SpringerWienNewYork

Margit Pavelka
Jürgen Roth

Functional Ultrastructure

Atlas of Tissue Biology and Pathology

Second, Revised and Enlarged Edition

SpringerWienNewYork

o. Univ. Prof. Dr. med. Margit Pavelka
Medical University of Vienna, Center for Anatomy and Cell Biology, Department of Cell Biology and Ultrastructure Research, Vienna, Austria
(margit.pavelka@meduniwien.ac.at)

Prof. Dr. med. Dr. sc. Dr. h. c. Jürgen Roth
Yonsei University Graduate School, World Class University Program, Department of Biomedical Science, Yonsei University, Seoul, Korea
(jurgen.roth@yonsei.ac.kr)

Softcover re-print of the Hardcover 2nd edition 2010

SpringerWienNewYork is part of Springer Science + Business Media
springer.at

Typesetting: Leo Mahler, 1150 Wien, Austria
Printing: Holzhausen Druck GmbH, 1140 Wien, Austria

Printed on acid-free and chlorine-free bleached paper
SPIN: 12634620

With 176 (partly coloured) Figures

Additional material to this book can be downloaded from http://extras.springer.com.

ISBN 3-211-83564-4 1st edn. SpringerWienNewYork
ISBN 978-3-7091-1661-6 (with CD-ROM) SpringerWienNewYork

This book is dedicated to

Michaela and Ernst
Verena, Raphael, Julia and David

FOREWORD

The period between 1950 and 1980 were the golden years of transmission electron microscopy and produced a plethora of new information on the structure of cells that was coupled to and followed by biochemical and functional studies. TEM was king and each micrograph of a new object produced new information that led to novel insights on cell and tissue organization and their functions. The quality of data represented by the images of cell and tissues had been perfected to a very high level by the great microscopists of that era including Palade, Porter, Fawcett, Sjostrand, Rhodin and many others. At present, the images that we see in leading journals for the most part do not reach the same technical level and are not prepared with the same attention to detail as in the golden era of TEM nor do they have the same information content and sheer artistic beauty.

This Atlas by Jürgen Roth and Margit Pavelka is a major exception, as it presents electron micrographs whose image quality and information content is uncompromised and unsurpassed. It has been prepared with great care and attention to detail. It depicts the remarkable diversity of specialized cell types such as those of the exocrine pancreas, intestinal epithelium, muscle and nervous system, for example. It reminds the reader that although each cell has the usual complement of organelles, their organization is quite different and distinctive and is recognizable to the trained eye. It reminds the cell biologist, biochemist, molecular biologist, geneticist and pathologist alike, who all too frequently work on cultured cells that lack differentiated features, of the diversity of cells in mammals and how their structure and organization reflect their specialized functions. Thus this atlas provides unique insights on how the architecture of cells, tissues and cell organelles mirror their functions. It also provides unique insights into how pathological processes affect cell organization.

This information is vital to current work in which the emphasis is on integrating approaches from proteomics, molecular biology, genetics, genomics, molecular imaging and physiology and pathology to understand cell functions and derangements in disease. In this current era, there is a growing tendency to substitute modern light microscopic techniques for electron microscopy, because it is less technically demanding and is more readily available to researchers.- This atlas reminds us that the information obtained by electron microscopy is invaluable and has no substitute. The increased insights obtained are comparable to the superior resolution (1000 x greater) obtained by the two methods. In fact, this atlas reminds us that these approaches are complementary and neither can substitute for the other.

Careful perusal of the images in this atlas makes one realize how many details are visible which go beyond those already known as far as even normal cell architecture is concerned. There is still a gold mine to be discovered for those wishing to put forth the effort. When it comes to cellular pathology in particular, the surface has barely been scratched and the potential is limitless.

It can be anticipated that this atlas may stimulate readers to undertake further ultrastructural studies coupled with functional studies on both normal and diseased cells and tissues to harvest the key insights this excursion will provide. In the age of harvesting the cell genome and proteome, we should also not forget to pay attention to harvesting the cell "structurome". This atlas provides the reader with the opportunity to delve into this world.

La Jolla, November 2009 *Marilyn G. Farquhar*

PREFACE OF THE SECOND EDITION

It provides us with both a pleasure and satisfaction to prepare the Preface for the Second Edition of this book. Foremost, it indicates to us, the authors, that the book was well received by the science community and that a strong interest exists for what it covers: modern biological and medical electron microscopy. Notwithstanding the enormous progress of light microscopy, it is simply the electron microscope which provides unmatched high resolution with the consequent detailed structural information.

For the second edition, we have kept the principal outline of the book. It is made up of two parts. Part 1, *The Cell,* which contains as much as possible a complete documentation of the various cellular organelles including the cytoskeleton, cell-cell contacts and cell-extracellular matrix interactions of normal cells in tissues or grown in culture. This part also includes a representative compilation of characteristic changes of organelles produced by experimental conditions commonly used to study membrane traffic and protein transport, or altered processes which arise from human diseases. Part 2, *Principles of Tissue Organisation,* is not intended to be complete in terms of all tissue structures. Rather, it is deliberately selective and explores the fine structural basis of specific functions performed by particular tissues and their constituent cells. Like the first part, it comprises a representative documentation of normal tissues and of changes that are characteristic and therefore diagnostic for particular diseases. For both parts, however, emphasis was placed on an integrated view of structure and function to illustrate and discuss the concept that cellular organelles provide the structural foundation for the fundamental processes of living organisms.

As stated for the first edition, this book was prepared for a broad readership and is intended to provide the reader with first-hand information about the major role that ultra-structural research continues to play in the various fields of cell and tissue biology and pathology. With this goal in mind, it was compiled to provide a stimulating entré, and hopefully to create a desire for more, for the novice in the field and an enjoyable dessert for the Connoisseur.

So, what makes the second edition different? The text part of the second edition of this book was thoroughly revised to account for the progress made in molecular cell biology during the last five years. Furthermore, both parts of the book were enlarged with several new plates to provide information we thought that needed to be added, and to document recent progress in techniques and what it provided as new information. Last but not least, the second edition comes with a CD-ROM, which should be helpful when teaching classes or for enhancing other presentations. Although this book is an atlas and by classical definition meant to be held in the hand and appreciated in a rich visual environment, it will be published as an e-book as well. The electronic version of the book thus will ensure that an even broader readership will have access to the knowledge it contains.

We would appreciate hearing from you, the readers, about the book to incorporate your suggestions in future editions. You can reach us at *margit.pavelka@meduniwien.ac.at* and *jurgen.roth@yonsei.ac.kr.*

Vienna Seoul, November 2009

Margit Pavelka
Jürgen Roth

PREFACE OF THE FIRST EDITION

The present-day exciting era of genomics and proteomics, which provided new and revolutionary insights into the life of cells, has also led to a renewed interest and special appreciation of ultrastructure research. For the understanding of the functions of cells and tissues, it is mandatory to precisely know the structure of their macromolecular and supramolecular assemblies and essential to identify their sites of action with high resolution as well as to explore their dynamics in the life of cells and their organisation in higher systems. It is the today's top challenge and priority of all ultramicroscopic methods to visua-localise functional processes in cells and tissues in their correlation with subcellular organelles and their ultrastructurally recognizable domains. Major progress has become possible through the refinement of existing preparation techniques and the development of new ones as well as the development of new types of microscopes. Among others, high pressure cryofixation and cryoelectron microscopy applied for high resolution 3D structural analysis of isolated macromolecular complexes, electron tomography and 3D reconstruction of the inner architectures of cells, low temperature embedding resins and cryoultramicrotomy in combination with immunogold labelling and hybridisation techniques and atomic force microscopes have become fully integrated into the range of methods used in modern molecular cell biology.

Our principal aim in compiling this atlas was to provide the reader with first-hand information about the major role ultrastructure research continues to play in the various fields of cell and tissue biology and pathology. We hope it will be useful for investigators, both beginners and experienced researchers, not only of biology and medicine but also of molecular biology and biochemistry as an aid and guide for the evaluation and interpretation of electron micrographs. The plates of electron micrographs of this atlas illustrate the use of both classical and present-day electron microscopy in the study of normal and diseased cells and tissues. They are accompanied by brief explanatory texts, schemes and diagrams and selected classical as well as recent publications and key reviews for further reading. For those readers who want to up-date the references, a most useful on-line service is provided by Pubmed (http://www.ncbi.nlm.nih.gov/entrez/query.fcgi?db).

The first part of the atlas deals with the cell and its various constituents, cell-cell contacts and cell-matrix interactions. Here we aimed to be as complete as possible in the documentation of the various structures and their function in the context of molecular cell biology. In addition we included representative examples of characteristic organelle changes under various experimental conditions and under conditions of disease. The second part exemplifies principles of tissue organisation and is supplemented with selected examples of ultrastructural tissue pathologies. Here, we aimed not on completeness but particular emphasis was placed on morpho-functional aspects in order to demonstrate that the ultrastructure of cells and tissues mirrors their main tasks and reflects specific functions.

We hope that this atlas is not looked upon as a mere collection of striking electron microscopic pictures. Each of the electron micrographs is intended to convey a specific message related to the properties, functions, or pathologies of the tissues and cells shown. Last but not least we would like to hear from our readers and use these suggestions (mail to: juergen.roth@usz.ch and margit.pavelka@meduniwien.ac.at) to improve future editions.

Vienna Zurich, July 2004

Margit Pavelka
Jürgen Roth

ACKNOWLEDGEMENTS

Preparing a book is a formidable task and to arrange the second edition is by no means less demanding. All this would not have been possible without the help and dedication of many people who were involved in the various aspects of this venture. We would like to acknowledge and thank them for their precious time, valuable suggestions and encouragement. We are greatly indebted to those colleagues who have generously supplied us with exceptional electron micrographs.

A number of micrographs from the own archives represent the results of fruitful collaboration with present and past members of our groups and colleagues from abroad and these include Moise Bendayan, Eric G. Berger, Dieter Bitter-Suermann, Daniela Brada, Dennis Brown, Eric Carlemalm, Pierre M. Charest, Paul Debbage, Michel Deschuyteneer, Adolf Ellinger, Jing yu Fan, Richard M. Franklin, Alfred Gangl, Walter J. Gehring, Irwin J. Goldstein, Bruno Guhl, Michael Hess, Kiyoko Hirano, Robert L. Hill, Kristijan Jezernik, Eduard Kellenberger, Peter M. Lackie, Hans Lassmann, John M. Lucocq, Roberto Montesano, Josef Neumüller, Armando Parodi, James C. Paulson, Hanns Plenk, Rok Romih, Christian Schöfer, Robert G. Spiro, Douglas J. Taatjes, Kiyoteru T. Tokuyasu, Monika Vetterlein, Werner Villiger, Franz Wachtler, Winifred M. Watkins, Klara Weipoltshammer, Gary Hin-Fai Yam, Martin Ziak and Christian Zuber.

The members of our groups and the many colleagues who took their precious time to read the texts and scrutinize the illustrations are gratefully acknowledged for their valuable feedback and suggestions for improvement. We also would like to extend our gratitude to Ulrich Kaindl, Norbert Wey, Stefanie Sulzer and Klaus Schönheinz for their invaluable help in preparing the micrographs and their professional skills in transforming our amateur sketches into artistic schemes and diagrams. A special note of thanks is due to Dr. Amrei Strehl and Mag. Wolfgang Dollhäubl at Springer Wien NewYork and others for their professional skills and great support who made the production of the second edition a pleasant affair.

WE ARE GRATEFUL TO THE FOLLOWING COLLEAGUES FOR PROVIDING MICROGRAPHS

Dr. Ueli Aebi, Basel (Figs. 8D, 9, 143)

Dr. Wolfgang Baumeister, Martinsried (Fig. 159)

Dr. Thomas Bächi, Zurich (Fig. 77B)

Dr. Peter H. Burri, Berne (Fig. 123)

Dr. Saverio Cinti, Ascona (Figs. 144, 145)

Dr. Dusan Cmarko, Lausanne (Figs. 4, 7)

Dr. H. Dariush Fahimi, Heidelberg (Figs. 66–69)

Dr. Stanislav Fakan, Munich (Figs. 4, 7)

Dr. Daniel S. Friend, San Francisco (Fig. 172)

Dr. Hanns-Joachim Gerdes, Oslo (Fig. 85)

Dr. Micheal Hess, Innsbruck (Fig. 37)

Dr. Ernst B. Hunziker, Berne (Figs. 146A, B)

Dr. Françoise Jaunin, Lausanne (Figs. 4, 7)

Dr. Kristijan Jezernik, Ljubljana (Figs. 124, 125)

Dr. Brigitte Kaissling, Zurich (Figs. 101, 114)

Dr. Dontscho Kerjaschki, Vienna (Fig. 117)

Dr. Lars-Inge Larsson, Frederiksberg 104B)

Dr. Hans Lassmann, Vienna (Figs. 157, 158)

Dr. Josef Neumüller, Vienna (Fig. 126)

Dr. Hanns Plenck, jr., Vienna (Figs. 147, 148)

Dr. Charlotte Remé, Zurich (Figs. 115, 116)

Dr. Rok Romih, Ljubljana (Figs. 124, 125)

Dr. Christian Schöfer, Vienna (Fig. 6)

Dr. Max Spycher, Zurich (Figs. 22, 55–61, 70B, 71, 153)

Dr. Franz Wachtler, Vienna (Fig. 5)

Dr. Ewald R. Weibel, Berne (Fig. 123)

Dr. Klara Weipoltshammer, Vienna (Fig. 6)

Dr. Sadaki Yokota, Nagasaki (Fig. 63)

CONTENTS

THE CELL

The Cell

STRUCTURAL ORGANISATION OF A MAMMALIAN CELL

All eukaryotic cells present qualitatively a very similar structural organisation that can vary quantitatively depending on their degree of differentiation and specialisation as well as functional state. The electron micrograph shows at low magnification highly specialised acinar cells of rat pancreas, which represent the prototype of a polarised secretory cell type. Several such acinar cells form a functional unit, the secretory acinus.

Each cell consists of two major compartments, the nucleus and the cytoplasm, and these are the centre of attention of the first chapter. The cytoplasm of different cell types contains a common set of membrane-bounded organelles and the various structural components of the cytoskeleton. They are both embedded in the cytosol, which houses the intermediate metabolism and is the place of protein synthesis and proteasomal degradation of misfolded proteins. The rough endoplasmic reticulum (RER) is engaged in the synthesis and quality control of proteins, is the site of initiation of protein *N*-glycosylation, and is a major Ca^{2+} store. Lipids are also synthesised in the endoplasmic reticulum. Many proteins and lipids are transported to the Golgi apparatus, where they receive further post-translational modifications and are sorted to their final destinations. The protein *O*-glycosylation is initiated in the Golgi apparatus. The pre-Golgi intermediates (pre-GI) are engaged in anterograde and retrograde transport of cargo between the endoplasmic reticulum and the Golgi apparatus. In acinar pancreatic cells and other types of secretory cells, secretory proteins are sorted and packed into immature secretory granules, so-called condensing vacuoles (CV) forming in the *trans* Golgi apparatus. They mature into zymogen granules (ZG), which are stored in the apical cytoplasm and undergo exocytosis on stimulation. The endoplasmic reticulum and its transitional elements (TE), pre-Golgi intermediates, Golgi apparatus, and secretory granules constitute the secretory pathway. Further cellular organelles are the mitochondria (M), which generate energy, peroxisomes, which perform oxidative reactions, lysosomes, which have a degradative function, and endosomes, which are involved in cellular uptake of substances.

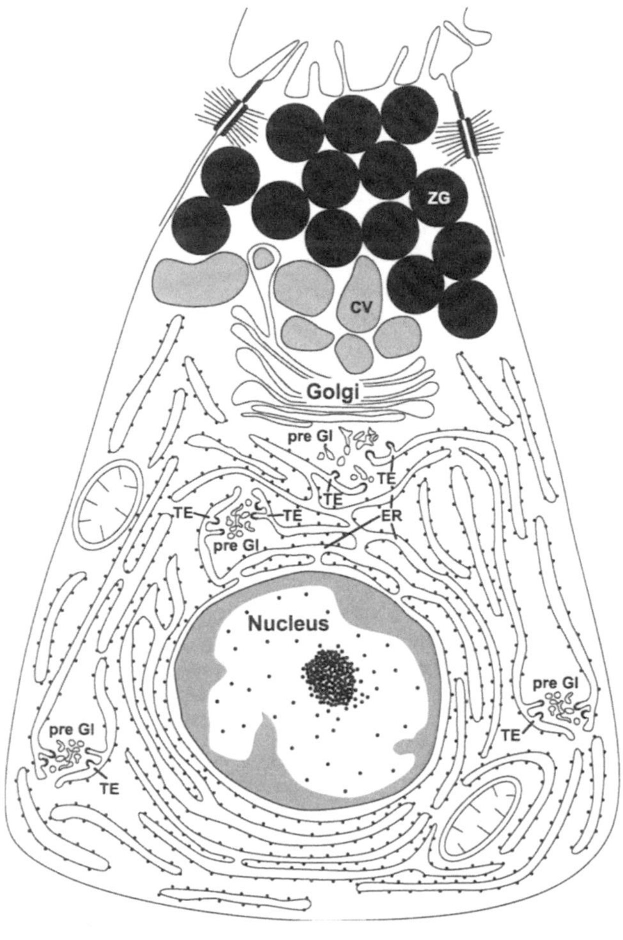

The plasma membrane is the cell's boundary with its environment. In epithelial cells such as the acinar cells, two plasma membrane domains can be distinguished. The apical plasma membrane forms the acinar lumen (AL) and is separated by junctional complexes (arrows) from the lateral (LPM) and basal (BPM) membrane domain.

The diagram presents a simplified version of a pancreatic acinar cell.

Magnification: x 11,250

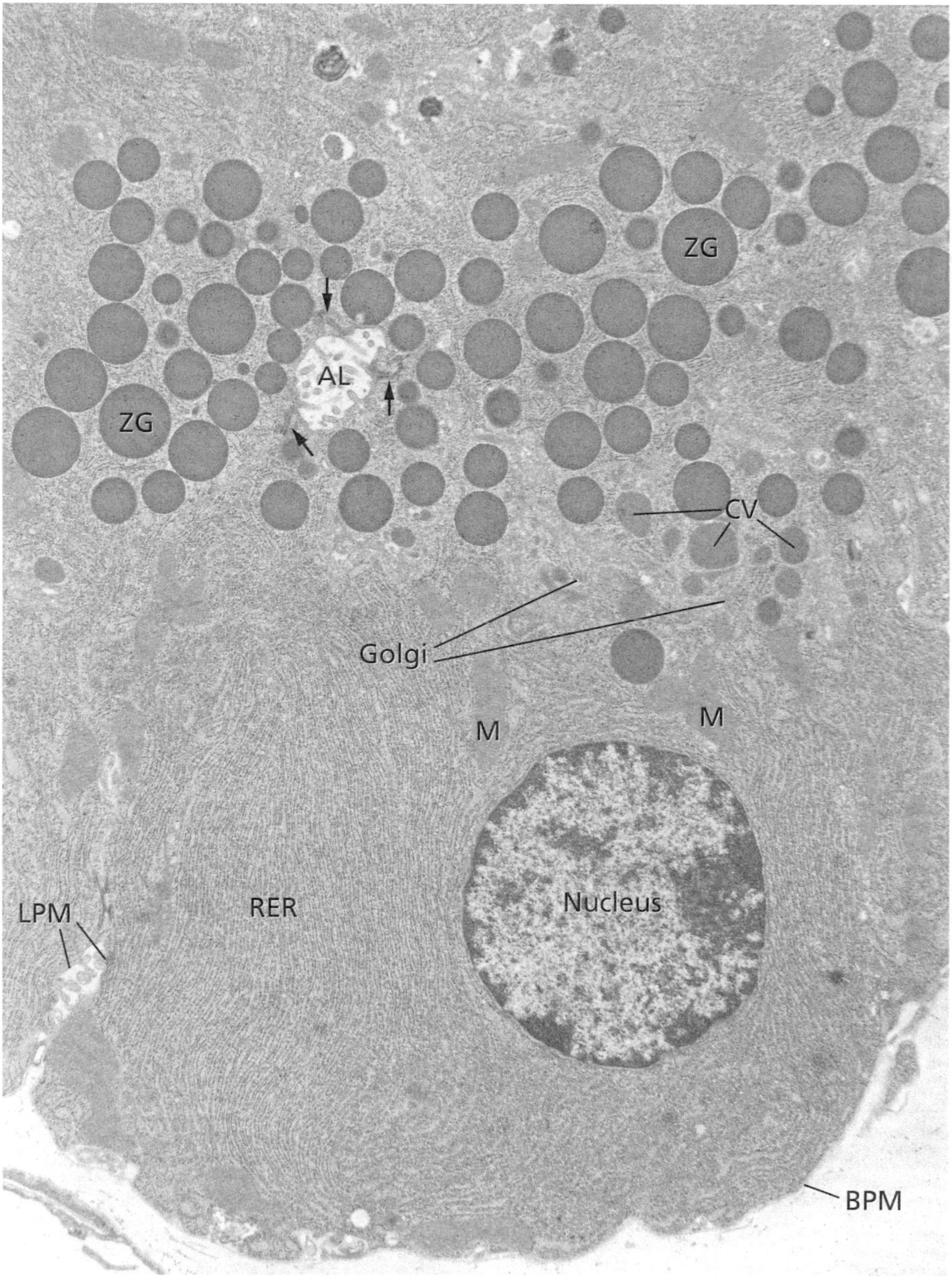
ZG
AL
ZG
CV
Golgi
M
M
LPM
RER
Nucleus
BPM

ARCHITECTURE OF THE CELL NUCLEUS

The nucleus is the largest of the cellular compartments, housing the chromosomes with the vast majority of the cellular genome, as well as multiple molecular machineries necessary for gene organisation and expression. During the past years contemporarily with the successful sequencing of the entire human genome, knowledge of nuclear architectures has also enormously increased and exciting results provide evidence that the longstanding dogma of a spatial separation of gene transcription in the nucleus and translation in the cytoplasm cannot be maintained.

The nuclear space is separated from the cytoplasm by the double membrane of the nuclear envelope (NE), in which the nuclear pores are embedded, forming routes for nuclear-cytoplasmic exchange (arrows in panel A, cf. Figs. 8 and 9). Panel B shows the outer and inner membranes enclosing the perinuclear cistern and a connection site, where the outer membranes of the nuclear envelope are continuous with those of the rough endoplasmic reticulum (RER) and the perinuclear cistern passes into the RER lumen (asterisk).

The nucleus is compartmentalised in chromosome territories and interchromatin spaces (diagrams A and B).

Genomic DNA condensed in chromosomes during mitosis (A-chromosomes 1,2,3), forms chromosome territories in the interphase nucleus (A-1,2,3). Genetic loci shown on mitotic chromosomes (B-1,2,3) are positioned in the interphase nucleus within or, presumably on decondensed chromatin fibre loops, outside the territories (B-1,2,3).

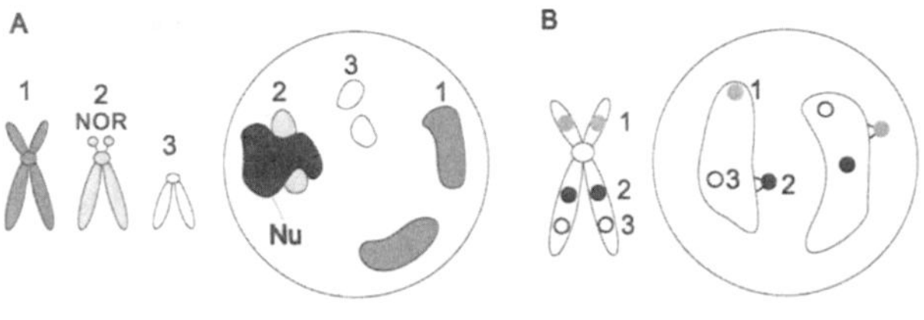

In the electron microscope due its intense staining, the condensed chromatin (heterochromatin, C) is clearly visible, whereas the decondensed euchromatic fibers are hardly discernible. Condensed chromatin dominates the peripheral nuclear areas corresponding to its attachment to the nuclear lamina (cf. Fig. 3). In the nucleus interior, condensed chromatin cords surround the interchromatin spaces, where numerous discrete domains related to gene expression exist. Among these, the nucleolus is the most prominent (Nu, cf. Figs. 5–6).

Interchromatin granule clusters (IG) represent the ultrastructural equivalent of the splicing speckles or splicing factor compartments, which are enriched in RNA and protein factors engaged in mRNA splicing. It is suggested that IGs represent sites of storage and/or assembly of splicing complexes. IGs also contain multiple RNA processing factors, transcription factors and potential structural proteins, such as lamins.

The Cajal body (CB), first described by Ramon y Cajal a century ago, is the best characterised of the small nuclear bodies. It corresponds to the "coiled body" and was recently renamed after its discoverer. In Cajal bodies, components of transcription and RNA processing machineries are transiently localised.

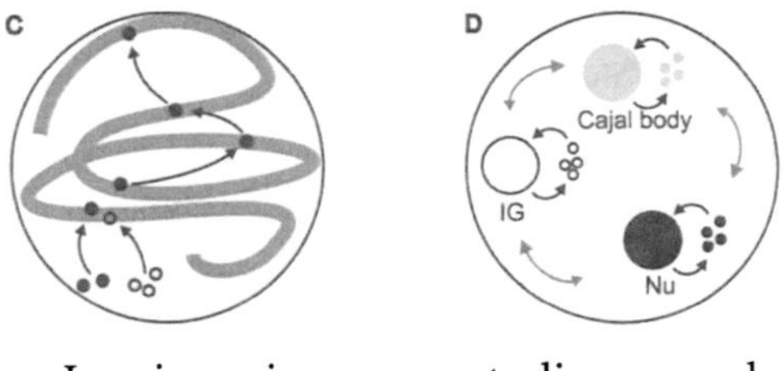

In vivo-microscopy studies proved that nuclear structures, although morphologically stable, are dynamic entities (diagrams C and D).

Both chromatin proteins (C) and proteins of interchromatin domains (D) reside only temporarily on chromatin and the respective compartments (Nu-nucleolus, IG-interchromatin granule clusters, Cajal bodies) and there is a continuous exchange with the nucleoplasm.

Diagrams A–D are modified from Roix J and Misteli T (2002) Histochem Cell Biol 118: 105–116.

References

Cremer T and Cremer C (2001) Chromosome territories, nuclear architecture and gene regulation in mammalian cells. Nat Rev Genetics 2: 292

Iborra FJ, Jackson DA, and Cook PR (2001) Coupled transcription and translation within nuclei of mammalian cells. Science 293: 1139

Lamond AI and Spector DL (2003) Nuclear speckles: a model for nuclear organelles. Nat Rev Mol Cell Biol 4: 605

Misteli-T (2008) Physiological importance of RNA and protein mobility in the cell nucleus. Histochem Cell Biol 129: 5

Pederson T (2002) Dynamics and genome-centricity of interchromatin domains in the nucleus. Nat Cell Biol 4: E287

Roix J and Misteli T (2002) Genomes, proteoms, and dynamic networks in the cell nucleus. Histochem Cell Biol 118: 105

Magnification: x 24,000 (A), x 106,000 (B)

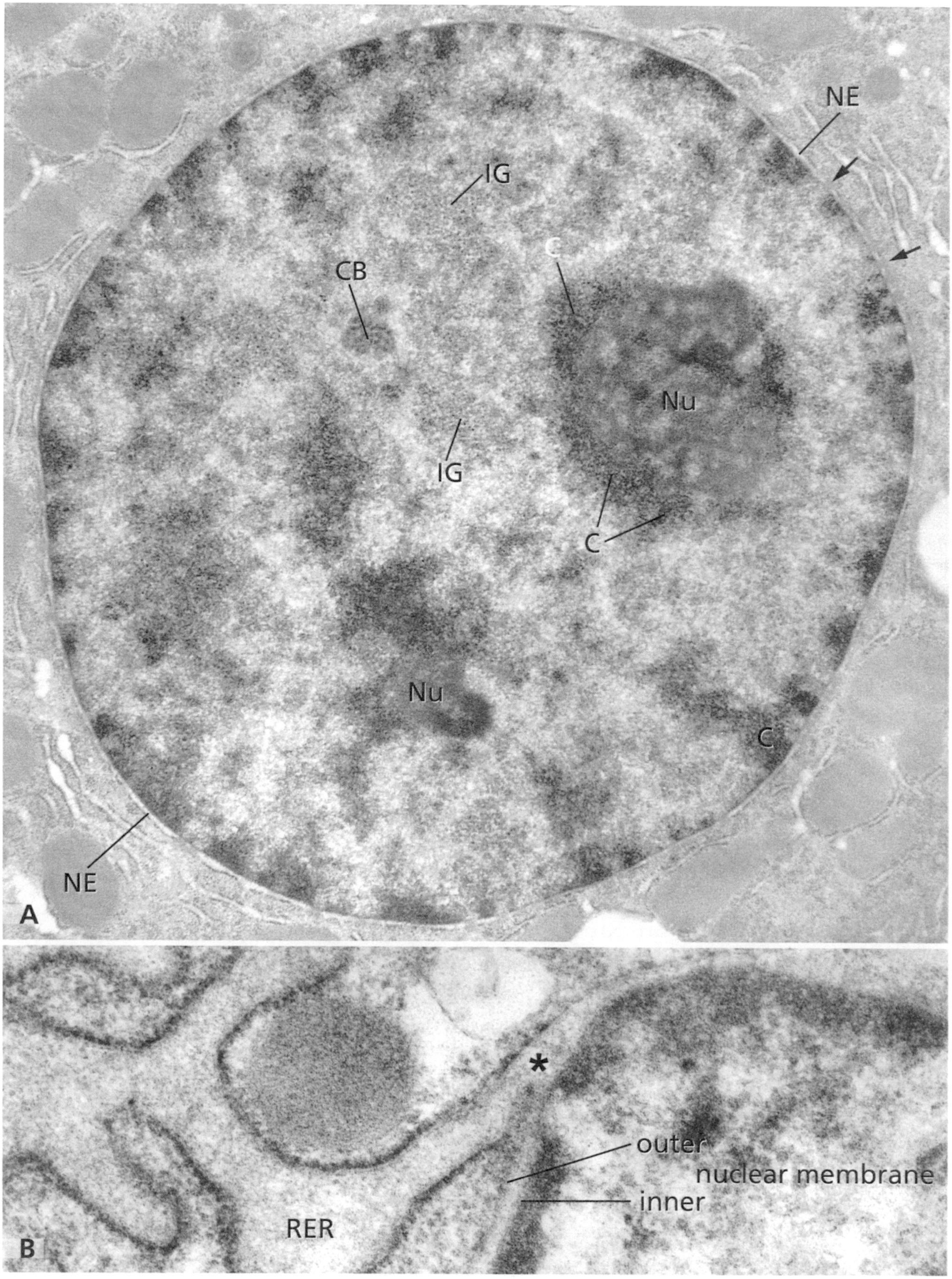
NE
IG
CB
C
Nu
IG
C
Nu
C
NE
A
*
outer
nuclear membrane
inner
RER
B

CYTOCHEMICAL DETECTION OF RIBONUCLEOPROTEINS

Various cytochemical techniques are available to study the distribution of DNA and RNA by electron microscopy. A classic technique for the detection of nuclear RNP RNA containing structures in ultrathin sections from tissues is the regressive contrasting procedure of Bernhard. In a first step, both DNA and RNA in Epon ultrathin sections of a tissue or cells are contrasted with uranyl acetate. Afterwards, the ultrathin sections are incubated with aqueous EDTA, which forms complexes with uranyl ions and results in the progredient removal of uranyl ions from DNA. The staining is of limited specificity and needs to be verified by enzyme digestion controls.

An example for the regressive RNA contrasting procedure is given in panel A. It shows an Epon ultrathin section from glutaraldehyde-fixed rat liver. In the nucleus of hepatocytes, the EDTA differential staining preferentially contrasts ribonucleoprotein-containing nuclear constituents. Perichromatin fibrils (arrowheads) are now visible in such a specimen, whose contrast after a conventional staining of sections would be too close to that of dispersed chromatin fibres. Arrows point to some perichromatin granules considered as nuclear storage and/or transport form of pre-mRNA particles. The DNA-containing peripheral and nucleolus-associated condensed chromatin (C) exhibits only light grey contrast. IG: interchromatin granules; Nu: nucleolus.

References

Bernhard W (1969) A new staining procedure for electron microscopical cytology. J Ultrastruct Res 27: 250

Gautier A (1976) Ultrastructural localization of DNA in ultrathin tissue sections. Int Rev Cytol 44: 113

Moyne G (1980) Methods in ultrastructural cytochemistry of the cell nucleus. Progr Histochem Cytochem 13: 1

NUCLEAR LAMINA

The inner nuclear membrane is covered by a 30 to 80 nm thick fibrous layer, the nuclear lamina. In panel B, the well developed nuclear lamina of a skin keratinocyte is seen at the inner nuclear membrane (IM). Outer nuclear membrane (OM); chromatin (C).

The nuclear lamina is composed of lamin A and variable amounts of lamins B and C, which are intermediate filaments. They differ from other intermediate filaments in that they form an orthogonal lattice. Nuclear lamins are higly organized fibrils with a central rod-like domain composed of α-helical coiled-coil and globular domains on either end. The C-terminal globular domain contains a nuclear localization sequence for nuclear import. Posttranslational modifications including farnesylation permit the assembly of lamin subunits at the inner nuclear membrane. The lamins are anchored to the inner nuclear membrane by the lamin B receptor and by lamina-associated proteins 1 and 2.

The nuclear lamina functions such as to give shape and provide mechanical support to the nuclear envelope and interacts with the chromatin thereby establishing a link between chromosomes and the nuclear envelope. During onset of mitosis, phosphorylation of lamins triggers the disassembly of the nuclear lamina and subsequently that of the nuclear envelope. Lamins become dephosphorylated when the nuclear envelope starts to reassemble in late anaphase.

Surprisingly, lamin mutations not only disrupt the nuclear envelope but also result in the development of certain types of muscular dystrophies, partial lipodystrophy syndromes, peripheral neuropathies and most impressive in progery syndromes.

References

Aebi U, Cohn J, Buhle L, and Gerace L (1986) The nuclear lamina is a meshwork of intermediate-type filaments. Nature 323: 560

Hutchinson C and Worman HJ (2004) A-type lamins: Guardians of the soma? Nat Cell Biol 6: 1062

Mattout A, Dechat T, Adam S, Goldman R, and Gruenbaum Y (2006) Nuclear lamins, diseases and aging. Curr Opin Cell Biol 18: 335

Scaffidi P and Misteli T (2006) Lamin A-dependent nuclear defects in human aging. Science 312: 1059

Stuurman N, Heins S, and Aebi U (1998) Nuclear lamina: their structure, assembly, and interactions. J Struct Biol 122: 42

Magnification: x 24,500 (A); x 70,000 (B)

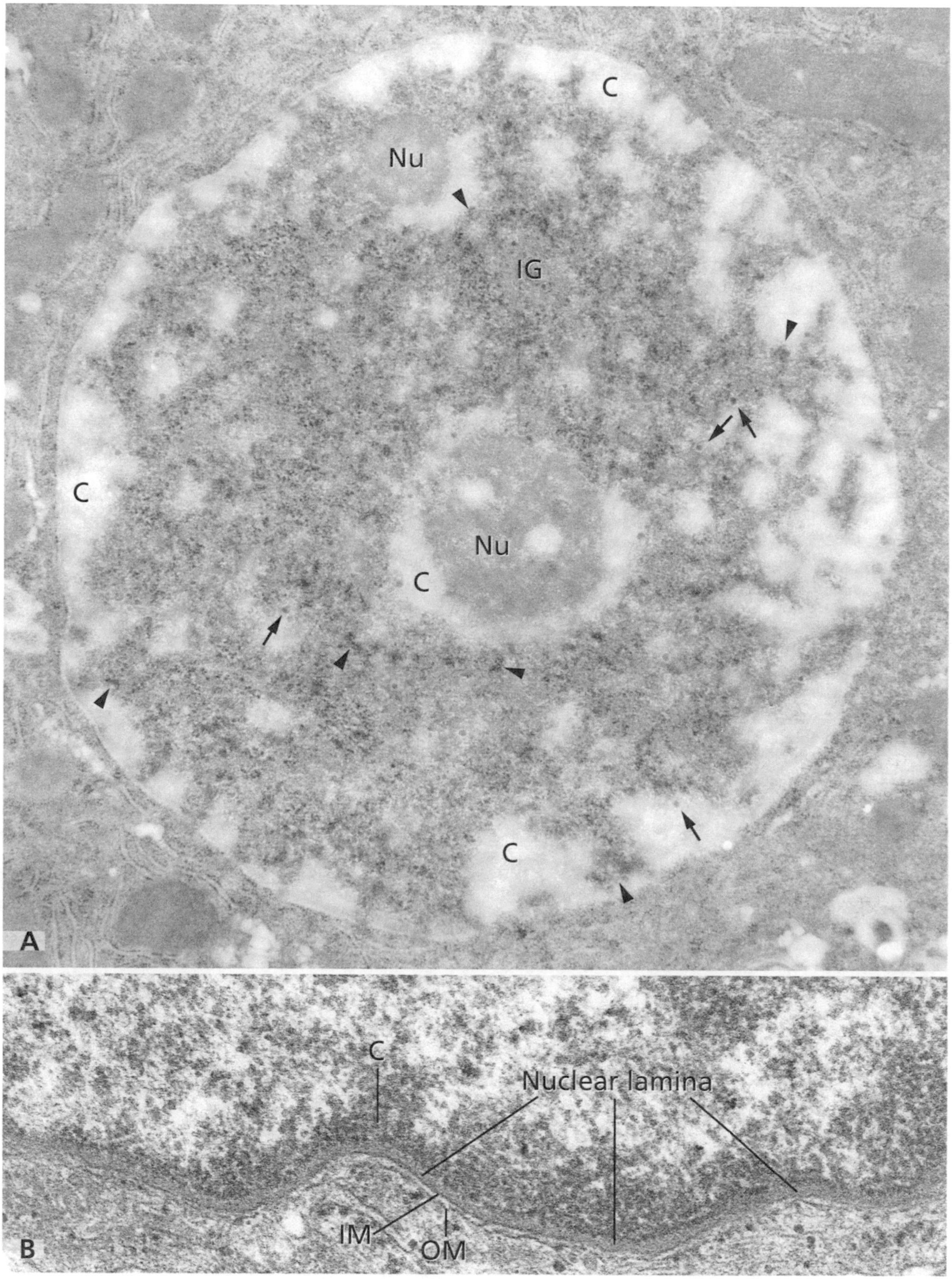
C
Nu
IG
C
Nu
C
C
A
C
Nuclear lamina
IM
OM
B

DETECTION OF SITES OF DNA REPLICATION AND OF INTERPHASE CHROMOSOME DOMAINS

DNA can be detected with the high precision and specificity of immunoelectron microscopy. For this purpose, antibodies raised experimentally or present in sera from patients who have autoimmune diseases as well as antibodies raised against nucleotide analogs have been applied. Formaldehyde or formaldehyde/glutaraldehyde fixation and embedding in Lowicryl K4M or LR white resins provides high detection sensitivity. High spatial resolution is obtained by the use of gold-labelled secondary antibodies. With these techniques, DNA has been observed consistently in the dispersed and condensed chromatin of interphase nuclei, chromosomes of mitotic cells, and the mitochondria.

In panels A and B, DNA has been revealed by the incubation of living V79 hamster cells with the thymidine analog bromodeoxyuridine (BrdU). The bromodeoxyuridine, which becomes incorporated into nuclear DNA during its replication, can be detected subsequently in ultrathin sections by a monoclonal anti-bromodeoxyuridine antibody and gold-labelled secondary antibody.

Panel A shows part of the nucleus of a cell that was incubated with bromodeoxyuridine for as short a period as 2 minutes. Sites of nascent DNA as detected with anti-bromodeoxyuridine antibody are indicated by the presence of gold particles. In addition, DNA in the Lowicryl K4M thin section was histochemically revealed by a Feulgen-type reaction using osmium ammine. Immunogold labelling showed DNA replication sites in association with individual dispersed chromatin fibers (arrowheads) on the periphery of a condensed chromatin area (arrow). In other experiments, these regions were shown to contain DNA polymerase alpha. Together with results from a 5 minutes bromodeoxyuridine pulse, alone or followed by a chase, it could be concluded that the sites of DNA replication corresponded essentially to perichromatin regions and that the newly replicated DNA moved rapidly from the replication sites towards the interior of condensed chromatin areas.

In panel B, V79 hamster cells were incubated with bromodeoxyuridine for 9 hours and then allowed to grow for an additional 62 hours. This time period allowed for about 5 cell divisions, during which the labelled chromosomes became segregated. Nuclei of these cells contained only 2–3 labelled chromosomes and were highly amenable to analyzing the presence of chromosome territories and subchromosomal domains. Such analysis indicated that chromosome domains were either separated from one another by interchromatin space or were in close contact with no or little intermingling of their DNA. Therefore, although chromosomes formed discrete territories, chromatin of adjacent chromosomes seemed to be in contact in limited regions, implying chromosome-chromosome interactions. Such an *in vivo* labelling approach allowed the analysis of the structural features of interphase chromosome domains as well as of their relations with the immediate neighbourhood. Arrows indicate RNP fibrils in perichromatin and interchromatin regions. NE: nuclear envelope.

References

Fakan S (2004) The functional architecture of the nucleus as analyzed by ultrastructural cytochemistry. Histochem Cell Biol 122: 83

Francastel C, Schubeler D, Martin DI, and Groudine M (2000) Nuclear compartmentalization and gene activity. Nat Rev Mol Cell Biol 1: 137

Gratzner HG (1982) Monoclonal antibody to 5-bromo- and 5-iododeoxyuridine: A new reagent for detection of DNA replication. Science 218: 474

Jaunin F, Visser AE, Cmarko D, Aten JA, and Fakan S (2000) Fine structural in situ analysis of nascent DNA movement following DNA replication. Exp Cell Res 260: 313

Liu DF, el-Alfy M, and Leblond CP (1995) DNA changes involved in the formation of metaphase chromosomes, as observed in mouse duodenal crypt cells stained by osmium-ammine. II. Tracing nascent DNA by bromodeoxyuridine into structures arising during the S phase. Anat Rec 242: 449

Tamatani R, Taniguchi Y, and Kawarai Y (1995) Ultrastructural study of proliferating cells with an improved immunocytochemical detection of DNA-incorporated bromodeoxyuridine. J Histochem Cytochem 43: 21

Thiry M (1999) Ultrastructural methods for nucleic acid detection by immunocytology. Progr Histochem Cytochem 34: 93

Visser AE, Jaunin F, Fakan S, and Aten JA (2000) High resolution analysis of interphase chromosome domains. J Cell Sci 113: 2585

Magnification: x 66,000 (A); x 63,000 (B)

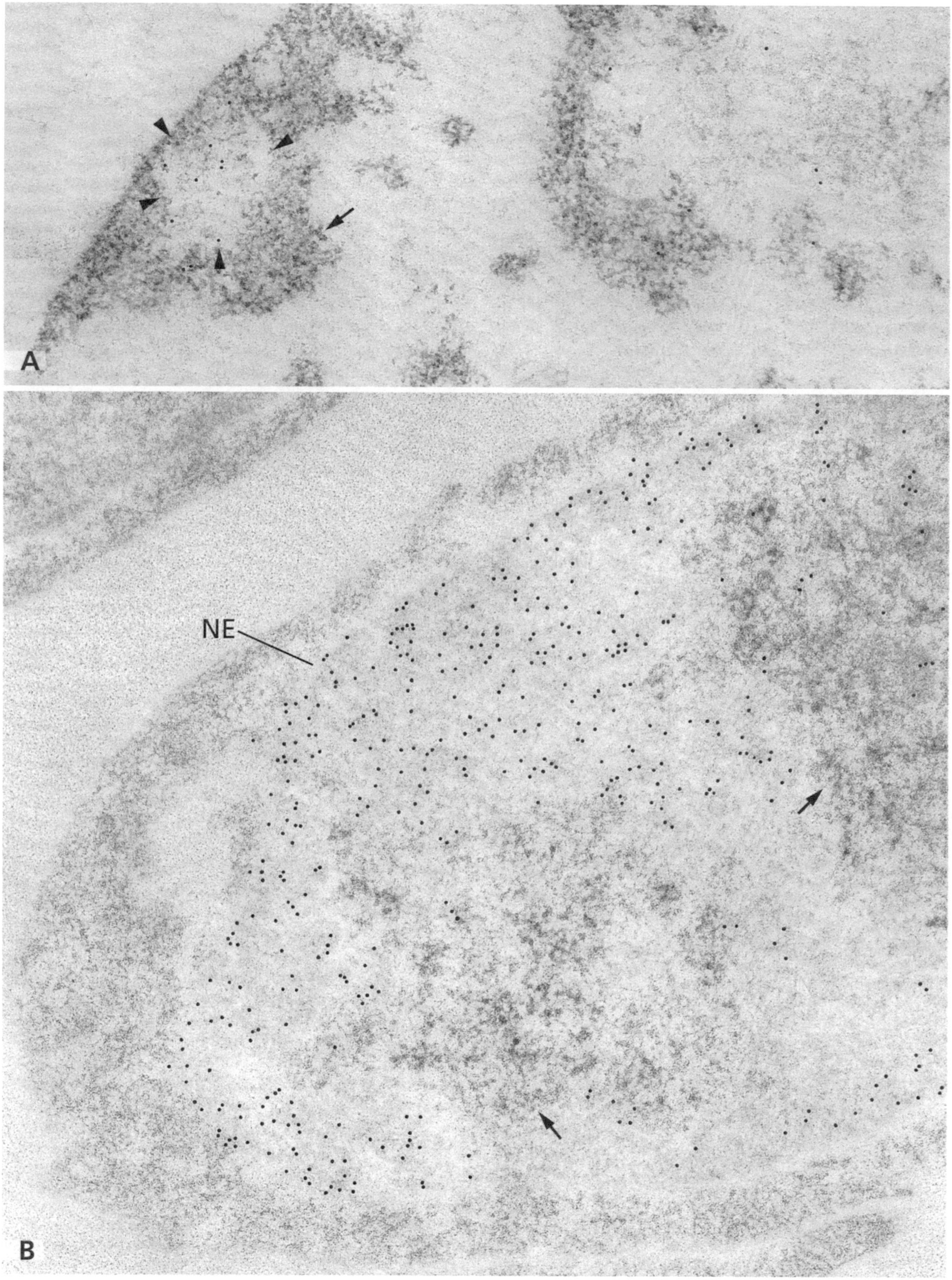
A
NE
B

NUCLEOLUS

Nucleoli are the most prominent of the nuclear bodies. They are formed around the nucleolus organiser regions (NORs), which are localised to the secondary constrictions of acrocentric chromosomes and bear numerous copies of genes coding for the preribosomal RNA (pre-rRNA). Production of ribosome precursors is the main task of nucleoli, although it is neither a privilege, nor is it the sole task, since preribosomal particles may also be built in the nucleus outside of nucleoli and, during the past decade, various additional functions have been ascribed to nucleoli.

In human cells, NORs are contained in five chromosomes. Hence in diploid cells, theoretically ten nucleoli could be organised, but nucleoli tend to fuse during interphase and mammalian cells show mostly one to four nucleoli.

The morphological appearance of nucleoli as shown electron microscopically in Fig. 5 and illustrated in the insert at the left lower corner reflects their main function in ribosome biogenesis. The structure of nucleoli is a result of the processes connected with transcription and processing of pre-rRNA and assembly of precursors of the small and large ribosome subunits. The nucleolar structure changes concomitantly with cell differentiation (cf. Fig. 6) and disappears with onset of mitosis, at which time transcription and processing of pre-rRNA are suppressed.

Nucleoli are composed of three different components, the dense fibrillar component (dfc), the granular component (gc), and the fibrillar centre (fc), and they are accompanied by masses of condensed chromatin (C), which mainly corresponds to chromosome territories of NOR-containing chromosomes (cf. Fig. 2). The fibrillar centre, which is assumed to be a protein storage site, appears as a distinct spherical body encircled by the dense fibrillar and granular components. These both compartments contain growing preribosomal particles, the dense fibrillar component at early, the granular component at late states of formation. The dense fibrillar component is enriched in newly synthesised pre-rRNA and a palette of proteins, the granular component consists preribosomal particles, which are nearly completed. Transcription is proposed to take place in the dense fibrillar component or at the border between fibrillar centre and dense fibrillar component. In another model, the fibrillar centre is suggested to be the main site of transcription.

The relation between nucleoli and Cajal bodies is close (cf. Fig. 2, CBs). CBs are important for the biogenesis and function of small nucleolar ribonucleoproteins. Several proteins, such as fibrillarin and Nopp40, are localised in both the nucleolus and the CB. There is also evidence that proteins move between the two bodies. CBs are found in direct contact with nucleoli and they are highly dynamic structures that have been shown travelling toward and away from nucleoli. Since CBs also have been shown to move from the nucleolar periphery into the nucleolus, the possibility is considered that transfer of the material takes place by direct contact between CBs and nucleoli.

During the past years, additional nucleolar functions not related to the assembly of preribosomes have been assumed. These include tRNA maturation, sequestration of regulatory molecules, roles in viral infection, export of viral RNAs and control of ageing. It has been proposed that for some of the nonconventional activities an immobile platform is necessary and thereby advantage is taken of the nucleolus.

References

Filipowicz W and Pogacic V (2002) Biogenesis of small nucleolar ribonucleoproteins. Curr Opin Cell Biol 14: 319

Gall JG (2003) The centennial of the Cajal body. Nat Rev Mol Cell Biol 4: 975

Hozak P (1995) Catching RNA polymerase I in flagranti: Ribosomal genes are transcribed in the dense fibrillar component of the nucleolus. Exp Cell Res 216: 285

Olson MOJ, Hingorani K, and Szeneni A (2002) Conventional and nonconventional roles of the nucleolus. Int Rev Cytol 219: 199

Raska I (2003) Oldies but goldies: Searching for Christmas trees within the nucleolar architecture. Trends Cell Biol 13: 517

Sirri V, Urcuqui-Inchima S, Roussel P, and Hernandez-Verdun D (2008) Nucleolus: the fascinating nuclear body. Histochem Cell Biol 129: 13

Tschochner H and Hurt E (2003) Pre-ribosomes on the road from the nucleolus to the cytoplasm. Trends Cell Biol 13: 255

Magnification: x 66,000

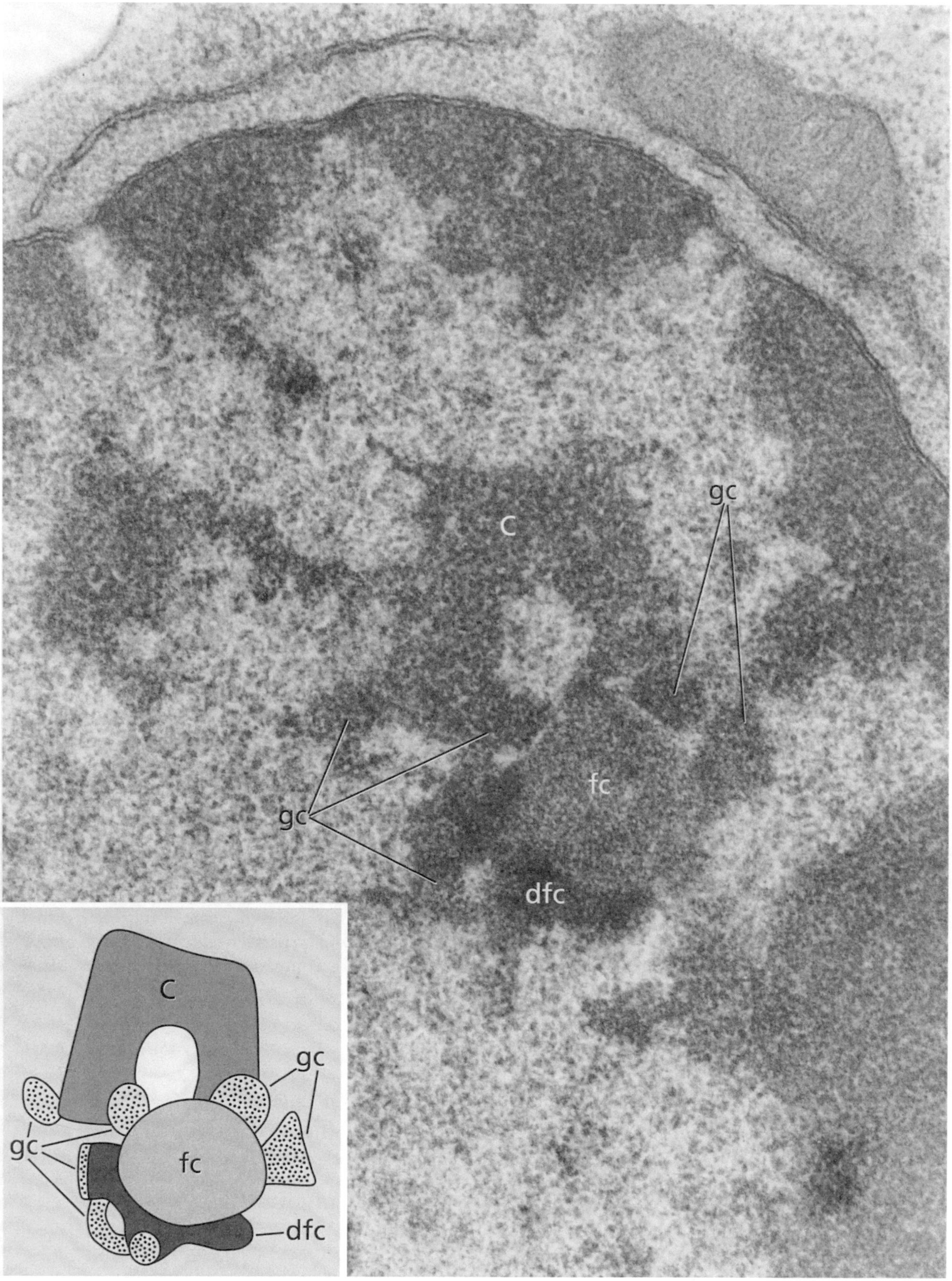
gc
C
fc
gc
dfc
C
gc
gc
fc
dfc

CHANGES OF THE NUCLEOLAR ARCHITECTURE

The nucleolar architecture depends on the functional state of cells and changes concomitantly with cellular changes. This has been studied in detail in human lymphocytes after stimulation with phytohaemagglutinin and in frog erythropoetic cells during maturation from proerythroblasts into erythrocytes.

Three types of nucleoli could be discriminated according to the appearance in the electron microscope:

1. "Ring-shaped" nucleoli (arrows in panel A) exhibiting a large fibrillar centre, which is surrounded by dense fibrillar and granular components. This type corresponds to the nucleolus shown in Fig. 5.
2. Nucleoli with "nucleolonema" (panel B) consisting of an extended network, which is composed of dense fibrillar and granular components. Fibrillar centers are small and inconspicuous.
3. "Compact" nucleoli (panel C), which usually contain several fibrillar centres, surrounded by thick circles of dense fibrillar material and large areas composed of granular nucleolar substance.

Panels A–C show a sequence of micrographs originating from studies with human lymphocytes analysed at different times after stimulation with phytohaemagglutinin (PHA) and showing a consecutive metamorphosis of the nucleoli.

About 75% of the unstimulated lymphocytes from the peripheral venous blood exhibit one ring-shaped nucleolus; the rest of 25% of the cells contains two or more nucleoli. The latter are usually smaller, but they are of the same ring-shaped type.

PHA-treatment leads to an increase of the cells with multiple ring-shaped nucleoli. An example is shown in panel A after 4 hours of PHA-stimulation with two ring-shaped nucleoli visible. By 12 hours, 34% of the lymphocytes contain more than one nucleolus. The increase in the number of nucleoli is interpreted as activation of additional NORs. Further PHA-treatment results in the appearance of one to two large nucleoli with nucleolonema (panel B). The multiple small nucleoli present earlier seem to fuse, thus forming the large nucleolar networks, in which dense fibrillar and granular components dominate. Fibrillar centres are inconspicuous and seem to be exhausted due to a decrease of stored proteins, which may have been used up during the assembly of ribosome precursors. This nucleolar type is most pronounced after 36 hours. An example after 16 hours is on display in panel B. With continued PHA-treatment, nucleoli are further transformed into compact types with multiple fibrillar centres and extended dense fibrillar and granular components, corresponding to a fresh highly active period. This nucleolus type is shown in panel C after 72 hours PHA-treatment.

The close connection between nucleolar structure and function is also shown by inhibition of protein- and RNA-biosynthesis. Inhibition of protein synthesis by treatment with puromycin causes a significant decrease in the size of fibrillar centres. Treatment with actinomycin D, which specifically inhibits rRNA-transcription, leads to in a general reduction of the nucleolar size and a segregation of the nucleolar components. Panel D shows a HeLa cell nucleolus after actinomycin D-treatment for 6 hours. The typical separation of the nucleolar subcompartments is further illustrated in the diagram shown below.

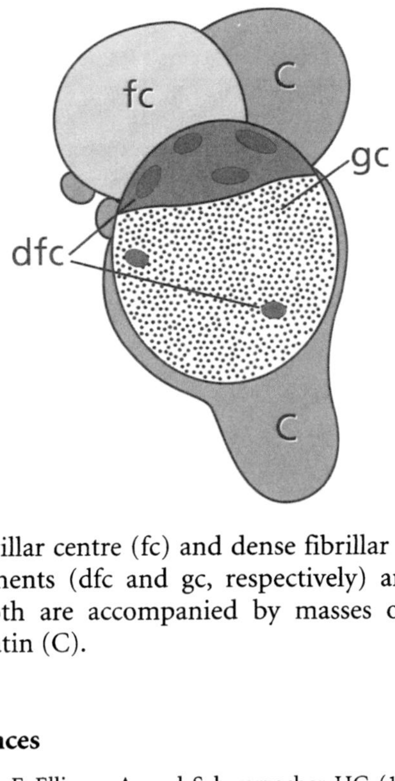

Fibrillar centre (fc) and dense fibrillar and granular components (dfc and gc, respectively) are segregated and both are accompanied by masses of condensed chromatin (C).

References

Wachtler F, Ellinger A, and Schwarzacher HG (1980) Nucleolar changes in human phytohaemagglutinin-stimulated lymphocytes. Cell Tiss Res 213: 351

Wachtler F, Popp W, and Schwarzacher HG (1987) Structural changes in nucleoli during inhibition of protein- and RNA-biosynthesis. Cell Tiss Res 247: 583

Magnification: x 7,000 (A), x 29,000 (B), x 14,000 (C) x 30,000 (D)

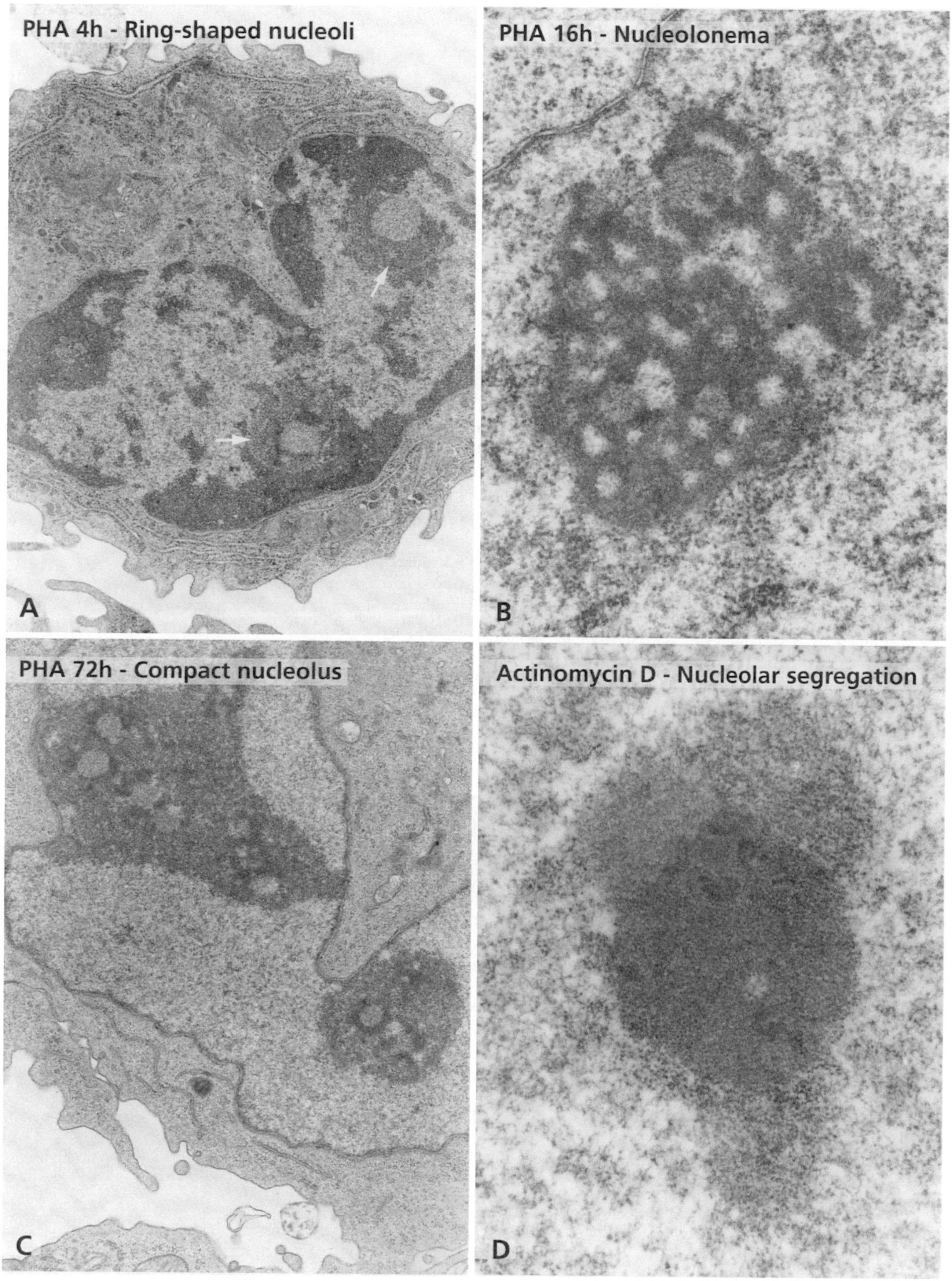
PHA 4h - Ring-shaped nucleoli
A
PHA 16h - Nucleolonema
B
PHA 72h - Compact nucleolus
C
Actinomycin D - Nucleolar segregation
D

DETECTION OF SITES OF RNA SYNTHESIS

The statement made for the visualization of DNA also fully applies to that of RNA: it can be readily detected by immunoelectron microscopy with high precision and specificity by labelling of ultrathin sections from cells and tissues. Antibodies against RNA can be applied, but the reagent most often used is the nucleotide analog 5'bromouridine-5-triphosphate (BrUTP), which, after its incorporation into RNA, can be detected with the monoclonal anti-bromodeoxyuridine antibody and gold-labelled secondary antibody with high spatial resolution. Formaldehyde or formaldehyde/glutaraldehyde fixation combined with embedding in Epon, Lowicryl K4M or LR white resins provides high detection sensitivity. The post-embedding immunogold labelling can be combined with other histochemical stains. Since 5'bromouridine-5-triphosphate is not membrane permeable, permeabilized cells were used or the nucleotide analog was microinjected into cells. Alternative approaches have been the use of the nucleoside analog BrU rather than BrUTP and of liposome transfection vectors.

In panel A, sites of RNA synthesis in the cell nucleus are visualized by means of brominated RNA precursor and immunoelectron microscopy. Cells of a human bladder carcinoma line T24 were microinjected in culture with 5'bromouridine-5-triphosphate and further incubated for 20 minutes. The immunogold labelling in the nucleoplasm was mostly localized on the periphery of condensed chromatin (C) areas, a nuclear compartment named the perichromatin region. In the nucleolus (Nu), newly synthesized pre-rRNA occurs mainly in the dense fibrillar component.

In panel B, a detail of a nuclear region after 10 minutes incubation following 5'bromouridine-5-triphosphate microinjection is shown. Here, a double immunogold labelling was carried out on the ultrathin section. Newly transcribed RNA was detected with anti-bromodeoxyuridine antibodies and 6 nm gold-labelled secondary antibody. In addition, hnRNP complexes were detected with anti-hnRNP core protein antibody and 15 nm gold-labelled secondary antibody. As can be readily seen, both signals often colocalized on perichromatin fibrils (arrowheads), which represent the *in situ* forms of pre-mRNA (hnRNA) transcripts. However, interchromatin granule clusters (IG) that are compartments of pre-mRNA splicing factor accumulation and storage were virtually devoid of any labelling.

From such experiments it could be concluded that the dense fibrillar component of the nucleolus is the site of pre-rRNA transcription and of initial steps of pre-rRNA processing. The perichromatin fibrils represent the *in situ* form of pre-mRNA transcripts and probably also the site of most pre-mRNA processing steps.

References

Cmarko D, Verschure P, Martin T, Dahmus M, Krause S, Fu X-D, van Driel R, and Fakan S (1999) Ultrastructural analysis of transcription and splicing in the cell nucleus after bromo-UTP microinjection. Mol Biol Cell 10: 211

Gorski SA, Dundr M, and Misteli T (2006) The road much traveled: trafficking in the cell nucleus. Curr Opin Cell Biol 18:284

Hobot J A, Bjornsti M A, and Kellenberger E (1987) Use of on-section immunolabeling and cryosubstitution for studies of bacterial DNA distribution. J Bacteriol 169: 2055

Hozak P, Cook PR, Schofer C, Mosgöller W, and Wachtler F (1994) Site of transcription of ribosomal RNA and intranucleolar structure in HeLa cells. J Cell Sci 107: 639

Jackson DA, Hassan AB, Errington RJ, and Cook PR (1993) Visualization of focal sites of transcription within human nuclei. EMBO J 12: 1059

Schul W, van Driel R, and de Jong L (1998) A subset of poly(A) polymerase is concentrated at sites of RNA synthesis and is associated with domains enriched in splicing factors and poly(A) RNA. Exp Cell Res 238: 1

Verheggen C, Le Panse S, Almouzni G, and Hernandez-Verdun, D (1998) Presence of pre-rRNAs before activation of polymerase I transcription in the building process of nucleoli during early development of *Xenopus laevis*. J Cell Biol 142: 1167

Sirri V, Urcuqui-Inchima S, Roussel P, and Vernandez-Verdun D (2008) Nucleolus: the fascinating nuclear body. Histochem Cell Biol 129:13

Magnification: x 32,000 (A); x 122,000 (B)

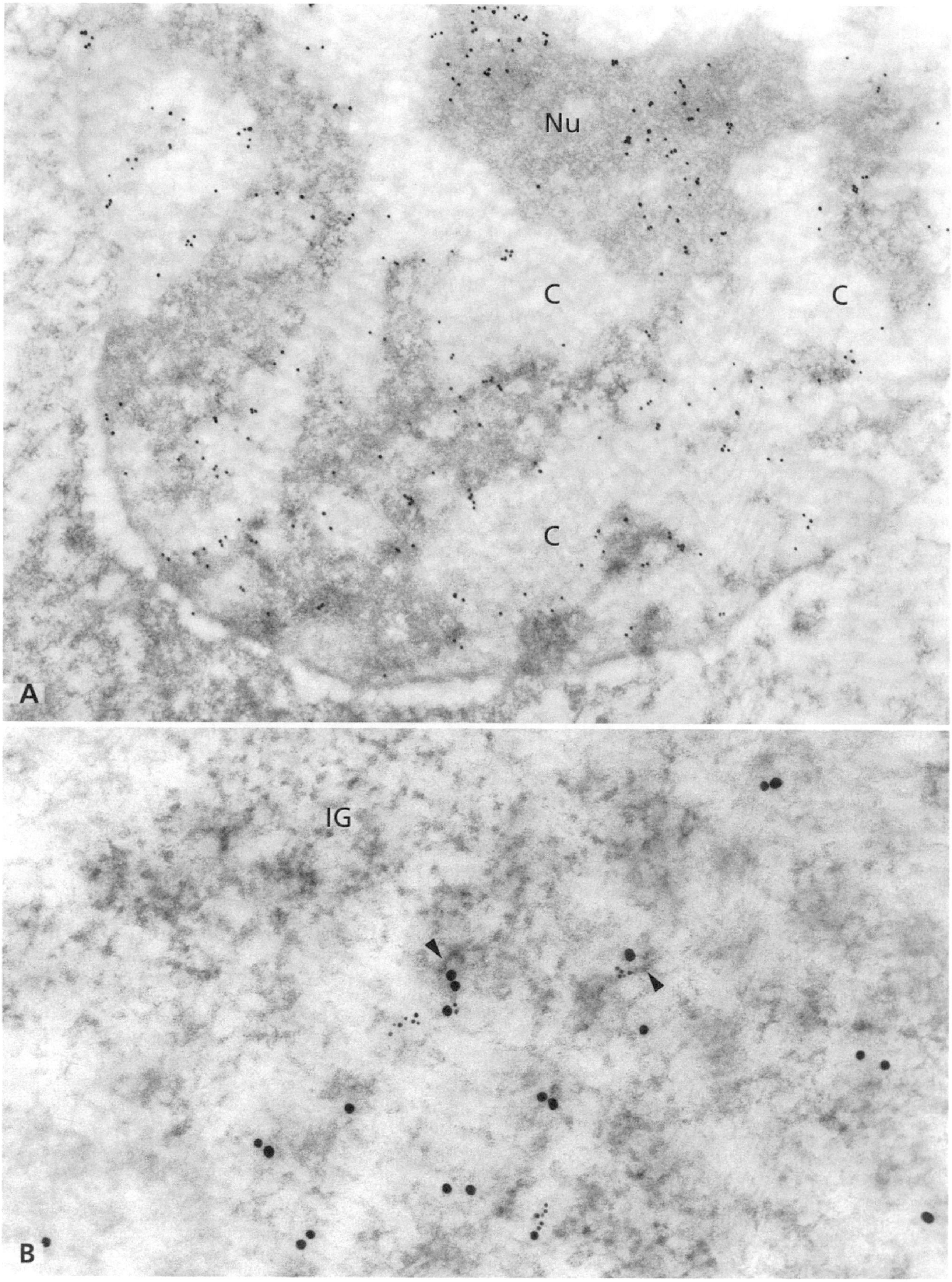
Nu
C
C
C
A
IG
B

NUCLEAR PORE COMPLEXES

Nuclear pore complexes (NPCs) are large macromolecular assemblies with an estimated mass of about 125 million Dalton. Depending on the activity of a cell, the number of NPCs varies but is usually around 3000-4000 per nucleus. NPCs are sites of bidirectional transport.

Panel A represents a freeze-fracture replica. Much of the nuclear envelope with the outer and inner nuclear membrane can be seen in a face-on view with a great number of NPCs. In panel B, a NPC is shown as it appears in a cross section from resin-embedded tissue by transmission electron microscopy. The pore is spanned by a diaphragm-like structure. The inner and outer nuclear membrane are continuous at the edges of the pore. At the surface of the inner nuclear membrane, the nuclear lamina (see Fig. 3) and condensed chromatin (CC) are present. In the cytoplasm next to the nuclear envelope (NE in B), a cisterna of the rough endoplasmic reticulum (ER in B) is seen. Panel C is a grazing section, which provides a face-on view of several NPCs, which often contain a central electron dense particle and ring-like arranged peripheral granular material (arrowhead). The electron dense material between the NPCs is condensed chromatin (CC). In panel D, isolated, detergent-extracted NPCs are shown by cryo-electron microscopy in the frozen-hydrated state. In favourable end-on views, one complex seems to be composed of eight peripheral particles and a single central one.

The three-dimensional structure of the NPC has been reconstructed from images of negative stained complexes and by cryo-electron microscopy (see diagrams in Figs. 8 and 9). The basic configuration of the NPC is a cylindrical channel with a diameter of about 125 nm. It is of a tripartite assembly composed of a large luminal ring embedded in the nuclear envelope flanked by a ring at its cytoplasmic and nuclear surface.

The barrel-like central ring is composed of eight spokes with protuberances toward the central opening. This is the structural basis for the characteristic eight-fold rotational symmetry of the NPC. Often a plug or particle is present in the central lumen. The mass of the spoke complex was determined to be 52 MDa by scanning transmission electron microscopy mass measurement. Between the adjacent eight spokes, eight longitudinal channels of about 10 nm diameter are formed. They are sites of passive diffusion of ions and water and small proteins (≤60 kDa). The transport of larger particles occurs by active transport involving specific receptor proteins. By using gold particles of various diameters, coated with peptides containing a nuclear localization signal, the maximal size of particles that could pass through the nuclear pore was about 26 nm. This indicates that the pore gate is a dynamic structure. Macromolecules larger than 26 nm may squeeze through the pores by deformation. From the nuclear and cytoplasmic ring, eight filaments extend outwards. The somewhat longer filaments of the nuclear ring converge and are capped by a terminal ring. This ensemble forms the nuclear basket which functions like an iris.

Many of the nucleoporins are glycoproteins with an *O*-glycosidically-linked *N*-acetylglucosamine residue on serine or threonine. In panel C, *O*-linked *N*-acetylglucosamine residues are detected by a gold labelling technique. In tangentially sectioned NPCs, the centre is labelled by gold particles (arrowheads). This type of glycosylation is found on many other proteins including the RNA polymerase II catalytic unit and a large number of transcription factors. It is an important regulatory modification, which, through a reciprocal relation with *O*-phosphorylation, modulates the function of many proteins.

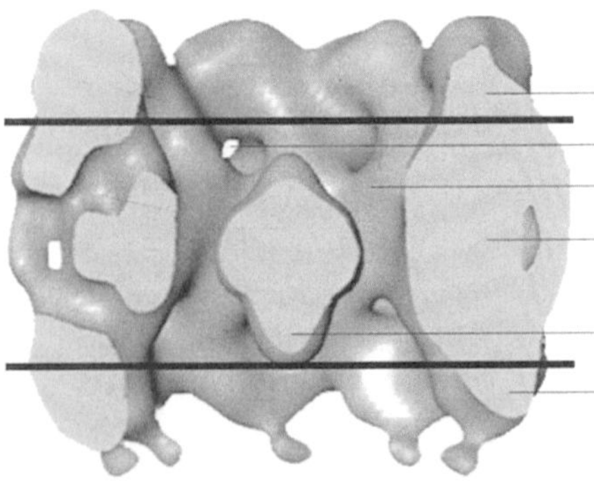

References

Beck M, Forster F, Ecke M, Plitzko JM, Melchior F, Gerisch G, Baumeister W, and Medalia O (2004) Nuclear pore complex structure and dynamics revealed by cryoelectron tomography. Science 306: 1387

Hart GW, Housley MP, and Slawson C (2007) Cycling of O-linked beta-N-acetylglucosamine on nucleocytoplasmic proteins. Nature 446: 1017

Lim RY, Aebi U, and Fahrenkrog B (2008) Towards reconciling structure and function in the nuclear pore complex. Histochem Cell Biol 129: 105

Webster M, Witkin KL, and Cohen-Fix O (2009) Sizing up the nucleus: nuclear shape, size and nuclear-envelope assembly. J Cell Sci 122: 1477

Magnification: x 30,000 (A); x 42,000 (B); x 36,000 (C); x 121,000 (D); x 17,000 (E)

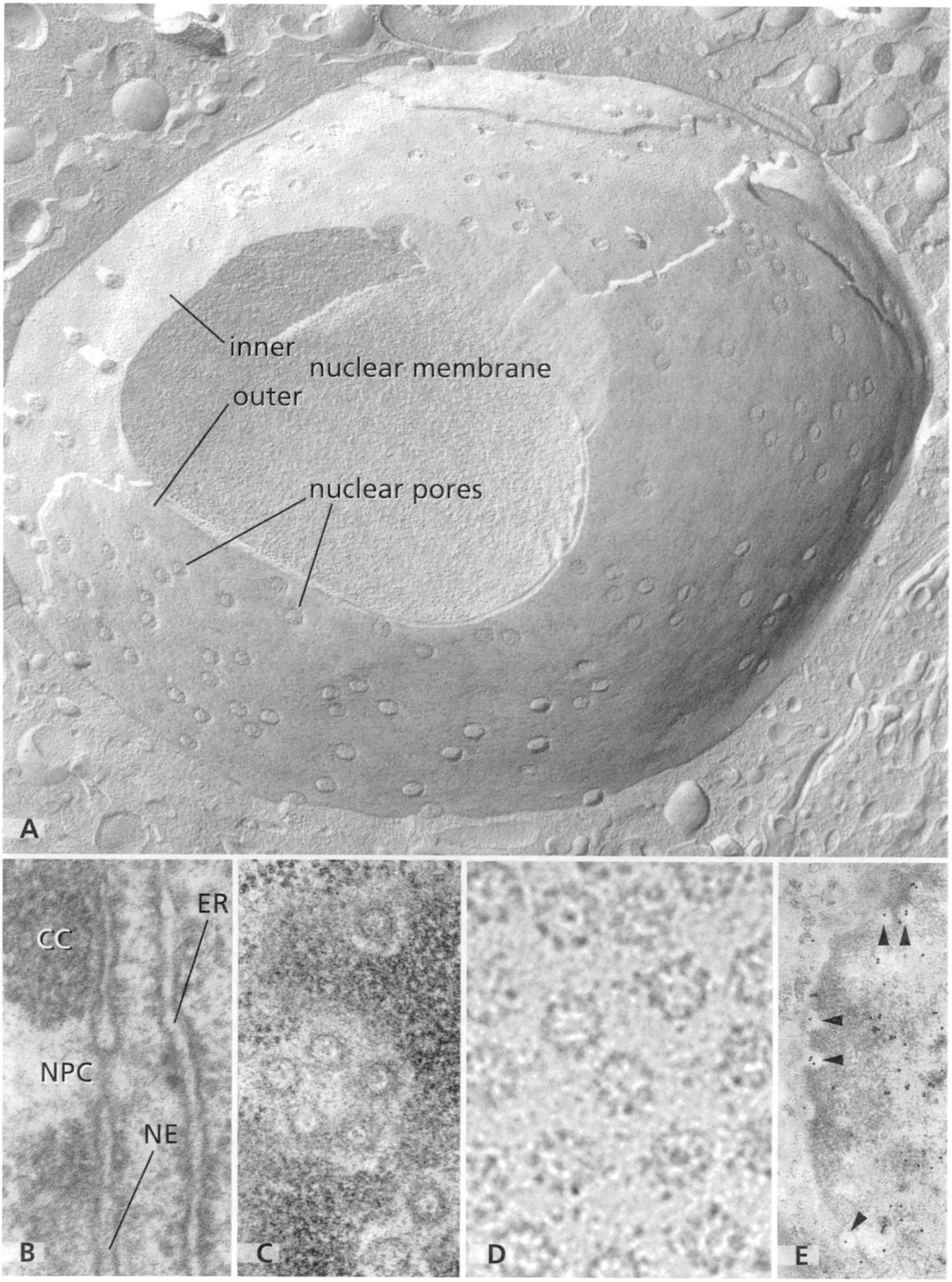
inner
nuclear membrane
outer
nuclear pores
A
CC
ER
NPC
NE
B
C
D
E

NUCLEAR PORE COMPLEXES: STRUCTURAL CHANGES AS MONITORED BY TIME-LAPSE ATOMIC FORCE MICROSCOPY

Atomic force microscopy is a unique imaging technique that allows high resolution studies of biological macromolecules in a native state under quasi-physiological conditions in an aqueous environment. In addition to static structural analysis, dynamic structural analysis by time lapse imaging in response to stimuli as well as of growth and growth kinetics of macromolecules has been achieved. The examples shown here concern the structure of native nuclear pore complexes and structural changes in response to calcium.

Panels A and B and their insets show native nuclear envelopes from *Xenopus* oocytes. Panel A shows the cytoplasmic surface of the nuclear envelope and panel B its nuclear surface. Clear-cut differences are visible since the nuclear pore complexes at the cytoplasmic surface appear doughnut-like (panel A) whereas those at the nuclear surface are dome-like (panel B) in appearance. Furthermore, the eightfold rotational symmetry of individual nuclear pore complexes is apparent (inset in A). For a quantitative analysis, nuclear pore complexes were aligned and averaged and their radial height profiles computed (panels C and D).

In panels E and F, nuclear pore complexes were observed by time lapse-atomic force microscopy from the nuclear surface in the absence of calcium (panel E) or in the presence of micromolecular concentrations (panel F). The arrowheads mark three corresponding nuclear pore complexes and the reversible calcium-mediated changes, closing in absence of calcium and opening in its presence, are visualised. In panels G and H, such conformational states were quantified by aligning and averaging a number of nuclear pore complexes and computing their average radial heights. From this analysis it became clear that 20–30 nm diameter openings occurred at the distal rings of the nuclear baskets without changes in the overall height of the baskets. These results fully support the proposal that the nuclear basket with its distal ring may act as an iris-like diaphragm. The ring is in a closed state in the absence of calcium and in an open state in the presence of micromolecular calcium concentrations as shown in the diagram.

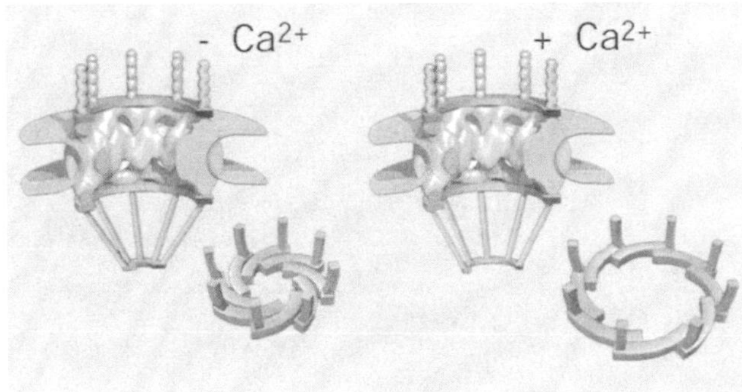

References

Binnig G, Quate C, and Gerber C (1986) Atomic force microscope. Phys Rev Lett 56: 930

Oberleithner H, Schneider S, and Bustamante JO (1996) Atomic force microscopy visualises ATP-dependent dissociation of multimeric TATA-binding protein before translocation into the cell nucleus. Pflügers Arch 432: 839

Perez-Terzic C, Gacy AM, Bortolon R, Dzeja PP, Puceat M, Jaconi M, Prendergast FG, and Terzic A (1999) Structural plasticity of the cardiac nuclear pore complex in response to regulators of nuclear import. Circ Res 84: 1292

Perez-Terzic C, Pyle J, Jaconi M, Stehno-Bittel L, and Clapham DE (1996) Conformational states of the nuclear pore complex induced by depletion of nuclear Ca^{2+} stores. Science 273: 1875

Ryan KJ and Wente SR (2000) The nuclear pore complex: a protein machine bridging the nucleus and cytoplasm. Curr Opin Cell Biol 12: 361

Stolz M, Stoffler D, Aebi U, and Goldsbury C (2000) Monitoring biomolecular interactions by time-lapse atomic force microscopy. J Struct Biol 131: 171

Wang H and Clapham DE (1999) Conformational changes of the in situ nuclear pore complex. Biophys J 77: 241

Magnification: x 75,000 (A, B); x 150,000 (upper insets in A and B); x 100,000 (lower insets in A); x 150,000 (E, F)

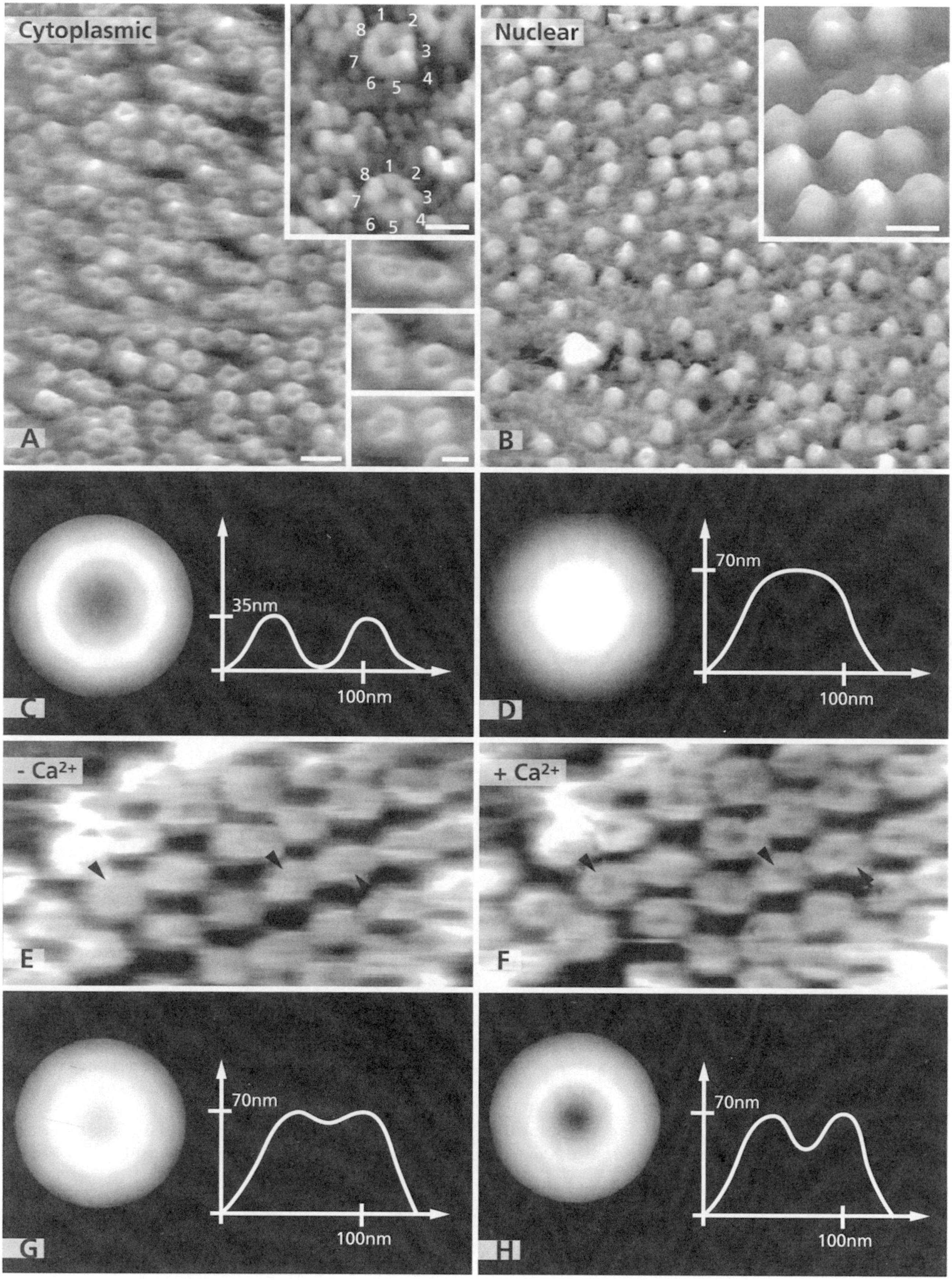
Cytoplasmic
1
2
3
4
5
6
7
8
A
Nuclear
B
35nm
100nm
C
70nm
100nm
D
- Ca2+
E
+ Ca2+
F
70nm
100nm
G
70nm
100nm
H

MITOSIS AND CELL DIVISION

For growth of tissue, development of organs, and maintenance of life functions, both production of new cells by cell division and elimination of cells by programmed cell death (cf. Fig. 11) are necessities. During cell division, triggered by the mitotic cyclin-Cdc2 protein complex at the end of the G2-phase of the cell cycle, mitosis serves to equally distribute all parts of the genome among two daughter cells. Duplication of the centrioles (cf. Fig. 72) and movement toward opposite poles of the cells precede mitosis, which can be followed during four well characterised phases, the prophase, metaphase (panels A and B), anaphase, and telophase. Mitosis is accompanied by reorganisation of the cytoplasm and followed by the actual cell division (cytokinesis).

After DNA-replication in the preceding S-phase of the cell cycle, each chromosome consists of two identical chromatids (sister chromatids) connected at the centromeric region. Two distinct protein complexes, the cohesin and the condensin complex, are required for cohesion of the chromatids and chromatin condensation leading to the typical thread-like mitotic transport structures of the chromosomes. The centrioles are critical for formation and maintenance of the mitotic spindle, which is composed of microtubules and multiple associated proteins. After breakdown of the nuclear envelope in prophase, spindle microtubules insert into the kinetochore at the chromosome centromere and chromosomes are aligned in the equatorial plate. Cohesin is necessary to ensure chromosome bi-orientation. At the metaphase-anaphase transition, destruction of the cohesin complex triggers the segregation of the sister chromatids, which are moved to the poles of the cell. In telophase, chromosomes decondense again, the nuclear envelope is rebuilt and cytokinesis takes place through contraction of a transient contractile actin-myosin ring formed around the equatorial region.

Cells undergo mitosis at different stages of development. Panel A shows an eosinophilic myelocyte from the human bone marrow, which represents a well differentiated cell of the myeloid haematopoietic lineage containing multiple specific eosinophilic granules (cf. Fig. 171). In panel B, a mitotic proerythroblast provides an example of a cell of the red blood cell lineage at an early stage of differentiation. Both cells are shown during metaphase. The homogenous, densely packed chromatin (C) comprises the most prominent structure in the cells. In panel B, a chromosome centromeric region (primary constriction) is visible. Nucleoli and other functional domains, typical for the interphase nucleus (cf. Fig. 2), are not existent, and the former nucleoplasm is intermixed with the cytoplasm in a common compartment. The breakdown of the nuclear envelope is closely connected with the disassembly of the nuclear lamina (cf. Fig. 3). Recent findings indicate that fragmentation of the nuclear envelope is initiated by a microtubule-dependent tearing mechanism, with dynein functioning as the motor protein and leading to the formation of gaps close to areas of maximal tension in the envelope.

In the cytoplasm of both cells, endoplasmic reticulum is still well organised, but the Golgi apparatus is not visible. In mammalian cells, the characteristic stacked Golgi architecture disassembles on onset of mitosis and is reorganised during telophase. The Golgi apparatus fate during mitosis has been a matter of intense debate over the past years. There exist two different models, persistence of distinct Golgi apparatus entities versus complete absorption into the endoplasmic reticulum. Moreover, recent findings imply that Golgi fragmentation not only accompanies, but is a prerequisite for entry into mitosis.

References

Beaudouin J, Gerlich D, Daigle N, Eils R, and Ellenberg J (2002) Nuclear envelope breakdown proceeds by microtubule-induced tearing of the lamina. Cell 108: 83

Lavoie BD, Hogan E, and Koshland D (2002) *In vivo* dissection of the chromosome condensation machinery: reversibility of condensation distinguishes contributions of condensin and cohesin. J Cell Biol 156: 805

Rabouille C and Jokitalo E (2003) Golgi apparatus partitioning during cell division (Review). Mol Membr Biol 20: 117

Schatten H (2008) The mammalian centrosome and its functional significance. Histochem Cell Biol 129: 667

Tanaka TU (2002) Bi-orienting chromosomes on the mitotic spindle. Curr Opin Cell Biol 14: 365

Wei JH and Seemann J (2009) Mitotic division of the mammalian Golgi apparatus. Seminars Cell Dev Biol 20: 810

Magnification: x 10,300 (A), x 13,500 (B)

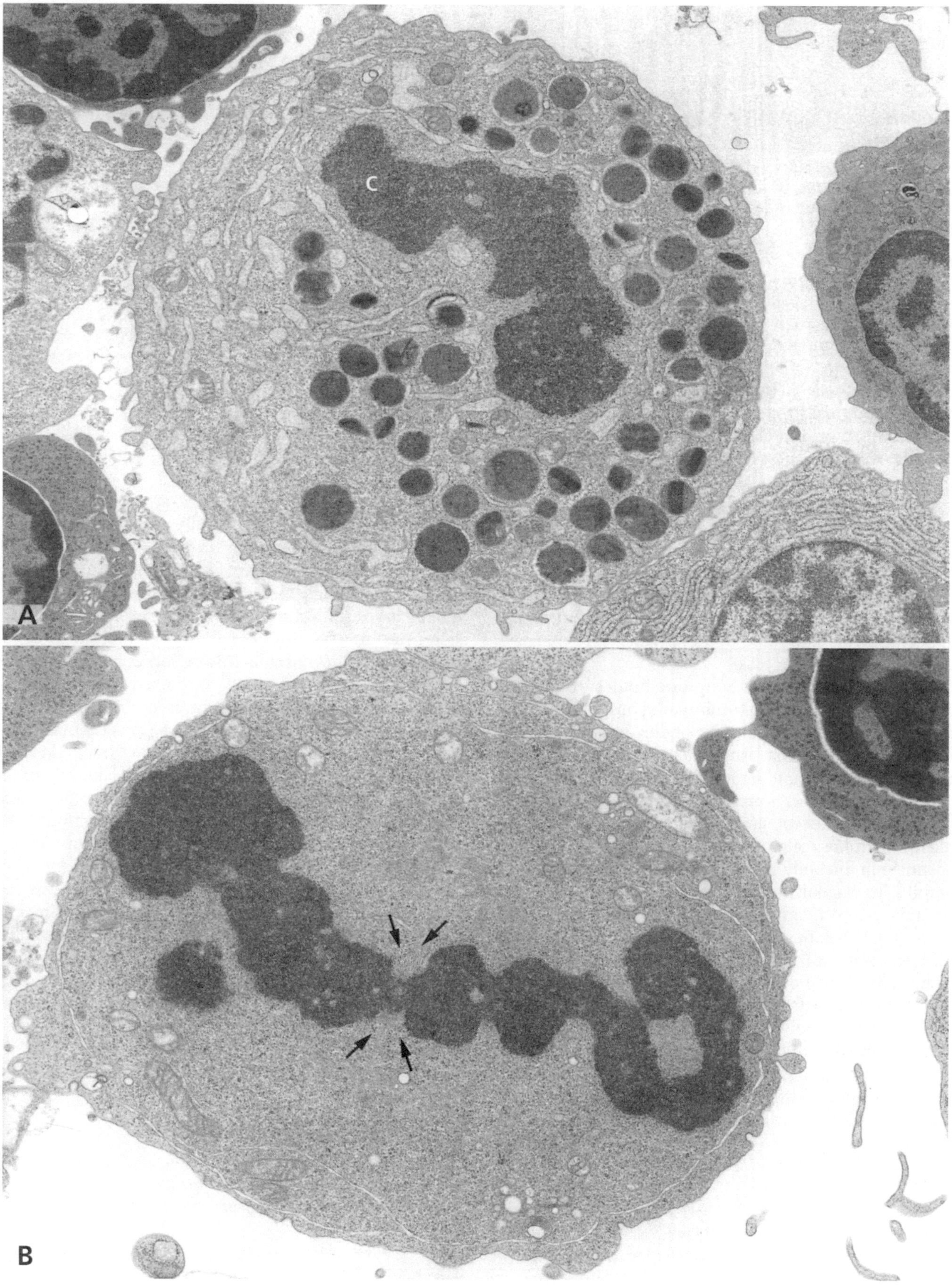
C
A
B

APOPTOSIS

Apoptosis is a physiologic type of programmed cell death that is essential for normal development and regular life of tissues and organs. It is used by multicellular organisms for elimination of "unwanted" cells, which may include excess and unnecessary cells, defective, senescent and harmful cells. Any apoptosis disorder potentially leads to diseases. By contrast to necrosis, which is a nonphysiologic accidental cell death resulting from irreversible cell injury, apoptosis takes place according a genetic cell-suicide programme and is an active process initiated by extern signals or by intrinsic events, such as DNA-damage or irreparable stress at cellular organelles. Both pathways that initiate and regulate apoptosis, the death receptor pathway and the mitochondrial pathway, lead to the activation of particular proteases called caspases (cysteine aspartic acid-specific proteases). A cascade of caspase-mediated cleavage processes takes place causing dramatic cellular changes, which give rise to the characteristic morphological apoptosis patterns. The cells loose plasma membrane asymmetry and surface differentiations and round up. DNA-fragmentation leads to a chromatin hypercondensation and a kind of "explosion" of the cells results in the appearance of "apoptotic bodies", which are phagocytosed by surrounding cells.

Panel A shows typical apoptotic bodies originating from human lymphocytes undergoing programmed cell death following irradiation. The bodies are rounded, but they show intact plasma membranes and contain remnants of the cell nucleus and organelles. The hypercondensed chromatin (asterisk) is present in multiple nuclear fragments and, due to chromatin collapse against the nuclear periphery often appears in the shape of a crescent, as shown in the body at the right hand side of the micrograph. Cleavage of the nuclear DNA hallmarks apoptosis and causes the most severe damage to the cells.

A large number of results indicate that cells use different ways for active self-destruction. Apoptosis is referred to a type I of programmed cell death. In another type (type II), autophagy has a major role, and cytoplasmic constituents are degraded before nuclear destruction. Recent data suggest that functional links exist between apoptosis and autophagic cell death. Both types may occur simultaneously in tissues, and can coexist in the same cell.

References

Abraham MC and Shaham S (2004) Death without caspases, caspases without death. Trends Cell Biol 14: 184

Bursch W, Ellinger A, Gerner Ch, and Schulte-Hermann R (2003) Caspase-independent and autophagic cell death. In: When cells die (Lockshin R, Zakeri Z, and Tilly JL, eds). New York Chichester Weinheim Brisbane Singapore Tokyo: Wiley-Liss

Cohen I, Castedo M, and Kroemer G (2002) Tantalizing Thanatos: unexpected links in death pathways. Trends Cell Biol 12: 293

Cuervo AM (2004) Autophagy: in sickness and in health. Trends Cell Biol 14: 70

Machamer CE (2003) Golgi disassembly in apoptosis: cause or effect? Trends Cell Biol 13: 279

Penninger JM and Kroemer G (2003) Mitochondria, AIF and caspases – rivalling for cell death execution. Nat Cell Biol 5: 97

Taatjes DJ, Sobel BE, and Budd RC (2008) Morphological and cytochemical determination of cell death by apoptosis. Histochem Cell Biol 129: 33

Zhang J and Xu M (2002) Apoptotic DNA fragmentation and tissue homeostasis. Trends Cell Biol 12: 84

VIRAL INCLUSIONS

Viruses account for a large number of acute infections and occur as a consequence of hereditary or acquired forms of immunodeficiency. Because of their small size (20–300 nm), single virus particles can only be detected by electron microscopy. Some viruses may form aggregates in the nucleus and/or cytoplasm of the infected cells. Such viral inclusion bodies may be visible by light microscopy and definitely by electron microscopy. Polyoma viruses are ubiquitous in nature. Infections can be often observed in kidney tubular cells of renal transplants. Panels B and C show viral inclusion bodies due to Polyoma viruses in a renal biopsy of a kidney transplant patient. The spherical virus particles have a diameter of 30–45 nm and are arranged in characteristic paracrystalline arrays, which occupy most of the nucleoplasm (B) and parts of the cytoplasm (C). The paracrystalline arrays formed by the viruses can be well appreciated at higher magnification (inset in B). The nuclear viral inclusions can be unequivocally distinguished from the nucleolus (Nu in B). Another virus causing combined nucleo-cytoplasmic inclusions is the cytomegalovirus which belongs to the Herpes virus family. Cytomegalovirus infections affect multiple organs and are often observed in immunosuppressed transplant recipients or AIDS patients.

References

Colvin RB (1998) Renal transplant pathology. In: Heptinstall's pathology of the kidney (Jenette J, Olson J, and Schwartz M, eds). Philadelphia: Lippincott-Raven, pp 1409–1540

Magnification: x 6,800 (A), x 10,000 (B, C), x 32,000 (inset)

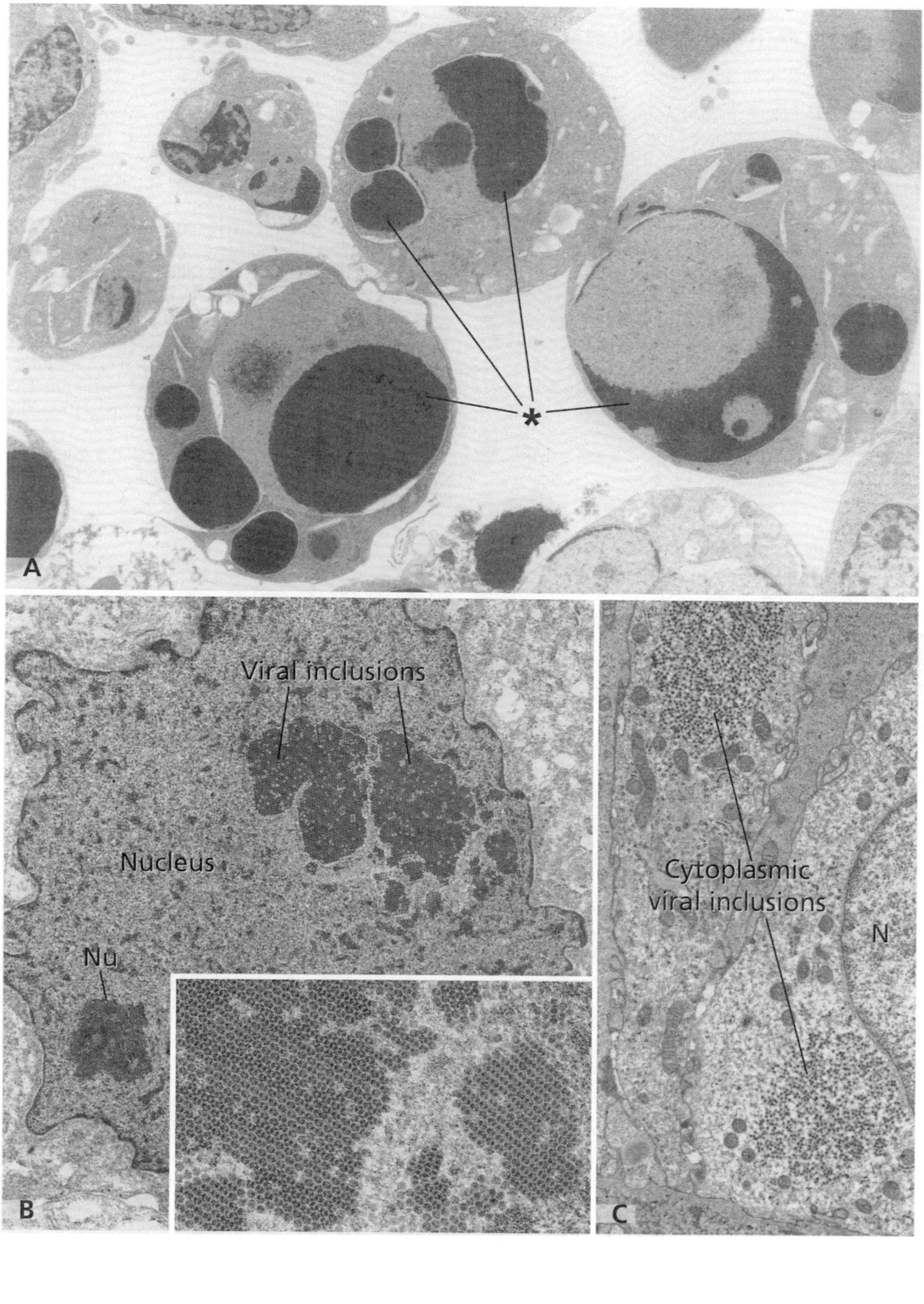
*
A
Viral inclusions
Nucleus
Nu
B
Cytoplasmic
viral inclusions
N
C

SECRETORY PATHWAY OF PANCREATIC ACINAR CELLS

In their classical Nobel prize winning studies on protein biosynthesis and secretion, Palade and coworkers used the exocrine pancreas as a model tissue. The low power electron micrograph from rat exocrine pancreas shows serous acinar cells and immunogold labelling for one of their major secretory products, the digestive enzyme amylase. As such, it provides a general idea about the main structural elements of the secretory pathway in a highly specialized cell. The immunoelectron microscopic demonstration of intracellular antigens in ultrathin sections has been critically improved by the introduction of the colloidal gold as a marker. The scheme shows the two-step protein A-gold technique to detect antigenic sites in an ultrathin section introduced by Roth and colleagues in 1978.

Several of serous cells assemble to form a structural and functional unit, a so called acinus, which is the main component of the gland parenchyma (cf. Figs. 1, 95 and 98).

The rough endoplasmic reticulum (RER), which is abundant in the basal and juxtanuclear portions of the acinar cells and other secretory cell types, represents the first structural element of the secretory pathway, also referred to as the endomembrane system. It shows abundant immunogold labelling for amylase and the gold particles appear as black spots. The next major element is represented by the Golgi apparatus, which can be seen in its typical supranuclear location and which is intensely labelled for amylase. The pre-Golgi intermediates cannot be recognized at this low magnification. The supranuclear cytoplasm contains a high number of secretory granules in which amylase and many other secretory products are stored. The secretory granules of the acinar cells are called zymogen granules (ZG) because of their content of proenzymes. On appropriate stimulation, the zymogen granules fuse with the apical plasma membrane and discharge their content into the acinus lumen (AL). This stimulus-mediated exocytosis is called regulated secretion. The acinus lumen represents the first part of the extracellular excretory duct system.

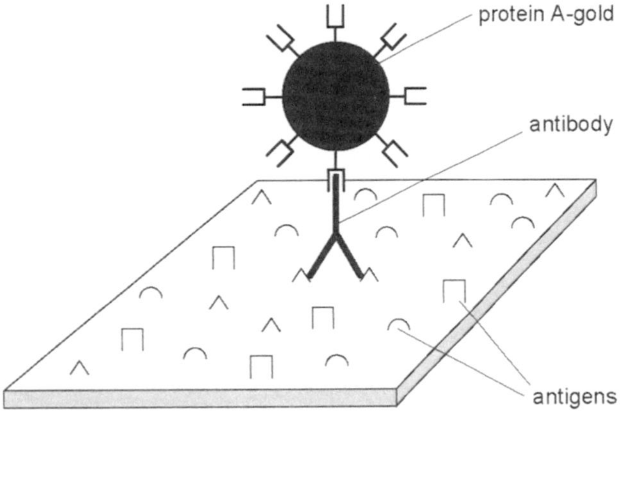

A detailed account of the structural and functional aspects of the secretory pathway is given in the following plates.

M: mitochondria.

References

Cantin M (1984) Cell biology of the secretory process. Basel: Karger

Palade G (1975) Intracellular aspects of the process of protein biosynthesis. Science 189: 347

Pfeffer S (2007) Unsolved mysteries in membrane traffic. Annu Rev Biochem 76: 629

Roth J (1981) The colloidal gold marker system for light and electron microscopic cytochemistry. In: Techniques in immunocyto-chemistry, vol 2 (Bullock GR, Petrusz P, eds). London: Academic Press, pp 217

Roth J (1989) Postembedding labeling on Lowicryl K4M tissue sections: detection and modification of cellular components. In: Meth Cell Biol (Tartakoff AM, ed). San Diego: Academic Press, pp 513

Roth J, Bendayan M, and Orci L (1978) Ultrastructural localization of intracellular antigens by the use of protein A-gold complex. J Histochem Cytochem 26: 1074

Steer C and Hanover J (1991) Intracellular trafficking of proteins. Cambridge: Cambridge University Press

Magnification: x 15,500

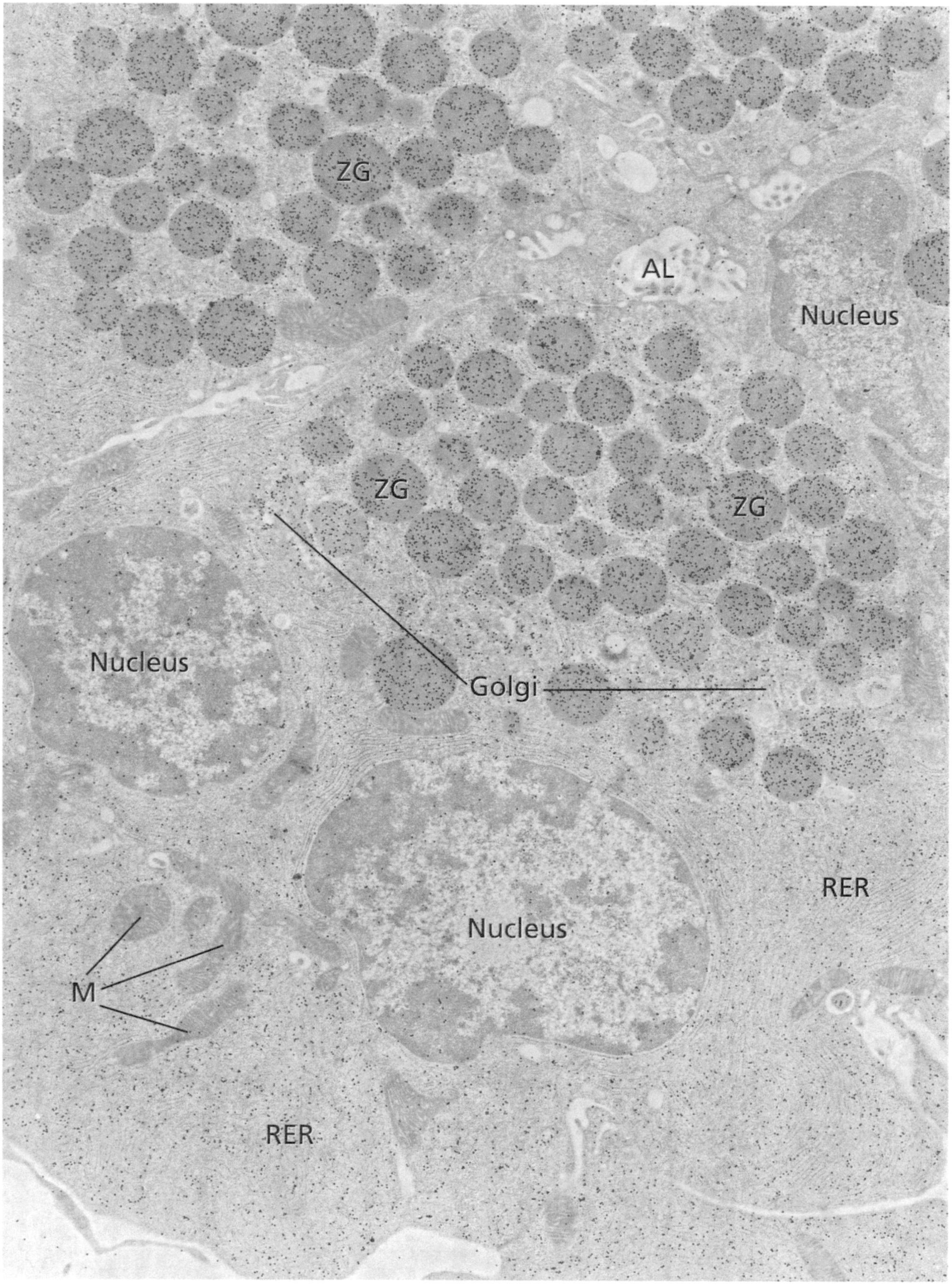
ZG
AL
Nucleus
ZG
ZG
Nucleus
Golgi
RER
Nucleus
M
RER

ENDOMEMBRANE SYSTEM OF DINOFLAGELLATES

Although the various organelles of mammalian cells are in the focus of this book, it should be noted that unicellular organisms have provided invaluable information about the biogenesis, molecular structure and function of various cellular organelles, since their principles have been found to be phylogenetically highly conserved. As a case in point, studies in the yeast *Saccharomyces cerevisiae* were of fundamental importance for the molecular dissection of the secretory pathway and of autophagy. Despite, in wild type *Saccharomyces cerevisiae,* elements of the rough endoplasmic reticulum are scarce and stacks of flat cisternae building a Golgi apparatus resembling the one of mammalian cells are rarely found. However, in *sec7* and *sec14* secretory mutants, Golgi apparatus composed of cisternal stacks could be readily observed at the non-permissive temperature. Other unicellular organisms such as the parasitic protists contain a comparatively well developed endomembrane system.

The marine dinoflagellates represent other unicellular organisms with a highly developed endomembrane system as shown in panel A. For high contrast and distinct appearance of cellular membranes, the cells were fixed in reduced osmium tetroxide after initial aldehyde fixation. However, by applying this fixation protocol, ribosomes can be barely recognized. In panel A, an equatorially sectioned dinoflagellate is shown, which contains a prominent nucleus and numerous plastids in its periphery. The many mitochondria are of tubulus type (cf. Fig. 64) and several cisternae of the endoplasmic reticulum can be seen. In panel B, a site of continuity between the nuclear envelope (NE) and endoplasmic reticulum is indicated by an arrowhead. In close relationship with a curved cisterna of the endoplasmic reticulum (ER), a membrane cluster composed of many tubules and vesicles is visible which most probably represents a peripheral pre-Golgi intermediate (cf. Fig. 23). In panel C, a Golgi apparatus cisternal stack is shown with numerous associated vesicles at either of its poles and the lateral extremities. Vesicles and short tubules are also located between a presumptive transitional element of the endoplasmic reticulum (ER) and a fenestrated *cis* Golgi cisterna. Such solid stacks of flattened cisternae appear to be linked by fenestrated elements and tubules, alike seen in mammalian cells. The arrowhead in panel C marks a (clathrin)-coated lateral extension of a *trans* Golgi cisterna.

References

Grell KG and Wohlfarth-Bottermann KE (1957) Light- and electron microscopy of the Dinoflagellate Amphidinium elegans n. sp. Z Zellforsch Mikrosk Anat 47: 7

Kaiser CA and Schekman R (1990) Distinct sets of SEC genes govern transport vesicle formation and fusion early in the secretory pathway. Cell 61: 723

Novick P, Field C, and Schekman R (1980) Identification of 23 complementation groups required for post-translational events in the yeast secretory pathway. Cell 21: 205

Rambourg A, Daraspe J, Képès F, and Verbavatz J-M (2008) Morphodynamics of the yeast Golgi apparatus. In: The Golgi apparatus. State of the art 110 years after Camillo Golgi's discovery (Mironov AA, Pavelka M, eds). Wien NewYork: Springer, pp 631

Sokolova YY and Mironov AA (2008) Structure and function of the Golgi organelle in parasitic protists. In: The Golgi apparatus. State of the art 110 years after Camillo Golgi's discovery (Mironov AA, Pavelka M, eds). Wien NewYork: Springer, pp 647

Taylor FJ (1980) On dinoflagellate evolution. Biosystems 13: 65

Wuestehube LJ, Duden R, Eun A, Hamamoto S, Korn P, Ram R, and Schekman R (1996) New mutants of Saccharomyces cerevisiae affected in the transport of proteins from the endoplasmic reticulum to the Golgi complex. Genetics 142: 393

Magnification: x 3,150 (A); x 50,000 (B); x 46,000 (C)

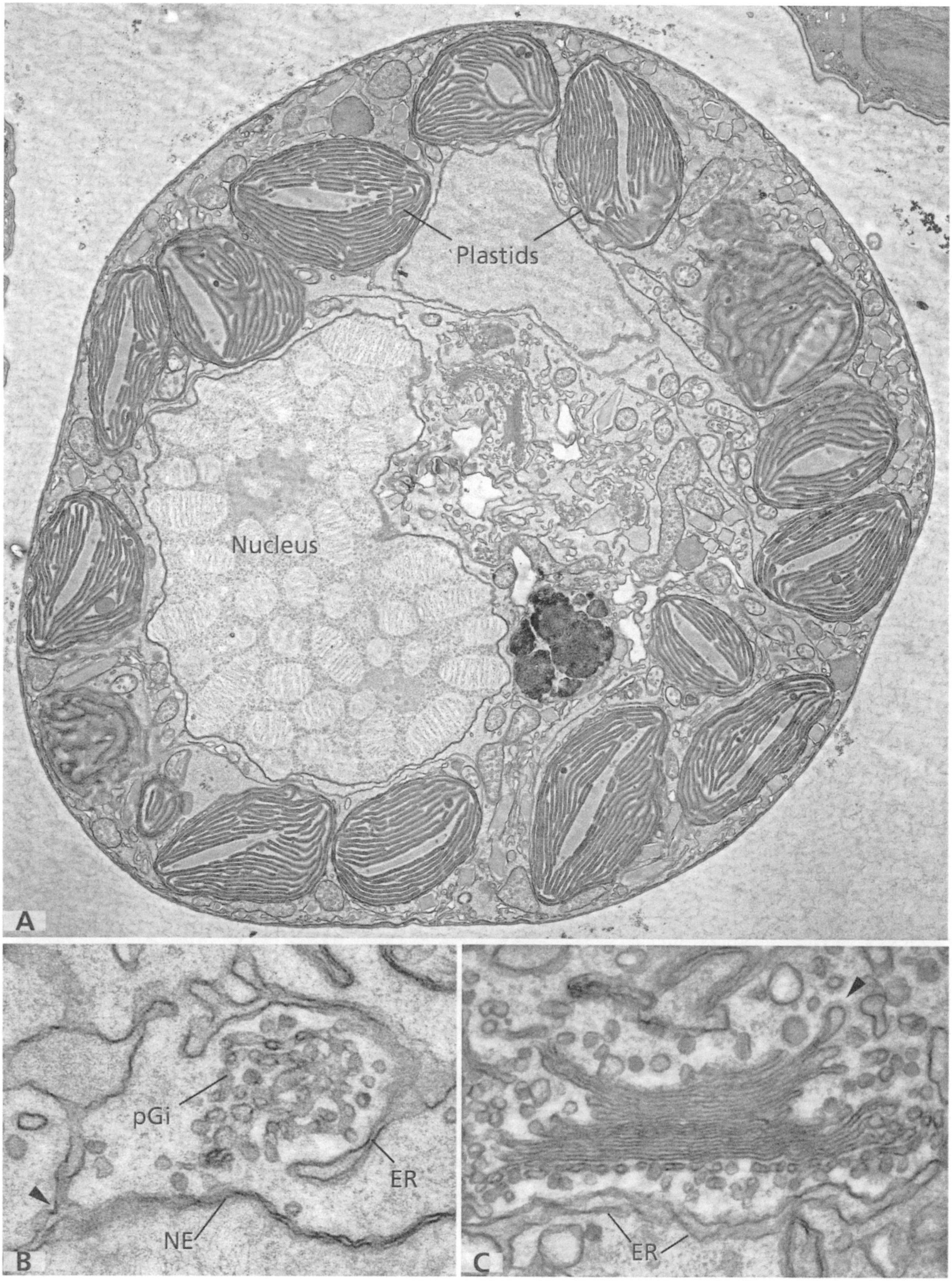
Plastids
Nucleus
A
pGi
ER
NE
B
ER
C

RIBOSOMES, ROUGH ENDOPLASMIC RETICULUM

Ribosomes are RNA-protein particles measuring 30nm at their largest dimensions, which function as machineries for protein synthesis. Within a ribosome, a messenger-RNA (mRNA) and transfer-RNAs (tRNA) are brought together, thus allowing base pairing between mRNA codons and tRNA anticodons, which drives the synthesis of a polypeptide with an amino acid sequence according to the model specified by the mRNAs codons.

Ribosomes are composed of two subunits, which are formed in the nucleolus. The diagram summarises the major steps in eukaryotic ribosome biogenesis.

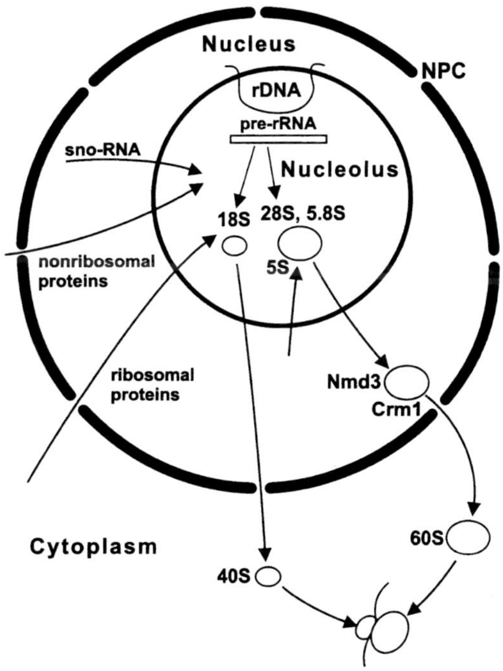

Co- and posttranscriptionally, the preribosomal RNA (pre-rRNA) encoded by the ribosomal DNA (rDNA) associates with small nucleolar RNAs (snoRNAs) and non-ribosomal proteins to be modified and processed. A series of cleavages takes place, leading to the 18S, 28S and 5.8S rRNAs, which together with ribosomal proteins and the 5S rRNA synthesised outside the nucleolus, assemble to form the small and large precursors of the 40S and 60S ribosomal subunits. They are separately exported out of the nucleus through the nuclear pore complex (NPC), the large precursor utilising the adaptor protein Nmd3 for binding to the Crm1 nuclear export factor. Ribosomes exist as distinct particles only during ongoing protein synthesis. The 40S and 60S subunits assemble during the protein synthesis initiation phase and again dissociate and enter in a common cytoplasmic pool after completion of the protein. The diagram is modified from Olson et al. (2002).

Polysomes are ribosome clusters attached to one mRNA, which simultaneously produce several copies of the same protein. Depending on the class of proteins synthesised, polysomes either reside "freely" in the cytoplasm ("free polysomes") or bind to membranes of the rough endoplasmic reticulum (RER, "bound polysomes"). The latter process is receptor-mediated and guided by a signal peptide, which is part of all proteins to be synthesised at the RER.

Panels A and B show details of cells extensively involved in synthesis of secretory proteins. The basal cytoplasm of a pancreatic acinar cell shown in panel A contains densely packed flat RER-cisternae forming a cellular compartment called ergastoplasm ("busy plasma"). The membranes are studded with ribosomes (arrowheads) and in the cisternal lumen fine filamentous materials are contained, mainly corresponding to newly synthesised secretory proteins. Polysomes appear as "rosettes", strings of beads in single or double rows (arrowheads in panel B) or serpent-like structures (arrows in panel B).

RER-bound ribosomes also produce luminal and membrane proteins of the entire secretory system, lysosomal enzymes and plasma membrane proteins. By contrast, ribosomes do not bind to the RER during synthesis of other proteins, such as proteins residing at the cytoplasmic side of membranes, cytoskeletal proteins and proteins destined to be transported into the nucleus, or inserted into mitochondria or peroxisomes. The RER is organised into subregions (cf. also Fig. 15) and recent findings suggest that particular mRNAs encoding for secretory or membrane proteins are confined to distinct areas of the RER. There is evidence for transport of mRNAs along either microtubules or actin filaments and several RER-associated proteins involved in anchoring mRNAs have been identified.

References

Baumann O and Walz B (2001) Endoplasmic reticulum of animal cells and its organization into structural and functional domains. Int Rev Cytol 205: 149

Nikonov AV, Snapp E, Lippincott-Schwartz J, and Kreibich G (2002) Active translocon complexes labeled with GFP-Dad1 diffuse slowly as large polysome arrays in the endoplasmic reticulum. J Cell Biol 158: 497

Olson, MOJ, Hingorani, K, and Szeneni, A (2002) Conventional and nonconventional roles of the nucleolus. Int Rev Cytol 219: 199

Magnification: x 52,000 (A), x 46,000 (B)

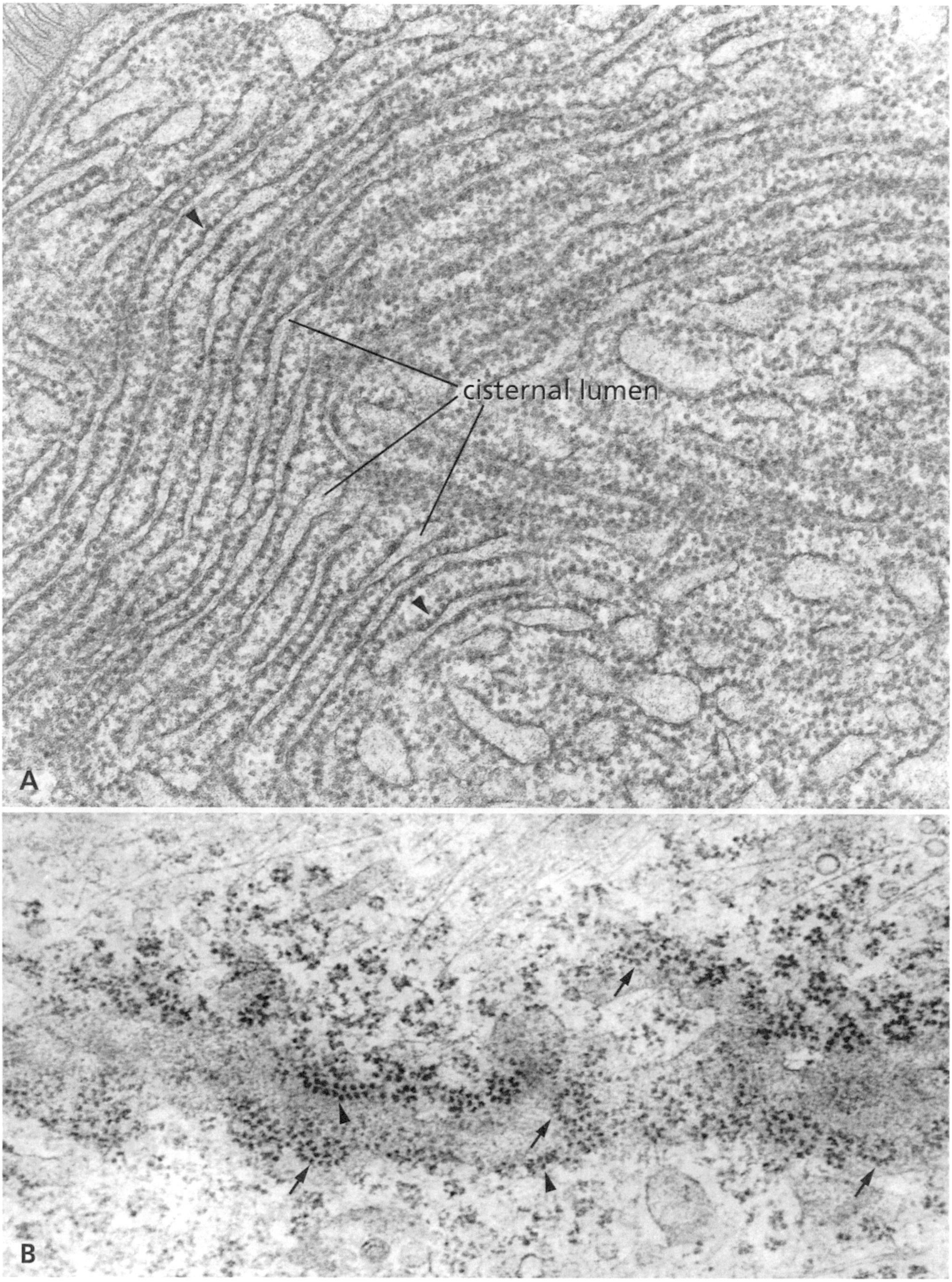
cisternal lumen
A
B

NUCLEAR ENVELOPE AND ROUGH ENDOPLASMIC RETICULUM

The nuclear envelope (NE) consists of an inner and an outer membrane enclosing the perinuclear cisterna or perinuclear space. The outer nuclear membrane is studded with ribosomes and continuous with the membrane of the rough endoplasmic reticulum (RER). Therefore, direct continuity exists between the lumen of the nuclear envelope and the rough endoplasmic reticulum, as marked by the lines in panel A. Evidence is ample that the outer nuclear membrane is engaged in protein synthesis and that *de novo* synthesized proteins are translocated in the perinuclear space. Post-translational modifications on proteins may occur here because of the presence of oligosaccharide-trimming glycosidases (cf. Fig. 18).

Although the rough endoplasmic reticulum and the nuclear envelope form a continuous network, functional domains seem to exist in this luminal interconnected system. As an example, the heterogeneous distribution of apomucin in mucus cells of submandibular glands is presented. Mucin is the main secretory product in this cell type and can be detected by immunoelectron microscopy. In panel A, immunogold labelling for apomucin in the cisternal lumen of the rough endoplasmic reticulum is evident. However, the perinuclear cisterna even at sites of continuity with the cisternal lumen of the rough endoplasmic reticulum is unlabelled. In panel B, intensely apomucin-positive parts of cisternae of rough endoplasmic reticulum can be seen alternating with parts that are completely unlabelled (marked by lines). This represents a remarkable example of segregation of a major secretory product in the lumen of the continuous network of rough endoplasmic reticulum cisternae. As for other reported examples, it is currently unknown how such domains become established and are maintained. It has been proposed that the endoplasmic reticulum is a site of mRNA localization which will associate for their life time with membrane-bound polyribosomes. As a results the synthesis of the respective proteins will be compartmentalized.

References

Baumann O and Walz B (2001) Endoplasmic reticulum of animal cells and its organization into structural and functional domains. Int Rev Cytol 205: 149

Deschuyteneer M, Eckhardt AE, Roth J, and Hill RL (1988) The subcellular localization of apomucin and nonreducing terminal N-acetylgalactosamine in porcine submaxillar glands. J Biol Chem 263: 2452

Ilnuma T, Aoki T, Arasaki K, Hirose H, Yamamoto A, Samata R, Hauri H, Arimitsu M, Tagaya M, Tani K (2009) Role of syntaxin 18 in the organization of endoplasmic reticulum subdomains. J Cell Sci 122: 1680

Nicchitta CV (2002) A platform for compartmentalized protein synthesis: protein translation and translocation in the ER. Curr Opin Cell Biol 14: 412

Nikonov AV, Snapp E, Lippincott-Schwartz J, and Kreibich G (2002) Active translocon complexes labeled with GFP-Dad1 diffuse slowly as large polysome arrays in the endoplasmic reticulum. J Cell Biol 158: 497

Pfeffer S (2003) Membrane domains in the secretory and endocytic pathways. Cell 112: 507

Prunuske AJ and Ullman KS (2006) The nuclear envelope: form and reformation. Curr Opin Cell Biol 18: 108

Sitia R and Meldolesi J (1992) Endoplasmic reticulum – a dynamic patchwork of specialized subregions. Mol Biol Cell 3: 1067

Spiliotis ET, Pentcheva T, and Edidin M (2002) Probing for membrane domains in the endoplasmic reticulum: Retention and degradation of unassembled MHC class I molecules. Mol Biol Cell 13: 1566

Voeltz GK, Prinz WA, Shibata Y, Rist JM, Rapoport TA (2006) A class of membrane proteins shaping the tubular endoplasmic reticulum. Cell 124:573

Magnification: x 98,000 (A); x 38,300 (B)

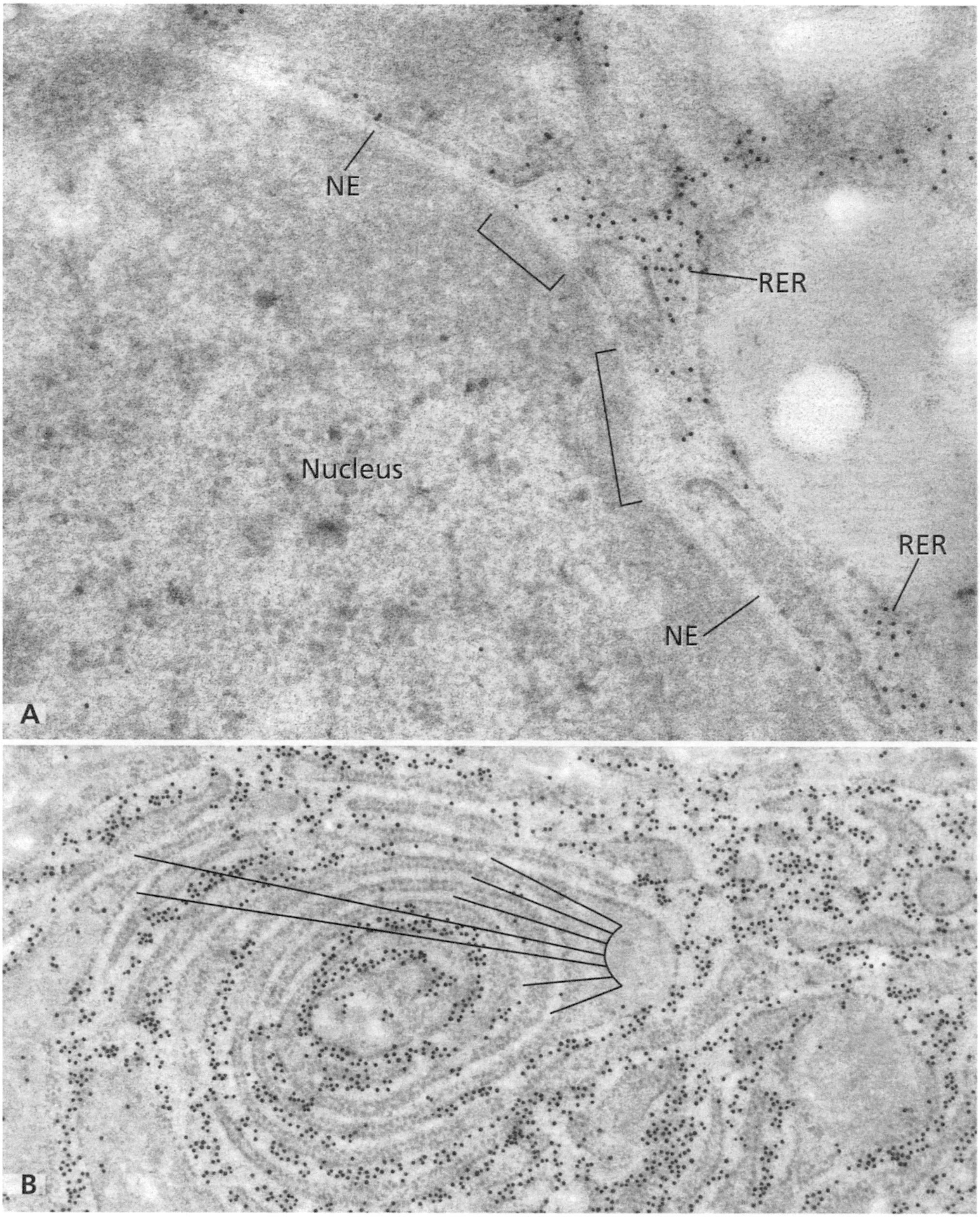
NE
RER
Nucleus
RER
NE
A
B

ANNULATE LAMELLAE

Annulate lamellae are peculiar cytoplasmic organelles composed of stacked sheets of endoplasmic reticulum-like membranes interrupted by annuli or pores. They are most commonly found in oocytes as well as embryonic and neoplastic cells, but they can be observed, albeit to a lesser degree, in virtually any eukaryotic cell type.

Both their origin and their function(s) are poorly understood. In many cell types, annulate lamellae seem to originate from the nuclear envelope, since they are often apposed to it as can be seen in the lower right corner of panel A. Here, a cross-sectioned single annulate lamellae structure with a pore (arrowhead) is located adjacent to the nuclear envelope with nuclear pore complexes (NPC). Thus, for embryonic cells, they have been proposed to transiently store or accumulate excess nuclear pore material and to be transitory organelles. Other investigators have concluded that annulate lamellae are derived from the rough endoplasmic reticulum, since they were found continuous with its cisternae as can also be seen in panel A (ER). In panel A, a cross-sectioned annulate lamellae is depicted, which consists of lamellae that are arranged in a parallel array. The lamellae are interrupted by pores (arrowheads), which are aligned. The pores are spanned by a diaphragm-like structure and exhibit small peripheral granules. Continuity between single lamella and cisternae of the rough endoplasmic reticulum (ER in A and B) is obvious. In panel B, a grazing section through an annulate lamellae is shown. Numerous pores (arrowhead) are seen in a face-on view, which reveals the presence of a large central granule and smaller annular granules in the pores. These are characteristic features of nuclear pore complexes. By morphological criteria, the pores of annulate lamellae are indistinguishable from nuclear pore complexes.

References

Beckhelling C, Chang P, Chevalier S, Ford C, and Houliston E (2003) Pre-M phase-promoting factor associates with annulate lamellae in Xenopus oocytes and egg extracts. Mol Biol Cell 14: 1125

Cordes VC, Reidenbach S, and Franke WW (1995) High content of a nuclear pore complex protein in cytoplasmic annulate lamellae of Xenopus oocytes. Eur J Cell Biol 68: 240

Cordes VC, Reidenbach S, and Franke WW (1996) Cytoplasmic annulate lamellae in cultured cells: composition, distribution, and mitotic behavior. Cell Tissue Res 284: 177

Franke WW, Scheer U, and Fritsch H (1972) Intranuclear and cytoplasmic annulate lamellae in plant cells. J Cell Biol 53: 823

Imreh G and Hallberg E (2000) An integral membrane protein from the nuclear pore complex is also present in the annulate lamellae: implications for annulate lamella formation. Exp Cell Res 259: 180

Kessel RG (1992) Annulate lamellae: a last frontier in cellular organelles. Int Rev Cytol 133: 43

Ralle T, Grund C, Franke WW, and Stick R (2004) Intranuclear membrane structure formations by CaaX-containing nuclear proteins. J Cell Sci 117: 6095

Scheer U and Franke WW (1969) Negative staining and adenosine triphosphatase activity of annulate lamellae of newt oocytes. J Cell Biol 42: 519

Stafstrom JP and Staehelin LA (1984) Are annulate lamellae in the Drosophila embryo the result of overproduction of nuclear pore components? J Cell Biol 98: 699

Sutovsky P, Simerly C, Hewitson L, and Schatten G (1998) Assembly of nuclear pore complexes and annulate lamellae promotes normal pronuclear development in fertilized mammalian oocytes. J Cell Sci 111: 2841

Magnification: x 53,500 (A); x 57,800 (B)

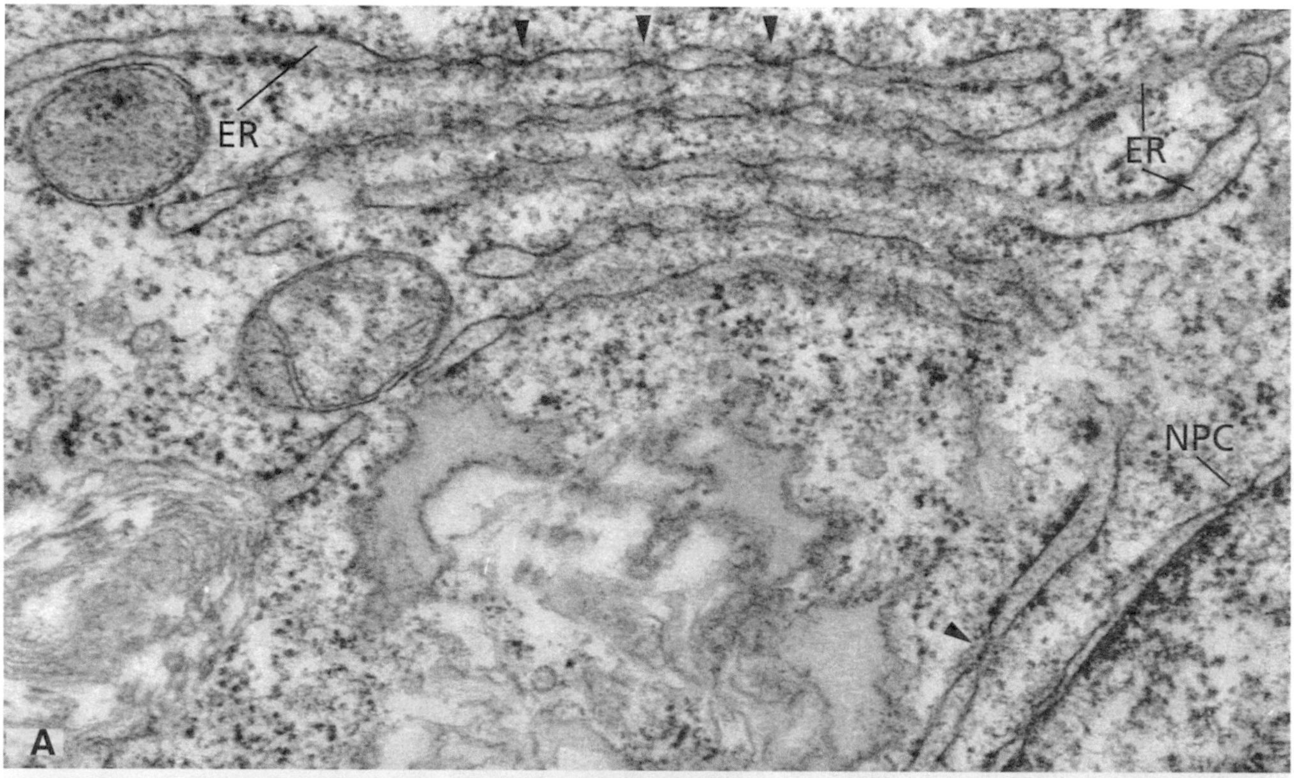
ER
ER
NPC
A

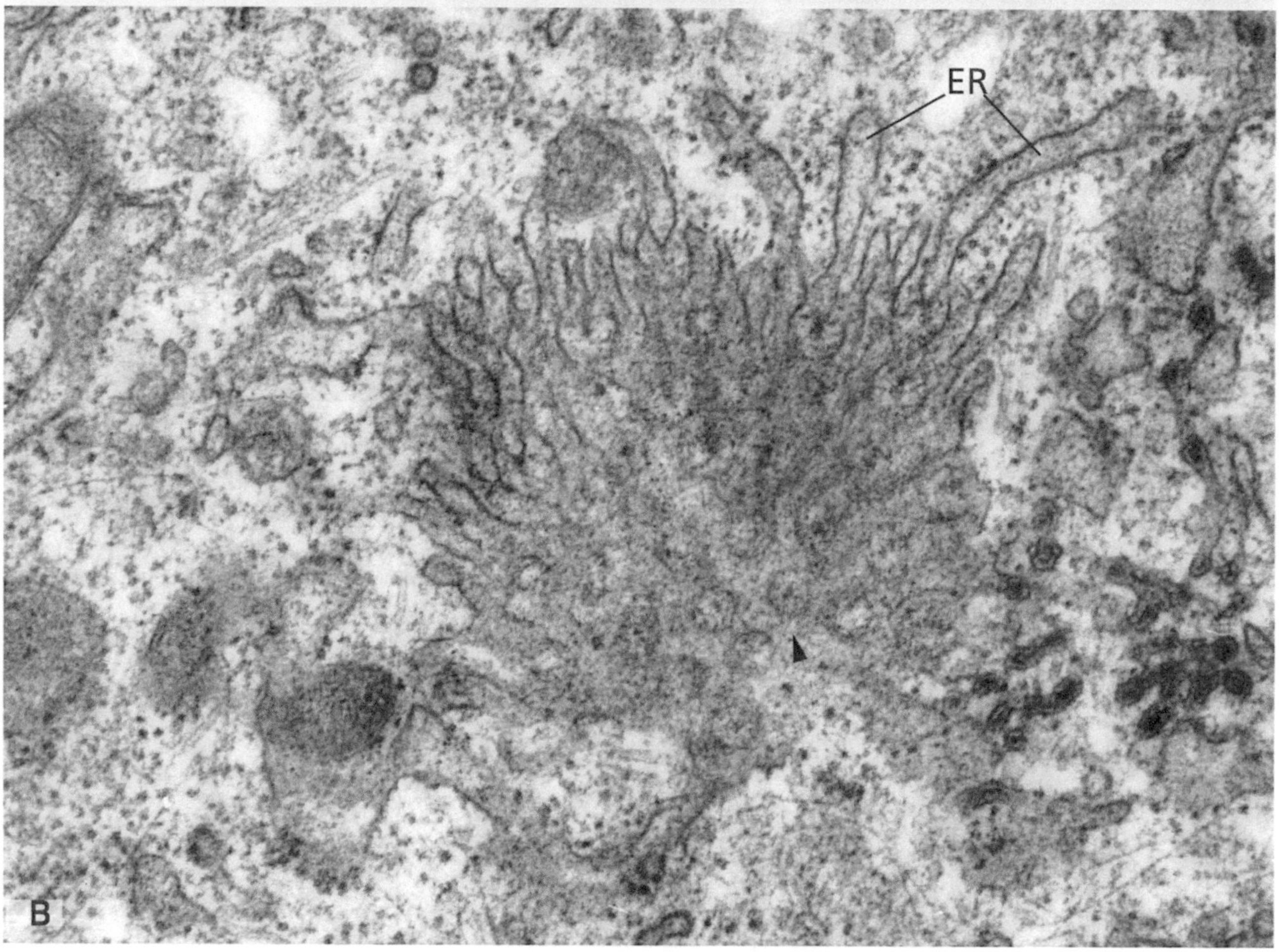
ER
B

ROUGH ENDOPLASMIC RETICULUM: SITE OF PROTEIN TRANSLOCATION AND INITIATION OF PROTEIN *N*-GLYCOSYLATION

The membranes of the endoplasmic reticulum (ER) represent the site of co-translational vectorial transfer of *de novo* synthesized proteins from the cytosol to the ER and of lipid biosynthesis. After translocation, post-translational protein modifications occur in the cisternal lumen.

In the rough endoplasmic reticulum (RER), soluble proteins become fully translocated into the cisternal lumen, whereas the other protein type, the transmembrane proteins, are translocated only partially and become inserted in the ER membrane to reside here or to become routed to other cellular destinations. The process of protein translocation is mediated by the translocon or Sec61 complex, which is a sophisticated modular protein machinery of which the heterotrimeric Sec61p, TRAM (translocation associated membrane protein) and TRAP (translocon associated protein) are the major components. Panel A depicts the immunogold localization of Sec61protein at the RER and outer nuclear membrane in a rat liver hepatocyte. General structural and functional aspects of the translocon seem to be highly conserved in prokaryotic and eukaryotic organisms. Based on high resolution structural analyses, the translocon consists of a doughnut-like structure with a central hydrophilic pore as the most striking feature. The pore diameter in a ribosome bound translocon is 40–60 Å and that of a ribosome free one 9–15 Å. A ribosome may be associated at the cytosolic side with its tunnel co-axially oriented over the Sec61 pore.

The oligosaccharyltransferase and the signal peptidase constitute major translocon-associated proteins not directly involved in the pore formation and in the translocation process. The oligosaccharyltransferase is a hetero-oligomeric complex composed of at least nine subunits showing significant homology from yeast to mammalian cells. For functional reasons it is neighbouring the translocon. It acts on nascent polypeptides while they are being translocated and catalyzes the *en bloc* transfer of a lipid-linked pre-assembled oligosaccharide (glucose$_3$mannose$_9$*N*-acetylglucosamine$_2$) to selected Asn-X-Ser/Thr sequences (X any amino acid except proline) of nascent polypeptide chains. This reaction occurs exclusively on the luminal side of the ER membrane and represents a key step in the pathway of protein *N*-glycosylation, which is an essential and highly conserved protein modification. The asparagine-linked oligosaccharide can be detected by the plant lectin Concanavalin A (Con A) because of its reactivity with oligosaccharides containing glucose or mannose. In panels B and C, Con A-gold labelling of the RER and of smooth endoplasmic reticulum (SER) of rat liver hepatocytes is shown. In regions containing SER, Con A labeling is also due to the presence of glycogen. The lumen of the liver sinus and bile capillary (BC) are unlabelled.

References

Alder NN, Shen Y, Brodsky JL, Hendershot LM, and Johnson AE (2005) The molecular mechanisms underlying BiP-mediated gating of the Sec61 translocon of the endoplasmic reticulum. J Cell Biol 168: 389

Burda P and Aebi M (1999) The dolichol pathway of N-linked glycosylation. Biochim Biophys Acta 1426: 239

Chavan M and Lennarz W (2006) The molecular basis of coupling of translocation and N-glycosylation. Trends Biochem Sci 31: 17

Johnson AE (2009) The structural and functional coupling of two molecular machines, the ribosome and the translocon. J Cell Biol 185: 765

Knauer R and Lehle L (1999) The oligosaccharyltransferase complex from yeast. Biochim Biophys Acta 1426: 259

Menetret JF, Hegde RS, Heinrich SU, Chandramouli P, Ludtke SJ, Rapoport TA, and Akey CW (2005) Architecture of the ribosome-channel complex derived from native membranes. J Mol Biol 348: 445

Osborne AR, Rapoport TA, and van den Berg B (2005) Protein translocation by the Sec61/SecY channel. Annu Rev Cell Dev Biol 21: 529

Silberstein S and Gilmore R (1996) Biochemistry, molecular biology, and genetics of the oligosaccharyltransferase. FASEB J 10: 849

van den Berg B, Clemons WM, Collinson I, Modis Y, Hartmann E, Harrison SC, and Rapoport TA (2004) X-ray structure of a protein-conducting channel. Nature 427: 36

Zuber Z and Roth J (2009) *N*-glycosylation. In: The Sugar Code. Fundamentals of Glycosciences (Gabius H-J, ed). Weinheim: Wiley/VCH, pp 87

Magnification: x 71,500 (A); x 11,000 (B); x 35,500 (C)

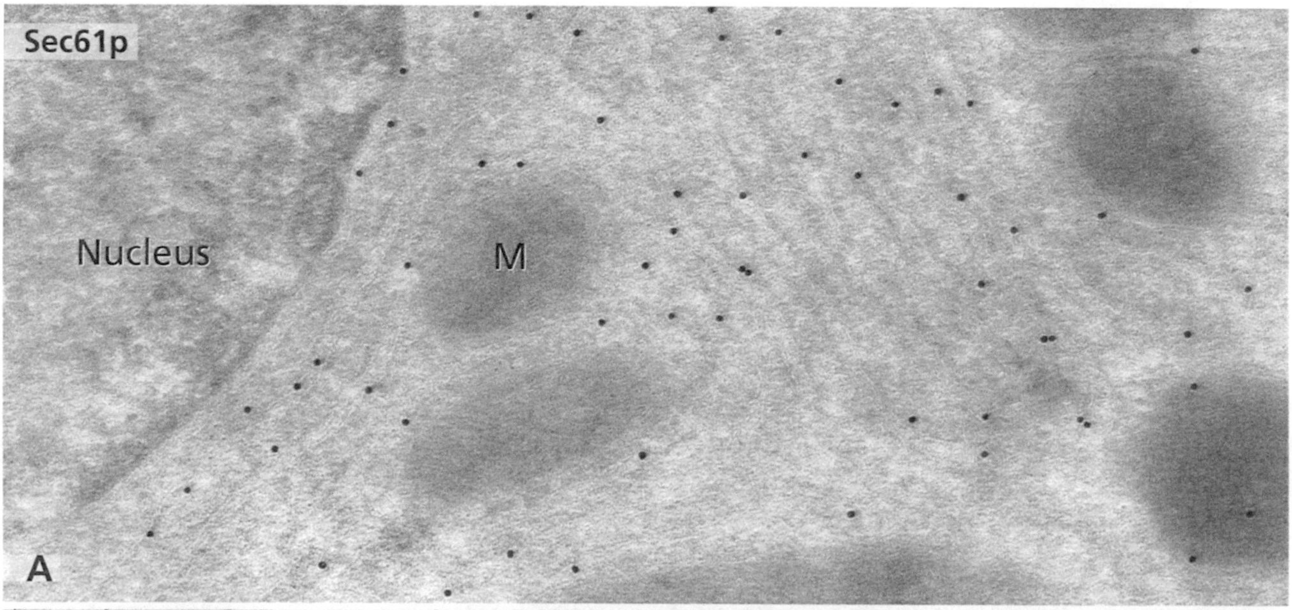
Sec61p
Nucleus
M
A

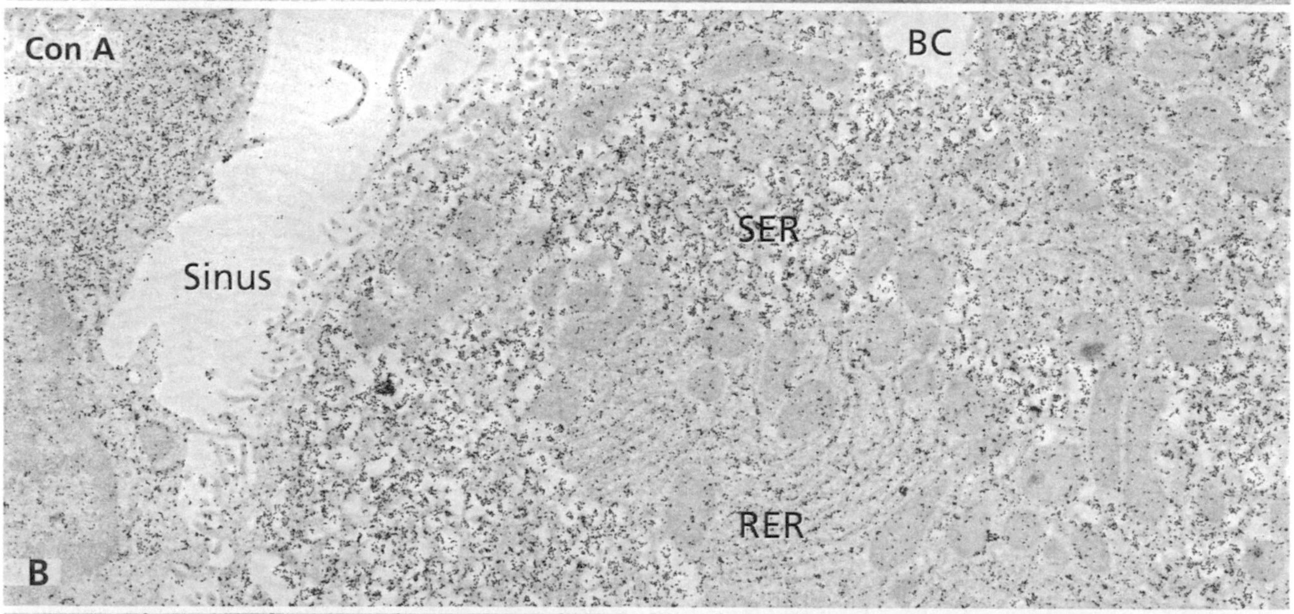
Con A
BC
SER
Sinus
RER
B

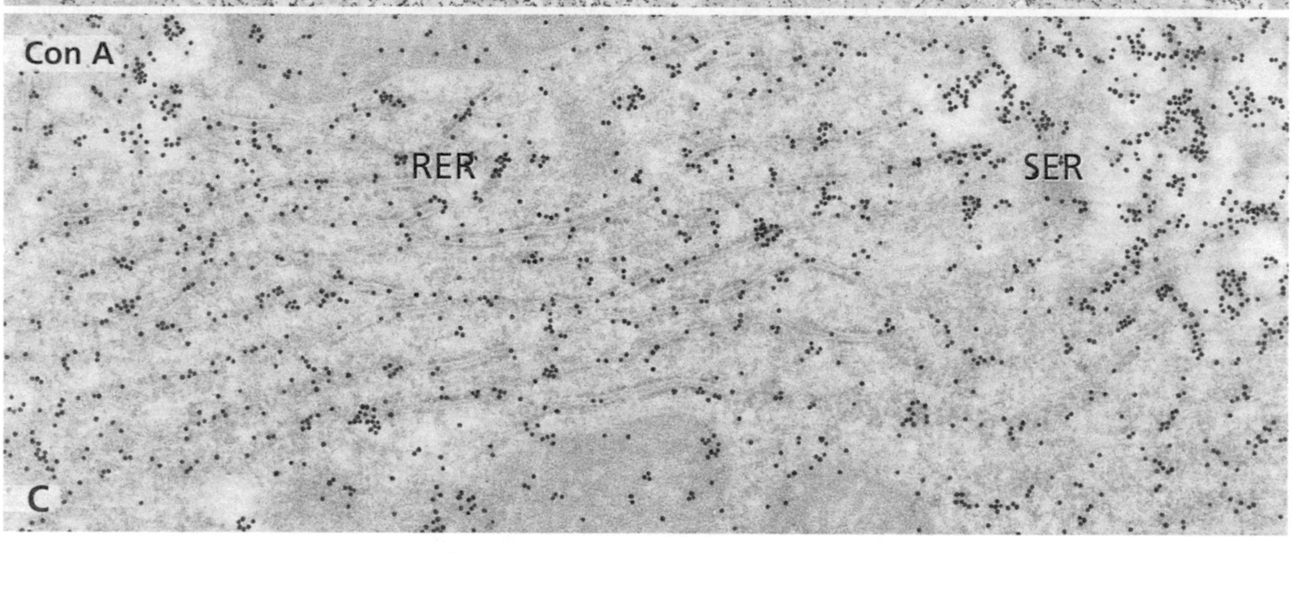
Con A
RER
SER
C

OLIGOSACCHARIDE TRIMMING, REGLUCOSYLATION, AND PROTEIN QUALITY CONTROL IN THE ROUGH ENDOPLASMIC RETICULUM

In the rough endoplasmic reticulum, first modifications on asparagine-linked oligosaccharides occur and include the removal of all three glucose residues by α-glucosidases I (Gls I) and II (Gls II) and some mannose residues by ER α-mannosidases I (ER Man I) and II (ER Man II), as illustrated in the scheme. These trimming reactions and the resulting specific oligosaccharides are important for protein quality control. Proper folding and correct assembly of glycoproteins are constantly monitored by a quality control machinery composed of chaperones, lectins such as calnexin and calreticulin, glucosidase II and UDP-glucose: glycoprotein glucosyltransferase (for additional details, cf. Fig. 24). On the other hand, defective oligosaccharide trimming has been recognized as the cause of congenital disorders of glycosylation.

In panel A, immunogold labelling demonstrates trimming glucosidase II in nuclear envelope (arrowheads) and the lumen of all endoplasmic reticulum cisternae in the form of microdomains. Both glucosidase I and II are present in the same endoplasmic reticulum cisternae as shown by double immunogold labelling (large gold particles for glucosidase I and small particles for glucosidase II). The importance of the trimming of the outer glucose residue by glucosidase I is unknown but that of the second glucose by glucosidase II provides a positive signal for protein folding. Glucosidase II also functions in the dissociation of complexes between calnexin/calreticulin and glycoproteins bearing monoglucosylated oligosaccharides.

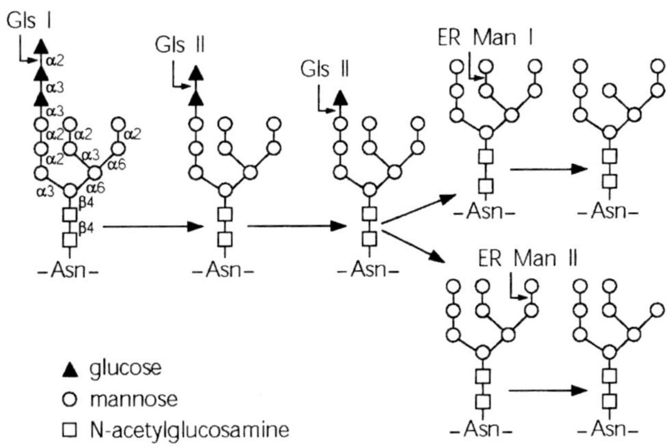

Panel B shows glucosyltransferase immunolabelling in nuclear envelope (arrowheads) and rough endoplasmic reticulum in a distribution similar to glucosidase II. Glucosyltransferase is unique in that it serves in a two-fold function as a protein folding sensor and as a glycosyltransferase. The enzyme recognizes not correctly folded glycoproteins and, by reglucosylating their oligosaccharides, tags the glycoprotein for entry in another calnexin/calreticulin cycle.

References

Ellgaard L and Helenius A (2003) Quality control in the endoplasmic reticulum. Nat Rev Mol Cell Biol 4: 181

Freeze HH (2006) Genetic defects in the human glycome. Nat Rev Genet 7: 537

Jakob CA, Burda P, Roth J, and Aebi M (1998) Degradation of misfolded endoplasmic reticulum glycoproteins in *Saccharomyces cerevisiae* is determined by a specific oligosaccharide structure. J Cell Biol 142: 1223

Hennet T (2009) Diseases of glycosylation. In: The Sugar Code. Fundamentals of Glycosciences (Gabius H-J, ed). Weinheim: Wiley/VCH, pp 365

Parodi AJ (2000) Protein glucosylation and its role in protein folding. Annu Rev Biochem 69: 69

Roth J, Yam GH, Fan J, Hirano K, Gaplovska-Kysela K, Le Fourn V, Guhl B, Santimaria R, Torossi T, Ziak M, and Zuber C (2008) Protein quality control: the who's who, the where's and therapeutic escapes. Histochem Cell Biol 129: 163

Vembar SS and Brodsky JL (2008) One step at a time: endoplasmic reticulum-associated degradation. Nat Rev Mol Cell Biol 9: 944

Magnification: x 53,000 (A); inset x 70,300; x 48,000 (B)

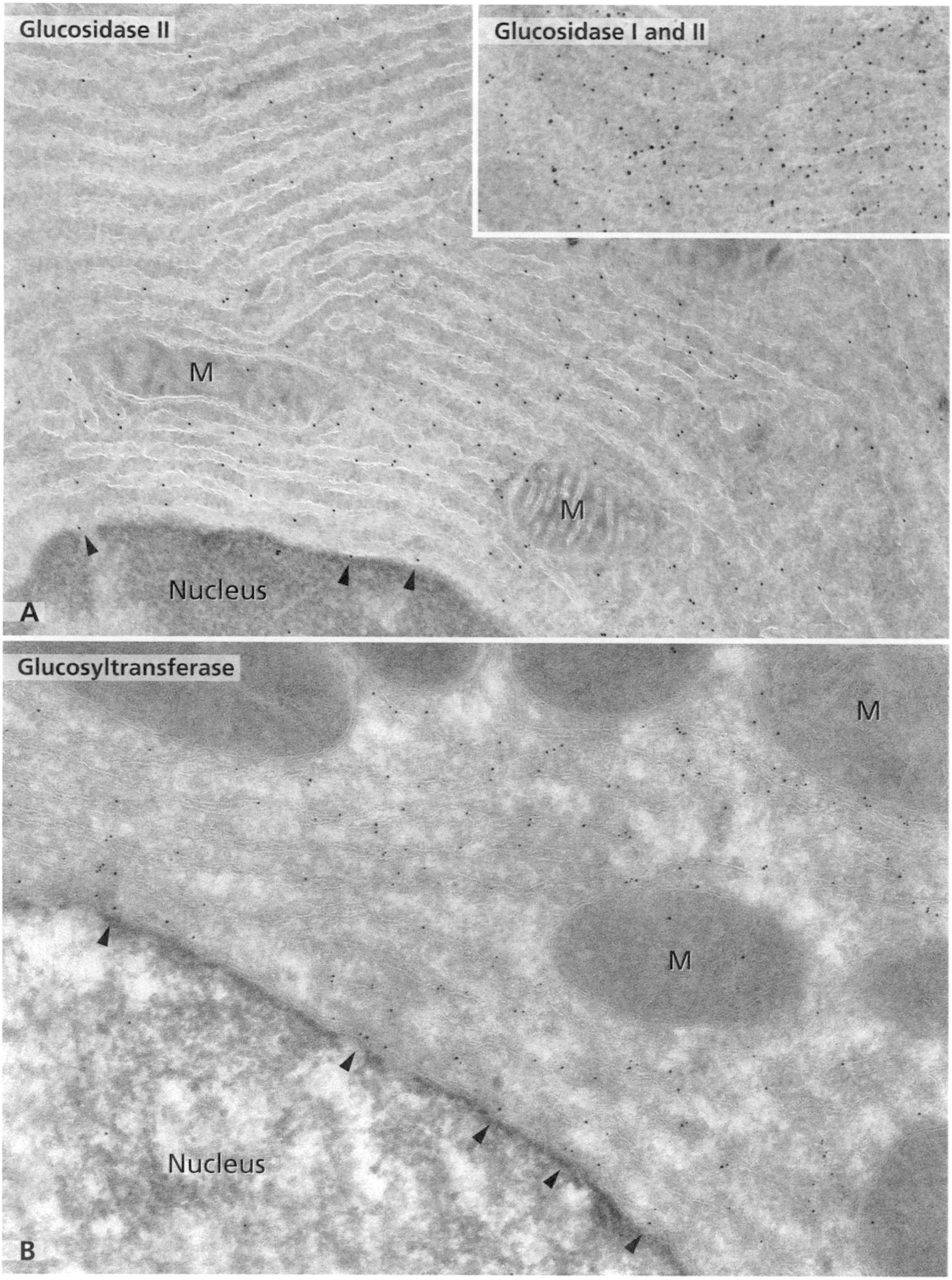
Glucosidase II
Glucosidase I and II
M
M
Nucleus
A
Glucosyltransferase
M
M
Nucleus
B

ROUGH ENDOPLASMIC RETICULUM: STORAGE SITE OF AGGREGATES OF MISFOLDED GLYCOPROTEINS

Protein misfolding occurs naturally at a high rate, or as a consequence of mutations and under various forms of cellular stress such as hyperthermia, hypoxia, starvation, experimental inhibition of protein synthesis by puromycin or of *N*-glycosylation by tunicamycin and inhibition of disulphide bridge formation by reducing agents. Accumulation of misfolded proteins also occurs in congenital human protein folding diseases, e.g. cystic fibrosis, α-1-antitrypsin deficiency and congenital goitrous hypothyreoidism.

The fate of misfolded proteins can be quite diverse. In the best case scenario, they will be dislocated and degraded by the ubiquitin-proteasome system without harmful consequences for the cells. Under unfavorable conditions, they may aggregate in the lumen of the rough endoplasmic reticulum. The classical manifestation of this form of protein aggregation are the Mott cells which represent plasma cells with stored immunoglobulins. This and other types of protein inclusions are generally called Russell bodies, which can also be observed in secretory cells other than plasma cells. Since such protein aggregates cannot be dislocated to the cytosol, the cisternal space of the entire rough endoplasmic reticulum may become greatly distended (asterisks in panel A showing part of a Mott cell from mouse spleen). Local protein aggregates represented by the intracisternal granules in the lumen of the ER can occur under conditions of starvation or treatment with puromycin aminonucleoside. In panel B, part of a pancreatic serous cell with intracisternal granules and a zymogen granule (ZG) with immunogold labelling for amylase is shown.

The functional consequences of protein accumulation in the ER are at least two-fold: induction of the unfolded protein response and an ER overload response. The unfolded protein response pathway, which consists of Ireα1, PERK and ATF6 controls the expression of various ER chaperones but may also induce apoptosis and inflammation. This signaling pathway is highly conserved from yeast to mammalian cells. The ER overload response involves another signaling pathway to induce the transcription factor NF-κB that finally results in the induction of cytokines followed by inflammation or in apoptosis.

References

Alanen A, Pira U, Lassila O, Roth J, and Franklin RM (1985) Mott cells are plasma cells defective in immunoglobulin secretion. Eur J Immunol 15: 235

Bernales S, Papa FR, and Walter P (2006) Intracellular signaling by the unfolded protein response. Annu Rev Cell Dev Biol 22: 487

Fan JY, Roth J, and Zuber C (2007) Expression of mutant Ins2^{C96Y} results in enhanced tubule formation causing enlargement of pre-Golgi intermediates of CHO cells. Histochem Cell Biol 128: 161

Jonikas MC, Collins SR, Denic V, Oh E, Quan EM, Schmid V, Weibezahn J, Schwappach B, Walter P, Weissman JS, and Schuldiner M (2009) Comprehensive characterization of genes required for protein folding in the endoplasmic reticulum. Science 323: 1693

Kim PS and Arvan P (1998) Endocrinopathies in the family of endoplasmic reticulum (ER) storage diseases: disorders of protein trafficking and the role of ER molecular chaperones. Endocr Rev 19: 173

Kopito RR (2000) Aggresomes, inclusion bodies and protein aggregation. Trends Cell Biol 10: 524

Palade G (1956) Intracisternal granules in the exocrine cells of the pancreas. J Biophys Biochem Cytol 2: 417

Ron D and Walter P (2007) Signal integration in the endoplasmic reticulum unfolded protein response. Nat Rev Mol Cell Biol 8: 519

Roth J, Yam GH, Fan J, Hirano K, Gaplovska-Kysela K, Le Fourn V, Guhl B, Santimaria R, Torossi T, Ziak M, and Zuber C (2008) Protein quality control: the who's who, the where's and therapeutic escapes. Histochem Cell Biol 129: 163

Valetti C, Grossi C, Milstein C, and Sitia R (1991) Russel bodies: a general response of secretory cells to synthesis of mutant immunoglobulin which can neither exit from, nor be degraded in the endoplasmic reticulum. J Cell Biol 115: 983

Yam GHF, Gaplovska-Kysela K, Zuber C, and Roth J (2007) Aggregated myocilin induces Russell bodies and causes apoptosis – Implications for the pathogenesis of myocilin-caused primary open-angle glaucoma. Amer J Pathol 170: 100

Zuber C, Fan JY, Guhl B, and Roth J (2004) Misfolded proinsulin accumulates in expanded pre-Golgi intermediates and endoplasmic reticulum subdomains in pancreatic beta cells of Akita mice. FASEB J 18: 917

Magnification: x 20,500 (A); x 95,000 (B)

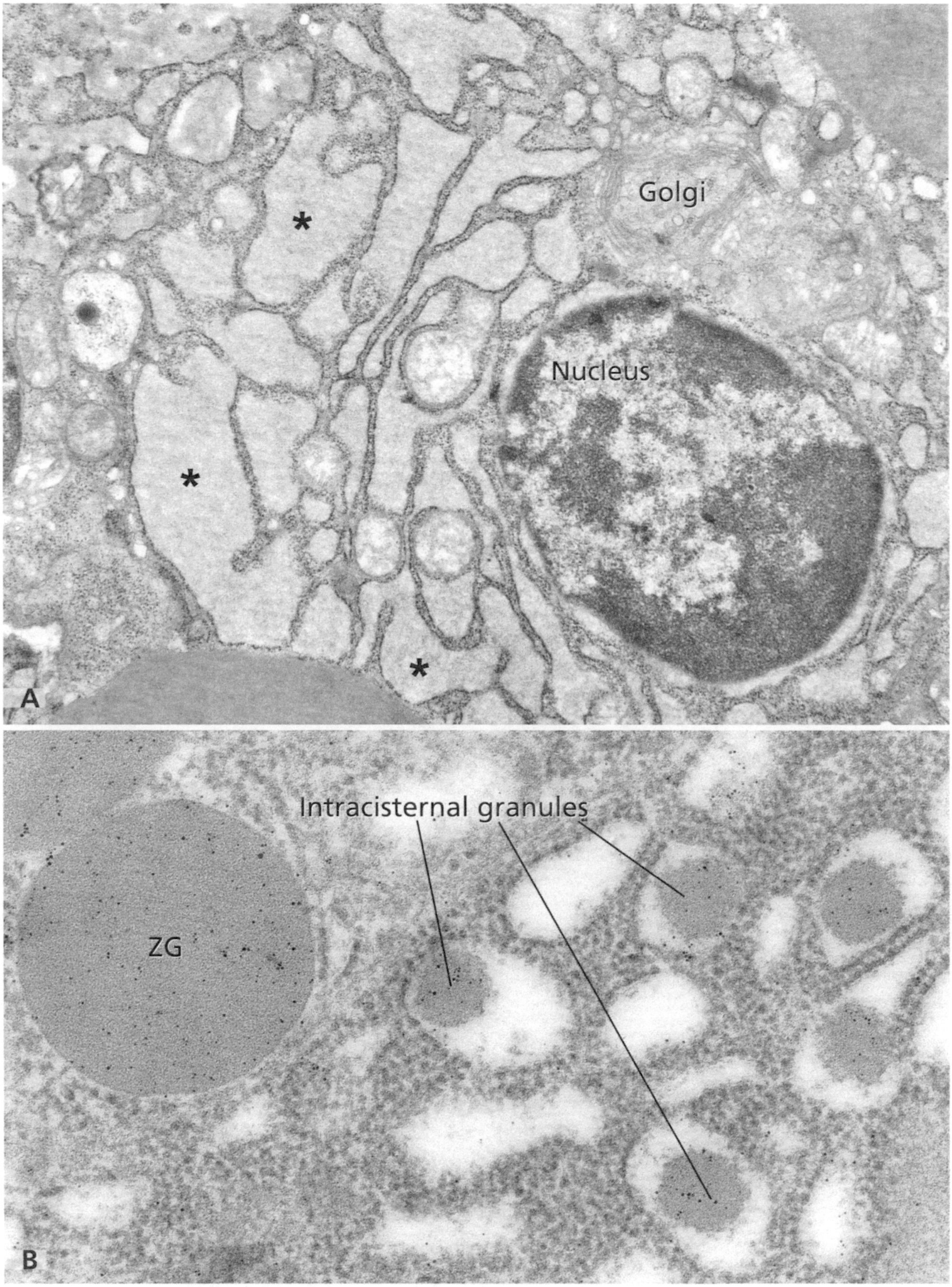
Golgi
*
Nucleus
*
*
A
Intracisternal granules
ZG
B

RUSSELL BODIES AND AGGRESOMES REPRESENT DIFFERENT TYPES OF PROTEIN INCLUSION BODIES

It has already been mentioned that Russell bodies result from the aggregation of misfolded glycoproteins in the lumen of the rough endoplasmic reticulum. A giant Russell body similar in size to the nucleus is seen in panel A with normal appearing rough endoplasmic reticulum (RER), mitochondria (M), and secondary lysosomes (Ly). A main structural feature is its limiting membrane studded with ribosomes, the endoplasmic reticulum membrane.

Another type of inclusion body that is also due to the production of misfolded glycoproteins are aggresomes. They differ from Russell bodies in that the proteinaceous aggregates, which can be observed in the electron microscope as amorphous dense material (asterisk), are located in the cytosol and have no limiting membrane (panel B). Furthermore, aggresomes in contrast to Russell bodies are composed of ubiquitylated protein aggregates, proteasomes, p62 and chaperones. Fully formed aggresomes are surrounded by intermediate filaments of the vimentin type (IF) and found close to the microtubule organizing centre. They are formed from smaller cytosolic protein aggregates present throughout the cytosol by active minus-end directed transport along microtubules (MT). Aggresomes can be experimentally induced by the inhibition of proteasome activity, or when the proteolytic capacity of the proteasomes is exhausted by the presence of aggregation-prone misfolded proteins. There is also evidence that protein aggregates can directly impair the proteasome activity.

Despite all these differences, both Russell bodies and aggresomes are morphological manifestations of cellular indigestion often observed under various diseased states. In particular, inclusion bodies are commonly found in association with chronic neurodegenerative diseases such as Alzheimer's and Parkinson's disease and familial amyotrophic lateral sclerosis.

References

Alanen, A, Pira U, Lassila O, Roth J, and Franklin R (1985) Mott cells are plasma cells defective in immunoglobulin secretion. Eur J Immunol 15: 235

Bence NF, Sampat RM, and Kopito RR (2001) Impairment of the ubiquitin-proteasome system by protein aggregation. Science 292: 1552

Fan JY, Roth J, and Zuber C (2007) Expression of mutant Ins2^{C96Y} results in enhanced tubule formation causing enlargement of pre-Golgi intermediates of CHO cells. Histochem Cell Biol 128: 161

Garcia-Mata R, Bebök Z, Sorscher EJ, and Sztul ES (1999) Characterization and dynamics of aggresome formation by a cytosolic GFP-chimera. J Cell Biol 146: 1239

Hirano K, Guhl B, Roth J, and Ziak M (2009) A cell culture system for the inducation of Mallory bodies: Mallory bodies and aggresomes represent different types of inclusion bodies. Histochem Cell Biol 132: 293

Johnston JA, Ward CL, and Kopito RR (1998) Aggresomes: a cellular response to misfolded proteins. J Cell Biol 143: 1883

Kawaguchi Y, Kovacs JJ, McLaurin A, Vance JM, Ito A, and Yao TP (2003) The deacetylase HDAC6 regulates aggresome formation and cell viability in response to misfolded protein stress. Cell 115: 727

Komatsu M, Waguri S, Koike M, Sou YS, Ueno T, Hara T, Mizushima N, Iwata J, Ezaki J, Murata S, et al. (2007) Homeostatic levels of p62 control cytoplasmic inclusion body formation in autophagy-deficient mice. Cell 131: 1149

Kopito RR (2000) Aggresomes, inclusion bodies and protein aggregation. Trends Cell Biol 10: 524

Kopito RR, and Sitia R (2000) Aggresomes and Russell bodies – Symptoms of cellular indigestion? EMBO Rep 1: 225

Roth J, Yam GH, Fan J, Hirano K, Gaplovska-Kysela K, Le Fourn V, Guhl B, Santimaria R, Torossi T, Ziak M, and Zuber C (2008) Protein quality control: the who's who, the where's and therapeutic escapes. Histochem Cell Biol 129: 163

Seibenhener ML, Babu JR, Geetha T, Wong HC, Krishna NR, and Wooten MW (2004) Sequestosome 1/p62 is a polyubiquitin chain binding protein involved in ubiquitin proteasome degradation. Mol Cell Biol 24: 8055

Valetti C, Grossi C, Milstein C, and Sitia R (1991) Russel bodies: a general response of secretory cells to synthesis of mutant immunoglobulin which can neither exit from, nor be degraded in, the endoplasmic reticulum. J Cell Biol 115: 983

Yam GHF, Gaplovska-Kysela K, Zuber C, and Roth J (2007) Aggregated myocilin induces Russell bodies and causes apoptosis – Implications for the pathogenesis of myocilin-caused primary open-angle glaucoma. Amer J Pathol 170: 100

Magnification: x 19,000 (A); x 48,000 (B)

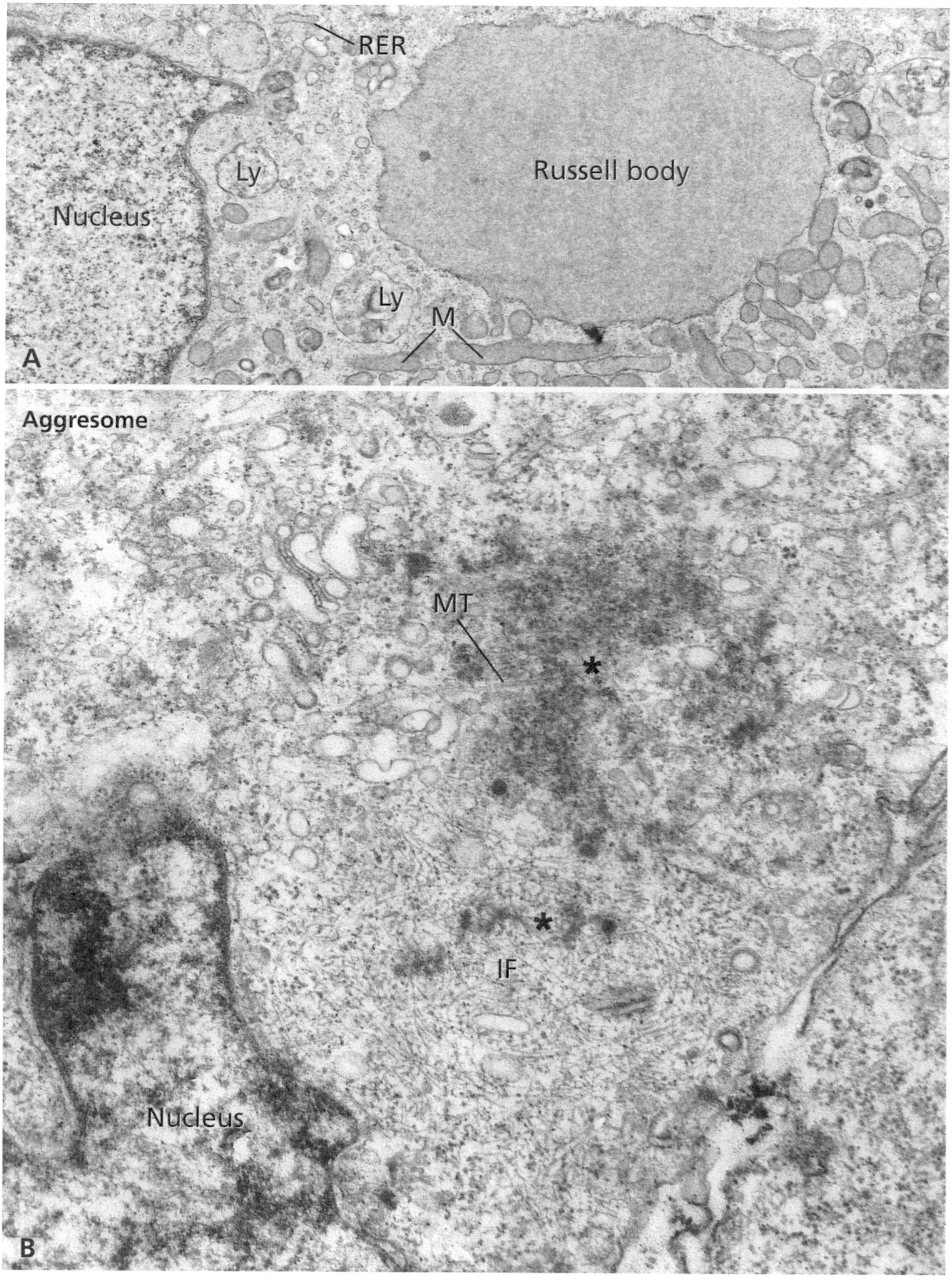
RER
Ly
Nucleus
Russell body
Ly
M
A
Aggresome
MT
*
*
IF
Nucleus
B

SMOOTH ENDOPLASMIC RETICULUM

Lacking attached ribosomes, the smooth endoplasmic reticulum (SER) forms a complex membranous system that is linked to various cellular functions. The main SER tasks include synthesis of lipids, maintenance of calcium homeostasis and detoxification reactions necessary for the conversion of harmful water-insoluble substances into water-soluble compounds that are more suitable for excretion through the kidney. Specialised SER regions are the membraneous networks and the terminal cisterns of the sarcoplasmic reticulum (cf. Fig. 150). Though continuous with the rough endoplasmic reticulum (see inset), the SER is organised in a different way. The differences are perfectly discernible in hepatocytes as shown in the detail presentation at the opposite page.

The rough endoplasmic reticulum (RER) visible in the upper left corner of the micrograph forms regular stacks of flat cisternae in parallel orientation, whereas the SER consists of interconnected tubular and finger-like architectures that appear tangled up, forming kinds of membranous convolutions. In hepatocytes and in steroid hormone-producing cells, extended SER membrane convolutes occupy wide areas of the cytoplasm. In the electron microscope, myriads of smooth ER membrane profiles characterise these cytoplasmic areas. There is a close spatial relationship of SER membranes and glycogen particles (asterisk), which may reflect functional connections underlined by enzymes of the glycogen metabolism that were found associated with SER membranes.

Membrane continuities shown in detail in the inset at the right lower corner (arrowhead) reflect the close functional cooperations and common tasks of smooth and rough endoplasmic reticulum. Synthesis of the different components of lipoproteins and formation of lipoprotein particles is one example, which also can be shown morphologically. Very low density lipoprotein particles (VLDLs) produced by the hepatocytes and secreted into the space of Disse (cf. Figs. 107 and 108) are visible as electron dense globules within the SER lumina (arrows). They are particularly frequent close to the SER-RER-transitions (inset).

Under physiological conditions, *de novo*-synthesis of ER-membranes and removal are in balance and export out of the ER is accompanied by input via recycling membrane constituents. The overall size of the endoplasmic reticulum is thought to remain constant, but ER-membranes and compartments are highly dynamic and change their shapes continuously (cf. Fig. 22). Sizes and shapes of ER-compartments are determined by opposing forces of assembly connected with membrane fusion and disassembly connected with membrane fragmentation.

For ER-organisation, cytoskeleton and microenvironments have important roles. Differences in the appearance of the cytoplasmic "matrices" of RER and SER are clearly discernible in the electron microscope. RER cisternae visible in the upper left quarter of the micrograph and some few cisternae in the lower part of the figure and in the inset, are embedded in dense, compact "matrices" that are lacking in the SER cytoplasmic regions (cf. also Figs. 107–108).

M – mitochondria, PO – peroxisome.

References

Baumann O and Walz B (2001) Endoplasmic reticulum of animal cells and its organization into structural and functional domains. Int Rev Cytol 205: 149

English AR, Zurek N, and Voeltz GK (2009) Peripheral ER structure and function. Curr Opin Cell Biol 21: 596

Fagone P and Jackowski S (2009) Membrane phospholipids synthesis and endoplasmic reticulum function. J Lipid Res 50: S311

Galteau MM, Antoine B, and Reggio H (1985) Epoxide hydrolase is a marker for the smooth endoplasmic reticulum in rat liver. Embo J 4: 2793

Iinuma T, Aoki T, Arasaki K, Hirose H, Yamamoto A, Samata R, Hauri H-P, Arimitsu N, Tagaya M, Tani K (2009) Role of syntaxin 18 in the organization of the endoplasmic reticulum subdomains. J Cell Sci 122: 1680

Lavoie C and Paiement J (2008) Topology of molecular machines of the endoplasmic reticulum: a compilation of proteomics and cytological data. Histochem Cell Biol 129: 117

Paiement J and Bergeron J (2001) The shape of things to come: Regulation of shape changes in endoplasmic reticulum. Biochem Cell Biol 79: 587

Steegmaier M, Oorschot V, Klumperman J, and Scheller RH (2000) Syntaxin 17 is abundant in steroidogenic cells and implicated in smooth endoplasmic reticulum membrane dynamics. Mol Biol Cell 11: 2719

Van Meer G, Voelker DR, and Feigenson GW (2008) Membrane lipids: where they are and how they behave. Nat Rev Mol Cell Biol 9: 112

Magnification: x 32,800, x 47,500 (inset)

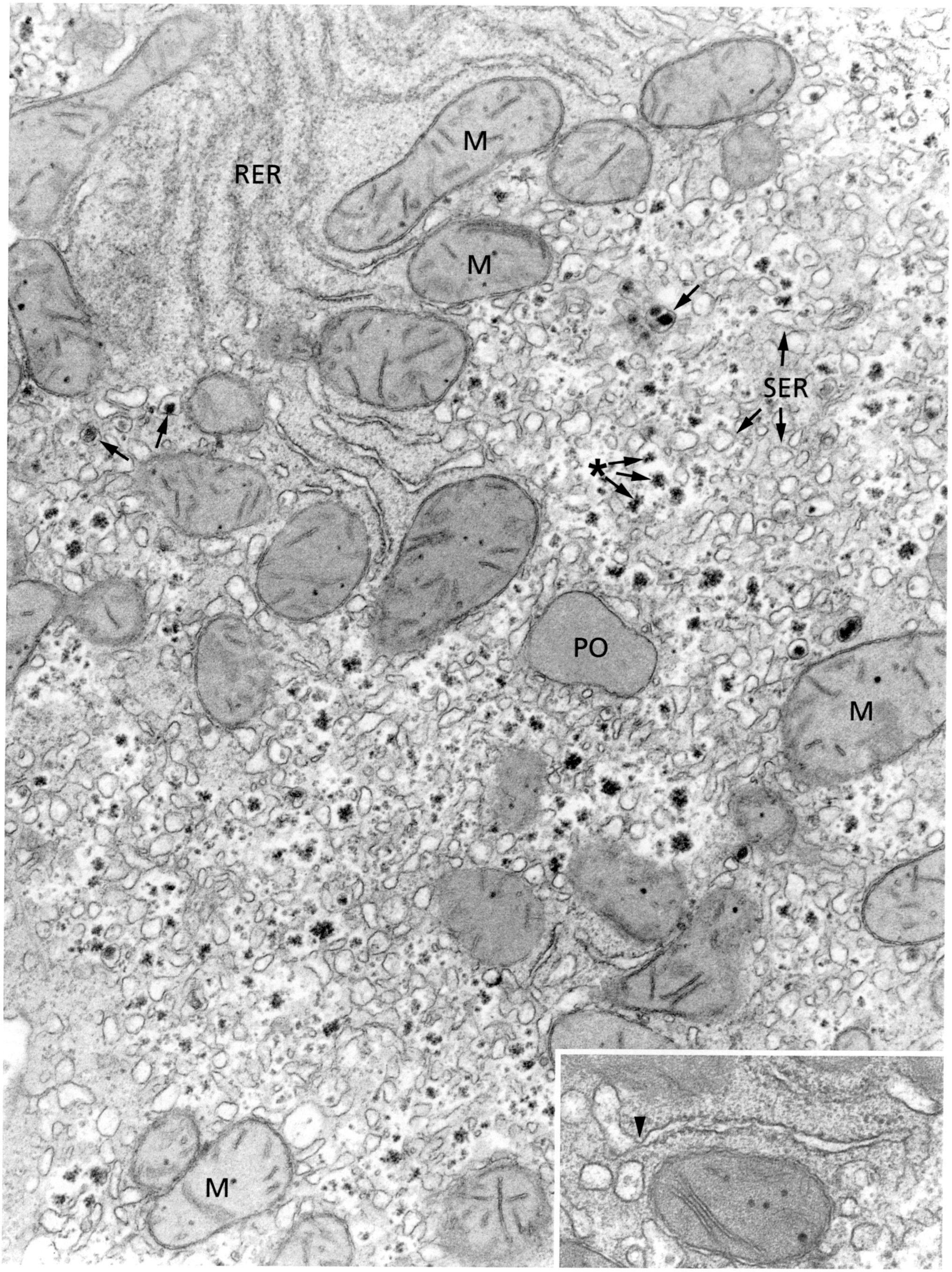
RER
M
M
SER
PO
M
M

PROLIFERATION OF THE SMOOTH ENDOPLASMIC RETICULUM

Hepatocytes and steroid-producing cells contain abundant smooth endoplasmic reticulum in contrast to most other cell types. The smooth endoplasmic reticulum membranes of liver hepatocytes amount to 16% of total cellular membranes whereas in pancreatic exocrine cells this value is less than 1%. The amount of smooth endoplasmic reticulum can increase rapidly and reversibly depending on functional demands, and its phenobarbital-induced proliferation represents a classical example for such a situation. After withdrawal of the drug, the excess smooth endoplasmic reticulum is removed by autophagy (cf. Fig. 62).

In hereditary disorders of bilirubin metabolism resulting in predominantly unconjugated hyperbilirubinaemia, the cytoplasm of liver hepatocytes can be filled with smooth endoplasmic reticulum (SER) at the expense of the rough endoplasmic reticulum and contains depositions of bilirubin (arrows and inset). These disorders include the Crigler-Najjar syndromes I and II and the Gilbert syndrome. The electron micrograph shows hepatocytes from a liver biopsy of a patient with Crigler-Najjar syndrome. Crigler-Najjar syndrome is inherited in an autosomal recessive mode. The molecular defect lies in the hepatic bilirubin UDP-glucoronosyltransferase A1 isoform, whose gene locus is located on chromosome 2q37. Genetic aberrations can occur in any of its five exons in the form of deletions, insertions, missense mutations, and premature stop codons. As a consequence, hepatic bilirubin UDP-glucoronosyltransferase activity can be undetectable or reduced and non-haemolytic icterus due to increased serum concentrations of unconjugated bilirubin occurs during the first days of life. When untreated, fatal kernicterus causing bilirubin encephalopathy results. However, long-term survival could be achieved by orthotopic liver transplantation.

References

Bosma P, Roy Chowdhury J, Huang T, Lahiri P, Oude Elferink R, Lederstein M, Whitington P, Jansen, P, and Roy Chowdhury N (1992) Mechanism of inherited deficiencies of multiple UDP-glucoronosyltransferase isoforms in two patients with Crigler-Najjar syndrome, type I. FASEB J 6: 2859

Bosma P, Roy Chowdhury N, Goldhoorn B, Hofker M, Oude Elferink R, Jansen P, and Roy Chowdhury J (1992) Sequence of exons and the flanking regions of human bilirubin-UPD-glucoronosyltransferase gene complex and identification of a genetic mutation in a patient with Crigler-Najjar syndrome, type I. Hepatology 15: 941

Bosma P, Seppen J, Goldhoorn B, Bakker C, Oude Elferink R, Roy Chowdhury J, Roy Chowdhury N, and Jansen P (1994) Bilirubin UDP-glucoronosyltransferase I is the only relevant bilirubin glucoronidating isoform in man. J Biol Chem 269: 17960

Crigler J and Najjar V (1952) Congenital familial non-hemolytic jaundice with kernicterus. Pediatrics 10: 169

Roy Chowdhury J, Wolkoff A, Roy Chowdhury N, and Arias I (2001) Hereditary jaundice and disorders of bilirubin metabolism. In: The metabolic and molecular bases of inherited diseases. New York: McGraw-Hill, pp 3063

Magnification: x 9,600; x 13,000 (inset)

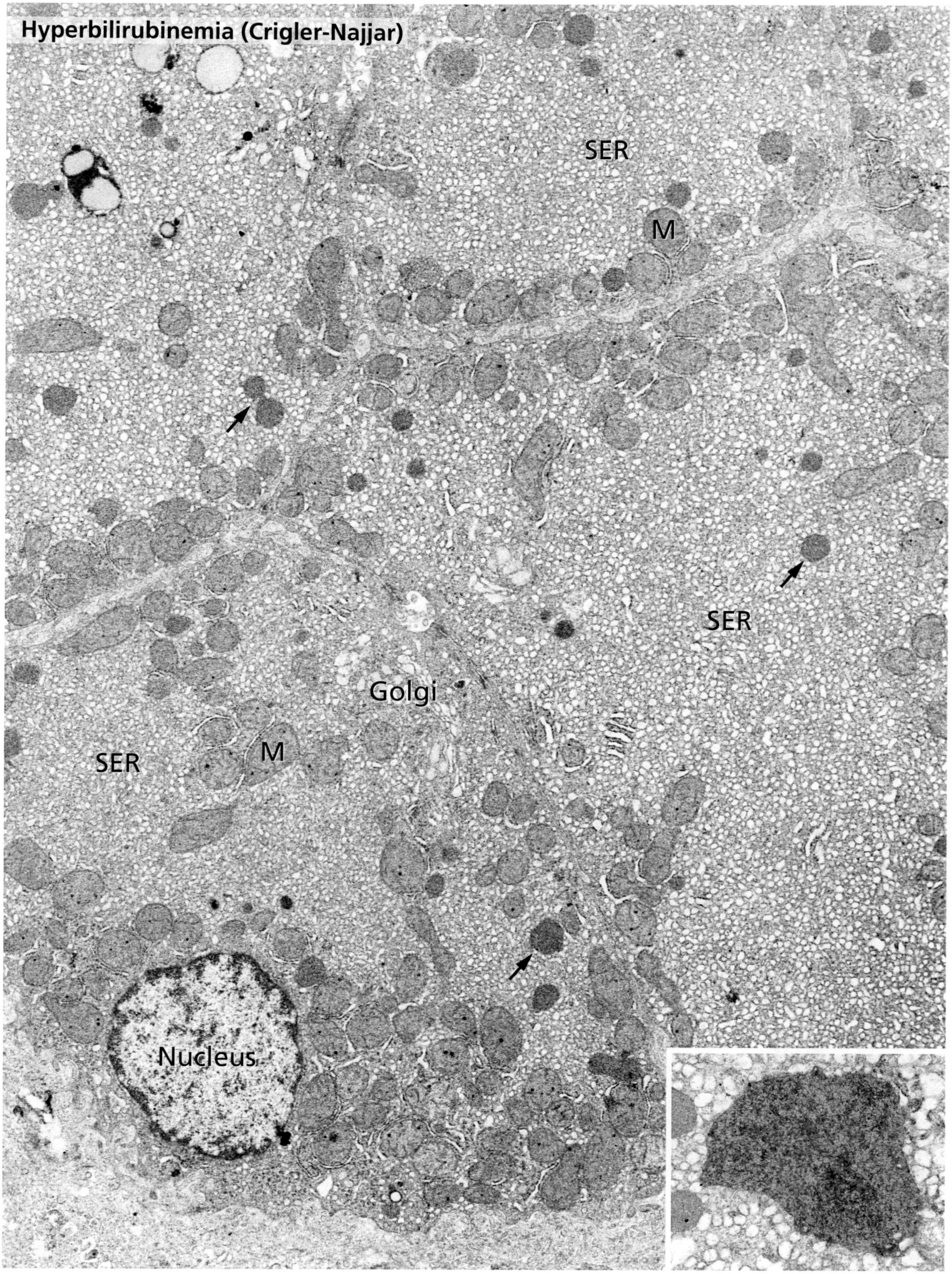
Hyperbilirubinemia (Crigler-Najjar)
SER
M
SER
Golgi
SER
M
Nucleus

PRE-GOLGI INTERMEDIATES

The interface between the endoplasmic reticulum and the Golgi apparatus consists of a complex and highly dynamic structure, the pre-Golgi intermediates. They are also referred to as intermediate (or salvage) compartment, ERGIC-53, or vesiculo-tubular clusters (VTC) and represent intermediates for transport out of the endoplasmic reticulum or for recycling to the endoplasmic reticulum.

The exit of cargo from the endoplasmic reticulum occurs at specialized, morphologically defined sites, the transitional elements of the rough endoplasmic reticulum, which are partly devoid of ribosomes (TE in panels A–C) and exhibit buds (arrowheads in panels A–C), which give rise to vesicles. Located between the transitional elements and the *cis* side of the Golgi apparatus are clusters of smooth membrane vesiculo-tubular structures, the pre-Golgi intermediates (pGI in panels A and B). Pre-Golgi intermediates together with transitional elements of rough endoplasmic reticulum exist also at multiple sites in the peripheral cytoplasm (panel C) and move from there along microtubules to the Golgi apparatus. Transitional elements and pre-Golgi intermediates are constituents not only of secretory cells (panel B shows part of an endocrine B-cell with secretory granules – SG) but also of less differentiated non-professional secretory cells such as Chinese hamster ovary (CHO) cells (panel B). The pre-Golgi intermediates are enriched in ERGIC-53 or p58 (rat homologue of human ERGIC-53) providing a marker for their identification by immunolabelling (panel D).

Intense studies have unravelled many aspects of the molecular mechanism of cargo selection in the transitional elements of the rough endoplasmic reticulum and of ER-to-Golgi transport and recycling. Cytosolic coat proteins (COP) such as Sar1p, Sec23p-Sec24p, and Sec13p-Sec31p bind to transitional elements resulting in the formation of COPII-coated buds into which cargo is actively sorted (see diagram). COPII-coated vesicles then carry out the anterograde transport. Such vesicles loose their coat and fuse with each other or neighbouring vesiculo-tubular clusters. There is also evidence for the direct *en bloc* formation of large tubular carriers from the endoplasmic reticulum to accomodate large size cargo. Pre-Golgi intermediates are not only functioning in anterograde COPII-mediated transport but also in retrograde transport, which is mediated by the COPI vesicular pathway. The COP coats are morphologically distinct from the clathrin coats of buds and vesicles at the *trans* side of the Golgi apparatus (arrow in panel B).

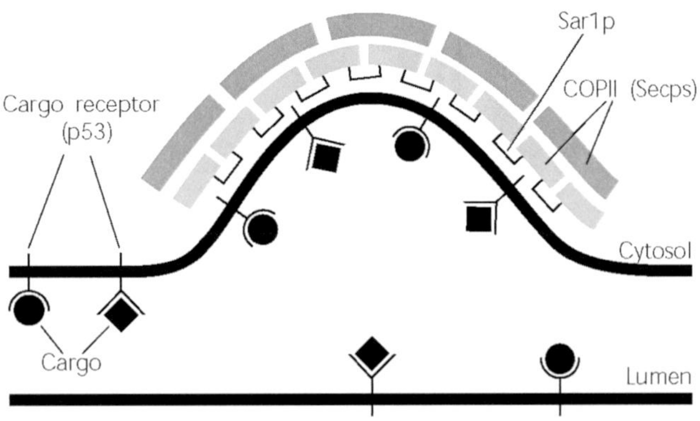

References

Appenzeller-Herzog C and Hauri HP (2006) The ER-Golgi intermediate compartment (ERGIC): in search of its identity and function. J Cell Sci 119: 2173

Bannykh SI and Balch WE (1997) Membrane dynamics at the endoplasmic reticulum Golgi interface. J Cell Biol 138: 1

Fan JY, Roth J, and Zuber C (2003) Ultrastructural analysis of transitional endoplasmic reticulum and pre-Golgi intermediates: a highway for cars and trucks. Histochem Cell Biol 120: 455

Gurkan C, Stagg SM, LaPointe P, and Balch WE (2006) The COPII cage: unifying principles of vesicle coat assembly. Nat Rev Mol Cell Biol 7: 727

Hammond AT and Glick BS (2000) Dynamics of transitional endoplasmic reticulum sites in vertebrate cells. Mol Biol Cell 11: 3013

Lee MC, Miller EA, Goldberg J, Orci L, and Schekman R (2004) Bi-directional protein transport between the ER and Golgi. Annu Rev Cell Dev Biol 20: 87

Liu W and Lippincott-Schwartz J (2005) Illuminating COPII coat dynamics. Nat Struct Mol Biol 12: 106

Simpson JC, Nilsson T, and Pepperkok R (2006) Biogenesis of tubular ER-to-Golgi transport intermediates. Mol Biol Cell 17: 723

Stagg SM, LaPointe P, Razvi A, Gurkan C, Potter CS, Carragher B, and Balch WE (2008) Structural basis for cargo regulation of COPII coat assembly. Cell 134: 474

Zeuschner D, Geerts WJC, vanDonselaar E, Humbel BM, Slot JW, Koster AJ, and Klumperman J (2006) Immuno-electron tomography of ER exit sites reveals the existence of free COPII-coated transport carriers. Nat Cell Biol 8: 377

Magnification: x 87,000 (A); x 75,000 (B); x 74,000 (C); x 66,500 (D)

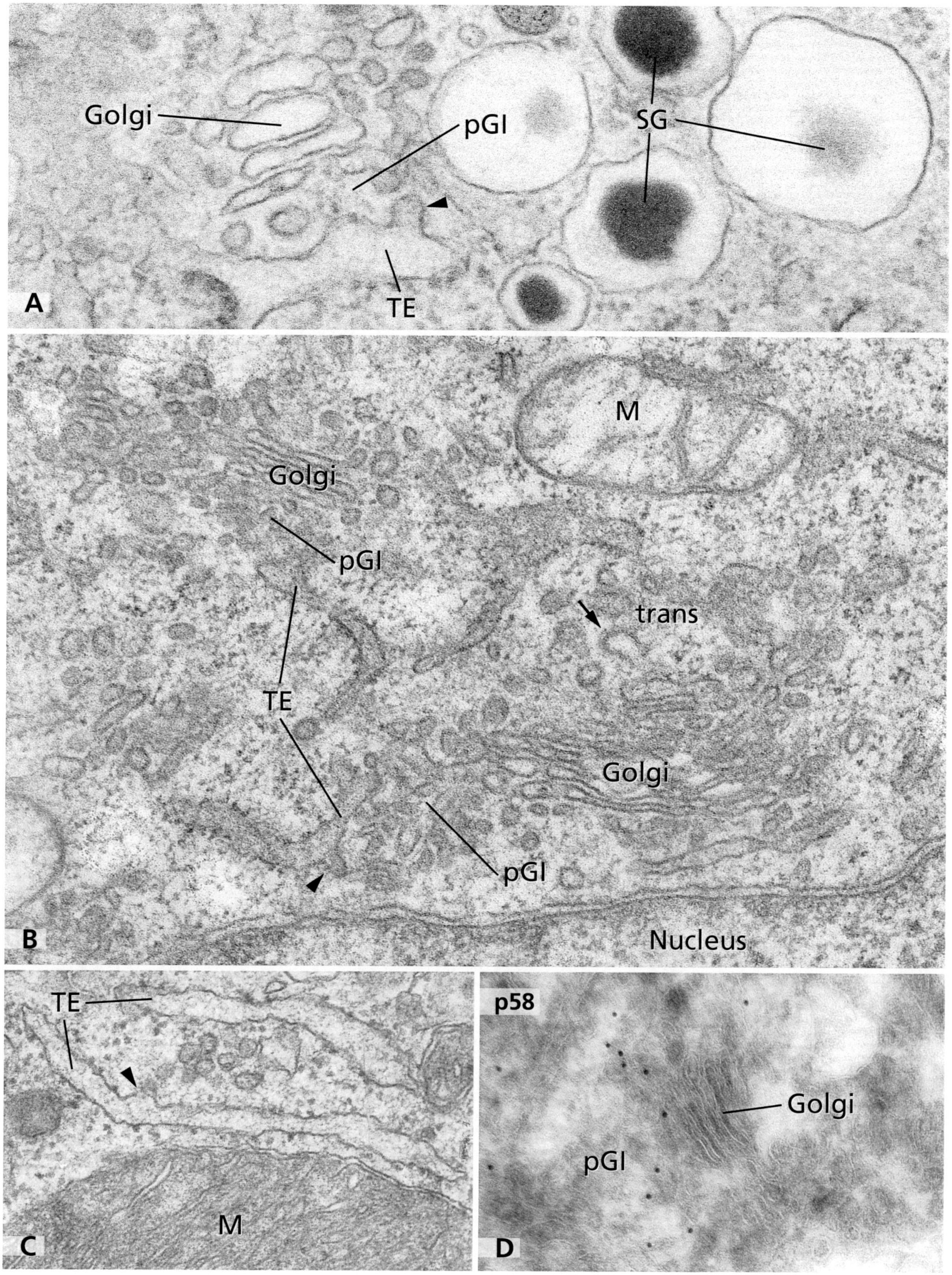
Golgi
pGI
SG
TE
A
M
Golgi
pGI
trans
TE
Golgi
pGI
B
Nucleus
TE
M
C
p58
Golgi
pGI
D

PRE-GOLGI INTERMEDIATES: OLIGOSACCHARIDE TRIMMING AND PROTEIN QUALITY CONTROL

Machinery components of the protein quality control such as chaperones, protein disulfide isomerase, lectins (calnexin and calreticulin), ERp57, glucosidase II, and glucosyltransferase are all functioning in the endoplasmic reticulum. The scheme depicts the major steps in the quality control of glycoprotein folding. Glycoproteins bearing monoglucosylated oligosaccharides either due to trimming by glucosidase II (Gls II) or through reglucosylation by glucosyltransferase (GT) will be bound by calnexin or calreticulin. Dissociation of such glycoprotein-lectin complexes will be achieved by glucosidase II, and correctly folded glycoproteins may be processed by ER-mannosidase I (ER Man I) and exported from the endoplasmic reticulum. However, not correctly folded glycoproteins will be recognized and reglucosylated by glucosyltransferase and enter a new calnexin/calreticulin cycle. Thus, the opposing actions of glucosidase II and glucosyltransferase provide the basis for the on-and-off-cycle. If correct folding cannot be achieved, trimming by ER-mannosidase I and II will be followed by interaction with EDEM 1 and OS9, respectively. Both direct misfolded glycoproteins for dislocation to the cytosol.

High resolution immunogold labelling has unequivocally shown the presence of glucosidase II and glucosyltransferase in pre-Golgi intermediates of various mammalian cell types (panel A: pig liver, panel B: rat pancreas, panel C: peripheral pGI in rat liver) and in the salivary glands of *Drosophila melanogaster*. Notably, quantification of the immunogold labelling for glucosyltransferase showed enrichment in pre-Golgi intermediates over rough endoplasmic reticulum. This indicates quality control of glycoprotein folding is not limited to the ER and additionally occurs in pre-Golgi intermediates. Pre-Golgi intermediates become greatly enlarged due to the accumulation of misfolded proteins.

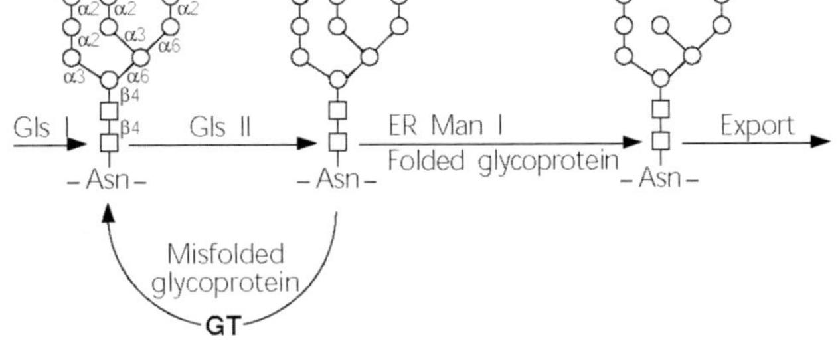

References

Christianson JC, Shaler TA, Tyler RE, and Kopito RR (2008) OS-9 and GRP94 deliver mutant alpha1-antitrypsin to the Hrd1/SEL1L ubiquitin ligase complex for ERAD. Nat Cell Biol 10: 272–282

Ellgaard L and Helenius A (2003) Quality control in the endoplasmic reticulum. Nat Rev Mol Cell Biol 4: 181

Fan JY, Roth J, and Zuber C (2007) Expression of mutant $Ins2^{C96Y}$ results in enhanced tubule formation causing enlargement of pre-Golgi intermediates of CHO cells. Histochem Cell Biol 128: 161

Lucocq JM, Brada D, and Roth J (1986) Immunolocalization of the oligosaccharide trimming enzyme glucosidase II. J Cell Biol 102: 2137

Roth J, Yam GH, Fan J, Hirano K, Gaplovska-Kysela K, Le Fourn V, Guhl B, Santimaria R, Torossi T, Ziak M, and Zuber C (2008) Protein quality control: the who's who, the where's and therapeutic escapes. Histochem Cell Biol 129: 163

Zuber C, Fan JY, Guhl B, Parodi A, Fessler JH, Parker C, and Roth J (2001) Immunolocalization of UDP-glucose: glycoprotein glucosyltransferase indicates involvement of pre-Golgi intermediates in protein quality control. Proc Natl Acad Sci USA 98: 10710

Vembar SS and Brodsky JL (2008) One step at a time: endoplasmic reticulum-associated degradation. Nat Rev Mol Cell Biol 9: 944

Zuber C, Fan JY, Guhl B, and Roth J (2004) Misfolded proinsulin accumulates in expanded pre-Golgi intermediates and endoplasmic reticulum subdomains in pancreatic beta cells of Akita mice. FASEB J 18: 917

Zuber C, Cormier JH, Guhl B, Santimaria R, Hebert DN, and Roth J (2007) EDEM1 reveals a quality control vesicular transport pathway out of the endoplasmic reticulum not involving the COPII exit sites. Proc Natl Acad Sci USA 104: 4407

Magnification: x 92,000 (A); x 71,500 (B); x 43,700 (C)

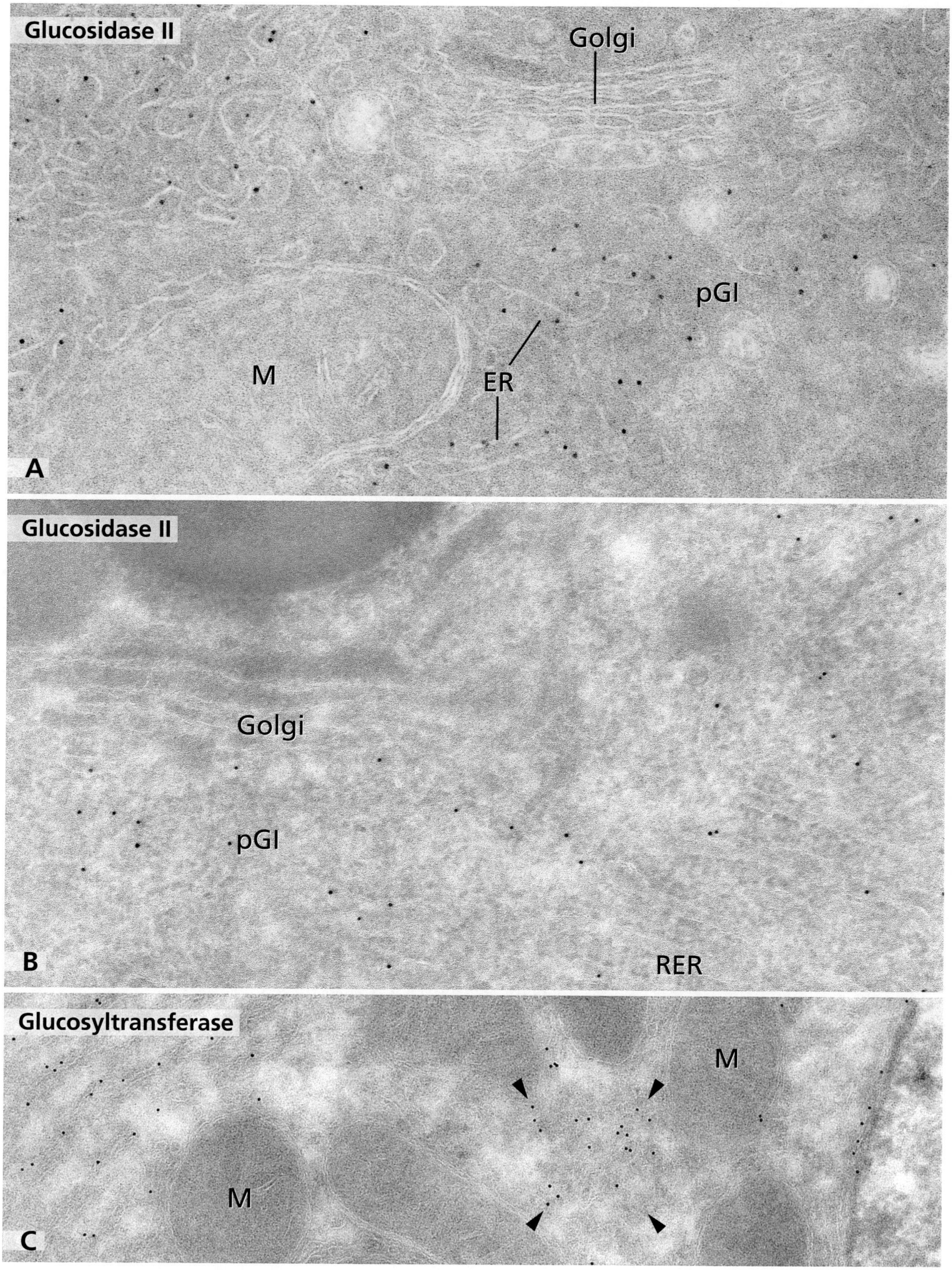
Glucosidase II
Golgi
pGI
M
ER
A
Glucosidase II
Golgi
pGI
B
RER
Glucosyltransferase
M
M
C

GOLGI APPARATUS: A MAIN CROSSROADS ALONG SECRETORY PATHWAYS

The Golgi apparatus is a central crossroads along the secretory pathways and has an important role also in endocytosis (cf. Figs. 33 and 48). Within the Golgi apparatus, multiple post-translational modifications of newly synthesised secretory and membrane proteins occur (cf. Figs. 27–29) and from here, molecules are sorted to various other destinations.

The Golgi apparatus is named after its discoverer Camillo Golgi, who first described the organelle in 1898. The original drawings of the osmium stained "apparato reticolare interno" in the somata of spinal ganglion cells published by Camillo Golgi already indicate that the entire apparatus is a continuous organelle and is composed of subunits that are connected among each other. The individual subunits are made up of stacks of flat cisternae, which are connected by non-compact, highly fenestrated or tubular regions (cf. Fig. 35 – diagram). Within each of the stacks, subcompartments are dedicated to special functions.

The micrograph presented at the opposite page shows one of the Golgi stacks in a rat pancreatic acinar cell. In this kind of cell, such as in other secretory cells of the regulated type, the Golgi stacks' organisation clearly mirrors the involvement in secretory routes. A clear polarity exists and the two different sides, the *cis* and *trans* sides, can readily be discriminated. Within the secretory pathway, the *cis* side is the import region of newly synthesised molecules coming from the endoplasmic reticulum and the *trans* side with the *trans* Golgi network (cf. Figs. 33–35) constitutes the region of export, where molecules modified within the Golgi stacks leave the organelle and are sorted to their final destinations. This clear *cis*-to-*trans* orientation of the secretory traffic is reflected morphologically, although the principles of traffic across the Golgi apparatus stacks are not definitely answered and different models are in discussion (cf. Fig. 40). Transitional elements (TE) of the rough endoplasmic reticulum (RER) and pre-Golgi intermediates (pGI, cf. Figs. 23 and 24) are located close to the *cis* side of the stack. Transitional elements correspond to the RER-export regions and exhibit ribosomes attached at one surface only. This is particularly visible in the lower right segment of the electron micrograph. Membranes of pre-Golgi intermediates are in close spatial relation to the *cis* Golgi side, where they are assumed to enter the stack and contribute to form a new Golgi cisterna. Both *cis* and medial cisternae are variably dilated and show fine flaky contents of secretory materials, which also are visible in the cisternae at the *trans* side and within the condensing vacuoles (CV). Condensing vacuoles correspond to immature secretory granules and are formed from particular dilations of *trans* Golgi cisternae (asterisk), shown in the upper left quarter of the micrograph. Several condensing vacuoles and mature secretory granules, in the pancreatic acinar cells termed zymogen granules (ZG), are accumulated in the *trans* Golgi area. The changes from fine flocculent contents in the Golgi cisternae and condensing vacuoles to greatly homogenous and dense contents visible in the mature zymogen granules correspond to condensation and maturation of the pancreatic secretion taking place in the condensing vacuoles (cf. Figs. 43 and 44).

M – mitochondria.

References

Dröscher A (1998) Camillo Golgi and the discovery of the Golgi apparatus. Histochem Cell Biol 109: 425

Farquhar MG and Hauri H-P (1997) Protein sorting and vesicular traffic in the Golgi apparatus. In: The Golgi apparatus (Berger EG and Roth J, eds). Basel Boston Berlin: Birkhäuser, p 63

Golgi C (1898) Sur la structure des cellules nerveuses. Arch Ital Biol 30: 60

Jackson CL (2009) Mechanisms of transport through the Golgi complex. J Cell Sci 122: 443

Patterson GH, Hirschberg K, Polishchuk RS, Gerlich D, Phair RD, and Lippincott-Schwartz J (2008) Transport through the Golgi apparatus by rapid partitioning within a two-phase membrane system. Cell 133: 1055

Mayinger P (2009) Regulation of Golgi function via phosphoinositide lipids. Seminars Cell Dev Biol 20: 793

Mironov AA and Pavelka M (2008) The Golgi apparatus as a crossroads in intracellular traffic. In: The Golgi apparatus. State of the art 110 years after Camillo Golgi's discovery (Mironov A, Pavelka M, eds). Wien New York: Springer, p 16

Rambourg A and Clermont Y (1997) Three-dimensional structure of the Golgi apparatus in mammalian cells. In: The Golgi apparatus (Berger EG and Roth J, eds). Basel Boston Berlin: Birkhäuser, p 37

Roth J (1997) Topology of glycosylation in the Golgi apparatus. In: The Golgi apparatus (Berger EG and Roth J, eds). Basel Boston Berlin: Birkhäuser, p 131

Short B, Haas A, and Barr FA (2005) Golgins and GTPases, giving identity and structure to the Golgi apparatus. Biochim Biophys Acta 1744: 383

Magnification: x 45,000

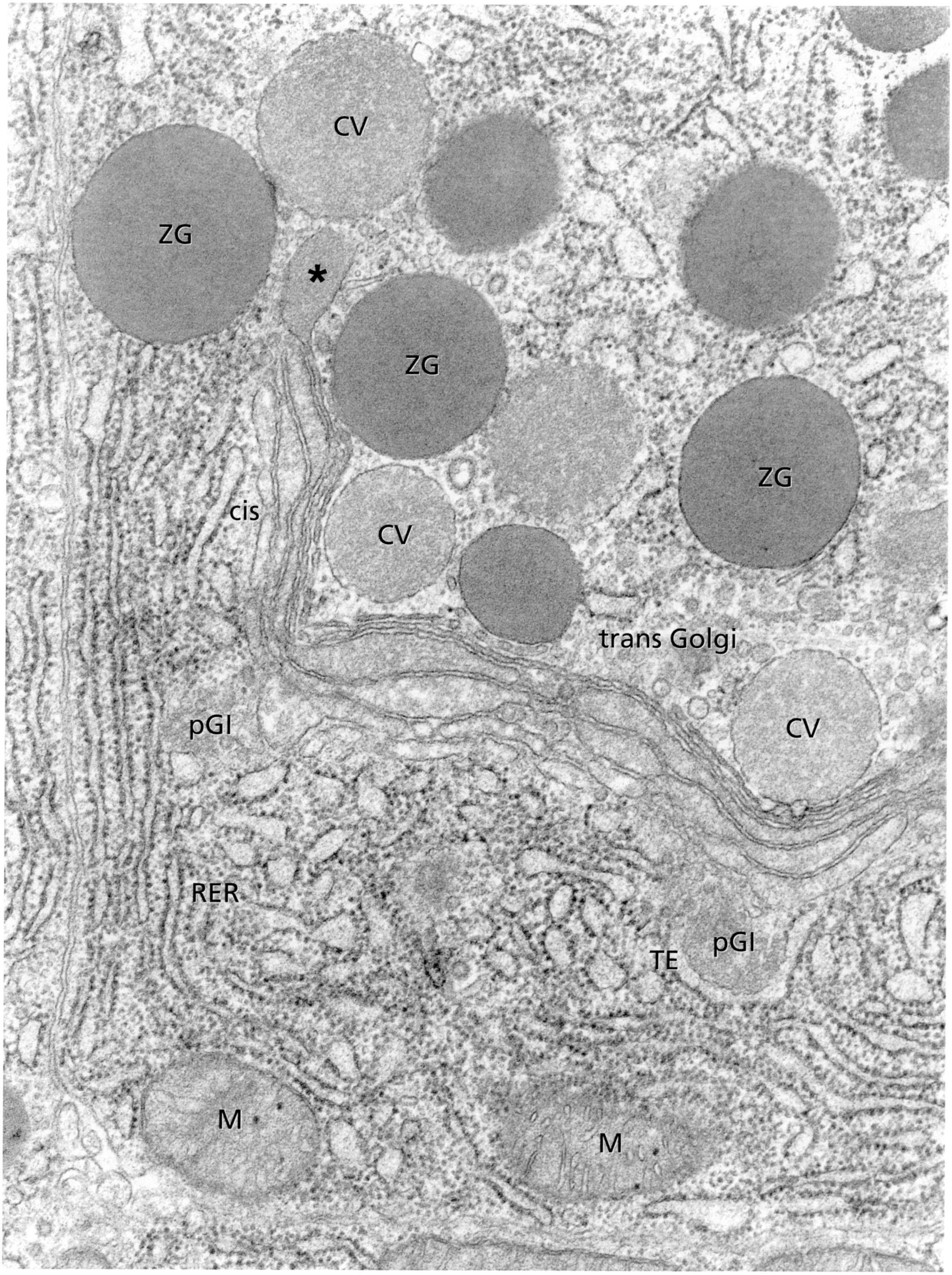
CV
ZG
*
ZG
ZG
cis
CV
trans Golgi
pGl
CV
RER
TE
pGl
M
M

PROTEIN SECRETION VISUALISED BY IMMUNOELECTRON MICROSCOPY

In general terms immunoelectron microscopy can be defined as a tool to reveal the relations of various cellular constituents of known function to subcellular organelles or parts of them, thereby providing information about organizational and functional principles of cellular compartments and subcompartments.

In panels A–C, the main compartments of the secretory pathway of a secretory cell prototype, the exocrine pancreatic cell (cf. Fig. 26), are shown together with immunogold labelling for amylase. The immunogold labelling of ultrathin sections showed amylase in the cisternal lumen of the rough endoplasmic reticulum (panel A). Labelling for amylase was also detectable in the transitional elements of the rough endoplasmic reticulum, the pre-Golgi intermediates and throughout the cisternal stack of the Golgi apparatus (panel B). Cisternae at the *trans* side of the Golgi apparatus exhibited local distensions filled with electron dense material (asterisks). They represent initial stages of zymogen granule formation, so called condensing vacuoles. Zymogen granules (ZG) are present above the *trans* side of the Golgi apparatus and in the apical cytoplasm. Here, secretion may occur by fusion of the zymogen granule membrane with the apical plasma membrane (arrows in panel C), which results in the exocytotic release of the zymogen granule content in the acinar lumen (cf. Fig. 45). Both the zymogen granule content and the acinar lumen are labelled for amylase.

Immunogold labelling can be quantified and provides a means to detect sites along the secretory pathway at which concentration of proteins or conversion of a prohormone into the mature hormone may occur. Quantitative immunogold labelling has provided evidence that the exit of proteins from the endoplasmic reticulum does not occur by bulk flow but is a selective process through which secretory and membrane proteins are sorted for packing into transport vesicles.

References

Balch WE, McCaffery JM, Plutner H, and Farquhar MG (1994) Vesicular stomatitis virus glycoprotein is sorted and concentrated during export from the endoplasmic reticulum. Cell 76: 841

Bannykh SI, Rowe T, and Balch WE (1996) The organization of endoplasmic reticulum export complexes. J Cell Biol 135: 19

Bendayan M, Roth J, Perrelet A, and Orci L (1980) Quantitative immunocytochemical localization of pancreatic secretory proteins in subcellular compartments of the rat acinar cell. J Histochem Cytochem 28: 149

Gurkan C, Stagg SM, LaPointe P, and Balch WE (2006) The COPII cage: unifying principles of vesicle coat assembly. Nat Rev Mol Cell Biol 7: 727

Malkus P, Jiang F, and Schekman R (2002) Concentrative sorting of secretory cargo proteins into COPII-coated vesicles. J Cell Biol 159: 915

Orci L, Ravazzola M, Amherdt M, Madsen O, Vassalli JD, and Perrelet A (1985) Direct identification of prohormone conversion site in insulin-secreting cells. Cell 42: 671

Orci L, Ravazzola M, Storch M-J, Anderson RGW, Vassalli J-D, and Perrelet A (1987) Proteolytic maturation of insulin is a post-Golgi event which occurs in acidifying clathrin-coated secretory vesicles. Cell 49: 865

Magnification: x 62,000 (A); x 55,000 (B); x 45,000 (C)

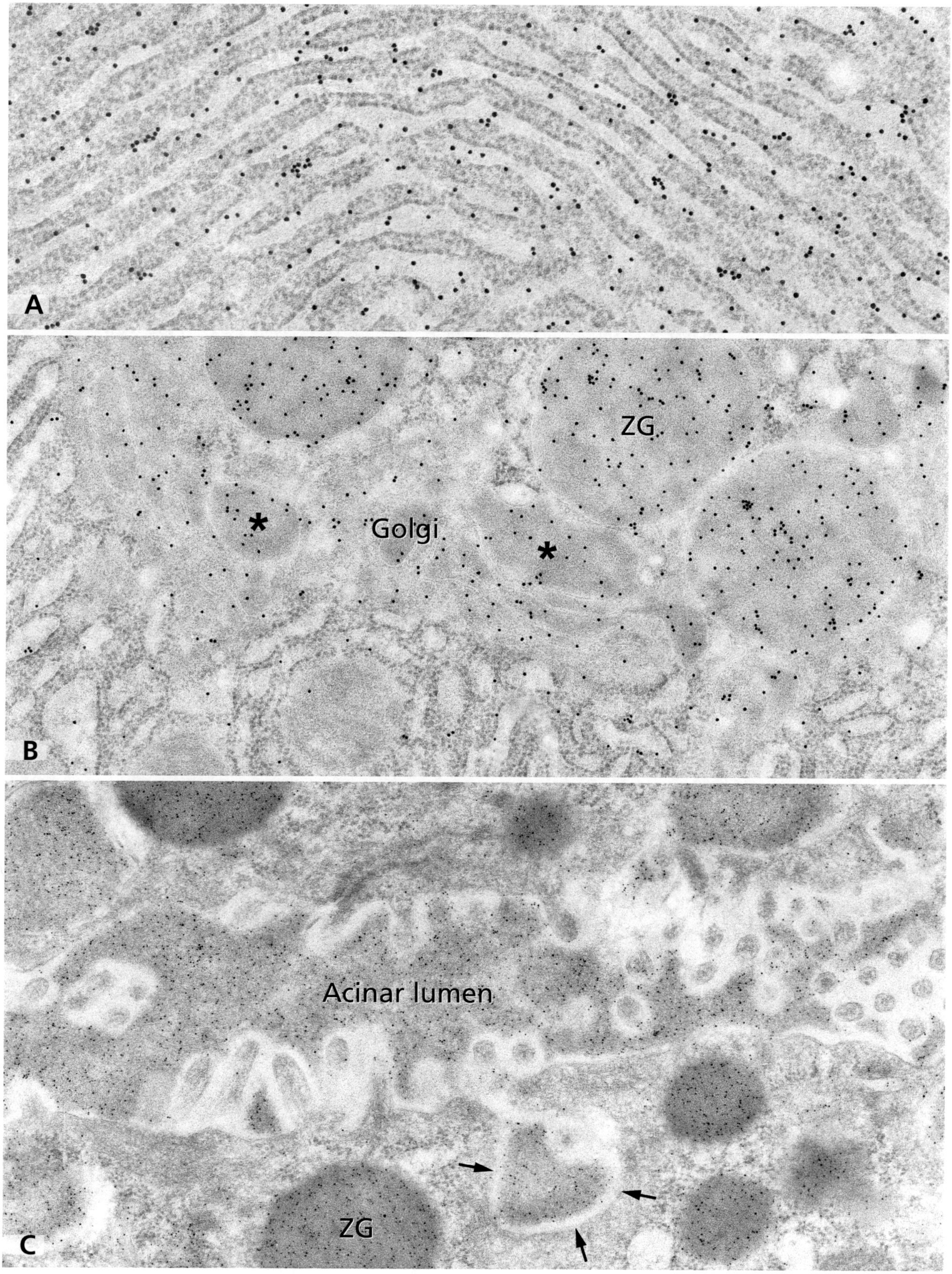
A
B
ZG
Golgi
*
*
C
Acinar lumen
ZG

PROTEIN *N*-GLYCOSYLATION: OLIGOSACCHARIDE TRIMMING IN THE GOLGI APPARATUS AND PRE-GOLGI INTERMEDIATES

Protein glycosylation continues in the Golgi apparatus by trimming and elongation of asparagine-linked oligosaccharides on glycoproteins to yield their mature forms. Trimming occurs by three mannosidases. Golgi mannosidase I removes the remaining α1,2-linked mannose residues followed by a first elongation reaction by *N*-acetylglucosaminyltransferase I and Golgi mannosidase II removes one α1,3 and one α1,6-linked mannose residue (scheme A). Endo-α-mannosidase trims monoglucosylated oligosaccharides of glycoproteins and has a substrate specificity similar to glucosidase II (scheme B). It provides a glucosidase-independent trimming pathway. In addition, certain mannosidase-trimmed monoglucosylated oligosaccharides are substrates for endomannosidase but not for glucosidase II (schemes C, D).

Golgi mannosidase I has been assumed to be compartmentalised in the *cis*-Golgi apparatus. However, depending on the cell type, it is detectable primarily in medial and *trans* Golgi cisternae. Likewise, Golgi mannosidase II shows a cell type-dependent Golgi distribution in medial (NRK and CHO cells), medial and *trans* (pancreatic cells) or *trans* cisternae (goblet cells), or across the entire cisternal stack (some enterocytes and hepatocytes; panel A).

Endomannosidase has a dual distribution: primarily in *cis* and medial Golgi cisternae with additional substantial labelling in p58-positive pre-Golgi intermediates (panels B and C). Panel C shows double immunogold labelling for endomannosidase (small gold particles) and p58 (large gold particles marked by arrowheads). This shows that glucose trimming occurs not only in the endoplasmic reticulum and pre-Golgi intermediates but also in the Golgi apparatus.

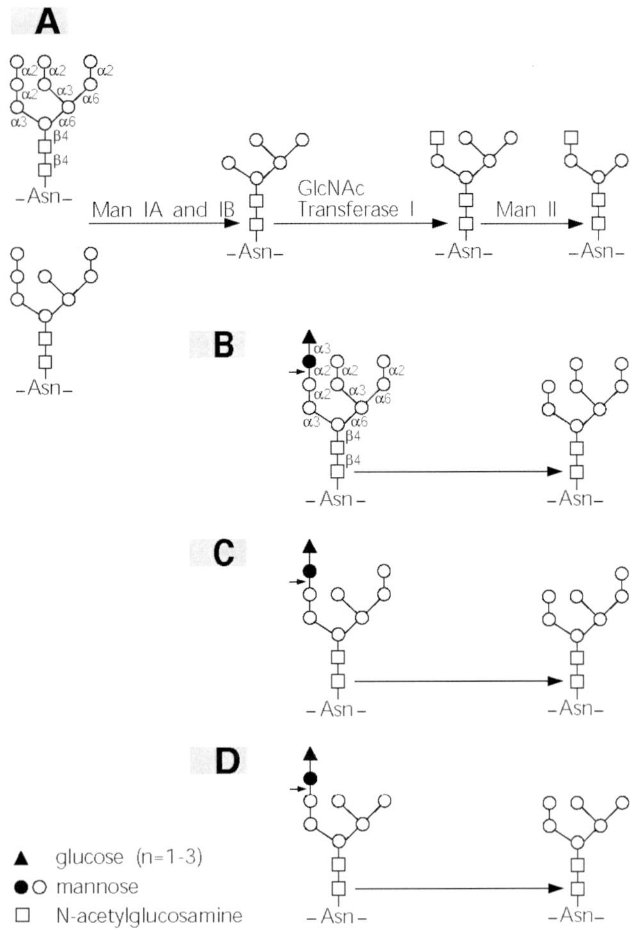

References

Herscovics A (1999) Importance of glycosidases in mammalian glycoprotein biosynthesis. Biochim Biophys Acta 1473: 96

Moore SE and Spiro RG (1990) Demonstration that Golgi endo-α-mannosidase provides a glucosidase-independent pathway for the formation of complex *N*-linked oligosaccharides of glycoproteins. J Biol Chem 265: 13104

Roth J (1997) Topography of glycosylation in the Golgi apparatus: In: The Golgi apparatus (Berger EG and Roth J, eds). Basel Boston Berlin: Birkhäuser, pp 131

Spiro R (2000) Processing enzymes involved in the deglucosylation of *N*-linked oligosaccharides of glycoproteins: glucosidases I and II and endomannosidase. In: Carbohydrate in chemistry and biology (Ernst B, Hart G, and Sinay P, eds). Weinheim: Wiley-VCH, pp 65

Velasco A, Hendricks L, Moremen KW, Tulsiani DRP, Touster O, and Farquhar MG (1993) Cell type-dependent variations in the subcellular distribution of alpha-mannosidase-I and alpha-mannosidase-II. J Cell Biol 122: 39

Zuber C, Spiro MJ, Guhl B, Spiro RG, and Roth J (2000) Golgi apparatus immunolocalisation of endomannosidase suggests post-endoplasmic reticulum glucose trimming: Implications for quality control. Mol Biol Cell 11: 4227

Zuber Z and Roth J (2009) *N*-glycosylation. In: The Sugar Code. Fundamentals of Glycosciences (Gabius H-J, ed). Weinheim: Wiley/VCH, pp 87

Magnification: x 83,000 (A); x 101,000 (B); x 94,000 (C)

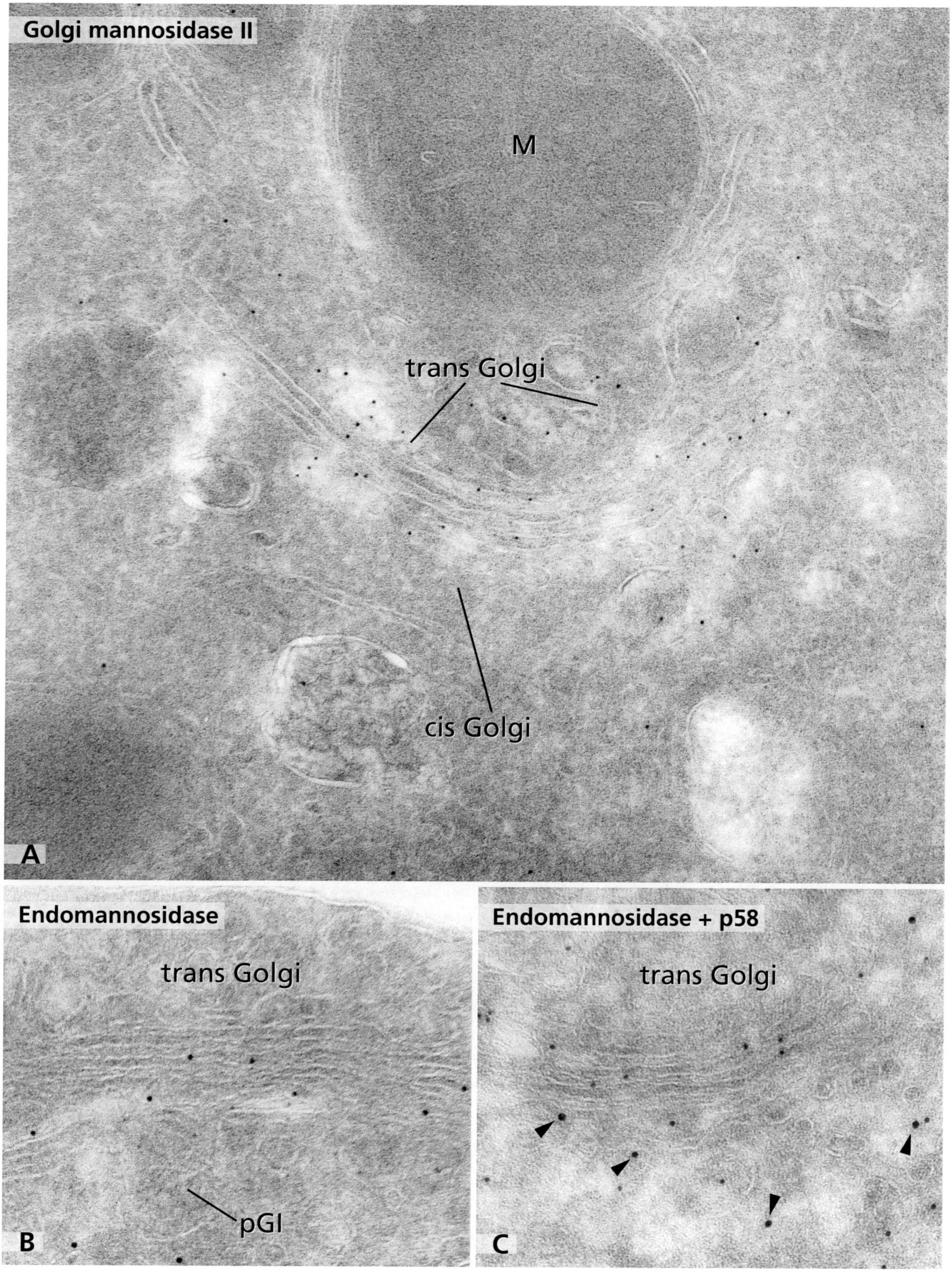
Golgi mannosidase II
M
trans Golgi
cis Golgi
A
Endomannosidase
trans Golgi
pGI
B
Endomannosidase + p58
trans Golgi
C

GOLGI APPARATUS: SITE OF MATURATION OF ASPARAGINE-LINKED OLIGOSACCHARIDES

The maturation of asparagine-linked oligosaccharides (or *N*-glycans) takes place in the Golgi apparatus and involves the families of *N*-acetylglucosaminyl-, galactosyl-, fucosyl-, and sialyltransferases, which elongate and diversify a mannose-GlcNAc$_2$ core oligosaccharide in a stepwise manner to yield so-called complex-type oligosaccharides. In it, elongation and mannose trimming reactions are intermingled. The complex-type oligosaccharides may exist as elaborate tetra-antennary structures with polylactosamine chains, which are repeats of galactose-*N*-acetylglucosamine disaccharides (scheme A). On the other hand, they may be rather simple bi-antennary structures, which are most commonly found (scheme B). These structures have to be distinguished from the high mannose type oligosaccharides composed of five to nine mannose residues and two *N*-acetylglucosamine residues and hybrid structures which are partly of complex and partly of high-mannose type.

The elongation reactions occur in a step-wise manner whereby the product of one glycosylation reaction provides the acceptor substrate for the next one. In line with this, sequentially acting glycosyltransferases such as *N*-acetylglucosaminyl-, galactosyl-, and sialyltransferase are located in medial cisternae, *trans* cisternae or in *trans* cisternae and the *trans* Golgi network, respectively (panels A and B) and serve as markers of these Golgi regions. However, although distinct, these glycosyltransferases partly overlap in the Golgi apparatus, and their distribution may vary depending on cell types (cf. Fig. 29).

Glycosyltransferases transfer a sugar from the respective nucleotide sugar (UDP-acetylglucosamine, GDP-fucose, UDP-galactose, CMP-sialic acid) to the acceptor. These glycosylation reactions take place in the lumen of the Golgi cisternae. Since sugar nucleotides are synthesised in the cytosol, except for CMP-sialic acid, which is synthesised in the nucleus, they must be transported into the lumen of Golgi cisternae. This is effected by nucleotide sugar transporters, which are multipassing Golgi membrane proteins. These transporters are antiporters because they function in the coupled cytosol-to-Golgi lumen transport of nucleotide sugars and in the Golgi lumen-to-cytosol transport of nucleotide derivatives (GMP, UMP) arising from glycosylation reactions.

References

Berninsone PM and Hirschberg CB (2000) Nucleotide sugar transporters of the Golgi apparatus. Curr Opin Struct Biol 10: 542

Hirschberg CB (1997) Transport of nucleotide sugars, nucleotide sulfate and ATP. In: The Golgi apparatus (Berger E and Roth J, eds). Basel Boston Berlin: Birkhäuser, pp 163

Perez M and Hirschberg CB (1987) Transport of sugar nucleotides into the lumen of vesicles derived from rat liver rough endoplasmic reticulum and Golgi apparatus. Meth Enzymol 138: 709

Rabouille C and Nilsson T (1995) Redrawing compartmental boundaries in the exocytic pathway. FEBS Lett 369: 97

Roth J and Berger E (1982) Immunocytochemical localisation of galactosyltransferase in HeLa cells: codistribution with thiamine pyrophosphatase in *trans* Golgi cisternae. J Cell Biol 93: 223

Roth J, Taatjes D, Lucocq J, Weinstein J, and Paulson J (1985) Demonstration of an extensive trans-tubular network continuous with the Golgi apparatus cisternal stack that may function in glycosylation. Cell 43: 287

Taniguchi N, Honke K, and Fukuda M (2002) Handbook of glycosyltransferases and related genes. Tokyo Berlin: Springer

Zuber Z and Roth J (2009) *N*-glycosylation. In: The Sugar Code. Fundamentals of Glycosciences (Gabius H-J, ed). Weinheim: Wiley/VCH, pp 87

Magnification: x 62,500 (A); x 46,200 (B)

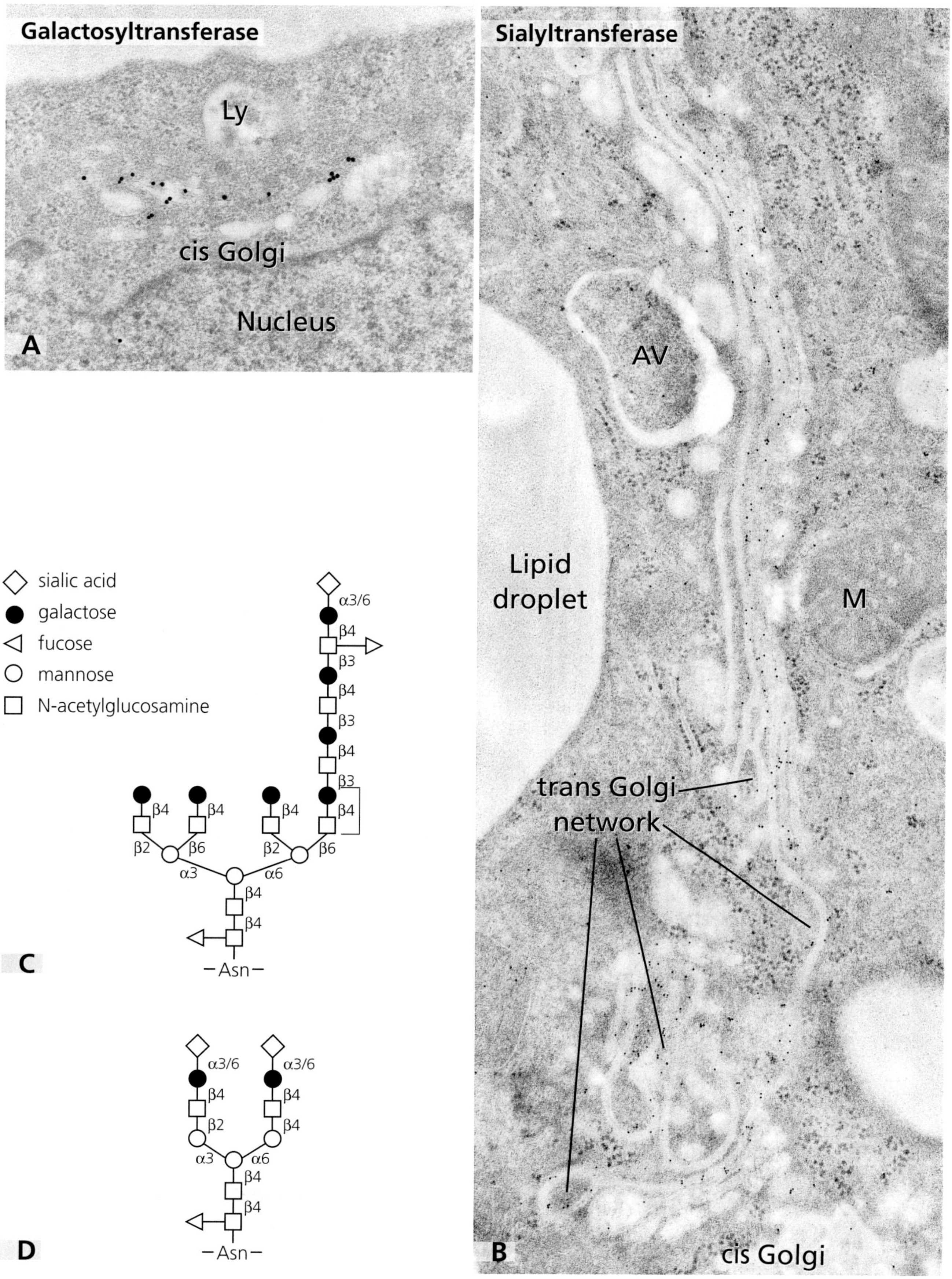
Galactosyltransferase
Ly
cis Golgi
Nucleus
A
Sialyltransferase
AV
Lipid droplet
M
trans Golgi network
B
cis Golgi
sialic acid
galactose
fucose
mannose
N-acetylglucosamine
α3/6
β4
β3
β2
β6
α3
α6
−Asn−
C
D

CELL TYPE-RELATED VARIATIONS IN THE TOPOGRAPHY OF GOLGI APPARATUS GLYCOSYLATION REACTIONS

Biochemical analyses have shown that glycosylation reactions in the Golgi apparatus occur in a step-wise manner by sequentially acting glycosyltransferases. Early subcellular fractionation and immunoelectron microscopic analyses showing enzyme activities in separate fractions and immunolabelling in different Golgi regions provided support for the concept of subdivision of the Golgi apparatus in functionally distinct *cis*, medial, and *trans* compartments. It is now evident that these are not distinct but overlapping compartments. Furthermore, there are cell type-related differences in Golgi distribution of glycosylation reactions. This is illustrated by the example of the β-galactoside α2,6 sialyltransferase (ST6Gal-I). In rat colon goblet cells, immunogold labelling for this sialyltransferase started to become detectable in *trans* Golgi cisternae (panel A). Its product, sialic acid as detected with a gold-labelled sialic acid specific lectin, also became first detectable in *trans* Golgi cisternae (panel B). It should be noted that the sialyltransferase immunolabelling in the mucus droplets of the goblet cells represents enzyme proteolytically cleaved in the Golgi apparatus. In contrast, labelling for the sialyltransferase and its product of action, sialic acid residues, was diffuse across the cisternal stack in the neighbouring absorptive enterocytes (panels C and D). The same restricted versus diffuse immunolabelling was observed for the blood group A transferase and blood group A substance in these cell types. Notably, both represent terminal glycosylation reactions in the synthesis of asparagine-linked or serine/threonine-linked oligosaccharides.

References

Dunphy WG, Brands R, and Rothman JE (1985) Attachment of terminal *N*-acetylglucosamine to asparagine-linked oligosaccharides occurs in central cisternae of the Golgi stack. Cell 40: 463

Dunphy WG, Fries E, Urbani LJ, and Rothman JE (1981) Early and late functions associated with the Golgi apparatus reside in distinct compartments. Proc Natl Acad Sci USA 78: 7453

Goldberg DE and Kornfeld S (1983) Evidence for extensive subcellular organisation of asparagine-linked oligosaccharide processing and lysosomal enzyme phosphorylation. J Biol Chem 258: 3159

Hedman K, Pastan I, and Willingham MC (1986) The organelles of the trans domain of the cell. Ultrastructural localisation of sialoglycoconjugates using Limax flavus agglutinin. J Histochem Cytochem 34: 1069

Rabouille C, Hui N, Hunte F, Kieckbusch R, Berger EG, Warren G, and Nilsson T (1995) Mapping the distribution of Golgi enzymes involved in the construction of complex oligosaccharides. J Cell Sci 108: 1617

Roth J and Berger EG (1982) Immunocytochemical localisation of galactosyltransferase in HeLa cells: codistribution with thiamine pyrophosphatase in *trans* Golgi cisternae. J Cell Biol 93: 223

Roth J, Taatjes DJ, Lucocq JM, Weinstein J, and Paulson JC (1985) Demonstration of an extensive trans-tubular network continuous with the Golgi apparatus cisternal stack that may function in glycosylation. Cell 43: 287

Roth J, Taatjes DJ, Weinstein J, Paulson JC, Greenwell P, and Watkins WM (1986) Differential subcompartmentation of terminal glycosylation in the Golgi apparatus of intestinal absorptive and goblet cells. J Biol Chem 261: 14307

Weinstein J, UE L, McEntee K, PH L, and Paulson J (1987) Primary structure of β-galactoside α2,6 sialyltransferase. Conversion of membrane bound enzyme to soluble form by cleavage of the N-terminal signal anchor. J Biol Chem 262: 17735

Magnification: x 49,500 (A); x 52,000 (B); x 51,000 (C, D)

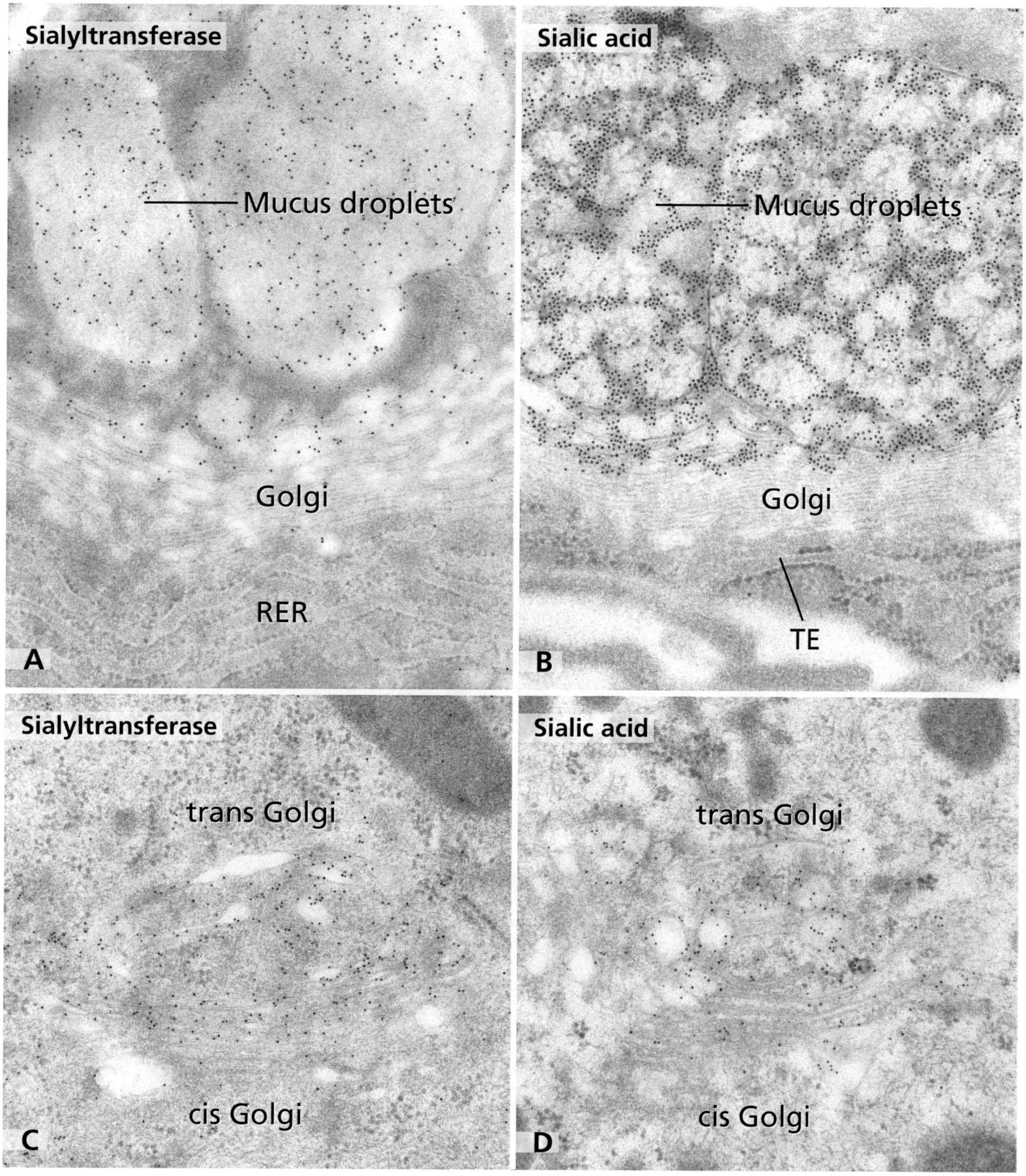
Sialyltransferase
Mucus droplets
Golgi
RER
A
Sialic acid
Mucus droplets
Golgi
TE
B
Sialyltransferase
trans Golgi
cis Golgi
C
Sialic acid
trans Golgi
cis Golgi
D

CELL TYPE-RELATED DIFFERENCES IN OLIGOSACCHARIDE STRUCTURE

Glycosyltransferases constitute large families of enzymes. About 15 sialyltransferases are currently known which all use CMP-sialic acid as donor substrate to covalently attach sialic acid to an oligosaccharide. However, the structure of the acceptor oligosaccharide may vary, and the sugar residues may be covalently attached in different ketosidic linkages. This represents one reason for the structural heterogeneity of asparagine-linked oligosaccharides. Hence, sialic acids may be found attached to galactose or *N*-acetylgalactosamine in α2,6, α2,3, α2,8 and α2,9 ketosidic linkage. Another reason for structural diversity of the oligosaccharides lies in differences in the cellular expression pattern of glycosyltransferases. Thus, sialyltransferases and other glycosyltransferases exhibit cell and tissue specific expression patterns. It should also be noted that sialyltransferases compete for a common nucleotide sugar and that differences in their tissue levels will influence the structure of the synthesised sialylated oligosaccharide.

It is now established that specific oligosaccharides of glycoproteins are involved in a variety of biological functions such as the modulation of the activity of proteins, cell-cell and cell-substratum interactions, protein quality control and trafficking. These are important aspects in the production of recombinant proteins such as erythropoetin, whose activity critically depends on correct glycosylation. It may therefore be necessary to modify the glycosylation machinery of host cells by introducing a specific glycosyltransferase so that they synthesise a particular oligosaccharide normally not made by them. For instance, Chinese hamster ovary (CHO) cells synthesise oligosaccharides terminated in α2,3-linked sialic acid but lack β-galactoside α2,6 sialyltransferase. A lectin specifically recognising α2,6-linked sialic acid therefore does not bind to the cell surface (PM) or intracellular structures such as Golgi apparatus and lysosomes (Ly) of CHO cells (panels A and B). After transfection, β-galactoside α2,6 sialyltransferase can be detected by immunoelectron microscopy in the Golgi apparatus of CHO cells (panel C) and oligosaccharides bearing α2,6-linked sialic in the plasma membrane (PM) (panel D) and the Golgi apparatus and lysosomes by lectin-gold labelling (panel E).

In the case of pig to human xenotransplantation, the hyperacute rejection is caused by oligosaccharides terminated in α1,3-linked galactose. To circumvent this, the respective galactosyltransferase has been inactivated by homologous recombination or competed for by the introduction of another fucosyltransferase using the same acceptor substrate.

References

Chen CG, Fisicaro N, Shinkel TA, Aitken V, Katerelos M, Vandenderen BJW, Tange MJ, Crawford RJ, Robins AJ, Pearse MJ, and Dapic AJF (1996) Reduction in Gal-alpha 1,3-Gal epitope expression in transgenic mice expressing human H-transferase. Xenotransplantation 3: 69

Galili U (1993) Evolution and pathophysiology of the human natural anti-α-galactosyl IgG(anti-Gal) antibody. Springer Semin Immunopathol 15: 155

Kitagawa H and Paulson JC (1994) Differential expression of five sialyltransferase genes in human tissues. J Biol Chem 269: 17872

Koike C, Kannagi R, Takuma Y, Akutsu F, Hayashi S, Hiraiwa N, Kadomatsu K, Muramatsu T, Yamakawa H, Nagai T, et al. (1996) Introduction of alpha(1,2)-fucosyltransferase and its effect on alpha-Gal epitopes in transgenic pig. Xenotransplantation 3: 81

Lee EU, Roth J, and Paulson JC (1989) Alteration of terminal glycosylation sequences on N-linked oligosaccharides of Chinese hamster ovary cells by expression of ß-galactoside α2,6 sialyltransferase. J Biol Chem 264: 13848

Osman N, McKenzie IFC, Ostenried K, Ioannou YA, Desnick RJ, and Sandrin MS (1997) Combined transgenic expression of alpha-galactosidase and alpha 1,2-fucosyltransferase leads to optimal reduction in the major xenoepitope Gal alpha(1,3)Gal. Proc Natl Acad Sci USA 94: 14677

Paulson J, Weinstein J, and Schauer A (1989) Tissue specific expression of sialyltransferases. J Biol Chem 264: 10931

Rosenberg A (1995) Biology of the sialic acids. New York: Plenum Press

Taniguchi N, Honke K, and Fukuda M (2002) Handbook of glycosyltransferases and related genes. Tokyo Berlin: Springer

Magnification: x 45,000 (A); x 55,500 (B); x 83,000 (C); x 51,000 (D); x 79,000 (E)

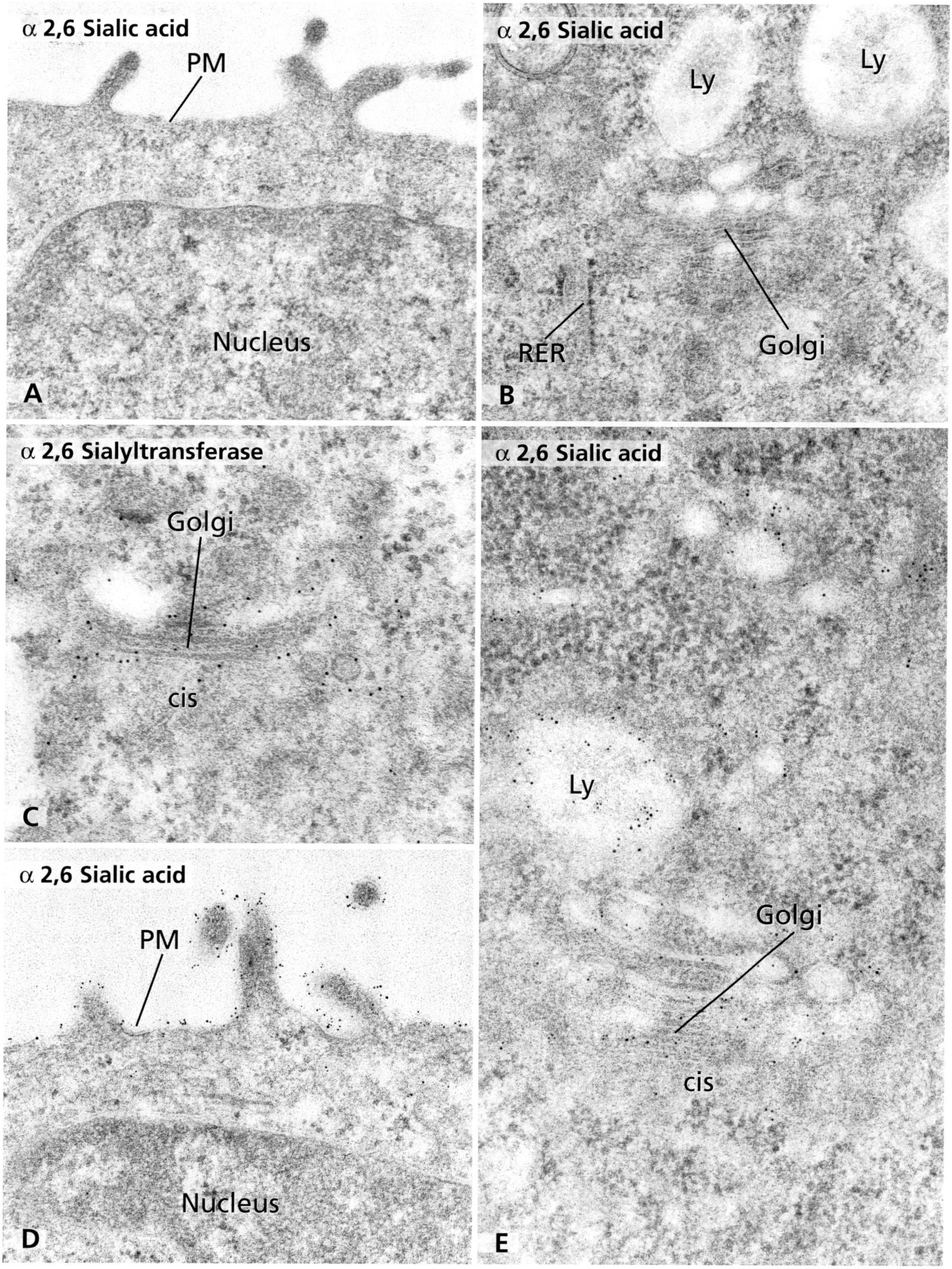
α 2,6 Sialic acid
PM
Nucleus
A
α 2,6 Sialic acid
Ly
Ly
RER
Golgi
B
α 2,6 Sialyltransferase
Golgi
cis
C
α 2,6 Sialic acid
PM
Nucleus
D
α 2,6 Sialic acid
Ly
Golgi
cis
E

TOPOGRAPHY OF BIOSYNTHESIS OF SERINE/THREONINE-LINKED OLIGOSACCHARIDES

In addition to asparagine-linked oligosaccharides (*N*-glycans), many glycoproteins bear serine-linked or threonine-linked oligosaccharides (*O*-glycans). The *O*-glycans considered here have as the core sugar *N*-acetylgalactosamine α-linked to the hydroxyl group of serine or threonine. Other *O*-glycans have xylose, fucose, or *N*-acetylglucosamine as the core sugar. Mucin glycoproteins that cover the digestive, respiratory, and genitourinary tract possess a high number of *O*-glycans that bind water and ions, serve as lubricants, and provide a defence against microbial invaders.

In contrast to the biosynthesis of *N*-glycans, which starts in the endoplasmic reticulum, those of *O*-glycans begins in the *cis*-Golgi apparatus, as illustrated in panels A–C. The mucus secreted by pig submaxillary glands is highly glycosylated by *O*-glycans. An antibody specifically recognising the non-glycosylated apomucin (panel A) marks the rough endoplasmic reticulum (RER) and pre-Golgi intermediates (pGI) but not the Golgi apparatus. An antibody raised against the polypeptide-GalNAc transferase catalysing the initial *O*-glycosylation reaction, the transfer of a GalNAc residue to serine or threonine, does not label the RER and the pGI but *cis* Golgi cisternae (panel B). This labelling pattern corresponds to that of lectin-gold labelling for polypeptide-linked *N*-acetylgalactosamine (panel C). The subsequent glycosylation reactions also take place in the Golgi apparatus. In yeast and fungi, *O*-glycosylation starts in the endoplasmic reticulum, but this is protein *O*-mannosylation.

The biosynthesis and structure of *O*-glycans with the serine/threonine-GalNAc core linkage is simpler than that of *N*-glycans. No lipid-linked oligosaccharide is preassembled, and no trimming reactions take place. There is no consensus *O*-glycosylation sequence, probably because of the numerous polypeptide-GalNAc transferases, which differ in their specificity for amino acid sequences. The biosynthetic pathway consists of a series of sequentially occurring classical glycoslytransfer reactions, yielding mainly bi-antennary structures. Some of the simple *O*-glycans are characteristic features of carcinoma mucins, such as the Tn antigen (GalNAc-ser/thre) and its sialylated form (sialic acid α2,6 GalNAc-ser/thre) or the Thomsen-Friedenreich antigen (Gal β1,3 GalNAc-ser/thre). Their occurrence seems to be of prognostic significance in some carcinomas.

References

Brockhausen I (1999) Pathways of O-glycan biosynthesis in cancer cells. Biochim Biophys Acta 1473: 67

Clausen H and Bennett EP (1996) A family of UDP-GalNAc: polypeptide *N*-acetylgalactosaminyltransferases control the initiation of mucin-type *O*-linked glycosylation. Glycobiology 6: 635

Deschuyteneer M, Eckhardt AE, Roth J, and Hill RL (1988) The subcellular localisation of apomucin and nonreducing terminal *N*-acetylgalactosamine in porcine submaxillar glands. J Biol Chem 263: 2452

Elhammer AP, Kezdy FJ, and Kurosaka A (1999) The acceptor specificity of UDP-GalNAc: polypeptide N-acetylgalactosaminyltransferases. Glycoconjugate J 16: 171

Hagen KGT, Fritz TA, and Tabak LA (2003) All in the family: the UDP-GalNAc:polypeptide *N*-acetylgalactosaminyltransferases. Glycobiology 13, 1R

Patsos G and Corfield A (2009) *O*-glycosylation: Structural diversity and functions. In: The Sugar Code. Fundamentals of Glycosciences (Gabius H-J, ed). Weinheim: Wiley/VCH, pp 111

Roth J, Wang Y, Eckhardt AE, and Hill RL (1994) Subcellular localisation of the UDP-*N*-acetyl-D-galactosamine: polypeptide *N*-acetylgalactosaminyltransferase-mediated *O*-glycosylation reaction in the submaxillary gland. Proc Natl Acad Sci USA 91: 8935

Strahl Bolsinger S, Gentzsch M, and Tanner W (1999) Protein *O*-mannosylation. Biochim Biophys Acta 1426: 297

Magnification: x 60,000 (A); x 76,000 (B); x 86,000 (C)

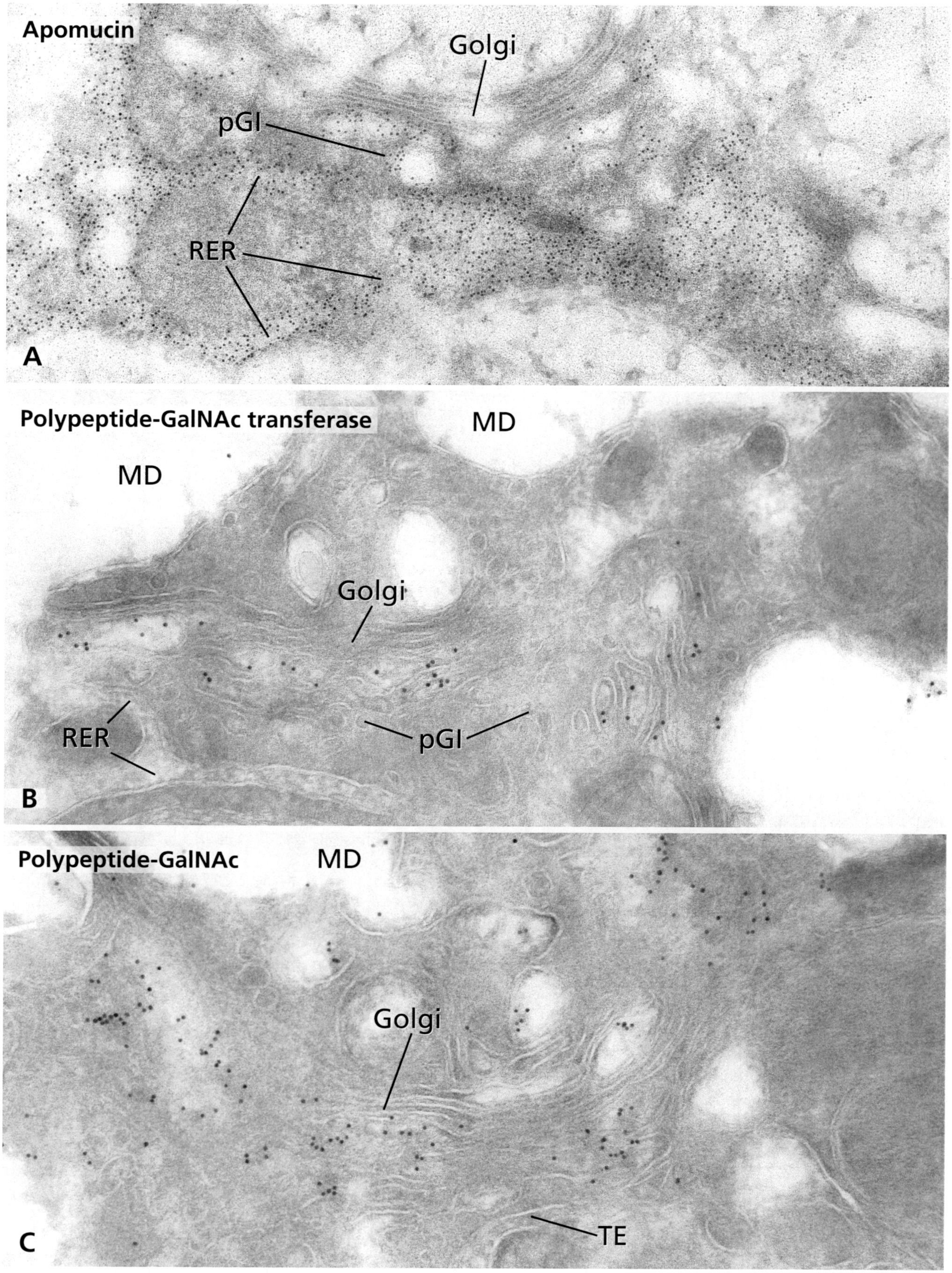
Apomucin
Golgi
pGI
RER
A
Polypeptide-GalNAc transferase
MD
MD
Golgi
RER
pGI
B
Polypeptide-GalNAc
MD
Golgi
TE
C

GOLGI APPARATUS AND TGN – STRUCTURAL CONSIDERATIONS

The pioneering work of Rambourg and colleagues laid the foundations for what is nowadays known as the *trans* Golgi network (TGN). As early as 1969, he reported an extensive system of anastomosing tubules at the *trans*-side of the Golgi apparatus of intestinal absorptive cells, named the *trans*-tubular network of the Golgi apparatus. Rambourg and colleagues, by analysing thick sections (0.5–1 μm) by high-voltage electron microscopy from resin embedded intestine (and other tissues) stained with phosphotungstic acid at low pH, pointed out the similarities of the *trans*-tubular network of the Golgi apparatus to the internal reticular apparatus discovered by Golgi, and to the system of GERL described by Novikoff and colleagues by using acid phosphatase cytochemistry. The stereoscopic analysis of thick sections showed that *trans*-tubular networks (and cis-networks) connect individual Golgi cisternal stacks to form a single ribbon-like organelle. The origin of the TGN from *trans* Golgi cisternae was demonstrated by its continuity with 2–3 *trans* Golgi cisternae. At the same time, specific Golgi apparatus functions such as terminal glycosylation reactions were unequivocally assigned to the TGN. The basic structure of the TGN has been confirmed by using high pressure freezing for fixation and electron microscopy tomography for three-dimensional analysis. The many studies have revealed cell type-related variability in the dimensions of the TGN and its highly dynamic nature.

The electron micrograph shown illustrates the Golgi apparatus and the TGN in the human hepatoma cell line HepG2. Cells were fixed with 2% formaldehyde in slightly hypertonic buffer solution. In ultrathin frozen sections, the cup-shaped cisternal stack of the Golgi apparatus is seen cross-sectioned together with the elaborate TGN. Because of the mild fixation, protein extraction as indicated by empty spaces in the cytoplasm has occurred. Therefore, the membranes appear highly distinct and the content of the Golgi cisternae and TGN is well contrasted. However, a specific element of the *trans*-side of the Golgi apparatus, the *trans* Golgi endoplasmic reticulum (cf. Figs. 34 and 35), usually escapes the observation since ribosomes cannot be easily discerned in ultrathin frozen sections. It is typically the *trans*-most Golgi cisterna of the HepG2 cells that gives raise to an elaborate TGN. As shown at higher magnification in the inset, numerous buds exist giving raise to clathrin-coated vesicles (arrow). The subsequent figures provide additional detail in regard to structure of the TGN and discuss its multiple functions in sorting as well exocytic and endocytic trafficking.

References

De Matteis MA and Luini A (2008) Exiting the Golgi complex. Nat Rev Mol Cell Biol 9: 273

Griffiths G and Simons K (1986) The *trans* Golgi network: sorting at the exit site of the Golgi complex. Science 234: 438

Hermo L and Smith CE (1998) The structure of the Golgi apparatus: a sperm's eye view in principal epithelial cells of the rat epididymis. Histochem Cell Biol 109: 431

Ladinsky MS, Mastronarde DN, McIntosh JR, Howell KE, and Staehelin LA (1999) Golgi structure in three dimensions: Functional insights from the normal rat kidney cell. J Cell Biol 144: 1135

Lieu ZZ, Derby MC, and Gleeson PA (2009) Structure and domain organization of the *trans* Golgi network. In: The Golgi Apparatus (Mironov A and Pavelka M, eds.) Wien New York: Springer, pp 358

Novikoff P, Novikoff A, Quintana N and Hauw J (1971) Golgi apparatus, GERL, and lysosomes of neurons in rat dorsal root ganglia, studied by thick section and thin section cytochemistry. J Cell Biol 50: 859

Rambourg A and Clermont Y (1997) Three-dimensional structure of the Golgi apparatus in mammalian cells. In: The Golgi apparatus (Berger E, Roth J, eds), Basel Boston Berlin: Birkhäuser, pp 37

Roth J, D JT, J ML, J W, and J CP (1985) Demonstration of an extensive trans-tubular network continuous with the Golgi apparatus cisternal stack that may function in glycosylation. Cell 43: 287

Roth J and Taatjes DJ (1998) Tubules of the trans Golgi apparatus visualized by immunoelectron microscopy. Histochem Cell Biol 109: 545

Taatjes D and Roth J (1986) The trans-tubular network of the hepatocyte Golgi apparatus is part of the secretory pathway. Eur J Cell Biol 42: 344

Magnification: x 64,000; x 75,000 (inset)

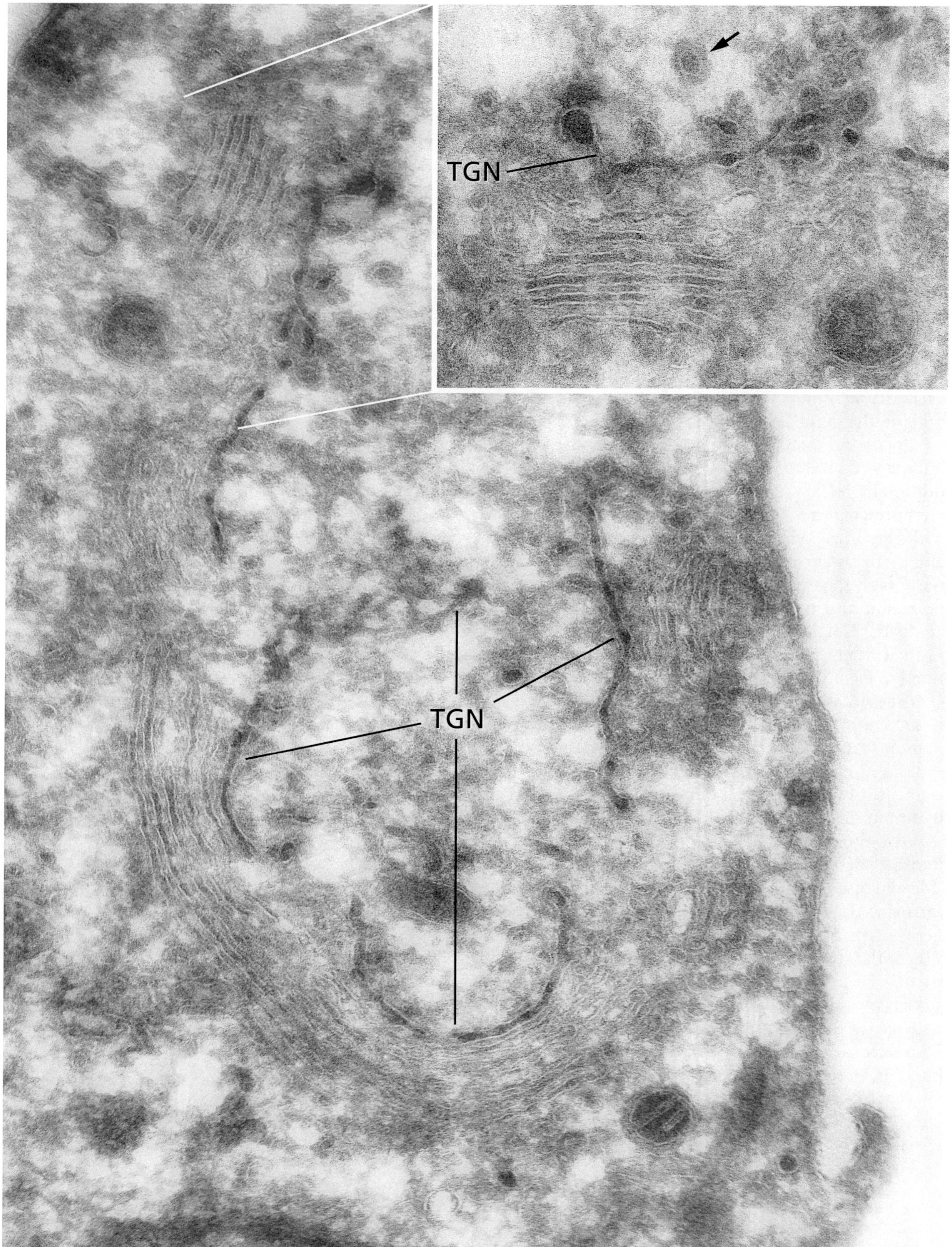
TGN
TGN

GOLGI APPARATUS AND TGN – SECRETION AND ENDOCYTOSIS

The Golgi apparatus is not only a pivotal organelle and main crossroads in the secretory pathway, but also has a central role during endocytosis (cf. Fig. 48). Studies with multiple cell types showed, that distinct endocytic routes exist from the plasma membrane to the Golgi apparatus, involving early and late endosomal compartments. For endocytosed materials, the *trans* Golgi side is the import region into the Golgi stacks and transport across the stacked cisternae appears to be directed from the *trans* to the *cis* side, which is opposite to the transport of secretory molecules that enter the Golgi stacks at the *cis* side and travel from *cis* to *trans* (cf. Figs. 23–25). Secretory and internalised molecules possibly meet each other in the same Golgi compartments.

Uptake of recycling membrane proteins and of internalised molecules into the Golgi apparatus plays an important role for reprocessing of plasma membrane glycoproteins and is of particular interest in connection with the retrograde pathways of toxins, such as ricin, and pertussis and cholera toxins, that take routes across the Golgi apparatus stacks back to the endoplasmic reticulum and to the cytosol. For these toxins, traffic across the Golgi apparatus is a prerequisite for exertion of their deleterious effects and Golgi apparatus disruption (cf. Figs. 37–40) prevents cells from intoxications.

Panels A and B show Golgi apparatus stacks engaged in secretion (panel A) and endocytosis (panel B), thus demonstrating that both secretory and endocytic molecules may visit both stacked Golgi cisternae and elements of the *trans* Golgi network.(TGN). In panel A by immunogold labelling, secretory albumin in shown in a hepatocyte of the rat liver. The gold marker is localised to all Golgi cisternae from the *cis* to the *trans* side and also labels all parts of the *trans* Golgi network (arrows). Typically, TGN elements form an extensive network that is continuous with cisternae at the *trans* side of the stacks. The TGN consists of two continuous parts, one being integrated component of the Golgi stacks and another turning away and building up a network distant from the stack (cf. also Figs. 34–36). Distended rims of both parts of the TGN correspond to budding secretory vesicles.

Panel B shows a rat pancreatic endocrine beta cell after endocytosis of concanavalin A (ConA). For visualisation of the endocytosis pathways ConA was conjugated with ferritin. Dense ConA-ferritin labelling is apparent in two prominent TGN-elements covered with extensive bristle coats, as well as in the *trans*most Golgi cisterna (arrows). Endocytosed ConA is also present in medial cisternae of the stack and in multiple tubular-vesicular elements in the stack's proximity. Detailed kinetic experiments showed an increased frequency of Golgi apparatus involved in endocytosis with increased incubation time. After 1 hour, Golgi stacks were occasionally labelled; after 3 hours, 10% of the stacks showed uptake of ConA-ferritin and 65% of the stacks were labelled after 14 hours.

Ly – lysosomes, M – mitochondria.

References

Barr FA (2009) Rab GTPase function in Golgi trafficking. Seminars Cell Dev Biol 20: 780

De Matteis MA, and Luini A (2008) Exiting the Golgi complex. Nat Rev Mol Cell Biol 9: 273

Griffiths G and Simons K (1986) The *trans* Golgi network: Sorting at the exit site of the Golgi complex. Science 234: 438

Luini A, Mironov AA, Polishchuk EV, and Polishchuk RS (2008) Morphogenesis of post-Golgi carriers. Histochem Cell Biol 129: 153

Mallard F, Antony C, Tenza D, Salamero J, Goud B, and Johannes L (1998) Direct pathway from early/recycling endosomes to the Golgi apparatus revealed through the study of Shiga toxin B-fragment transport. J Cell Biol 143: 973

Pavelka M, Neumüller J, and Ellinger A (2008) Retrograde traffic in the biosynthetic-secretory route. Histochem Cell Biol 129: 277

Sandvig K and van Deurs B (2000) Entry of ricin and Shiga toxin into cells: molecular mechanisms and medical perspectives. EMBO J 19: 1

Ramirez IB-R and Lowe M (2009) Golgins and GRASPs: Holding the Golgi together. Seminars Cell Dev Biol 20: 770

Roth J, Taatjes D, Lucocq J, Weinstein J, and Paulson J (1985) Demonstration of an extensive trans-tubular network continuous with the Golgi apparatus cisternal stack than may function in glycosylation. Cell 43: 287

Sallese M, Giannotta M, and Luini A (2009) Coordination of the secretory compartments via inter-organelle signalling. Seminars Cell Dev Biol 20: 801

Volz B, Orberger G, Porwoll S, Hauri H-P, and Tauber R (1995) Selective reentry of recycling cell surface glycoproteins to the biosynthetic pathway in human hepatocarcinoma HepG2 cells. J Cell Biol 130: 537

Magnification: x 45,000 (A), x 45,000 (B)

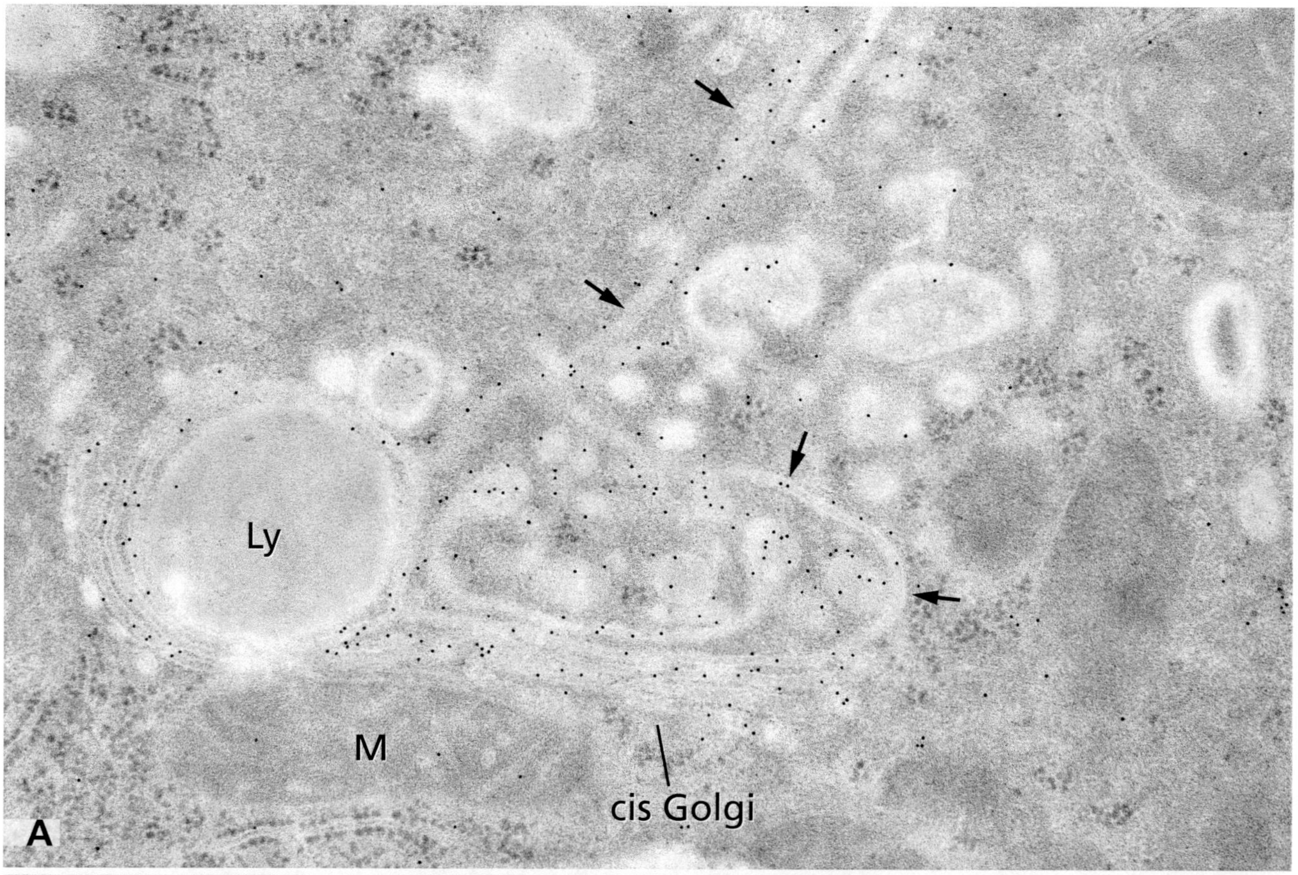
Ly
M
cis Golgi
A

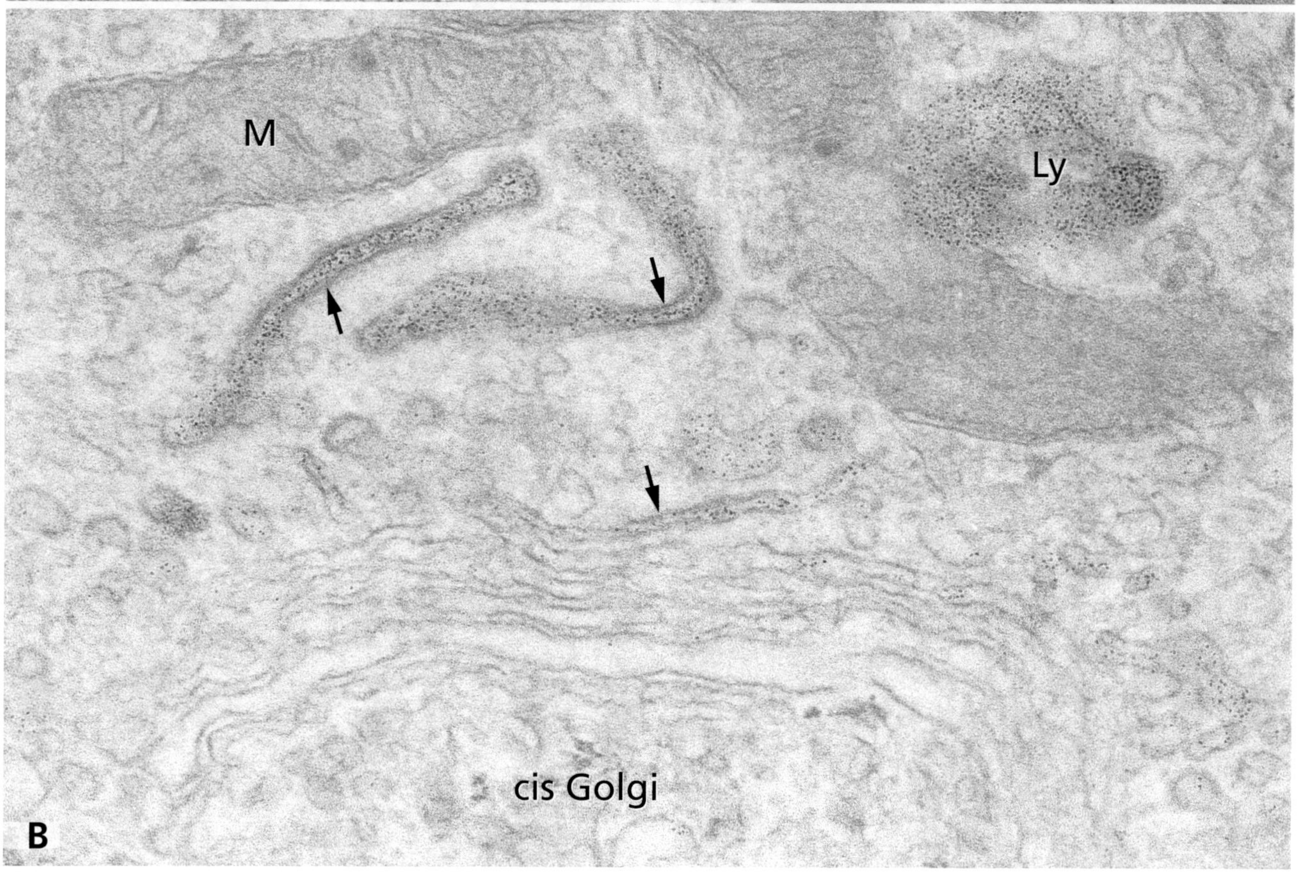
M
Ly
cis Golgi
B

GOLGI APPARATUS, TGN AND *TRANS* GOLGI-ER

The Golgi apparatus is a highly dynamic organelle that changes its shape and architecture concomitantly with the continuous flow and extensive traffic across its compartments. Chemical fixation is too slow to resolve properly most of the rapid shape changes and moreover in multiple cases, obscures the critical view of ultrastructural details due to artefacts occurring during fixation. These problems and restrictions may be overcome by using cryotechniques and in particular, by employing high pressure freezing for ultrafast immobilisation of cells and their subcompartments. This makes it possible to arrest cellular dynamics in less than half a second. High pressure cryoimmobilisation of membranes and compartments considerably improves the temporal resolution of cellular dynamics and, by preventing artefacts, improves spatial resolution and permits a view that is closer to the *in vivo*-state than that obtained by chemical fixation.

Panels A and B show Golgi stacks of HepG2-hepatoma cells, which were grown on carbon-coated sapphire disks, cryoimmobilised by high pressure freezing and subsequently freeze substituted and embedded in Epon. Both panels present a stack of Golgi cisternae with clear structural and architectural *cis* versus *trans* differences. Contrasting with the sights obtained with chemically fixed specimens such as shown in Fig. 25, in the cryoimmobilised cells clear differences of the Golgi cisternae membranes are evident. Membranes show a clearly three-lamellar and "frayed out" structure in cisternae localised in the medial part of the stacks. These membrane characteristics are lacking in the *trans* Golgi network (TGN), which in HepG2 hepatoma cells consists of two parts: one part is integrated in the stack and corresponds to the *trans*most Golgi cisterna (arrowheads in panels A and B). The other part turns away from the stack (the continuity is shown in panel B) and forms an extensive network with multiple budding vesicles (TGN in panels A and B). Both parts of the TGN show distinct, "non-frayed out"-membranes that clearly differ to those of the medial Golgi cisternae. Both parts of the TGN are closely associated with cisternae of the endoplasmic reticulum. This *trans* Golgi-ER (arrows) shows ribosomes attached to one side only; the other side is smooth and is in a close spatial relationship to the TGN.

These differentiated Golgi and TGN-patterns obtained after cryoimmobilisation underline the functional specialisations of the TGN, which is a central part of both the secretory pathway and the endocytic system (cf. Figs. 33 and 48).

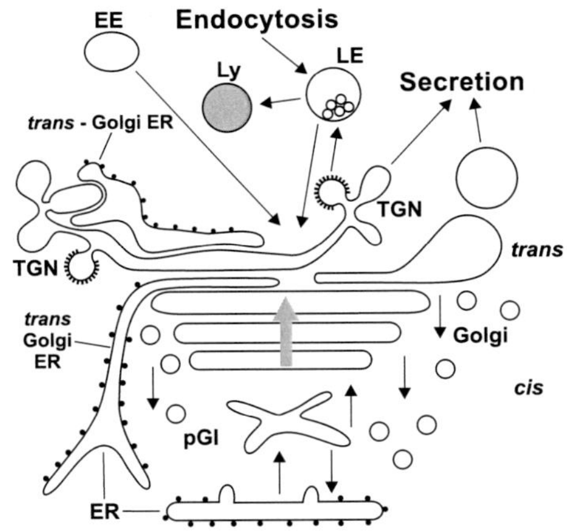

The diagram summarises the Golgi stack compartments and indicates the main directions of traffic into and across a Golgi apparatus stack. Models of traffic are addressed together with Fig. 40.

PGI – pre-Golgi intermediate, Ly – lysosome, EE – early endosome, LE – late endosome.

References

Bouchet-Marchis C, Starkuviene V, and Grabenbauer M (2008) Golgi apparatus studied in vitreous sections. J Microsc 230: 308

Griffiths G and Simons K (1986) The *trans* Golgi network: Sorting at the exit site of the Golgi complex. Science 234: 438

Hess MW, Müller M, Debbage PL, Vetterlein M, and Pavelka M (2000) Cryopreparation provides new insight into effects of Brefeldin A on the structure of the HepG2 Golgi apparatus. J Struct Biol 130: 63

Jackson CL (2009) Mechanisms of transport through the Golgi complex. J Cell Sci 122: 443

Ladinsky MS, Mastronarde DN, McIntosh JR, Howell KE, and Staehelin LA (1999) Golgi structure in three dimensions: Functional insights from the normal rat kidney cell. J Cell Biol 144: 1135

Leis A, Rockel B, Andrees L, and Baumeister W (2008) Visualizing cells at the nanoscale. Trends Biochem Sci 34: 60

Lučić V, Leis A, and Baumeister W (2008) Cryo-electron tomography of cells: connecting structure and function. Histochem Cell Biol 130:185

Mironov AA and Pavelka M (2008) The Golgi apparatus as a crossroads in intracellular traffic. In: The Golgi apparatus. State of the art 110 years after Camillo Golgi's discovery (Mironov A and Pavelka M, eds). Wien New York: Springer, p 16

Moor H (1987) Theory and practice of high pressure freezing. In: Cryotechniques in biological electron microscopy (Steinbrecht RA, and Zierold K, eds). Berlin Heidelberg: Springer, p 175

Magnification: x 74,000 (A, B)

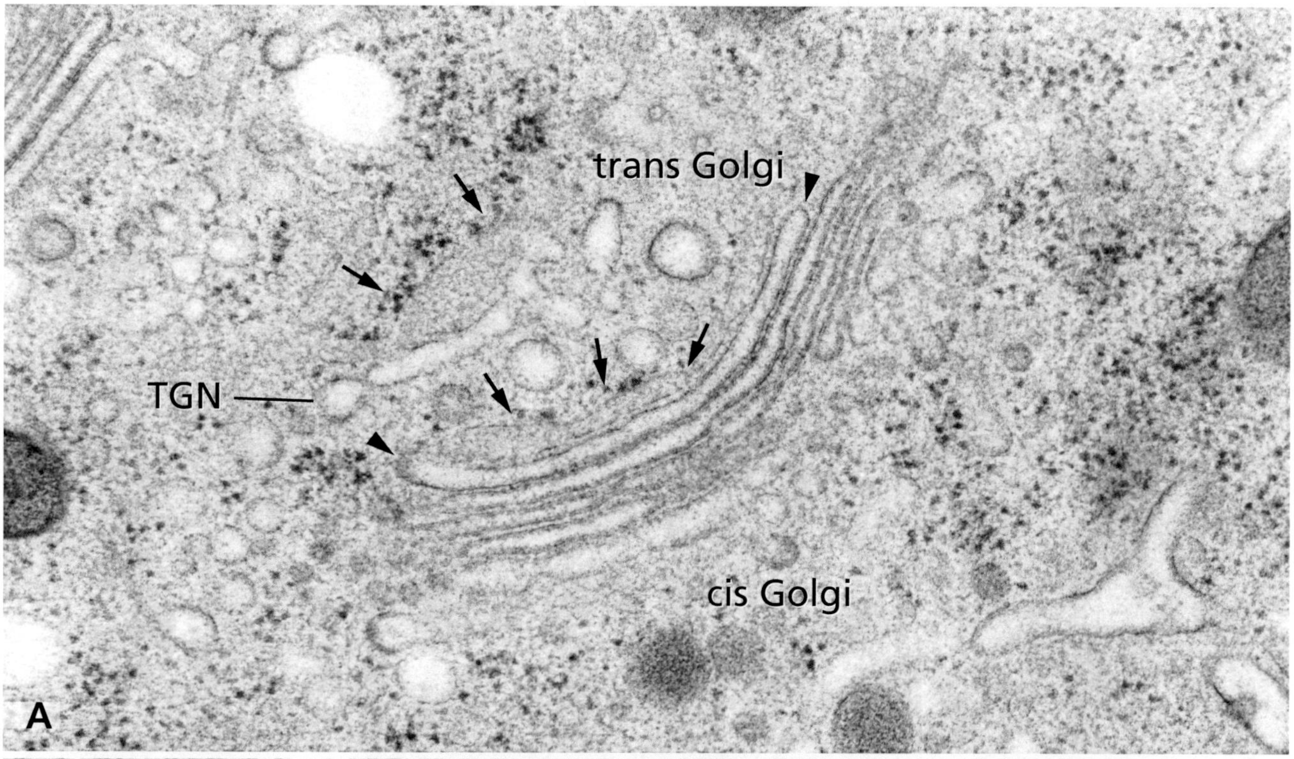
trans Golgi
TGN
cis Golgi
A

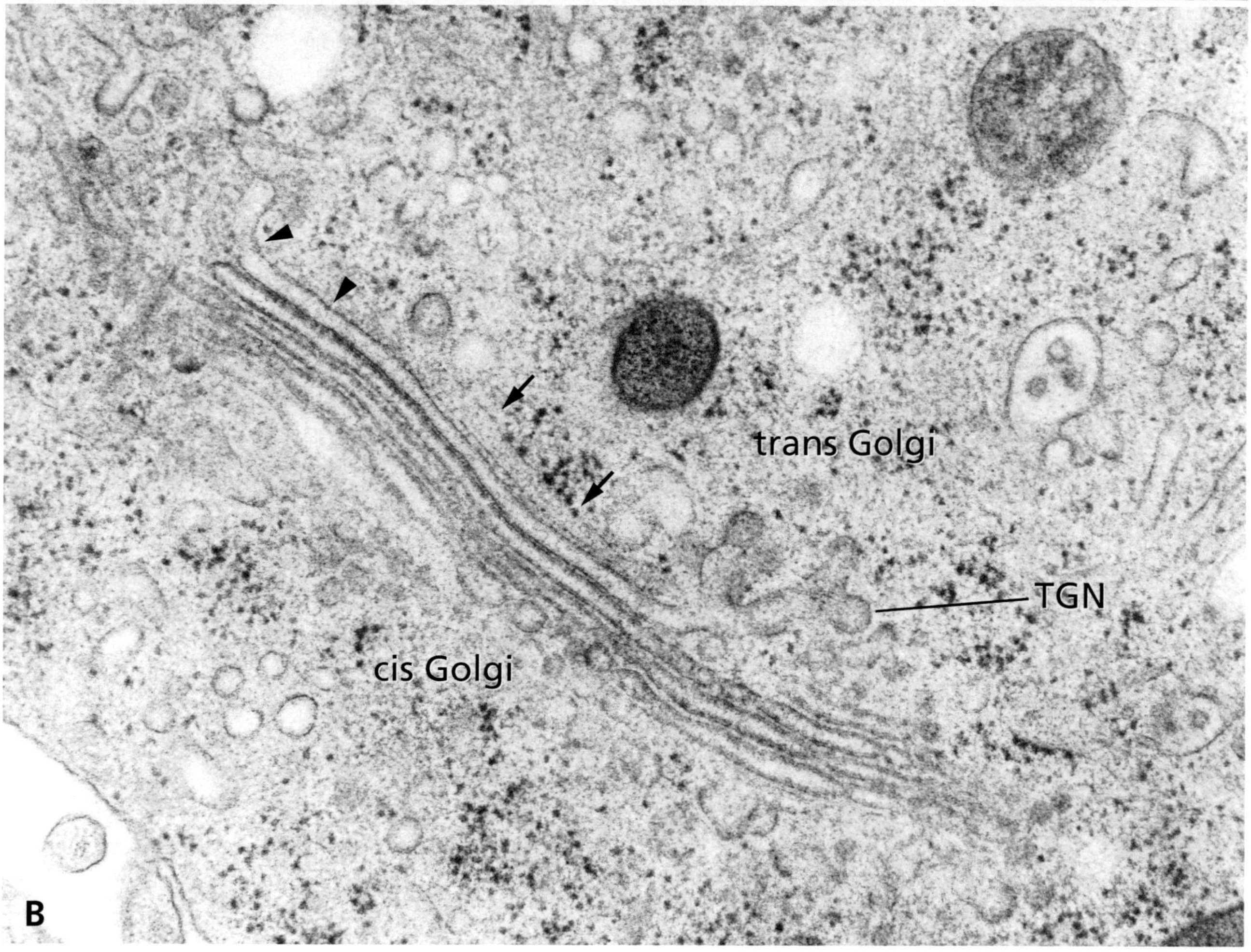
trans Golgi
TGN
cis Golgi
B

GOLGI APPARATUS, TGN AND *TRANS* GOLGI-ER: TILT SERIES

The enormous complexity of the Golgi apparatus is easily overlooked when electron microscopic studies are restricted to analyses of single thin sections. Additional studies of semithin and thick sections, analyses of section series, use of eucentric goniometer equipment for controlled tilting of the specimens within the electron microscope, and the introduction of electron tomography into the field of cell biology has tremendously improved the knowledge of the morpho-functional architecture of the Golgi apparatus and its dynamic organisation.

Panels A, B, and C show a series of micrographs of the same Golgi apparatus stack of high pressure frozen HepG2 hepatoma cells taken at different degrees of tilting, minus 43°, plus 3°, and +plus 33°. The series shows how many details of the complex structures would have been overlooked, if only one single angle had been analysed. Particular attention should be drawn to the endoplasmic reticulum (ER) exit site localised at a transitional ER-element (TE) and the heteromorphous compartments of the pre-Golgi intermediates (pGI) at the left hand side of the micrograph, the *cis* Golgi architecture, the small densely coated buds at *cis*- and medial Golgi compartments presumably corresponding to sites, where coat protein I (COP I)-coated vesicles are formed (arrowheads, cf. also Fig. 23), and the close appositions of the *trans* Golgi-endoplasmic reticulum (arrows) to cisternae of the *trans* Golgi networks.

The diagram below shows a reconstruction of a Golgi apparatus stack of a principal cell of the rat

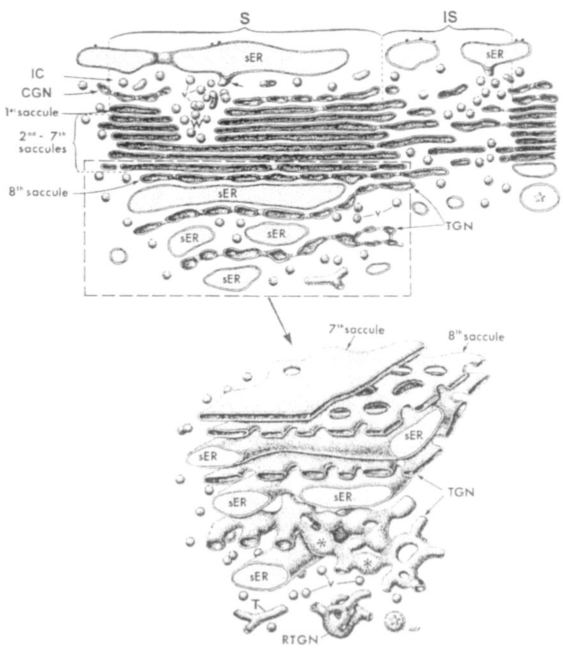

epididymis (from Hermo and Smith, 1998), which includes a cisternal (saccular, S) and an intercisternal connecting region (intersaccular, IS).

From the sparsely granulated endoplasmic reticulum (sER) buds project toward the pre-Golgi intermediate (pGI, intermediate compartment, IC). The stack shows a network at the *cis* side (*cis* Golgi network, CGN), eight flattened cisternae (saccules), and *trans* Golgi networks (TGN). Large perforations within several of the *cis* localised cisternae are in register with each other and form wells (w), which contain abundant small transport vesicles. At the *trans* side, sparsely granulated endoplasmic reticulum (sER) is closely apposed to the *trans* cisternae and to the TGN. A face view of the *trans* side shows that the eight cisterna is highly fenestrated. The *trans* Golgi networks show dilated areas (asterisks), which presumably represent prosecretory granules. It is suggested that the *trans* Golgi networks fragment and give rise to tubular elements (T), small vesicles (v), secretory granules (open star), and small compacted residual *trans* Golgi networks (RTGN).

The Golgi stacks presented in the diagram and those in the tilt series, show similarly prominent *trans* Golgi-ER apposed to the TGN. These close ER-TGN associations are assumed to be involved in non-vesicular lipid transfer.

References

Baumeister W, Grimm R, and Walz J (1999) Electron tomography of molecules and cells. Trends Cell Biol 9, 81

D'Angelo G, Vicinanza M, and de Matteis MA (2008) Lipid transfer proteins in biosynthetic pathways. Curr Opin Cell Biol 20: 360

Hanada K, Kumagai K, Tomishige N, and Yamaji T (2009) CERT-mediated trafficking of ceramide. Biochim Biophys Acta 1791: 684

Hermo L and Smith CE (1998) The structure of the Golgi apparatus: a sperm's eye view in principal epithelial cells of the rat epididymis. Histochem Cell Biol 109, 431

Ladinsky MS, Mastronarde DN, McIntosh JR, Howell KE, and Staehelin LA (1999) Golgi structure in three dimensions: functional insights from the normal rat kidney cell. J Cell Biol. 144: 1135

Marsh M (2005) Lessons from tomographic studies of the mammalian Golgi. Biochem Biophys Acta 1744: 273

Munro S (2003) Earthworms and lipid couriers. Nature 426: 775

Polishchuk RS and Mironov AA (2004) Structural aspects of Golgi function. Cell Mol Life Sci 61: 146

Magnification: x 62,000 (A, B, C)

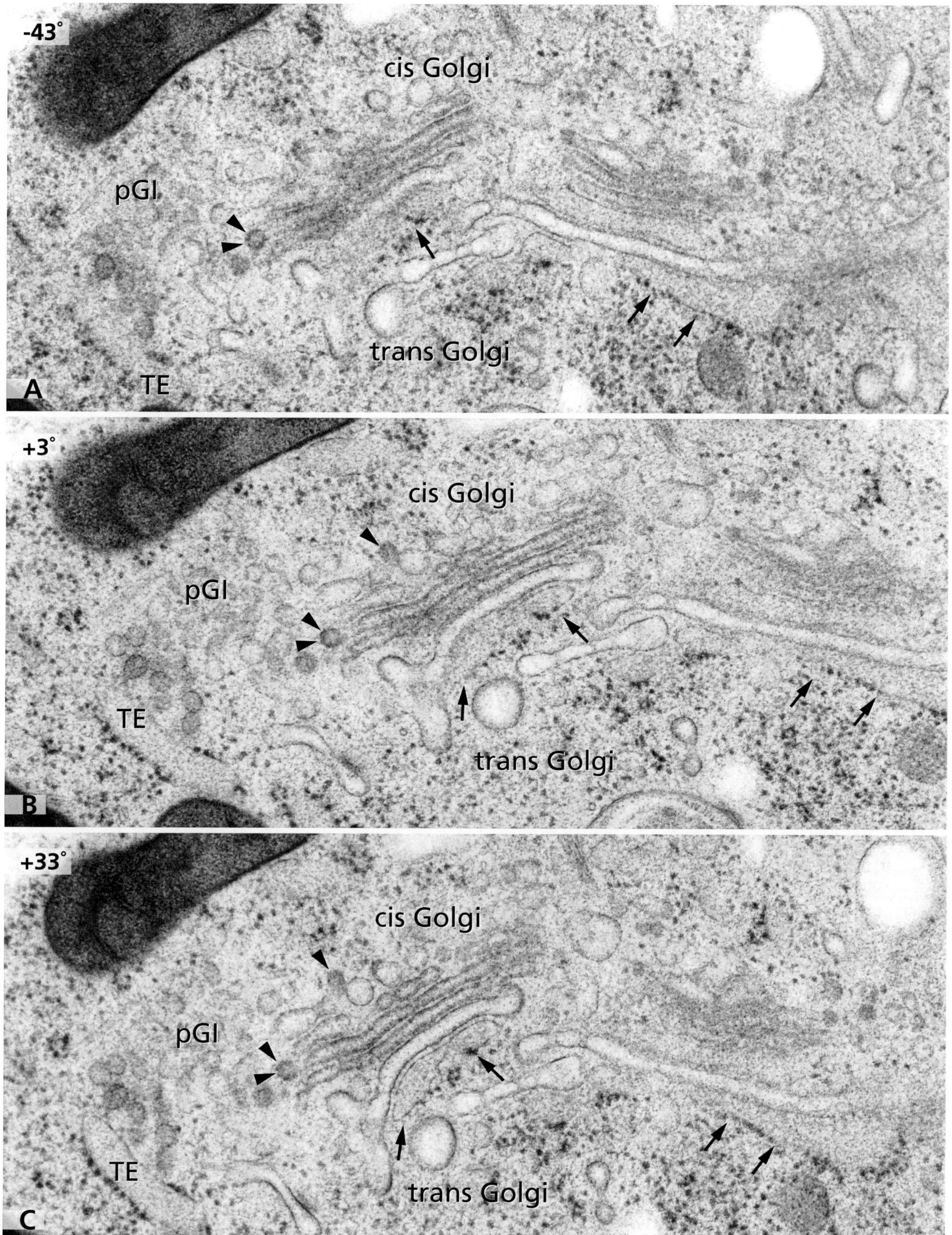
-43°
cis Golgi
pGI
trans Golgi
A
TE
+3°
cis Golgi
pGI
TE
trans Golgi
B
+33°
cis Golgi
pGI
TE
trans Golgi
C

STRUCTURE OF THE TGN

The term "*trans* Golgi network" (TGN) is used for complex and dynamic membrane compartments located at the *trans* Golgi side (cf. Figs. 33–35, 48). Because of its complexity and highly dynamic behaviour, the TGN belongs to the most interesting but also enigmatic and challenging structures of the cell. The name is solely descriptive, not taking into account different functions and organisations, and different relationships to the Golgi apparatus and its stacked cisternae. The TGN is variously extended and organised in different types of cells, and its organisation changes concomitantly with the differentiation of cells and functional states. In some cells, the TGN, as its name indicates, forms an extended network, whereas in other cells a TGN is hardly visible. Furthermore, its location is variable. The TGN often resides at some distance to the stacked Golgi cisternae, but, on the other hand, parts of it may be attached to the *trans*most Golgi cisterna, making it an integrated component of the Golgi stack. According to these varying appearances, the TGN is seen differently by different authors, either as a component of the Golgi apparatus and *trans*most constituent of the Golgi stacks, or as an individual organelle and entity. These contradictory interpretations may result from the dual tasks of the TGN; on the one hand, it has a key role in the biosynthetic-secretory system for the export of newly synthesised molecules out of the Golgi region and their targeting to different destinations, and on the other hand, it is involved in endocytic pathways; the TGN-organisation is influenced by the retrograde transport in the endocytosis system and changes concomitantly with endocytic flow.

Panels A–B and C–F in this figure display tomographic slices and different aspects of 3D-models of the TGN in a cell of the human hepatoma HepG2 line showing function-dependent differences in the organisation. In the models, *cis* and medial cisternae are coloured yellow, the TGN is red, and the *trans* Golgi-ER is blue. Panels A and B show the TGN-organisation in a cell 48 hours after seeding without further treatments; in panels C-F, the TGN after 30 min uptake of peroxidase-labelled wheat germ agglutinin (WGA) is displayed, corresponding to the pattern shown in Fig. 48A. In both cases, a part of the TGN is attached to the *trans*most cisterna of the Golgi stack (arrows in C–F), and in both cases, *trans* Golgi-endoplasmic reticulum (ER) is closely associated. The attached TGN is continuous with other parts located apart from the stack. Although there is a comparable general order of arranged Golgi cisternae, attached and non-attached TGN, and *trans* Golgi-ER, major differences are visible in the architecture of the TGN. In the cells not further treated, the TGN consists of a simple cisterna with attached and non-attached parts, whereas the endocytic TGN apparent after WGA-endocytosis exhibits a highly complex, rugged architecture. Both the Golgi-attached parts (arrows in C–F) and the non-attached parts of the endocytic TGN are composed of multiple different elements building large cylindrical and column-like structures, interconnected by ridges and bridging arches. These complex architectures are reduced at later times after onset of WGA-endocytosis, and the TGN again shows a more simple architecture. Multiple questions are open as to the mechanisms and the functional background of these endocytosis-connected TGN transformations, and their relationships to the tasks of the TGN in the secretory system.

References

Clermont Y, Rambourg A, Hermo L (1995) *Trans* Golgi network (TGN) of different cell types: three-dimensional structural characteristics and variability. Anat Rec 242: 289

De Matteis MA and Luini A (2008) Exiting the Golgi complex. Nat Rev Mol Cell Biol 9: 273

Griffiths G and Simons K (1986) The *trans* Golgi network: Sorting at the exit site of the Golgi complex. Science 234: 438

Luini A, Mironov AA, Polishchuk EV, and Polishchuk RS (2008) Morphogenesis of post-Golgi carriers. Histochem Cell Biol 129: 153

Ladinsky MS, Mastronarde DN, McIntosh JR, Howell KE, Staehelin LA (1999) Golgi structure in three dimensions: Functional insights from the normal rat kidney cell. J Cell Biol 144: 1135

Mollenhauer HH and Morré DJ (1998) The tubular network of the Golgi apparatus. Histochem Cell Biol 109: 533

Pavelka M, Neumüller J, and Ellinger A (2008) Retrograde traffic in the biosynthetic-secretory route. Histochem Cell Biol 129: 277

Roth J and Taatjes DJ (1998) Tubules of the *trans* Golgi apparatus visualized by immunoelectron microscopy. Histochem Cell Biol 109: 545

The pictures shown in panels C–F are modified from Pavelka M and Ellinger A (2008) Retrograde plasma membrane-to-Golgi apparatus transport. In: The Golgi apparatus. State of the art 110 years after Camillo Golgi's discovery (Mironov A and Pavelka M, eds). Wien New York: Springer, p 459

Magnification: x 59,000 (A,B), x 50,000 (C), x 60,000 (D), x 45,000 (E), x 70,000 (F)

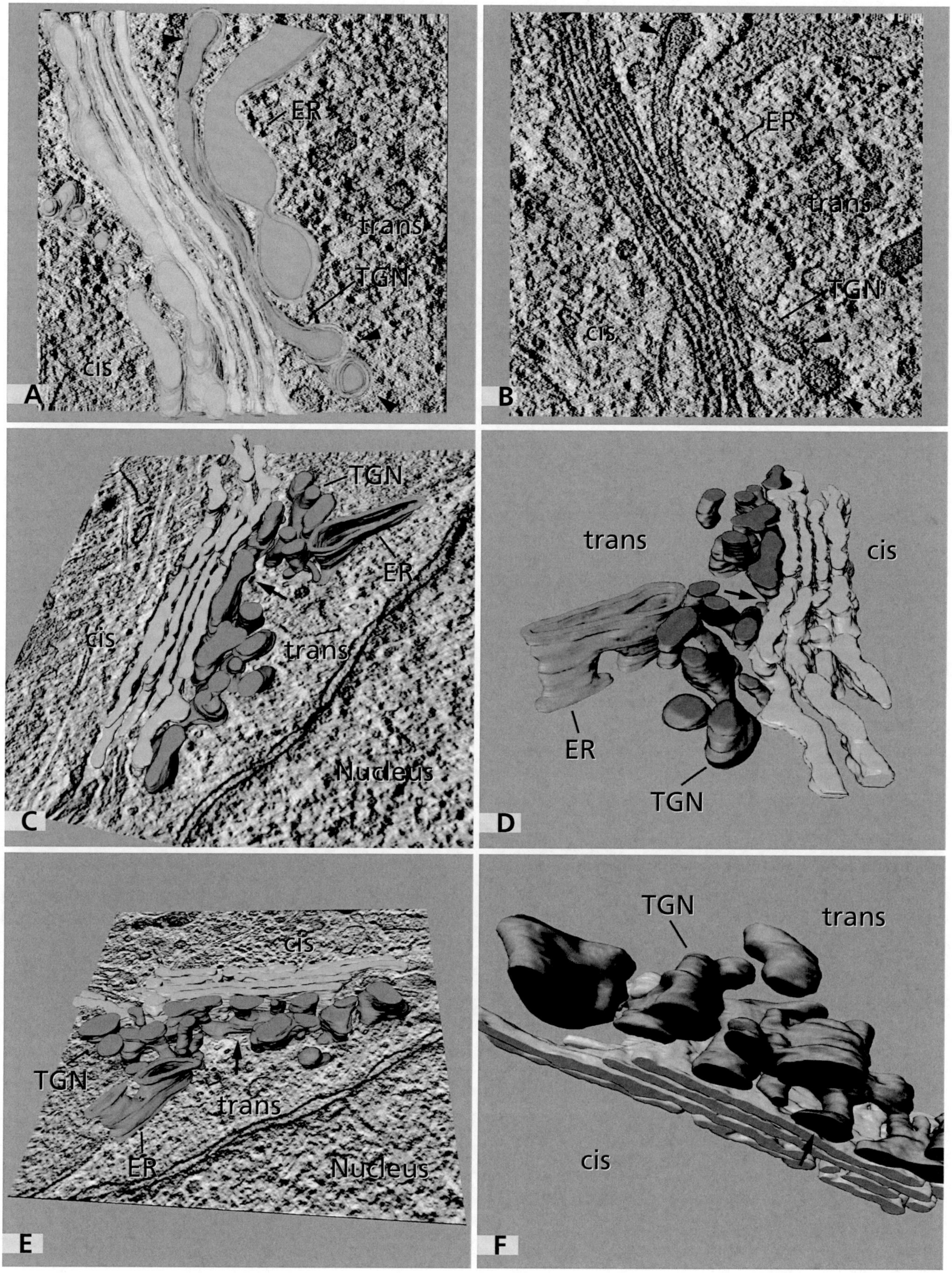
A
ER
trans
TGN
cis
B
ER
trans
TGN
cis
C
TGN
ER
cis
trans
Nucleus
D
trans
cis
ER
TGN
E
cis
TGN
trans
ER
Nucleus
F
TGN
trans
cis

BREFELDIN A-INDUCED GOLGI APPARATUS DISASSEMBLY

Among the agents known to disturb the functions and organisation of the Golgi apparatus, Brefeldin A (BFA) is one of the most commonly used. BFA is a fungal metabolite well characterised to prevent assembly of cytoplasmic coat proteins onto membranes by inhibiting nucleotide exchange onto ADP-ribosylation factor. As a consequence, formation of COP I-coated vesicles (coat protein I; cf. Fig. 23), involved in multiple steps of traffic between ER and Golgi apparatus and within the Golgi apparatus stacks, is inhibited. Within a few minutes, the Golgi apparatus disintegrates, the stacked Golgi cisternae transform into tubular-reticular compartments, long tubules grow out of the Golgi region, and both endogenous Golgi molecules, such as Golgi enzymes, and itinerant molecules, such as secretory proteins, are redistributed into the endoplasmic reticulum.

In panel A, the perinuclear Golgi region in a high-pressure frozen HepG2 hepatoma cell with one of the Golgi apparatus stacks is on display. The micrograph shows all characteristic features of the Golgi apparatus in these cells, a complex *cis*-Golgi side, coated vesicles budding from the rims of the cisternae (arrowheads), a prominent *trans* Golgi network (TGN), and extensive endoplasmic reticulum interposed between *trans* Golgi cisternae and TGN (arrow). The small densely coated vesicles (arrowheads) presumably correspond to COP I-vesicles known to be involved in retrograde traffic between Golgi apparatus and endoplasmic reticulum. In another concept, a morphogenetic function has been ascribed to the COP I vesicles. When monitoring the very early effects of BFA-treatment by using high-pressure freezing with a considerably improved temporal resolution in comparison to chemical fixation (cf. Figs. 34 and 35), it could be shown that BFA-induced Golgi breakdown occurs in distinct steps. As early as 30 sec after BFA-administration coated membrane regions and coated buds disappear from the Golgi membranes, but the disassembly of the stacks does not start before 90-120 seconds of BFA-treatment have been completed. At this time, the Golgi stacks start to transform into tubules and networks (cf. Fig. 38). At 3–5 minutes of BFA-treatment, stacked Golgi cisternae are no longer visible.

Panel B shows the perinuclear cytoplasm with a centriole (C) marking the cytocentre of a hepatoma cell after 5 min of BFA-treatment. Golgi cisternae are entirely missing, instead multiple long and, in part, branched membrane tubules prominently occupy the Golgi region, which further shows an accumulation of small and large vesicles. Some of the larger vesicles are multivesiculated, and their distinct thick membranes are similar to those of some of the network-forming tubules (cf. Fig. 38).

References

Dinter A and Berger EG (1998) Golgi-disturbing agents. Histochem Cell Biol 109: 571

Hess MW, Müller M, Debbage PL, Vetterlein M, and Pavelka M (2000) Cryopreparation provides new insight into Brefeldin A-effects on the structure of the HepG2 Golgi apparatus. J Struct Biol 130: 63

Klausner RD, Donaldson JG, and Lippincott-Schwartz J (1992) Brefeldin A: insights into the control of membrane traffic and organelle structure. J Cell Biol 116: 1071

Lippincott-Schwartz J, Yuan L, Tipper C, Amherdt M, Orci L, and Klausner RD (1991) Brefeldin A's effects on endosomes, lysosomes, and TGN suggests a general mechanism for regulating organelle structure and membrane traffic. Cell 67: 601

Pavelka M and Ellinger A (1993) Early and late transformations occurring at organelles of the Golgi area under the influence of Brefeldin A. An ultrastructural and lectinocytochemical study. J Histochem Cytochem 41: 1031

Wagner M, Rajasekaran AK, Hanzel DK, Mayor S, and Rodriguez-Boulan E (1994) Brefeldin A causes structural and functional alterations of the *trans* Golgi network of MDCK cells. J Cell Sci 107: 933

Wood SA, Park JE, and Brown WJ (1991) Brefeldin A causes a microtubule-mediated fusion of the *trans* Golgi network and early endosomes. Cell 67: 591

Magnification: x 44,000 (A), x 27,000 (B)

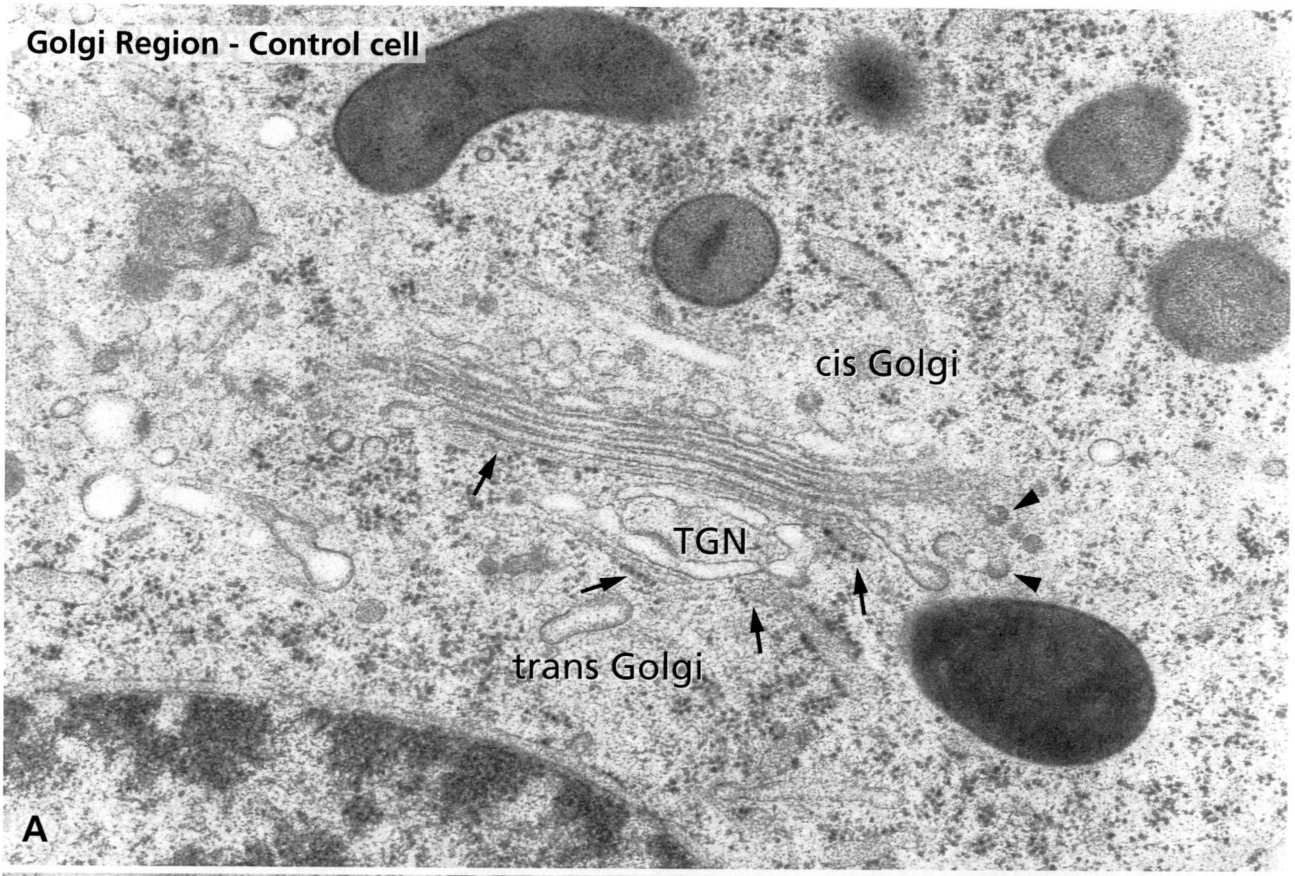
Golgi Region - Control cell
cis Golgi
TGN
trans Golgi
A

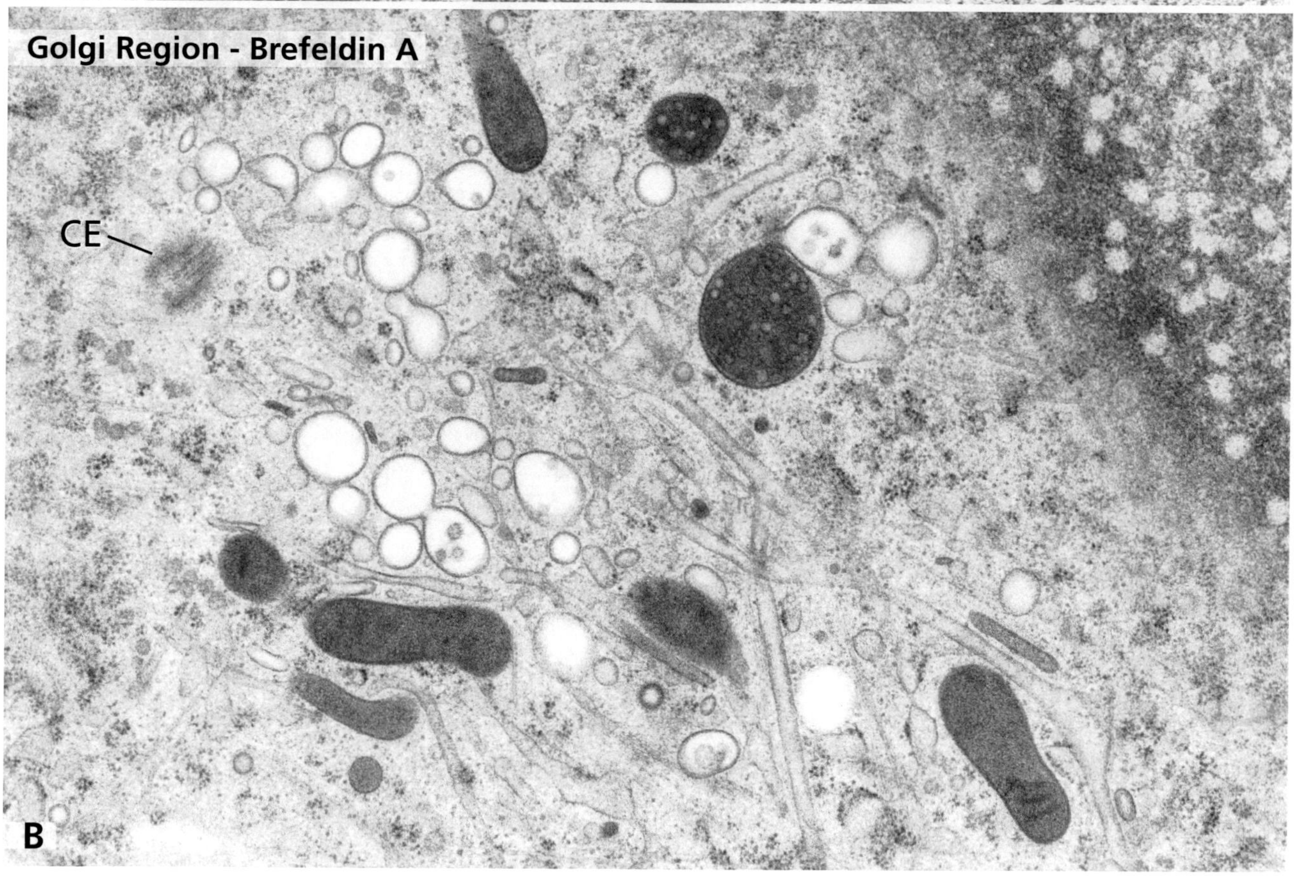
Golgi Region - Brefeldin A
CE
B

BREFELDIN A-TREATMENT: TUBULATION OF GOLGI APPARATUS AND ENDOSOMES

Golgi apparatus changes start at 90 to 120 seconds following to the onset of Brefeldin A (BFA)-administration and after coated regions had already disappeared at 30 to 60 seconds. The Golgi apparatus stacks disassemble and long tubules grow out of the Golgi region. After 3 to 5 min, a Golgi apparatus is missing in the cells and instead, the cytoplasm is occupied by a network of membrane tubules being interwoven with membranes of the endoplasmic reticulum. At places, glomerular membrane figures form conspicuous "organelles" in BFA-treated cells (cf. Figs. 39 and 40).

Panels A, B, and C show segments of the cytoplasm of high-pressure frozen HepG2 hepatoma cells 3 min (panels A and C) and 5 min (panel B) after BFA-treatment. Stacked Golgi cisternae are not visible. The BFA-induced tubules occupying wide cytoplasmic areas show different dimensions and delineating membranes, indicating their origin from different cellular compartments. BFA induces tubulation not only of the Golgi apparatus and *trans* Golgi network, but also transformation into tubules of endosomes and lysosomes. The voluminous tubules delineated by distinct thick membranes shown in panel A, are dilated at places and contain interior vesicles resembling multivesiculated endosomal compartments. They differ from the more densely stained tubules with less prominent membranes that are shown in panels B and C. The latter are commonly found in connection with or growing out from prominent networks (panel C) that resemble *trans* Golgi networks in untreated cells.

In panel B, the close spatial relation between a membrane tubule (arrowhead) and an accompanying microtubule (arrow) reflects the role of the microtubular cytoskeleton for the rapid formation of the BFA-induced tubular networks.

BFA-induced tubulation is considered to be a consequence of disturbed vesicular traffic because of inhibited formation of coat protein I (COP I)-coated transport vesicles owing to failed activation of adenosine ribosylation factor (ARF). Although fusion of tubules with membranes of the endoplasmic reticulum have not yet been shown, Golgi membrane tubulation is considered to be involved in the redistribution of Golgi molecules into the endoplasmic reticulum.

References

Hess MW, Müller M, Debbage PL, Vetterlein M, and Pavelka M (2000) Cryopreparation provides new insight into Brefeldin A-effects on the structure of the HepG2 Golgi apparatus. J Struct Biol 130: 63

Knoll G, Verkleij AJ, and Plattner H (1987) Cryofixation of dynamic processes in cells and organelles. In: Cryotechniques in biological electron microscopy (Steinbrecht RA, and Zierold K, eds). Berlin Heidelberg: Springer, pp 258

Lippincott-Schwartz J, Yuan L, Tipper C, Amherdt M, Orci L, and Klausner RD (1991) Brefeldin A's effects on endosomes, lysosomes, and TGN suggests a general mechanism for regulating organelle structure and membrane traffic. Cell 67: 601

Sciacky N, Presley J, Smith C, Zaal KJ, Cole N, Moreira JE, Terasaki M, Siggia E, and Lippincott-Schwartz J (1997) Golgi tubule traffic and the effects of Brefeldin A visualised in living cells. J Cell Biol 139: 1137

Wood SA, Park JE, and Brown WJ (1991) Brefeldin A causes a microtubule-mediated fusion of the *trans* Golgi network and early endosomes. Cell 67: 591

Magnification: x 48, 000 (A), x 50,000 (B), x 46,000 (C)

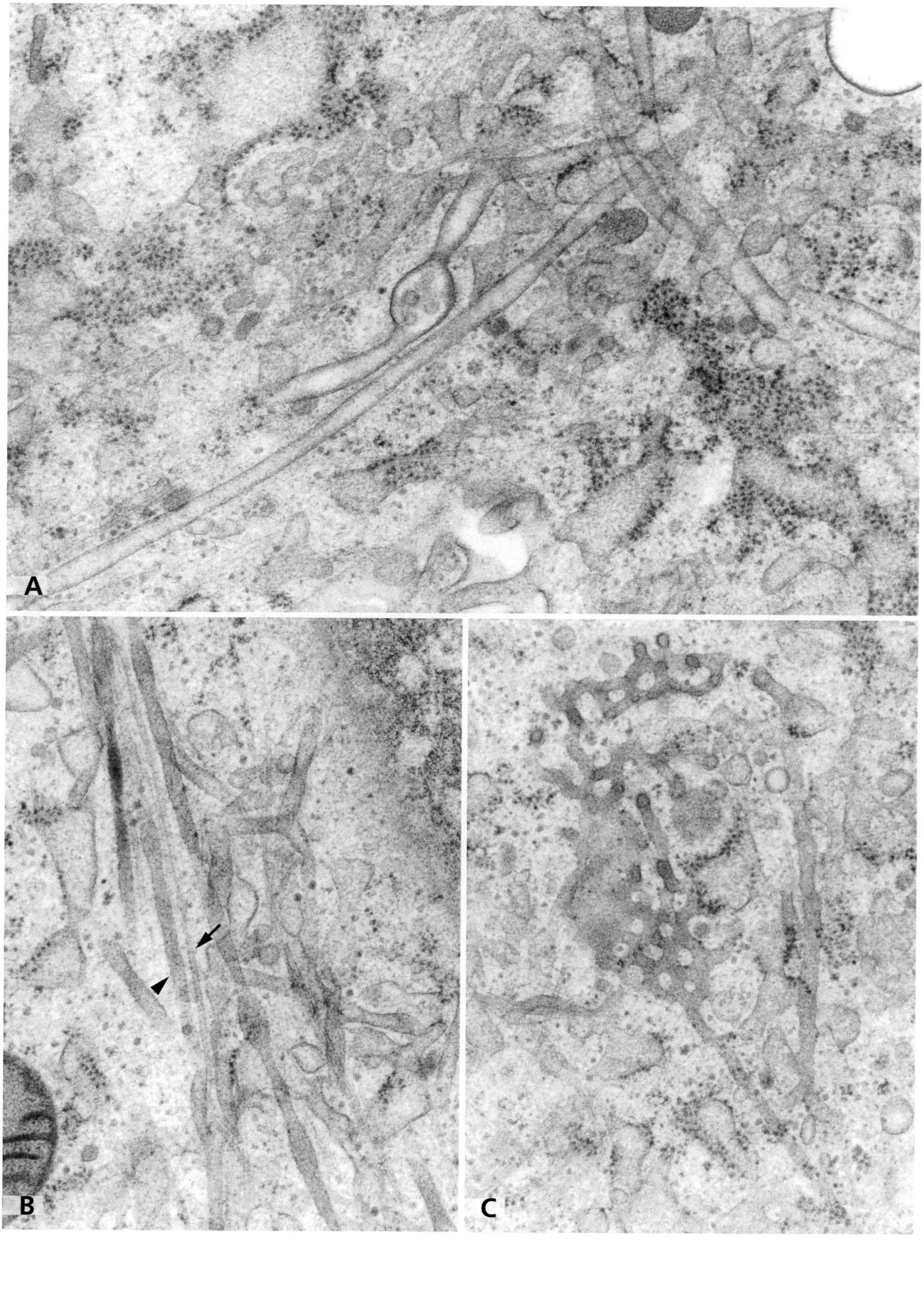
A
B
C

BREFELDIN A-TREATMENT: EFFECT ON RETROGRADE TRANSPORT OF INTERNALISED WGA

Treatment with Brefeldin A (BFA) leads to a protection of cells from the harmful effects of several toxins. These include ricin, as well as Shiga and cholera toxins. The mechanisms of this protective BFA-effect presumably are related to the BFA-induced breakdown of the Golgi apparatus, an event that leads to an interruption of retrograde pathways from the plasma membrane to the endoplasmic reticulum (ER) involving the Golgi apparatus (cf. Fig. 48). Such pathways have to be travelled by those toxins in order to finally reach the cytosol, where they exert their toxic effects.

Recent results have revealed that BFA not only can be used as an agent that disturbs retrograde pathways, but can also be used to induce retrograde transport into the endoplasmic reticulum and to control these routes. A precondition is detailed knowledge of the dynamics occurring during the transport of internalised molecules into the Golgi apparatus and to determine the precise time for BFA-application. Studies using wheat germ agglutinin (WGA) in HepG2 hepatoma cells showed that endocytic traffic into the Golgi apparatus is a complex multi-step process, during which three stages can be discriminated (cf. Figs. 47 and 48).

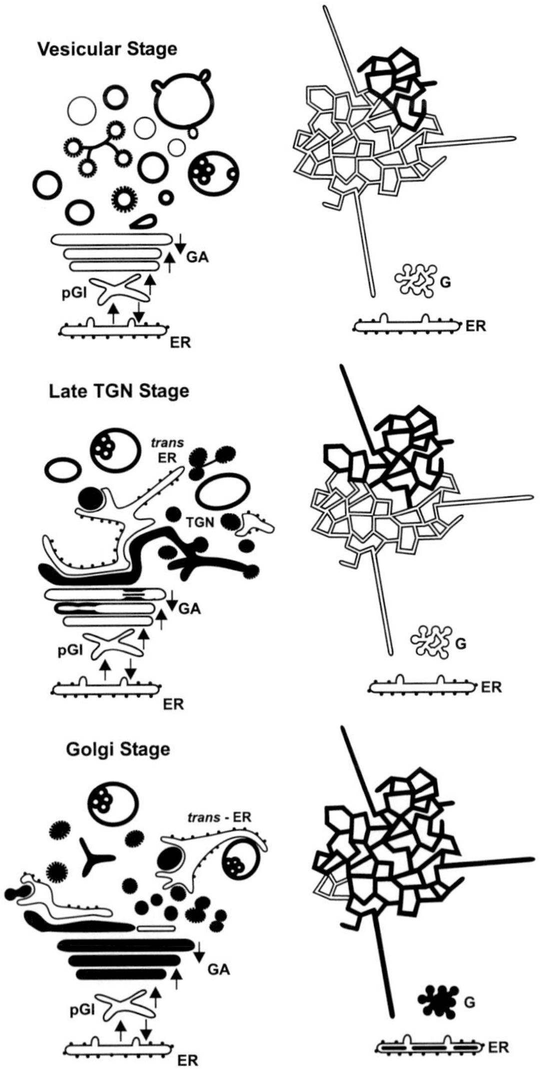

The stages are summarised in the diagram at the left hand side: In stage I ("Vesicular Stage"), vesicular endosomes accumulate in the *trans* Golgi region. During stage II ("TGN Stage"), an extended endocytic *trans* Golgi network is formed and, in part, is integrated into the Golgi stacks. In stage III, ("Golgi Stage"), prominent amounts of internalised WGA are taken up into the stacked Golgi cisternae.

With respect to the further fate of internalised WGA, the time of BFA-administration is most critical. This is summarised at the right hand side of the diagram. Administration of BFA during the stages I and II, though resulting in tubular-reticular Golgi transformations (cf. Fig. 38), does not lead to an uptake of internalised WGA into cisternae of the ER (upper and middle parts of the diagram). In constrast, when BFA is administered during stage III, internalised WGA is redistributed into ER cisternae (lower part of the diagram). This is most conspicuous after 10 minutes of BFA-treatment.

This particular situation is shown in the micrograph on the opposite page, which presents a perinuclear segment of a BFA-treated HepG2 hepatoma cell. Stacked Golgi cisternae are absent, and multiple glomerular structures (G), considered to represent pre-Golgi intermediates or Golgi remnants, dominate (cf. also Fig. 40). Most of the cisternae of the ER are labelled with reactions for the internalized WGA, which also is contained in the perinuclear cistern (double arrows).

Abbreviations: ER – endoplasmic reticulum, GA – Golgi apparatus, pGI – pre-Golgi intermediate, G – BFA induced glomerular structures.

References

Sandvig K and van Deurs B (2000) Entry of ricin and Shiga toxin into cells: molecular mechanisms and medical perspectives. EMBO J 19: 1

Vetterlein M, Niapir M, Ellinger A, Neumüller J, and Pavelka M (2003) Brefeldin A-regulated retrograde transport into the endoplasmic reticulum of internalised wheat germ agglutinin. Histochem Cell Biol 120: 121

Magnification: x 24,000

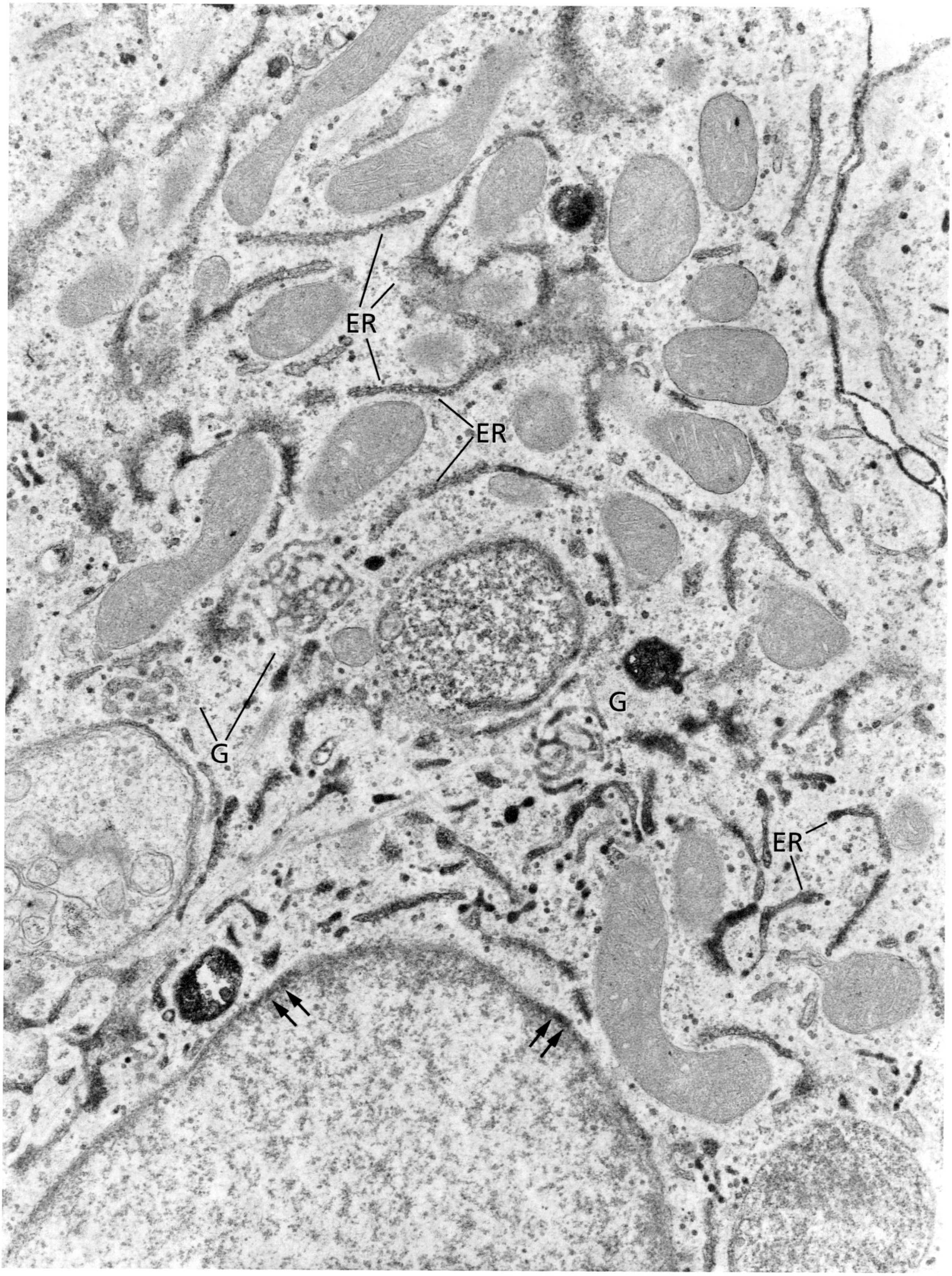
ER
ER
G
G
ER

BREFELDIN A-TREATMENT: TRANSITIONAL ER-ELEMENTS AND PRE-GOLGI INTERMEDIATES

Although multiple parts of cellular membranes and organelles, including Golgi apparatus, *trans* Golgi network, endosomes and lysosomes, are transformed under the influence of Brefeldin A, and Golgi apparatus constituents are redistributed into the endoplasmic reticulum (ER), ER-export sites are not conspicuously changed and multiple buds with fine cytoplasmic membrane coats are visible under the electron microscope. This is evident in high-pressure frozen HepG2 hepatoma cells shown in panel A, as well as in chemically fixed cells of the same types, shown in panel B, although the fine coats are much better visible in the cryo-immobilised cells in panel A. Both single buds projecting from ER-cisternae (double arrowheads in panel A) and buds associated with typical cup-shaped transitional elements (TE) of the ER (arrowhead in panel B) are frequent, as are pleomorphic tubular-vesicular structures nearby, that resemble pre-Golgi intermediates (pGI) in control cells (cf. Figs. 23, 25, and 35). The ER is moderately dilated and, at places, shows constrictions, which are particularly well visible in the cryo-immobilised cell in panel A.

Accumulated and particularly assembled pre-Golgi intermediates are considered to be the functional and structural bases for formation of the conspicuous glomerular-like structures that are prominent in BFA-treated cells and increase in number from 5–10 minutes after BFA application. One example is shown in panel C, and several of these glomerular structures are visible and marked with G in the survey micrograph shown in Fig. 39. They also have been considered to correspond to Golgi remnants and are involved in Golgi recovery. The micrograph in Fig. 39 shows that the lumina of the glomerular organelles contain reactions for internalised WGA, which could derive from residual Golgi materials on the one hand, but, on the other hand, could take an indirect route via Golgi redistribution into the ER and subsequent transfer to the glomerular structures via vesicle budding at ER export sites (arrowheads in panels A and B).

Although many details about mechanisms and machineries of vesicle budding, membrane fusion, protein sorting, progression, and maturation of compartments are known and the molecules involved are partly well characterised, the principles of operation of traffic across the Golgi apparatus from the *cis* side to the *trans* side are still unresolved. Research in this field focuses mainly on two models with multiple variations including the temporal opening of transport gates: The "vesicular transport model" proceeds on the assumption that the individual Golgi cisternae are stable entities and transport across the stacks of cisternae occurs via budding and fusion of small vesicles. In the "cisternal progression-maturation model" cargo remains confined within the lumen of the Golgi cisternae and the cisternae move through the stack and gradually "mature" concomitantly with movement from the *cis* to the *trans* side. The anterograde progression of cisternae is accompanied by retrograde vesicular traffic and recycling of endogenous Golgi molecules. There are more recent results in favour of the latter model or a combination of both, and it becomes increasingly evident that cargo contained in the lumen of Golgi compartments is an important modulator of Golgi structure and function, and traffic across the stacks of cisternae.

References

Hendricks LC, McClanahan SL, McCaffery M, Palade GE, and Farquhar MG (1992) Golgi proteins persist in the tubulovesicular remnants found in Brefeldin A-treated pancreatic acinar cells. Eur J Cell Biol 58: 202

Jackson CL (2009) Mechanisms of transport through the Golgi complex. J Cell Sci 122: 443

Patterson GH, Hirschberg K, Polishchuk RS, Gerlich D, Phair RD, and Lippincott-Schwartz J (2008) Transport through the Golgi apparatus by rapid partitioning within a two-phase membrane system. Cell 133: 1055

Mironov AA and Pavelka M (2008) The Golgi apparatus as a crossroads in intracellular traffic. In: The Golgi apparatus. State of the art 110 years after Camillo Golgi's discovery (Mironov A and Pavelka M, eds). Wien New York: Springer, p 16

Mironov AA, Beznoussenko GV, Nicoziani P, Martella O, Trucco A, Kweon H-S, Di Giandomenico D, Polishchuk RS, Fusella A, Lupetti P, Berger EG, Geerts WJC, Koster AJ, Burger KNJ, and Luini A (2001) Small cargo proteins and large aggregates can traverse the Golgi by a common mechanism without leaving the lumen of cisternae. J Cell Biol 155: 1225

Polishchuk RS and Mironov AA (2004) Structural aspects of Golgi function. Cell Mol Life Sci 61: 146

Ward TH, Polishchuk RS, Caplan S, Hirschberg K, and Lippincott-Schwartz J (2001) Maintenance of Golgi structure and function depends on the integrity of ER export. J Cell Biol 155: 557

Magnification: x 50,000 (A), x 50,000 (B), x 50,000 (C)

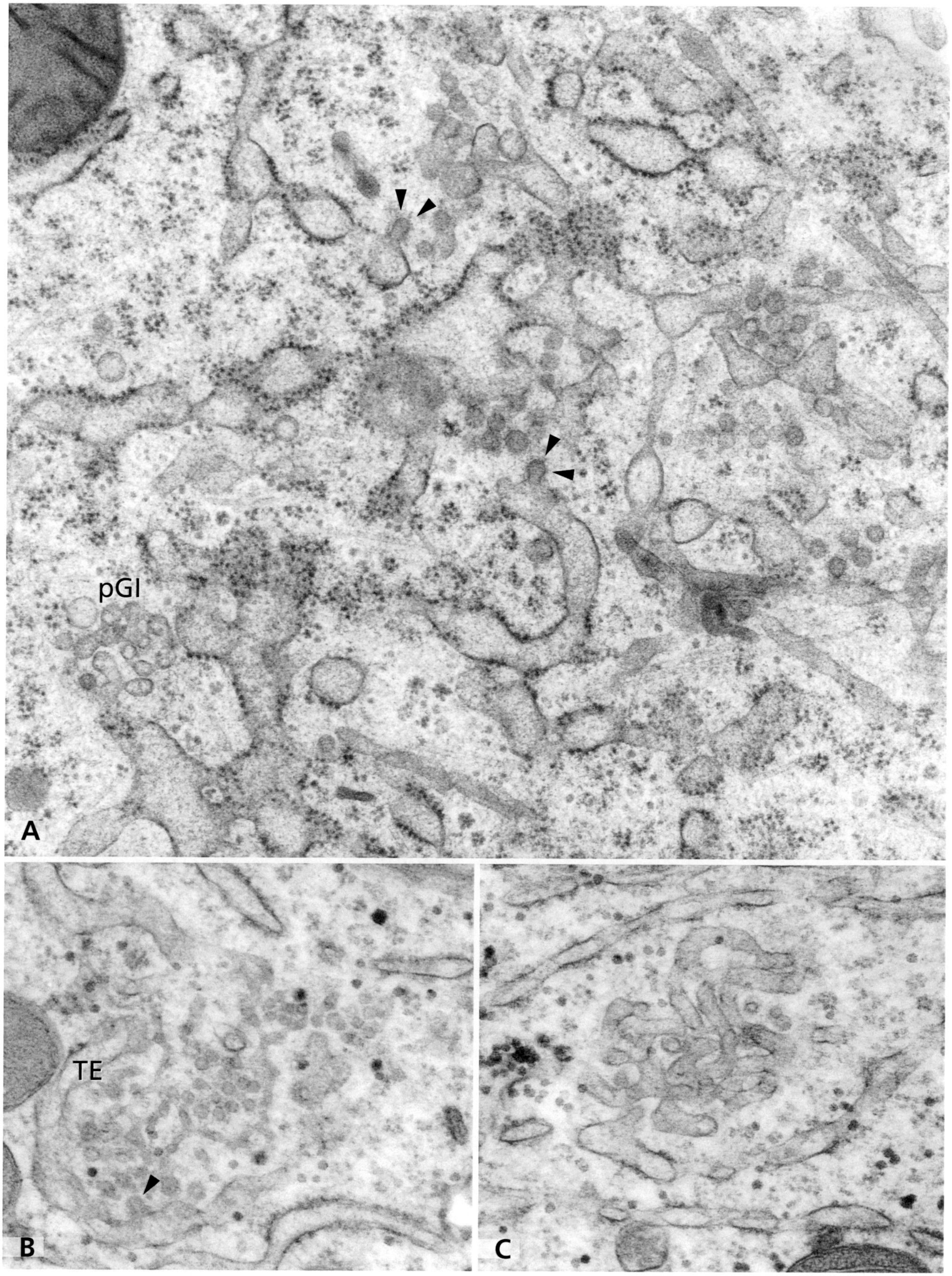
pGI
A
TE
B
C

HEAT SHOCK RESPONSE OF THE GOLGI APPARATUS

Cells in living organisms or cultured *in vitro* respond in a specific mode to elevated temperature (heat shock) or other stressful conditions. They rapidly and transiently accelerate the rate of expression of specific genes, the heat shock genes. This results in increased levels of constitutively expressed heat shock proteins fulfilling protective functions in that they act as chaperones to assist polypeptides and proteins to fold properly. The other group of stress proteins are the glucose-regulated proteins. Concomitantly with the activation of heat shock genes, most other cellular genes are supressed under conditions of stress. In cell culture, this results in arrest of cell growth

Structural changes of cytoplasmic organelles and the nucleus as well as rearrangements of the cytoskeleton can be observed under heat shock conditions by electron microscopy. The well organised stacks of cisternae of the Golgi apparatus as seen in cultured CHO cells (panel A) are dramatically changed upon heat shock. They consist of dispersed membrane vesicles of various sizes and remnants of cisternae (panel B). Furthermore, the cisternal lumen of the rough endoplasmic reticulum became distended. A member of the glucose-regulated protein family, BiP, is an abundant endoplasmic reticulum luminal chaperone. Other heat shock proteins and glucose-regulated proteins are present in the endoplasmic reticulum, Golgi apparatus, and mitochondria. Mitochondria in heat shocked cells are swollen, their intracristal spaces are distended, and their oxidative phosphorylation impaired. Due to changes of the cytoskeleton, mitochondria accumulate in the perinuclear region. In response to heat shock, the actin filaments and stress fibres formed by them are increased. The intermediate filament network collapses and becomes concentrated around the nucleus. No apparent changes of the microtubuli seem to occur. Heat shocks affect the structure and function of the nucleus as well. Bundles of intranuclear actin filaments have been observed. The granular component of the nucleolus is either lost or becomes aggregated. Accumulation and aggregation of the perichromatin granules is also observed.

Heat shock response is a universal reaction, and the heat shock proteins are among the most highly conserved proteins. Several families of heat shock proteins exist in prokaryotes and eukaryotes: the large molecular mass (83 to 90 kDa) hsp90 family, the 66 to 78 kDa hsp70 family, the hsp60 family, and a diverse group of small (15 to > 30 kDa) heat shock proteins such as DnaJ and GrpE. Although many of the heat shock proteins are present in the cytoplasm and nucleoplasm, some are in membrane-bounded organelles such as the endoplasmic reticulum, Golgi apparatus, and mitochondria.

If the heat shock is severe, cells may die. However, thermotolerance can be acquired by mild thermal stress before normally lethal thermal stress. The hsp70 seems to be particularly involved in the induction of thermotolerance. Of medical relevance is that heat shock proteins are elevated in various human diseases involving ischaemia, fever, inflammation and reactive oxygen species, and neoplasia.

References

Cervera J (1978) Effects of thermic shock on HEp-2 cells. An ultrastructural and high-resolution autoradiographic study. J Ultrastruct Res 63: 51

Iida K, Iida H, and Yahara I (1986) Heat shock induction of intranuclear actin rods in cultured mammalian cells. Exp Cell Res 165: 207

Morimoto R, Tissières A, and Georgopoulos C (1990) Stress proteins in biology and medicine. New York Cold Spring Harbor: Cold Spring Harbor Laboratory Press

Pelham HR (1984) Hsp70 accelerates the recovery of nucleolar morphology after heat shock. EMBO J 3: 3095

Welch WJ (1992) Mammalian stress response – cell physiology's structure/function of stress proteins, and implications for medicine and disease. Physiol Rev 72: 1063

Welch WJ and Suhan JP (1985) Morphological study of the mammalian stress response: characterisation of changes in cytoplasmic organelles, cytoskeleton, and nucleoli, and appearance of intranuclear actin filaments in rat fibroblasts after heat-shock treatment. J Cell Biol 101: 1198

Welch WJ and Suhan JP (1986) Cellular and biochemical events in mammalian cells during and after recovery from physiological stress. J Cell Biol 103: 2035

Magnification: x 25,500(A); x 23,500 (B)

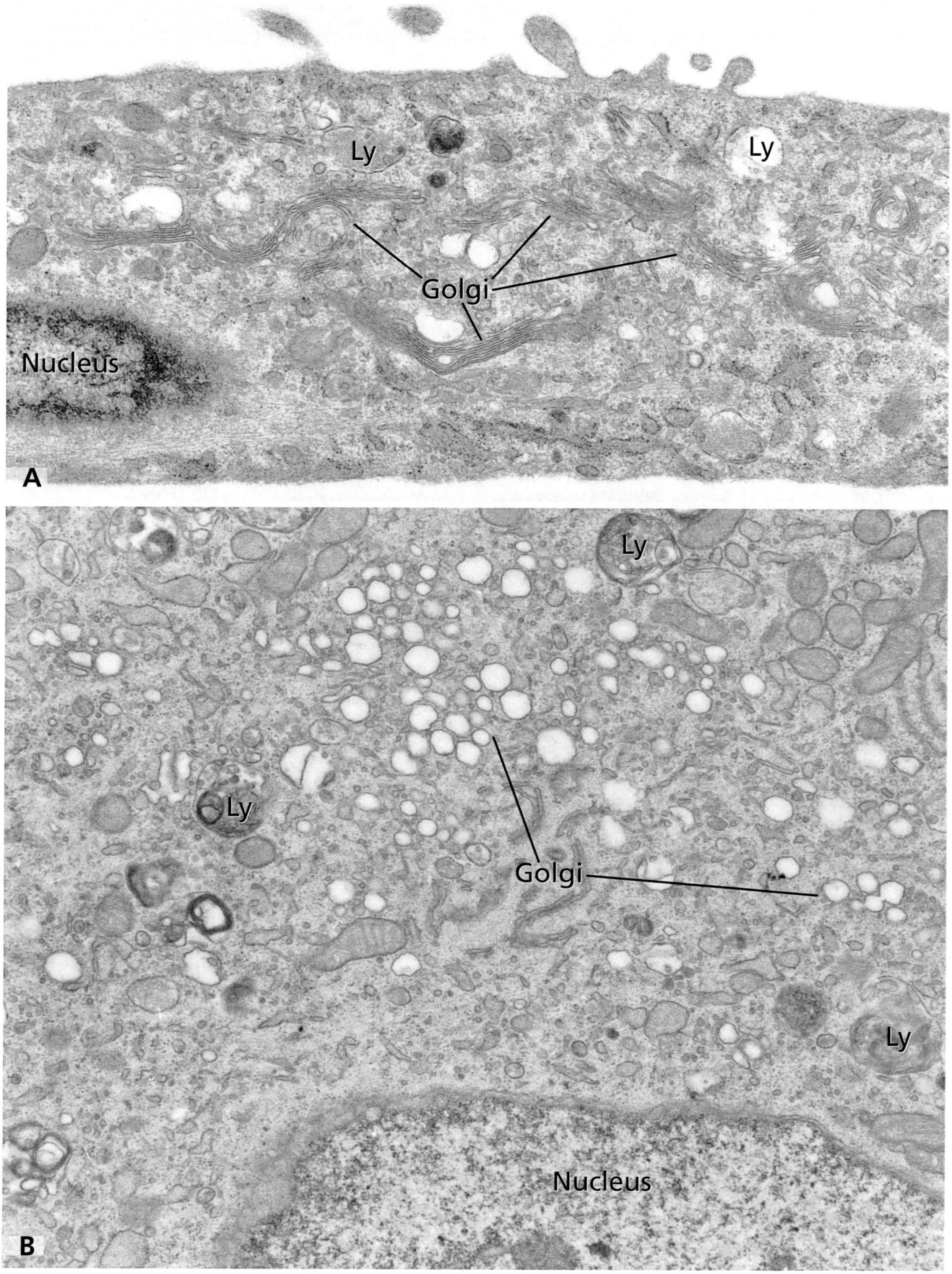
Ly
Ly
Golgi
Nucleus
A
Ly
Ly
Golgi
Ly
Nucleus
B

GOLGI APPARATUS CHANGES UPON ATP-DEPLETION AND ATP-REPLENISHMENT

In addition to Brefeldin A, the effects of which are shown in detail in the preceding pages, multiple other substances are known to disturb Golgi functions and interfere with membrane and cargo transport in the pre-Golgi area, across the Golgi apparatus stacks, and from the *trans* Golgi side to other destinations. Although basing on different mechanisms, most of these substances lead to massive changes of the Golgi apparatus architecture. Intracellular traffic also is interrupted when energy supply is turned down. Protein export from the endoplasmic reticulum (ER) is inhibited at 65% of the normal cellular adenosine triphosphate (ATP) pool. By means of controlled ATP depletion and replenishment, the effects on more distal compartments of the secretory pathway of impeded and restored traffic from ER to the Golgi apparatus can be analysed. Studies with various types of cells show tubulation of Golgi membranes, disintegration of *cis*-Golgi compartments, transformations into long tubules and redistribution of *cis*-Golgi constituents into the ER. A normal Golgi architecture is restored after ATP replenishment.

Since a reduction of the energy supply leads to a Golgi disintegration, which is reversible after normalisation of cellular ATP contents, controlled depletion and replenishment of the cellular ATP pools also provide a valuable tool for detailed analyses of distinct steps during the new formation of Golgi apparatus compartments and stacks. Panel A shows the perinuclear Golgi region of a HepG2 human hepatoma cell upon ATP-depletion effected by incubation of the cells in a glucose- and pyruvate-free medium containing 50 mM 2-deoxy-D-glucose (DOG medium). Panels B and C show the Golgi region of respective cells on ATP replenishment by incubation in medium supplemented with 100 mM glucose (GLUC medium). In the human hepatoma cells used, ATP depletion for 45 minutes results in a disassembly of the Golgi apparatus (panel A). Regular Golgi stacks are absent. They are replaced by prominent accumulations of structures of diverse shapes, some forming halves of small rings or honeycomb-like structures, some being club-like and rod-like. This images change rapidly during recovery from reduced energy supply. After 15 minutes incubation in GLUC medium, some cells already contain small reformed Golgi stacks (panel C). Most of the cells show intermediates that are characteristic for this stage of ATP-replenishment (panel B). Delicate networks and honeycomb-like structures are visible. Frequently, small tubular structures are terminated by buds (small arrows). Similar buds are visible, terminating the short cisternae in the reconstituted Golgi apparatus stack shown in panel C (small arrows).

References

Banta M, Polizotto RS, Wood SA, de Figueiredo P, and Brown WJ (1995) Characterization of a cytosolic activity that induces the formation of Golgi membrane tubules in a cell-free reconstitution system. Biochemistry 34: 13359

Cluett EB, Wood SA, Banta M, and Brown WJ (1993) Tubulation of Golgi membranes *in vivo* and *in vitro* in the absence of Brefeldin A. J Cell Biol 120: 15

Del Valle M, Robledo Y, and Sandoval IV (1999) Membrane flow through the Golgi apparatus: specific disassembly of the *cis*-Golgi network by ATP depletion. J Cell Sci 112: 4017

Dinter A and Berger EG (1998) Golgi-disturbing agents. Histochem Cell Biol 109: 571

Magnification: x 37,000 (A), x 43,000 (B, C)

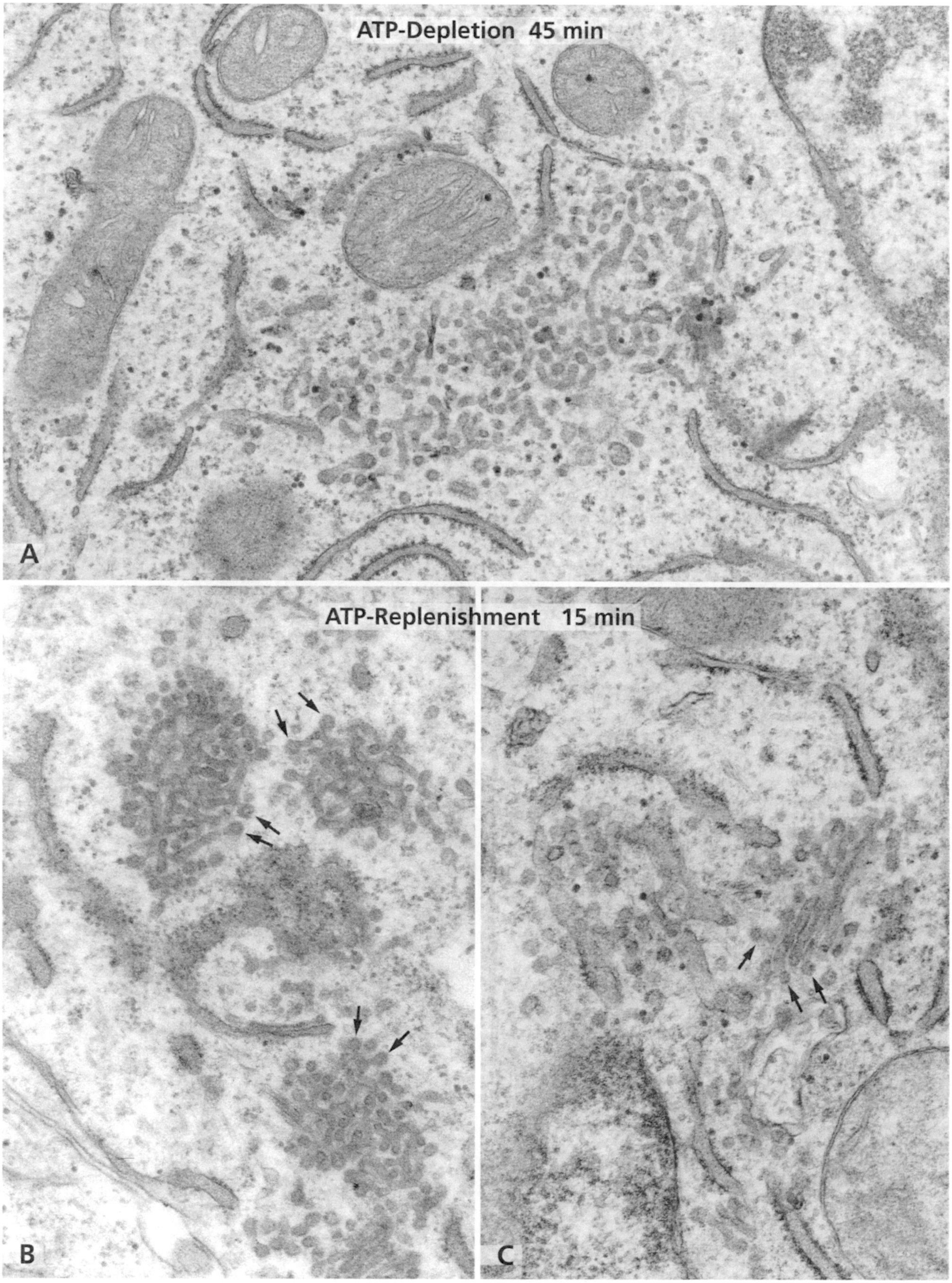
ATP-Depletion 45 min
A
ATP-Replenishment 15 min
B
C

SECRETORY GRANULES

Secretory granules function as storage compartments for secretory products and are the main organelles involved in regulated secretion. Their transport from the site of formation to the site of their export out of the cell is a multiple step process that involves budding from the TGN, microtubule- and actin-based transport, tethering, docking, and fusion with the plasma membrane. Discharge of the granule content occurs under precise local requirements after external, either neuronal or hormonal, stimulation. The micrographs in panels A and B show mature secretory granules, the zymogen granules (ZG), in the apical cytoplasm of acinar cells of the rat pancreas. The ultrastructural details shown in an ultrathin section in panel A and in a freeze-fracture replica in panel B correspond to each other. The acinar lumen is visible in the centres of the micrographs (AL). Arrows indicate the apical junctional complexes, by which the adjacent acinar cells are joined to each other. In the freeze-fracture replica in panel B, tight junction strands are visible in the form of prominent networks (arrows) that build up a broad occluding zone, which prevents reflux of the secretions from the acinar lumen into the intercellular spaces (cf. Figs. 83, 84 and 86). In both micrographs, the lateral plasma membranes of the neighbouring cells are labelled by arrowheads.

The zymogen granules mature from condensing vacuoles. These are the immature secretory granules (CV in Figs. 25 and 96) formed at the *trans* Golgi side within the *trans*most cisterna and the *trans* Golgi network (TGN). The biogenesis of zymogen granules involves several consecutive steps, which include aggregation and sorting of the secretory proteins to the *trans* Golgi and TGN membranes, budding of the immature granules (condensing vacuoles) from the TGN, homotypic fusion of immature secretory granules, and remodelling of membranes and contents. For sorting of proteins into the regulated pathway, their aggregation at the mildly acidic pH in the TGN lumen is an important event and interactions with accessory molecules, such as proteoglycans, sulphated glycoproteins, and lectin-like proteins, in the membranes of TGN and condensing vacuoles have crucial roles. Complex protein-protein and lipid-protein interactions are required, and recent results show that particular lipid microdomains (lipid rafts) are involved.

The endoplasmic reticulum, although accumulated in the basal cytoplasm of the acinar cells (cf. Figs. 1 and 95), is not confined to this cellular area but is present also between the zymogen granules in the apex of the cells, as shown in panel A. Ca^{2+}, released from the endoplasmic reticulum is a trigger for controlled exocytosis of the zymogen granule contents. The apical cell surface in part is occupied by microvilli. Microvilli are lacking at bay-like invaginations of the acinar lumen (upper part of the lumen in panel A), which possibly correspond to exocytosis domains. In the apical cytoplasm, in addition to small coated vesicles possibly involved in endocytosis and membrane internalisation, small secretory granules with contents less dense compared with the zymogen granules are present close to the plasma membrane (open arrows in A). They may be part of the constitutive-like and minor-regulated pathways assumed to be sources for basal secretion. Detailed studies of the secretory pathways in parotid acinar cells have indicated that the minor-regulated pathway provides surface docking/fusion sites for zymogen granule exocytosis.

References

Arvan P, Zhang B-Y, Feng L, Liu M, and Kuliawat R (2002) Luminal protein multimerization in the distal secretory pathway/secretion granules. Curr Opin Cell Biol 14: 448

Castle AM, Huang AY, and Castle JD (2002) The minor regulated pathway, a rapid component of salivary secretion, may provide docking/fusion sites for granule exocytosis at the apical surface of acinar cells. J Cell Sci 115: 2963

De Lisle RC (2002) Role of sulphated O-linked glycoproteins in zymogen granule formation. J Cell Sci 115: 2941

Kraemer J, Schmitz F, and Denckhahn D (1999) Cytoplasmic dynein and dynactin as likely candidates for microtubule-dependent apical targeting of pancreatic zymogen granules. Eur J Cell Biol 78: 265

Park JJ and Loh P (2009) How peptide hormone vesicles are transported to the secretion site for exocytosis. Mol Endocrin 22: 2583

Schmidt K, Dartsch H, Linder D, Kern H-F, and Kleene R (2000) A submembranous matrix of proteoglycans on zymogen granule membranes is involved in granule formation in rat pancreatic acinar cells. J Cell Sci 113: 2233

Tooze SA, Martens GJM, and Huttner WB (2001) Secretory granule biogenesis: rafting to the SNARE. Trends Cell Biol 11: 116

Magnification: x 31,000 (A); x 22,000 (B)

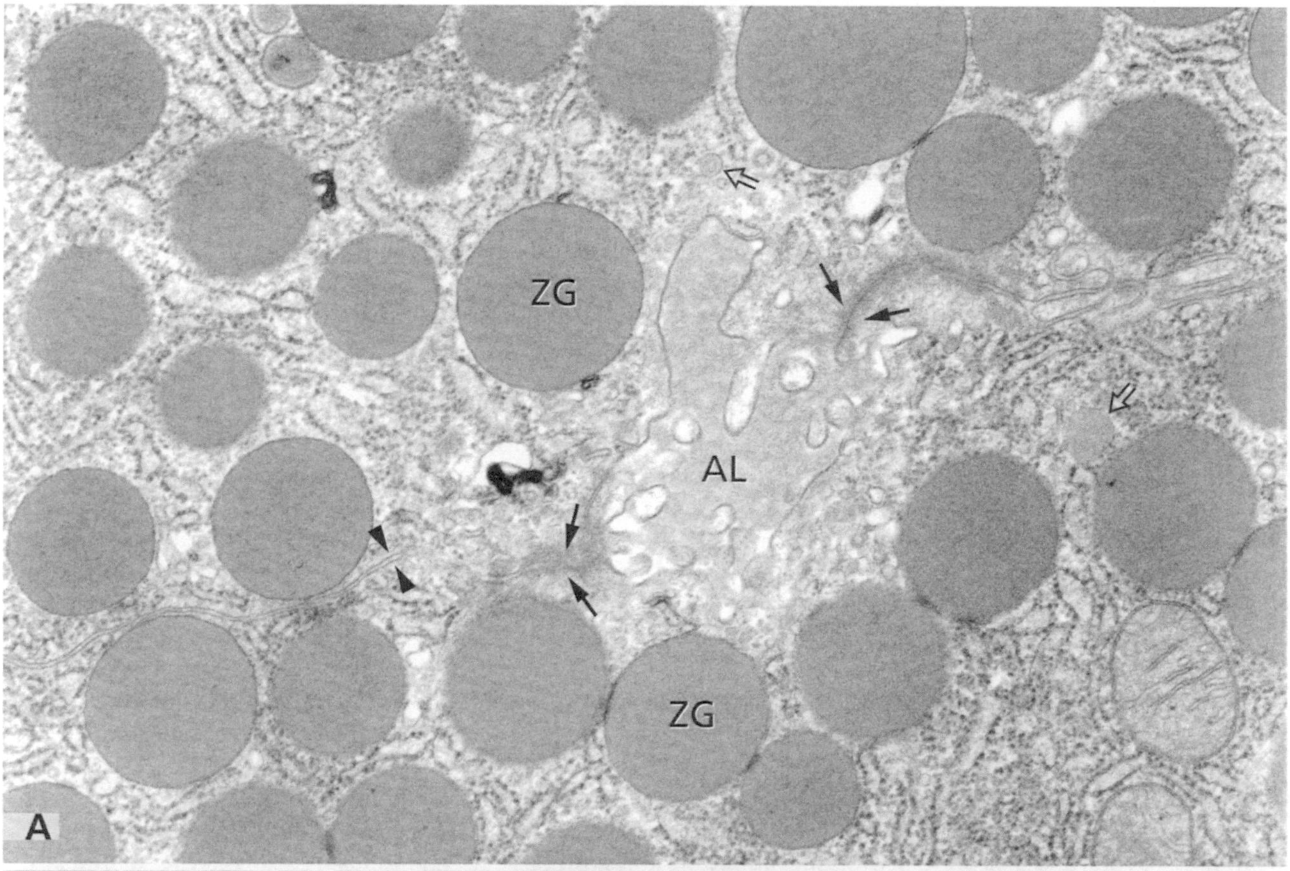
ZG
AL
ZG
A

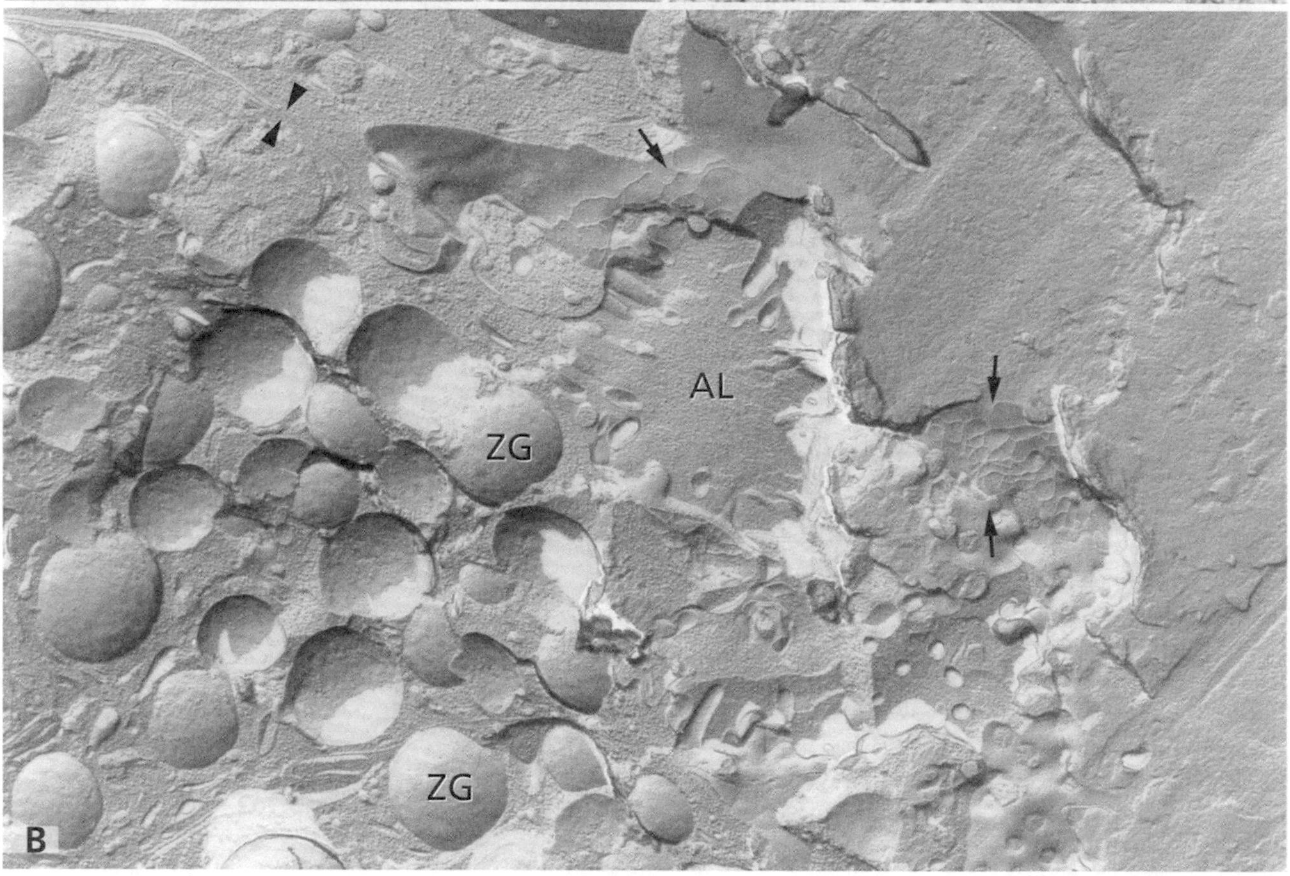
AL
ZG
ZG
B

SECRETORY GRANULE TYPES

The various exocrine and endocrine glands as well as single exocrine, endocrine and neuroendocrine cells scattered in different organs, and some neurons store their excretory products in secretory granules. In addition, a wide range of cells including the different types of leukocytes, macrophages, mast cells, platelets, T lymphocytes, melanocytes, type II alveolar epithelial cells, and endothelial cells contain secretory granules. All types of secretory granules are sealed off by a membrane. But their size and shape, and the appearance of their content differ greatly by electron microscopy. The latter largely depends on the chemical nature of the stored products (which can be identified by immunoelectron microscopy) and a few examples are presented here.

In submandibular gland (cf. Fig. 98), mucous and serous cells exist. Mucous cells synthesize mucins, which are highly glycosylated glycoproteins. By electron microscopy, mucous cell granules appear lucent with some granular material (panel A; see. also the goblet cell granules in Fig. 99). The serous cell granules contain an electron dense material with a tight fitting limiting membrane (panel B; see also the zymogen granules in Figs. 43 and 96). The secretion of the large salivary glands forms the saliva, which is composed of mucins, water, ions, digestive enzymes, and other proteins. Secreted mucins tightly bind water and form an insoluble gel on the mucosal surface of many organs, which functions as a lubricant and protects against physical and chemical injury. Saliva in addition to digestive function has also a protective antibacterial function through its content of IgA, lysozyme, and lactoferrin.

Endocrine and neuroendocrine cells and certain neurons contain granules with a dense core, which, depending on the cell type, may vary greatly in size. Panel C shows part of a human insulin-producing pancreatic beta cell with numerous secretory granules. They typically have a dense core and a lucent halo (arrows). The dense core can become crystalloid when insulin forms complexes with zinc (arrowhead). By immunoelectron microscopy, the dense core contains insulin (as indicated by the gold particles). C-peptide is present predominantly in the halo. Panel D shows part of growth hormone-producing cells from the human pituitary with immunogold labelling for this hormone. In this and many other endocrine cell types, the dense core granules have a tight fitting membrane.

The secretory granules are the organelles of the regulated secretory pathway that is tightly regulated in order to permit the controlled release of granule content in response to physiological signals. Secretion occurs by exocytosis, a multistage process that may result in the release of the entire pool of secretory granules of a given cell or only in a small number of them. In many cell types, a raise in intracellular Ca^{2+} is the main trigger for granule exocytosis, and synaptogamin is the Ca^{2+} sensor for regulated exocytosis. This is followed by ATP-depending priming steps, which include a reorganization of the cortical actin cytoskeleton so that granules can be recruited to the plasma membrane and modification of the SNARE proteins. Afterwards, ATP depending secretory granule tethering and docking takes place, followed by not ATP-requiring fusion of the granule membrane with the plasma membrane and subsequent release of content. The membrane fusion involves the formation and expansion of a fusion pore. The expansion of the fusion pore may be transient and followed by its rapid closure. This is the kiss-and-run exocytosis, which results in partial release of secretory granule content. Full fusion results in release of entire granule content and subsequent retrieval of the secretory granule membrane.

References

Borgonovo B, Ouwendijk J, and Solimena M (2006) Biogenesis of secretory granules. Curr Opin Cell Biol 18: 365

Burgoyne RD and Morgan A (2003) Secretory granule exocytosis. Physiol Rev 83: 581

Bennett M (1997) Ca^{2+} and the regulation of neurotransmitter secretion. Curr Opin Cell Biol 7: 316

Burgess TL and Kelly RB (1987) Constitutive and regulated secretion of proteins. Ann Rev Cell Biol 3: 243

Freeman JA (1966) Goblet cell fine structure. Anat Rec 154: 121

Gundelfinger E, Kessels M, and Qualmann B (2003) Temporal and spatial coordination of exocytosis and endocytosis. Nat Rev Molec Cell Biol 4: 127

Morvan J and Tooze S (2008) Discovery and progress in our understanding of the regulated secretory pathway in neuroendocrine cells. Histochem Cell Biol 129: 243

Verhage M and Toonen R (2007) Regulated exocytosis: merging ideas on fusing membranes. Curr Opin Cell Biol 19: 402

Magnification: x 20,000 (A); x 30,200(B); x 22,000 (C); x 20,500 (D)

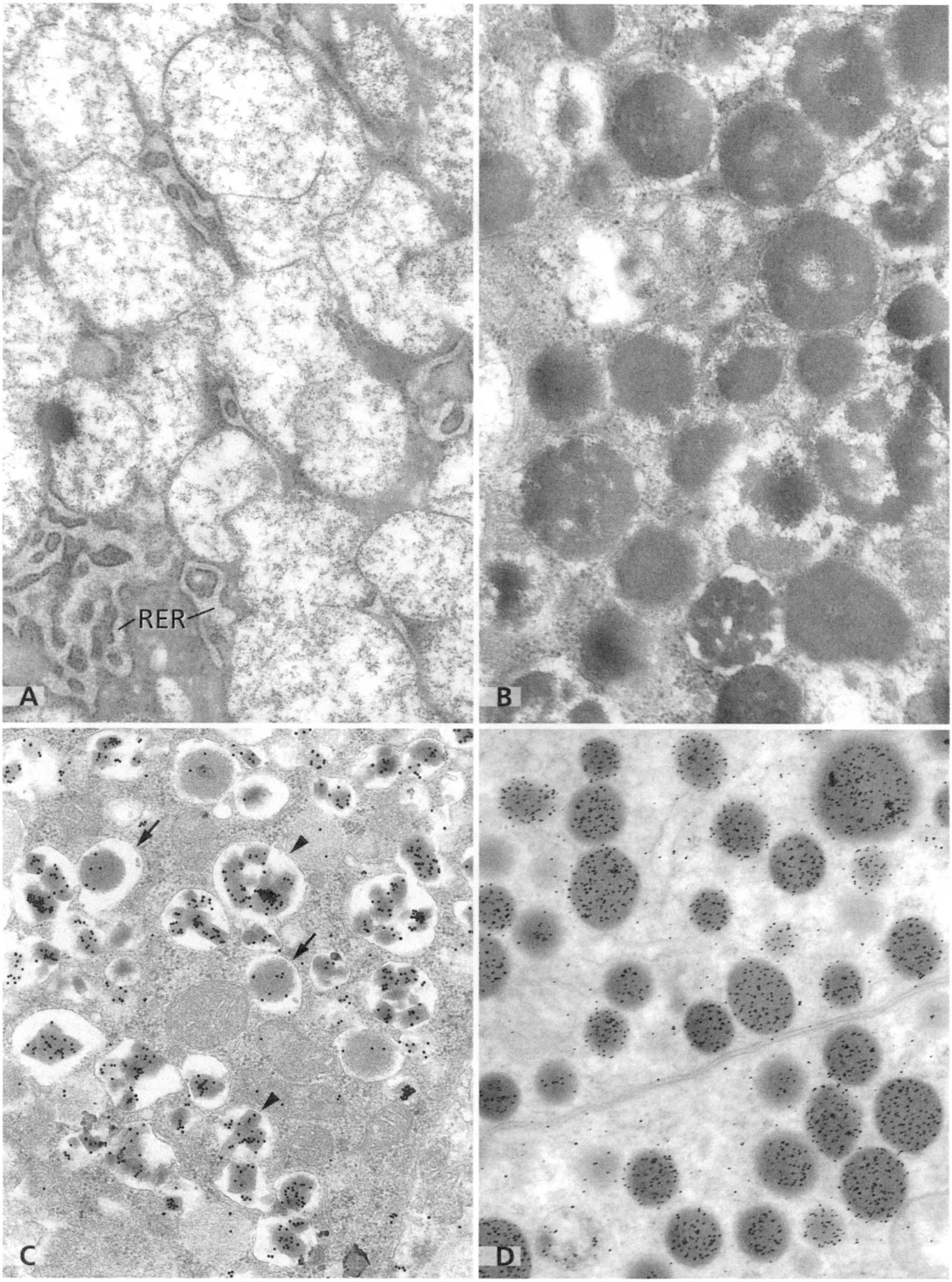
RER
A
B
C
D

GOBLET CELL – COMPOUND EXOCYTOSIS

In the goblet cells, the mucin droplets are the storage sites of the major secretory product, mucin. In contrast to the acinar pancreatic cells (cf. Fig. 43), the apical cytoplasm of the goblet cells is completely filled with the mucin droplets. The release of the mucin by secretion (exocytosis) is tightly regulated and occurs upon different stimuli such as parasympathetic stimulation or chemical and physical irritation.

The micrograph shows the apical portions of goblet cells and of absorptive cells together with their characteristic brush border (BB) from human duodenum. In contrast to the the kiss-and-run exocytosis, which results in partial release of secretory granule content, the ongoing compound exocytosis results in bulk secretion of the mucin of one of the goblet cells. This will eventually result in the almost complete release of the mucus, after which the mucus stores of the goblet cell will be rapidly replenished. The impressive, volcano-like look of the released mucus represents the prototype of compound exocytosis. Upon hydration, the secreted mucus will form a blanket protecting the underlying cells from mechanical damage and bacterial invasion.

Compound exocytosis can be observed in various other exocrine and endocrine cell types and involves both the fusion of secretory granules with each other and with the plasma membrane.

References

Burgoyne RD and Morgan A (2003) Secretory granule exocytosis. Physiol Rev 83: 581

Bennett M (1997) Ca^{2+} and the regulation of neurotransmitter secretion. Curr Opin Cell Biol 7: 316

Burgess TL and Kelly RB (1987) Constitutive and regulated secretion of proteins. Ann Rev Cell Biol 3: 243

Davis CW and Dickey BF (2008) Regulated airway goblet cell mucin secretion. Annu Rev Physiol 70: 487

Pickett JA and Edwardson JM (2006) Compound exocytosis: mechanisms and functional significance. Traffic 7: 109

Perez-Vilar J and Hill RL (1999) The structure and assembly of secreted mucins. J Biol Chem 274: 31751

Verhage M and Toonen R (2007) Regulated exocytosis: merging ideas on fusing membranes. Curr Opin Cell Biol 19: 402

Magnification: x 10,500

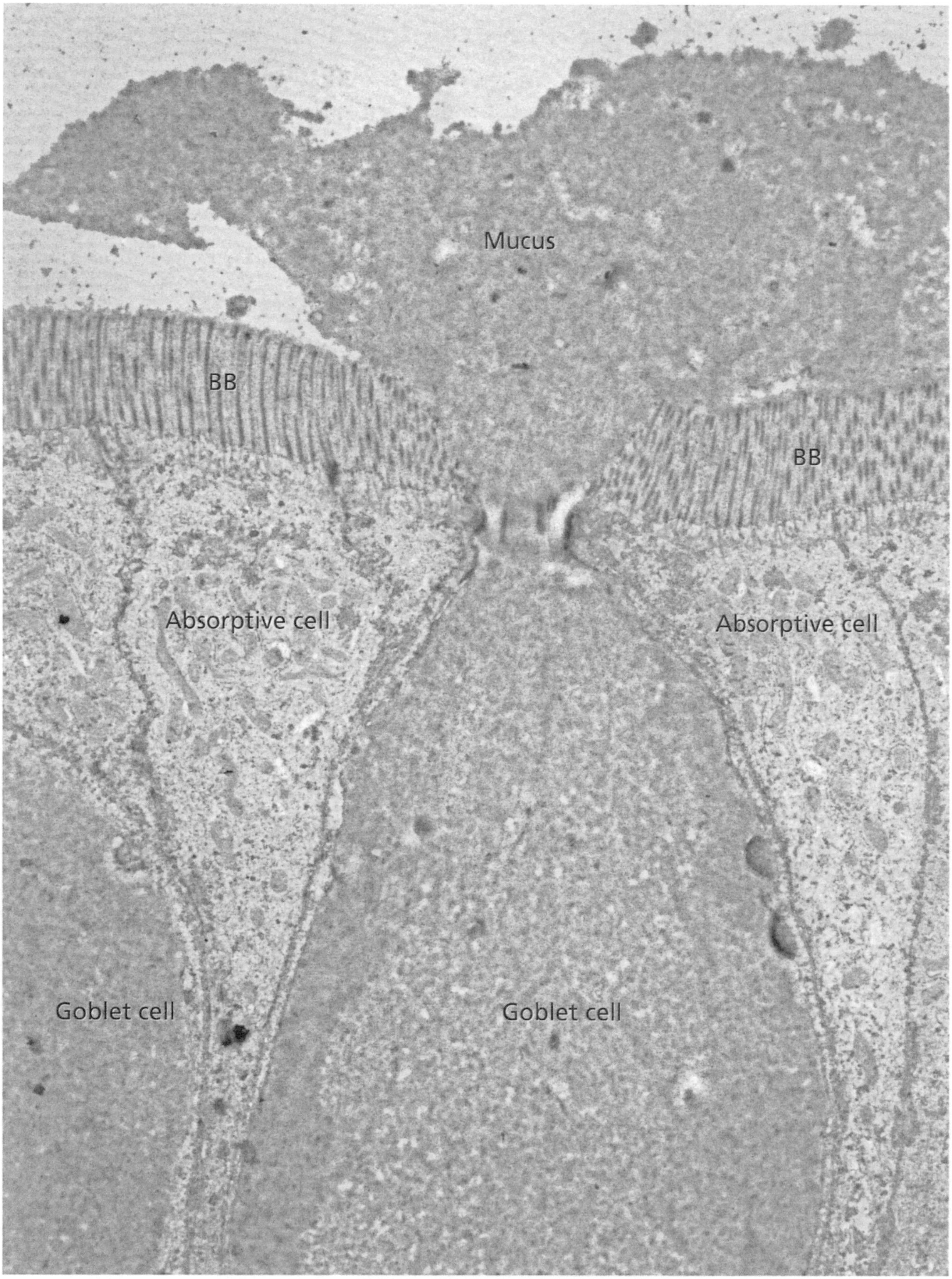
Mucus
BB
BB
Absorptive cell
Absorptive cell
Goblet cell
Goblet cell

RECEPTOR-MEDIATED ENDOCYTOSIS VIA CLATHRIN-COATED VESICLES AND VIRUS ENDOCYTOSIS

Adsorptive and receptor-mediated endocytosis via clathrin-coated vesicles provides the major and best characterised portal for uptake of multiple molecules and particles into cells. Via clathrin-dependent endocytosis, cells receive nutrients, regulate receptors and other plasma membrane constituents, take up antigens, and remove senescent, excess, and potentially harmful substances from the extracellular fluid.

Because of their typical bristle-like coats, the pits, buds, and vesicles formed during clathrin-dependent endocytosis can easily be differentiated under the electron microscope (panels A and C). The coat consists mainly of clathrin and adaptor proteins. The individual cytosolic clathrin molecules are triskelions, which assemble and recruit to the cytoplasmic face of the plasma membrane in concerted interactions with adaptor proteins and the lipid bilayer which results in the formation of a mainly hexagonal lattice. Through rapid disassembly and reassembly of the clathrin triskelions, the membrane deforms into pits and deeply invaginated buds coated by a clathrin basketwork of polygons in which pentagons and heptagons are juxtaposed to hexagons. The basket seems to function as a stabilising coat for the molecular machineries necessary to concentrate plasma membrane proteins into the endocytic pits and vesicles. Adaptors, such as the AP-2 adaptor protein complex, facilitate selection of the cargo for uptake into the vesicles and connect the clathrin coat with the transmembrane receptors. Detachment of the buds from the plasma membrane involves the action of dynamins, a class of scission-molecules essential for the pinching-off of the buds to create vesicles. Shortly after a vesicle is formed, the clathrin coat disassembles and detaches from the membrane, and the coat-free endocytic vesicle is capable of fusing with other endosomes.

Panels A and C show coated pits during internalisation of *Ricinus communis I* lectin-ferritin in cultured pancreatic beta cells. They are covered at the cytoplasmic face with the characteristic bristle-like clathrin coat. The ferritin particles concentrated in the coated pits and along the outer surface of the plasma membrane reflect the lectin binding to the galactose-bearing oligosaccharides in the plasma membrane. In panel C, endocytic vesicles containing internalised lectin-ferritin are already devoid of a clathrin coat. In the freeze fracture replica in panel B, filipin-sterol complexes are visible as small protuberances. They are absent from the pits indicating that they have a lower cholesterol content than the surrounding plasma membrane.

In panels D and E, human immunodeficiency virus (HIV) particles (arrowheads) bound to the surface of a macrophage are shown. They are seen attached to the plasma membrane (PM) contained in large endocytic vacuoles or surface invaginations (arrows in panel E). Viruses use different endocytic pathways; besides internalisation by clathrin-coated vesicles, viruses can be taken up into cells via caveolae (cf. Fig. 51), lipid-raft-mediated endocytosis routes, macropinocytosis, and presumably other mechanisms as well. HIV1 enters cells by direct fusion with the plasma membrane and by fusion with the membrane of macropinosomes. The large vacuoles in panel E containing HIV particles possibly correspond to macropinosomes formed by deep plasma membrane invaginations and closure of membrane ruffles.

SG – secretory granule.

References

Grant BD and Donaldson JG (2009) Pathways and mechanisms of endocytic recycling. Nat Rev Mol Cell Biol 10: 597

Heuser J (1980) Three-dimensional visualization of coated vesicle formation in fibroblasts. J Cell Biol 84: 560

Lafer EM (2002) Clathrin-protein interactions. Traffic 3: 513

Montesano R, Perrelet A, Vassalli P, and Orci L (1979) Absence of filipin-sterol complexes from large coated pits on the surface of culture cells. Proc Natl Acad Sci USA 76: 6391

Mousavi SA, Malerod L, Berg T, and Kjeken R (2004) Clathrin-dependent endocytosis. Biochem J 377: 1

Mukherjee S, Gosh RN, and Maxfield FR (1997) Endocytosis Physiol Rev 77: 759

Pelkmans L and Helenius A (2003) Insider information: what viruses tell us about endocytosis. Curr Opin Cell Biol 15: 414

Praefcke GJK and McMahon HT (2004) The dynamin superfamily: Universal membrane tubulation and fission molecules? Nat Rev Mol Cell Biol 5: 133

Robinson MS (2004) Adaptable adaptors for coated vesicles. Trends Cell Biol 14: 167

Schmid SL (1997) Clathrin-coated vesicle formation and protein sorting: an integrated process. Annu Rev Biochem 66: 511

Traub LM (2003) Sorting it out: AP-2 and alternate clathrin adaptors in endocytic cargo selection. J Cell Biol 163: 203

Magnification: x 35,000 (A, B); x 102,000 (C); x 26,500 (D); x 71,000 (E)

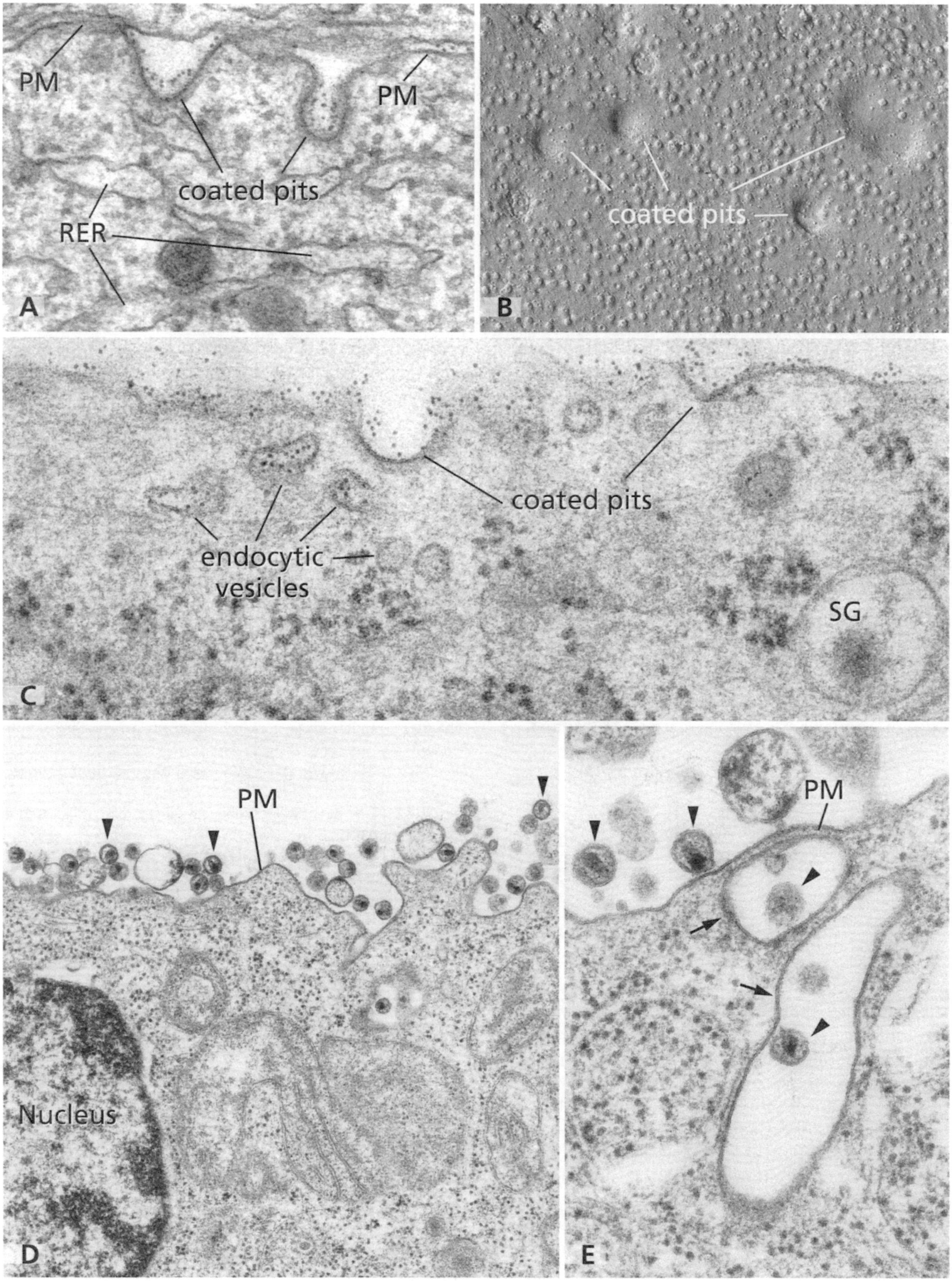
PM
PM
coated pits
RER
A
coated pits
B
coated pits
endocytic
vesicles
SG
C
PM
Nucleus
D
PM
E

ENDOSOMES AND ENDOCYTIC PATHWAYS

Different endocytic portals and pathways are known in mammalian cells (see diagram), roughly classified as clathrin-dependent (1, cf. Fig. 46) and clathrin-independent ones. The latter include endocytosis by budding of "smooth" vesicles lacking a distinct morphologically visible coat (2, cf. Fig. 52), uptake via pits containing lipid rafts (3), traffic via caveolae (4, cf. Fig. 51) and caveosomes (C), phagocytosis (5, cf. Fig. 52) and uptake via macropinocytosis (6) leading to the formation of large endocytic vacuoles, termed macropinosomes (MP, cf. Fig. 46). The routes travelled by endocytosed molecules involve complex endosomal compartments with mosaics of specialised structural and functional domains. Endosomes are highly dynamic and transform in response to endocytosis-connected signals. Early endosomes, classified in early sorting endosomes (SE) and recycling endosomes (RE), represent first stations from where proteins and lipids are sorted to different routes. These can be either routes to late endosomes (LE) and lysosomes (LY) for degradation, or recycling pathways to the plasma membrane (PM), or routes to other destinations involving *trans* Golgi networks (TGN), and the Golgi apparatus (GA). The endoplasmic reticulum (ER) is reached by retrograde pathways across the Golgi apparatus and pre-Golgi intermediates (pGI), or via more direct routes, as is known for caveolae-derived endosomes.

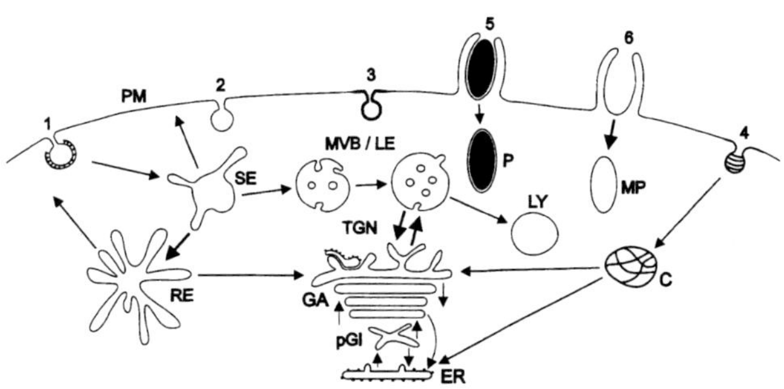

Panels A–E show early endosomal compartments after internalisation of peroxidase-labelled wheat germ agglutinin (WGA, panels A, B, and E) and a multivesicular body containing internalised *Ricinus communis I* lectin-ferritin (panel C). Different domains on early endosomes, such as large vacuolar parts (asterisks in panels A and E), small vesicles budding outward from the limiting membranes, and long tubular appendices reflect the sorting to different destinations. Inward budding of vesicles leads to the formation of "vesiculated" endosomes and multivesicular bodies (MVB, panels B and C). The early endosome in panel B contains some interior vesicles and a long outward tubule, and its limiting membrane exhibits a conspicuous coated area (arrows). A similar region is visible at one of the multivesicular bodies in panel C (arrow). Specialised bilayered clathrin coats on vacuolar endosomes have been suggested to have a role in the targeting of proteins to lysosomes.

In HepG2 hepatoma cells, internalised WGA is taken up into the Golgi apparatus by a complex multistep process (cf. also Fig. 48), which starts with an accumulation of early vesicular endosomes at the *trans* Golgi side (panel D). The Golgi apparatus is small and inconspicuous in the control cells. It expands during endocytosis, and an endocytic TGN starts to be formed. Initial endocytic networks labelled with internalised WGA (TGN) are visible in panel D (cf. also Fig. 48).

References

Barysch SV, Aggarwal S, Jahn R, and Rizzoli SO (2009) Sorting in early endosomes reveales connections to docking- and fusion-associated factors. Proc Natl Acad Sci USA 106: 9697

Doherty GJ and McMahon HT (2009) Mechanisms of endocytosis. Annu Rev Biochem 78: 857

Gruenberg J (2001) The endocytic pathway: A mosaic of domains. Nat Rev Mol Cell Biol 2: 721

Johannes L and Lamaze C (2002) Clathrin-dependent or not: Is it still the question? Traffic 3: 443

Kerr MC and Teasdale RD (2009) Defining macropinocytosis. Traffic 10: 364

Maxfield FR and McGraw TE (2004) Endocytic recycling. Nat Rev Mol Cell Biol 5: 121

Nichols B (2003) Caveosomes and endocytosis of lipid rafts. J Cell Sci 116: 4707

Pelkmans L and Helenius A (2002) Endocytosis via caveolae. Traffic 3: 311

Sachse M, Ramm G, Strous G, and Klumperman J (2002) Endosomes: multipurpose designs for integrating housekeeping and specialized tasks. Histochem Cell Biol 117: 91

Sandvig K, Torgersen ML, Raa HA, and van Deurs B (2008) Clathrin-independent endocytosis: from nonexisting to an extreme degree of complexity. Histochem Cell Biol 129: 267

Sorkin A and von Zastrow M (2009) Endocytosis and signalling: interwining molecular networks. Nat Rev Mol Cell Biol 10: 609

Woodman PG and Futter CE (2008) Multivesicular bodies: co-ordinated progression to maturity. Curr Opin Cell Biol 20: 408

Magnification: x 35,000 (A); x 41,000 (B); x 51,000 (C); x 33,800 (D); x 42,000 (E)

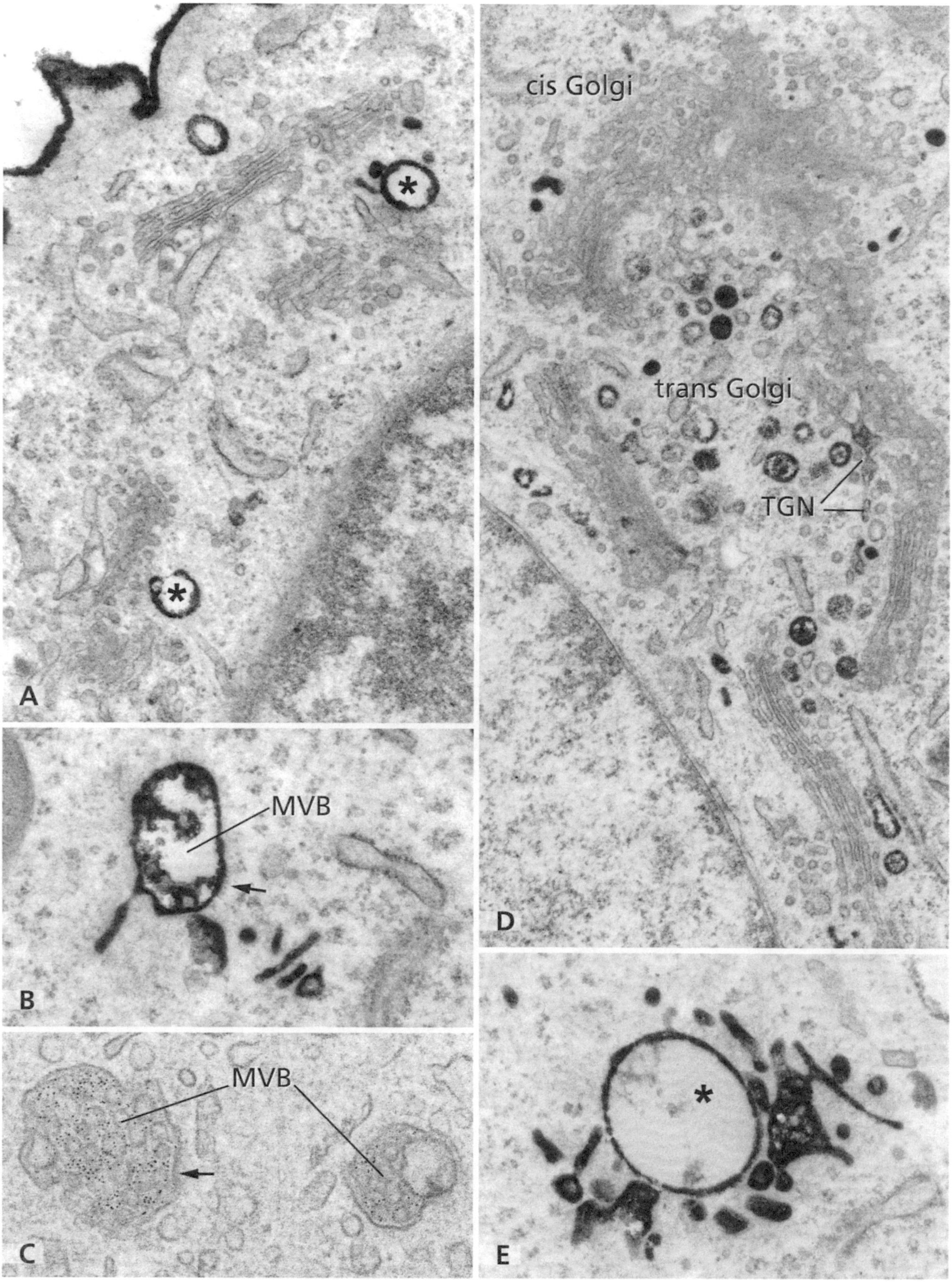
A
B
MVB
C
MVB
D
cis Golgi
trans Golgi
TGN
E
*

ENDOCYTIC *TRANS* GOLGI NETWORK AND RETROGRADE TRAFFIC INTO THE GOLGI APPARATUS

Different itineraries from the plasma membrane and early endosomes to the Golgi apparatus have been characterised, either or not involving late endosomes (cf. diagram in Fig. 47). A direct pathway travelled by the Shiga toxin B-fragment, TGN38, and the cation-independent mannose-6-phosphate receptor in part leads from the early recycling compartment en route to the *trans* Golgi network (TGN).

Studies with peroxidase labelled wheat germ agglutinin (WGA) in HepG2 hepatoma cells have shown early endosomes a few minutes after onset of endocytosis appearing in close neighbourhood to the Golgi stacks (cf. Fig. 47, panel A), which are small and inconspicuous at that time. Uptake of internalised WGA into the Golgi apparatus is a complex multi-step process. After accumulation of early vesicular endosomes at the *trans* Golgi side, endocytic networks are formed (panel D in Fig. 47), expand in size and closely associate with the Golgi stacks within 15-30 minutes. Panels A and B show Golgi stacks in HepG2 cells at 30 and 60 min of WGA-endocytosis, respectively. At 30 minutes, internalised WGA is localised within cisternae in *trans*most position, parts of which turn away from the stack and are continuous with extended *trans* Golgi endocytic networks, from where clathrin-coated vesicles are budding (arrowhead). Endocytic cisternae and networks are closely associated with cisternae of *trans* Golgi-endoplasmic reticulum (*trans*-ER). The endocytic *trans* Golgi networks decrease in size during the next thirty minutes and internalised WGA is taken up into the stacked Golgi cisternae, as shown in panel B. A TGN is not visible, but many clathrin-coated vesicles (arrowheads) occupy the *trans* Golgi region and most of the stacked Golgi cisternae contain internalised WGA (arrows). At both times 30 and 60 min, large vacuolar endosomes are prominent close to the Golgi cisternae (asterisks in panels A and B). The successive stages of endocytic traffic into the Golgi apparatus, the vesicular stage, the early and late TGN-stages, and the Golgi stage are additionally outlined in the diagram.

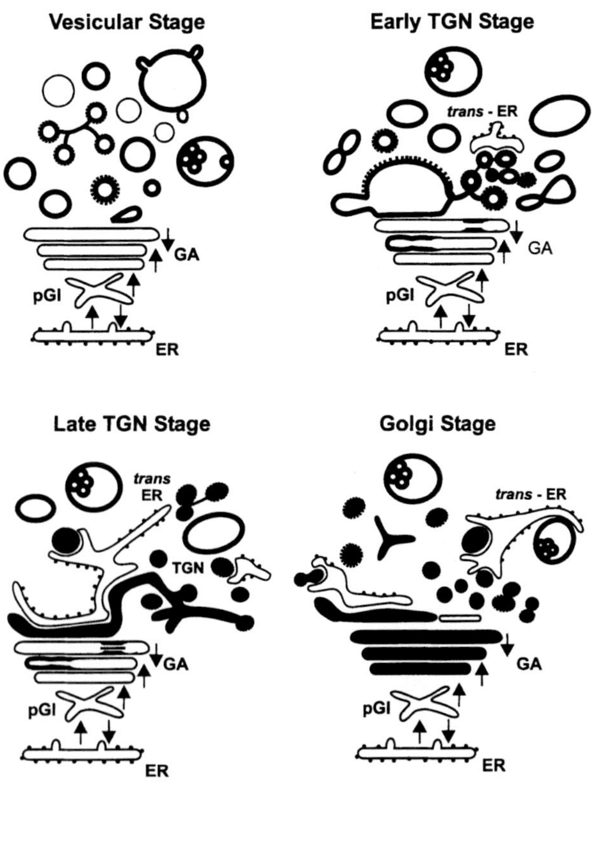

Endocytic traffic into and across the Golgi apparatus stacks is of particular interest in connection with the retrograde routes of toxins, and with developments of drug targeting systems.

References

Bonifacino JS and Rojas R (2006) Retrograde transport from endosomes to the *trans* Golgi network. Nat Rev Mol Cell Biol 7: 568

Johannes L and Popoff V (2008) Tracing the retrograde route in protein trafficking. Cell 135: 1175

Maxfield FR and McGraw TE (2004) Endocytic recycling. Nat Rev Mol Cell Biol 5: 121

Pavelka M, Neumüller J, and Ellinger A (2008) Retrograde traffic in the biosynthetic-secretory route. Histochem Cell Biol 129: 277

Sandoval RM and Molitoris BA (2004) Gentamycin traffics retrograde through the secretory pathway and is released in the cytosol via the endoplasmic reticulum. Am J Renal Physiol 286: F617

Sandvig K and van Deurs B (2005) Delivery into cells: lessons learned from plant and bacterial toxins. Gene Ther 12: 865

Sannerud R, Saraste J, and Goud B (2003) Retrograde traffic in the biosynthetic-secretory route: pathways and machinery. Curr Opin Cell Biol 15: 438

Tarrago-Trani MT and Storrie B (2007) Alternate routes for drug delivery to the cell interior: pathways to the Golgi apparatus and endoplasmic reticulum. Adv Drug Deliv Rev 59: 782

Vetterlein M, Ellinger A, Neumüller J, and Pavelka M (2002) Golgi apparatus and TGN during endocytosis. Histochem Cell Biol 117: 143

Weissenböck A, Bogner E, Wirth M, and Gabor F (2004) Binding and uptake of wheat germ agglutinin-grafted PLGA-nanospheres by Caco-2 monolayers. Pharmac Res 21: 1917

Magnification: x 43,000 (A); x 56,000 (B)

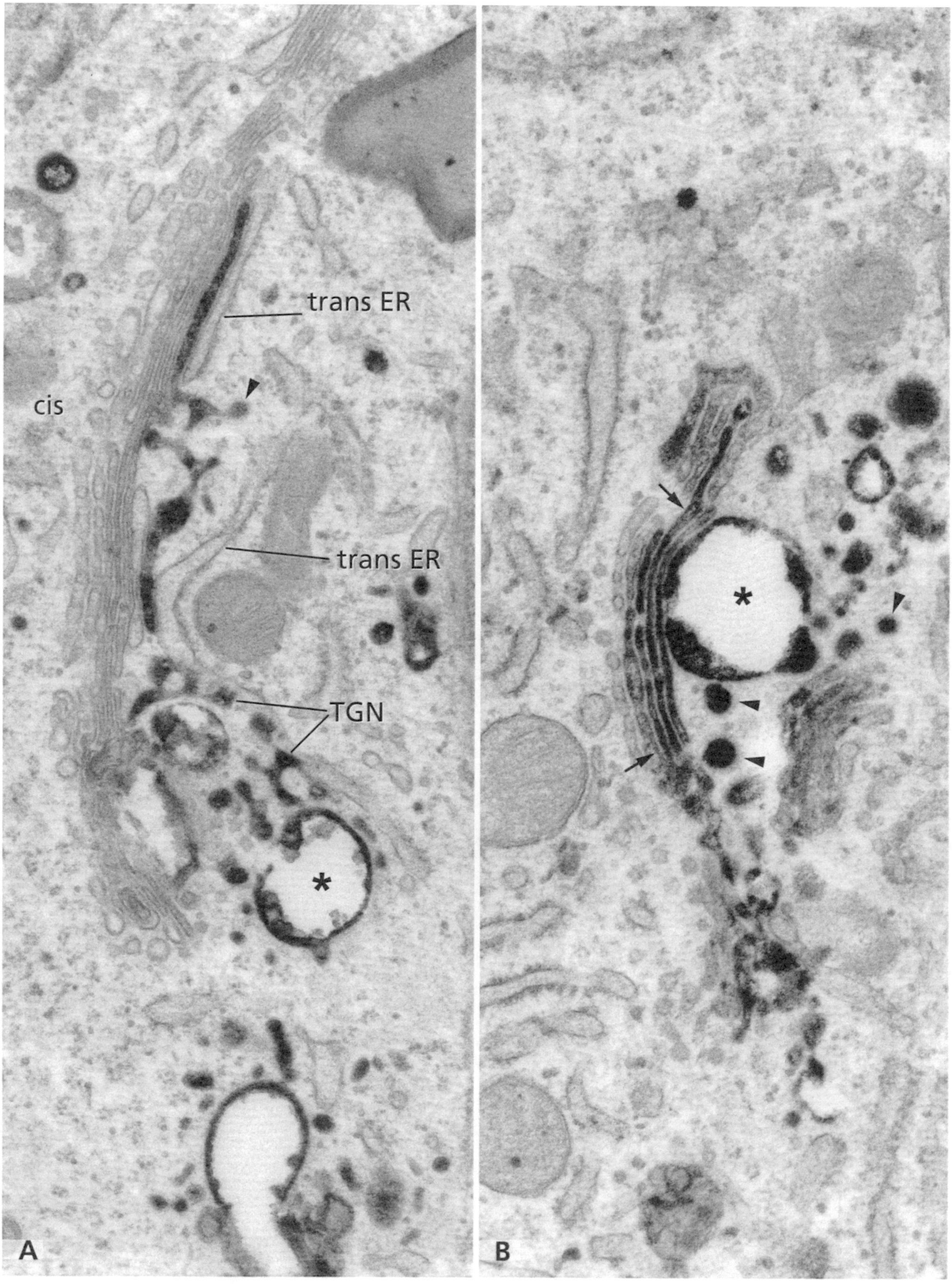

trans ER
cis
trans ER
TGN
*
A
*
B

TUBULAR PERICENTRIOLAR ENDOSOMES

Endosomes consist of pleiomorphic vacuoles of 0.5-1 μm diameter and tubules of 250 nm diameter (cf. Fig. 47). In a large variety of cultured cell types, including fibroblasts, CHO cells, epithelial cell types, and several endocrine cell types, an additional structural component has been identified that consists of tubules of approximately 60 nm diameter with a length of up to 2 μm, hence their name, tubular endosomes. These tubules can be loaded with the fluid phase marker horseradish peroxidase and with markers of receptor-mediated endocytosis (transferrin, α-2-macroglobulin, lectins). These multibranching tubules form local networks that are preferentially found in the perinuclear cytoplasm in association with centrioles but also in patches somewhere else in the cell body and are positive for Rab11 and unreactive with the early endosome marker EEA1.

In panel A, a cluster of tubular endosomes (outlined by a broken line) is present in the perinuclear region of Ehrlich tumour cells grown in monolayer culture. Since the cells had been fixed and embedded *in situ* in the Petri dishes, this cell monolayer is preserved. Note also the presence of large endocytotic vacuoles in adjacent parts of the cytoplasm. As shown at higher magnification in panel B, the tubular endosomes surround a centriole (arrow). Although many profiles of tubular endosomes can be observed in this 60 nm ultrathin section, the networks formed by them can be appreciated only by the electron microscopic investigation of thicker (1.0 μm) sections. In the electron micrograph shown here, endocytic vacuoles and lysosomes are also present. The labelling is due to a gold-labelled lectin. In panel C, a detail of the cytoplasm of a cultured pancreatic beta cell is shown that had endocytosed ferritin-conjugated *Ricinus communis* lectin. The ferritin label can be seen in typical tubular endosomes (arrowheads), which surround the centriole (double arrow), and in addition is present in lysosomes (Ly). Sometimes, a close spatial relation between a tubular endosome and a microtubulus is obvious (upper right corner of the micrograph). SG: insulin secretory granule.

From the results of uptake studies, tubular endosomes represent a specialised distal segment of early endosomes involved in receptor recycling and are distinct from late endosomes. Transferrin receptors seem to pass sequentially from early endosomes to pericentriolar tubular endosomes, followed by their return to the cell surface. However, passage through tubular endosomes seems to be not obligatory and followed only by a fraction of recycling transferrin receptors. With regard to other functional aspects of the pericentriolar endosomes, it has been proposed that they direct recycling receptors along microtubules radiating from the centrioles to particular cell surface regions of the cultured and migrating cells. It has been shown that recycling transferrin receptors are preferentially delivered to the leading lamella of migrating fibroblasts.

References

Hopkins C, Gibson A, Shipman M, and Miller K (1990) Movement of internalised ligand-receptor complexes along a continuous endosomal reticulum. Nature 346: 335

Hopkins C, Gibson A, Shipman M, Strickland D, and Trowbridge, I (1994) In migrating fibroblasts, recycling receptors are concentrated in narrow tubules in the pericentriolar area, and then routed to the plasma membrane of the leading lamella. J Cell Biol 125: 1265

Lin SX, Gundersen GG, and Maxfield FR (2002) Export from pericentriolar endocytic recycling compartment to cell surface depends on stable, detyrosinated (glu) microtubules and kinesin. Mol Biol Cell 13: 96

Sheff D, Pelletier L, OConnell CB, Warren G, and Mellman I (2002) Transferrin receptor recycling in the absence of perinuclear recycling endosomes. J Cell Biol 156: 797

Tooze J and Hollinshead M (1991) Tubular early endosomal networks in AtT20 and other cells. J Cell Biol 115: 635

Ullrich O, Reinsch S, Zerial M, and Parton R (1996) Rab11 regulates recycling through the pericentriolar recycling endosome. J Cell Biol 135: 913

Verges R, Havel RJ, and Mostov KE (1999) A tubular endosomal fraction from rat liver: Biochemical evidence of receptor sorting by default. Proc Natl Acad Sci USA 96: 10146

Magnification: x 7,300 (A); x 28,000 (B); x 19,250 (C)

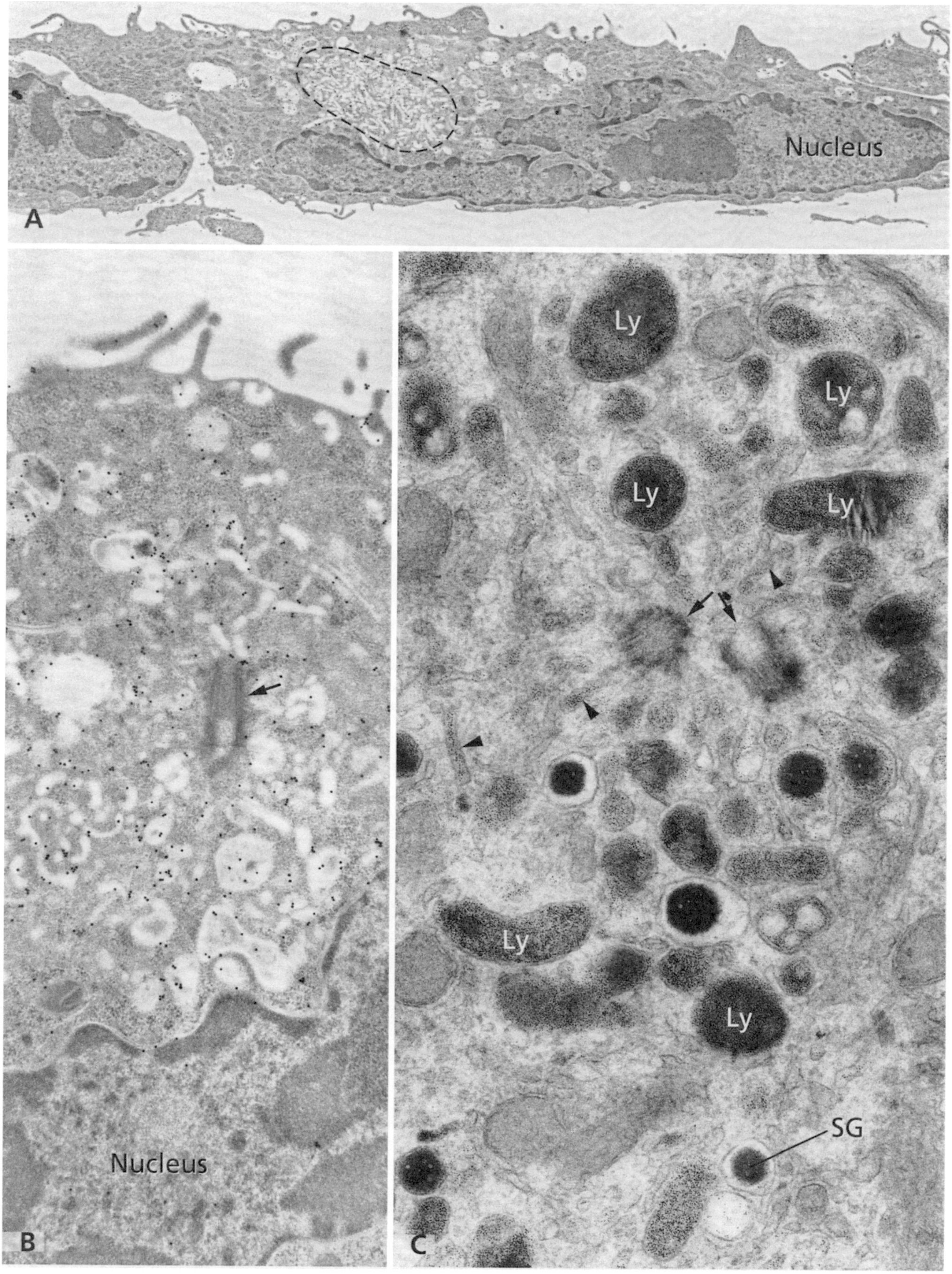
A
Nucleus
B
Nucleus
C
Ly
Ly
Ly
Ly
Ly
Ly
SG

LANGERHANS CELLS AND BIRBECK GRANULES: ANTIGEN PRESENTING DENDRITIC CELLS OF THE EPIDERMIS

Langerhans cells are antigen trapping dendritic cells of the epidermis. They capture and endocytose antigens and thereafter migrate from the epidermis to draining lymph nodes. Before arrival at the lymph nodes, Langerhans cells undergo maturation into activated dendritic cells and as a consequence present antigens to naïve T-cells. They are involved in cutaneous immune responses.

In panel A, a Langerhans cell of human epidermis is shown, which contains a large number of Birbeck granules in the neighbourhood of the Golgi apparatus. A few of its typical cytoplasmic processes that extend among keratinocytes can be seen partially (arrows). Birbeck granules are highly characteristic, rod-shaped structures of variable length with an internal periodically striated lamella corresponding to Langerin (panel B and inset). Often, Birbeck granules exhibit coated buds at their tips (arrowhead in inset), and this is the morphological manifestation of receptor-mediated endocytic processes. Furthermore, they characteristically exhibit local distentions (asterisks in panel B and arrow in inset).

Langerhans cells differ from dendritic cells in other locations by the presence of Langerin at their cell surface and of Birbeck granules in their cytoplasm. Birbeck granules are formed as a consequence of Langerin endocytosis and are due to the accumulation of Langerin in tubular endosomes. Thus, Birbeck granules represent endocytic structures that are formed during receptor-mediated endocytosis. Since Langerin colocalises with Rab11, the Langerin endocytic structures including the Birbeck granules represent the equivalent of tubular pericentriolar endosomes found in other cell types (cf. Fig. 49). Langerin itself recycles to the surface of Langerhans cells. However, on maturation of Langerhans cells the capacity of Langerin to recycle is gradually lost, and this explains the dramatic reduction in the number and size of their Birbeck granules.

References

Hanau D, Fabre M, Schmitt D, Stampf J, Garaud J, Bieber T, Grosshans E, Benezra C, and Cazenave J (1987) Human epidermal Langerhans cells internalise by receptor-mediated endocytosis T6 (CD1 "NA1/34) surface antigen: Birbeck granules are involved in the intracellular traffic of the T6 antigen. J Invest Dermatol 89: 172

McDermott R, Ziylan U, Spehner D, Bausinger H, Lipsker D, Mommaas M, Cazenave JP, Raposo G, Goud B, delaSalle H, et al. (2002) Birbeck granules are subdomains of endosomal recycling compartment in human epidermal Langerhans cells, which form where langerin accumulates. Mol Biol Cell 13: 317

Mellman I and Steinman RM (2001) Dendritic cells: specialised and regulated antigen processing machines. Cell 106: 255

Moll H, Fuchs H, Blanck C, and Röllinghoff M (1993) Langerhans cells transport *Leishmania major* from the infected skin to the draining lymph node for presentation to antigen-specific T-cells. Eur J Immunol 23: 1595

Moommas A, Wijsman M, Mulder A, van Praag M, Vermeer B, and Koning F (1992) HLA class II expression on human epidermal Langerhans cells in situ: up-regulation during the elicitation of allergic contact dermatitis. Hum Immunol 34: 99

Saudrais C, Spehner D, de la Salle H, Bohbot A, Cazenave JP, Goud B, Hanau D, and Salamero J (1998) Intracellular pathway for the generation of functional MHC class II peptide complexes in immature human dendritic cells. J Immunol 160: 2597

Schuler G, Romani N, Stössel H, and Wolff K (1991) Structural organisation and biological properties of Langerhans cells. In: Epidermal Langerhans cells (Schuler G, ed). Boca Raton, CA: CRC Press, pp 87

Stössel H, Koch F, Kämpgen E, Stöger P, Lenz A, Heufler C, Romani N, and Schuler G (1990) Disappearance of certain acidic organelles (endosomes and Langerhans cell granules) accompanies loss of antigen processing capacity upon culture of epidermal Langerhans cells. J Exp Med 172: 1471

Valledeau J (2000) Langerin, a novel C-type lectin specific to Langerhans cells, is an endocytic receptor that induces the formation of Birbeck granules. Immunitiy 12: 71

Magnification: x 11,500 (A); x 80,000 (B and inset)

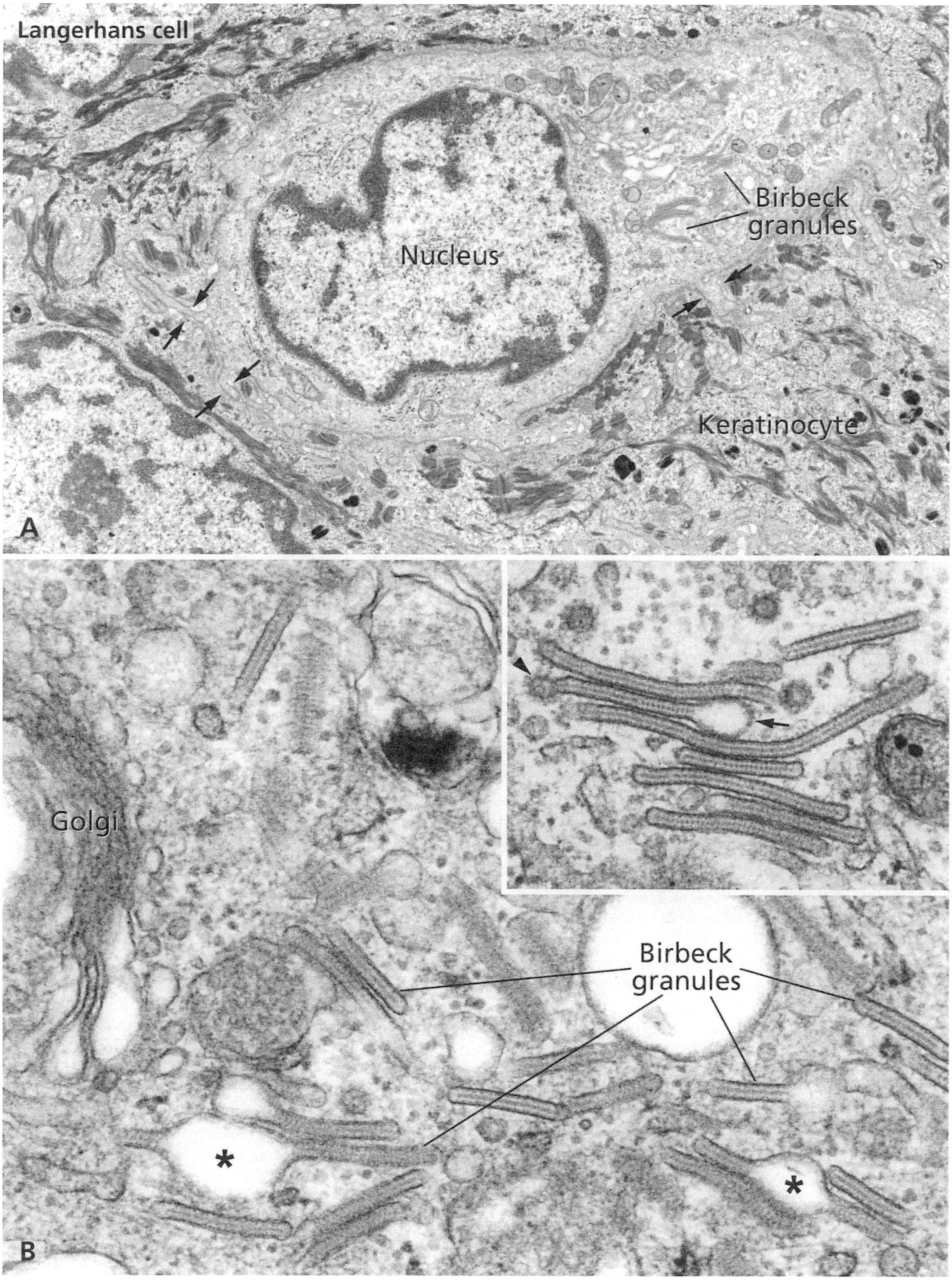
Langerhans cell
Birbeck granules
Nucleus
Keratinocyte
A
Golgi
Birbeck granules
*
*
B

CAVEOLAE

Caveolae are flask shaped, invaginated domains of the plasma membrane in various kinds of cells and measure 50-100 nm in diameter. They are known for about 50 years since the early days of electron microscopy but up to now there is a question as to their vital functions and a controversery exists about to their properties and capabilities. The caveolar membrane is enriched with cholesterol and glycosphingolipids. A characteristic coat at the cytoplasmic face of the membrane is composed of caveolins, the major structural proteins of caveolae. Caveolins bind cholesterol with high affinity and are necessary for forming the small caves and stabilise them at the plasma membrane. Caveolae are particularly abundant in adipocytes, endothelial cells (cf. Figs. 126 and 128), and muscle cells (panel A), but are missing in some other types of cells, such as lymphocytes and cells of the nerve system.

Panel A shows segments of smooth muscle cells. Extended areas of the cell surface are occupied by caveolar invaginations (arrows). In sections perpendicular to the cell surface, the flask shape and the narrow necks are apparent but the small superficial pores are discernible only rarely. Neighbouring caveolae may be continuous forming multicaveolated spaces (arrowheads). The cytoplasm beneath "caveolated" areas is free of myofibrils, and both perpendicular and cross sections show profiles of smooth membranes closely apposed to the membranes of caveolae (asterisks). These observations raise questions as to whether connections may exist to one of the tasks of caveolae in Ca^{2+} signalling and regulation of Ca^{2+} homeostasis. Caveolae in smooth muscle cells have been proposed to represent the counterpart of the transverse tubules in skeletal muscle cells (cf. Fig. 150).

Multiple other functions have been ascribed to caveolae, although differences may exist between different kinds of cells. Caveolae are important in relation with several types of signalling. They are involved in cholesterol transport and homeostasis, transendothelial traffic, bacterial entry, and development of the transverse tubule system in muscle cells. Caveolae are connected with endocytosis pathways and responsible for potocytosis, a specialised triggered uptake of small molecules across the caveolar membrane, and they have an important role in virus internalisation leading to the formation of special endosomes, the caveosomes. However, the opinions are divided with respect to several properties and functions of caveolae, in particular concerning their mobility, budding and internalisation. Although they are rather seen as immobile, stabilised platforms within the plasma membrane, caveolae can become internalised in special situations and subsequent to special stimuli. Distinct caveolae-mediated pathways have been characterised, which either involve the Golgi apparatus and from there, lead to one of the retrograde routes to the endoplasmic reticulum or, bypassing the Golgi apparatus, directly target the endoplasmic reticulum.

Panels B and C show cholera toxin-coated gold particles associated with the surface entrances into caveolae (arrows in panel B) and located within the small caves (arrows in panel C). The B subunit of cholera toxin binds to a glycolipid receptor, the G_{M1} ganglioside, present in caveolae. Since cholera toxin-gold complexes are concentrated within the caveolar invaginations, they have been used as a marker for caveolae. However, caveolae are not the sole structures for uptake of cholera toxin. It has been shown that internalisation of cholera toxin is dependent on lipid rafts and the receptor, G_{M1}, is also present in clathrin-coated pits.

References

Isshiki M and Anderson RGW (2003) Function of caveolae in Ca^{2+} entry and Ca^{2+}-dependent signal transduction. Traffic 4: 717

Mineo C and Anderson RGW (2001) Potocytosis. Histochem Cell Biol 116: 109

Montesano R, Roth J, Robert A, and Orci L (1982) Non-coated membrane invaginations are involved in binding and internalisation of cholera and tetanus toxins. Nature 296: 651

Nabi IR and Le PU(2003) Caveolae/raft-dependent endocytosis. J Cell Biol 161: 673

Nichols B (2003) Caveosomes and endocytosis of lipid rafts. J Cell Sci 116: 4707

Parton RG and Richards AA (2003) Lipid rafts and caveolae as portals for endocytosis: New insights and common mechanisms. Traffic 4: 724

Pelkmans L and Helenius A (2002) Endocytosis via caveolae. Traffic 3: 311

Van Deurs B, Roepstorff K, Hommelgaard AM, and Sandvig K (2003) Caveolae: anchored, multifunctional platforms in the lipid ocean. Trends Cell Biol 13: 92

Magnification: x 55,000 (A); x 96,000 (B, C)

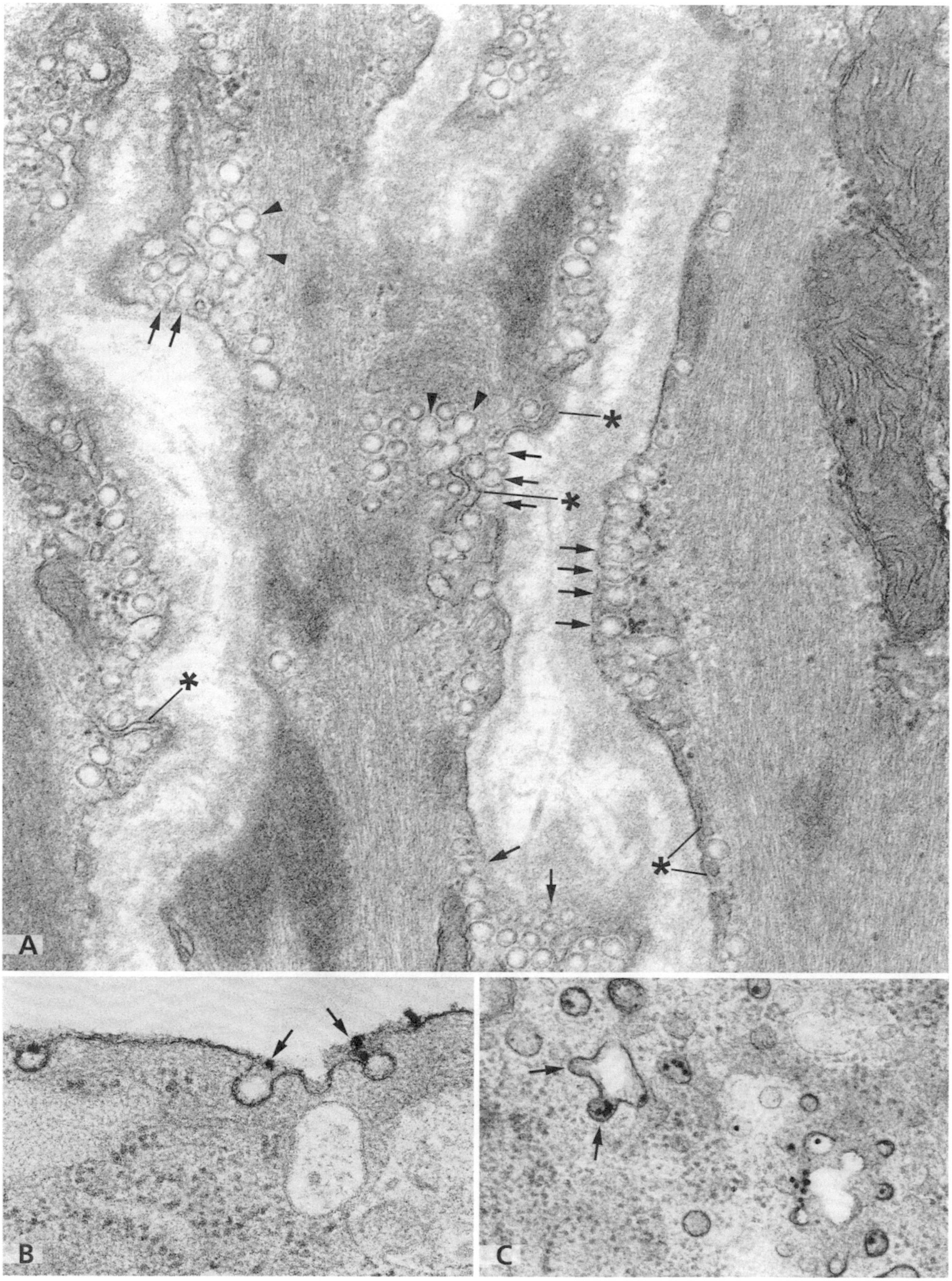
A
B
C

FLUID-PHASE ENDOCYTOSIS AND PHAGOCYTOSIS

Cells continuously take up fluid from their neighbourhood by a process designated as fluid-phase endocytosis. Unlike to other types of uptake, this kind of endocytosis is not preceded by a specific binding to the plasma membrane or concentration of molecules at special sites. It is considered that cells use this continuous unspecific uptake of fluid for a kind of surveillance of the surroundings or for the purpose of "cleaning" the plasma membrane in the acidic interior milieu of endosomes on the passage through endocytic compartments. Different ports have been proposed for unspecific fluid-phase uptake and different types of vesicles suggested to be involved, including clathrin-coated vesicles, macropinosomes, and other, poorly characterised, clathrin-independent vesicles. Small amounts of fluid from the surroundings of the cells are internalised together with the cargo concentrated in clathrin-coated pits (cf. Fig. 46) and together with substances internalised via caveolae. By macropinocytosis, large volumes of fluid are internalised in some cells, such as in dendritic cells that use this mechanism for immune surveillance. Horseradish peroxidase (HRP) has often been used for visualisation of fluid-phase endocytosis. Panel A shows intravenously applied HRP in the rat small intestinal mucosa, visualised by the electron dense reaction products after cytochemical oxidation of diaminobenzidine. HRP is distributed within all extracellular spaces surrounding smooth muscle cells and the endothelium of a lymph capillary (open and black arrowhead, respectively) and is transported across the endothelium by vesicular transport carriers (arrows). At places, the vesicles seem to fuse forming channel-like compartments. The apical surface of the endothelial cells is lined by multiple HRP-filled pits (inset) and abundant HRP-containing caveolae (arrows) occupy the surface of smooth muscle cells. Lipoprotein particles (asterisk) are contained within the lumen of the lymph capillary, appearing negatively stained by the dense HRP-reactions.

By phagocytosis, large particles, such as senescent and dead cells, apoptotic bodies, bacteria, inhaled carbon particles and other non-biological foreign materials are eliminated from the body. The process involves specialised motile cells, microphages (cf. also Figs. 170 and 171) and macrophages (cf. also Figs. 136-139). Uptake of a particle via phagocytosis is initiated by attachment to the plasma membrane and activation of receptors. A signal transduction cascade triggers rearrangement of the actin cytoskeleton that leads to an extension of the plasma membrane along the sides of the particle, which increasingly becomes internalised and finally is enclosed in a phagocytic vacuole. The phagocytic vacuole matures by multiple fusion and fission processes involving mainly endosomes, in specialised cases also the endoplasmic reticulum, and becomes prepared for fusion with lysosomes. Phagocytosis has a main role in the immune system, in the remodelling of tissue and programmed elimination of cells and matrix constituents. In panel B, a Kupffer cell present in the lumen of a rat liver sinus is shown after phagocytosis of a red blood cell (RBC). The engulfed erythrocyte is contained in a huge phagosome that occupies an extended part of the intracellular space of the macrophage. Gold particles show labelling with the galactose-binding *Ricinus communis I* lectin being particularly abundant in a lysosome (Ly).

References

Dejardins M and Griffiths G (2003) Phagocytosis: latex leads the way. Curr Opin Cell Biol 15: 498

Gagnon E, Duclos S, Rondeau C, Chevet E, Cameron PH, Steele-Mortimer O, Paiement J Bergeron JJM, and Desjardins M (2002) Endoplasmic reticulum-mediated phagocytosis is a mechanism of entry into macrophages. Cell 110: 119

Garin J, Diez R, Kieffer S, Dermine J-F, Duclos S, Gagnon E, Sadoul R, Rondeau C, and Desjardin M (2001) The phagosome proteom: Insight into phagosome functions. J Cell Biol 152: 165

Gillooly DJ, Simonsen A, and Stenmark H (2001) Phosphoinositides and phagocytosis. J Cell Biol 155: 15

Lefkir Y, Malbouyres M, Gotthardt D, Ozinsky A, Cornillon S, Bruckert F, Aderem AA, Soldati T, Cosson P, and Letourneur F (2004) Involvement of the AP-1 adaptor complex in early steps of phagocytosis and macropinocytosis. Mol Biol Cell 15: 861

Russell DG (2003) Phagosomes, fatty acids and tuberculosis. Nat Cell Biol 5: 776

Magnification: x 34,500 (A); x 47,000 (inset); x 6,400 (B)

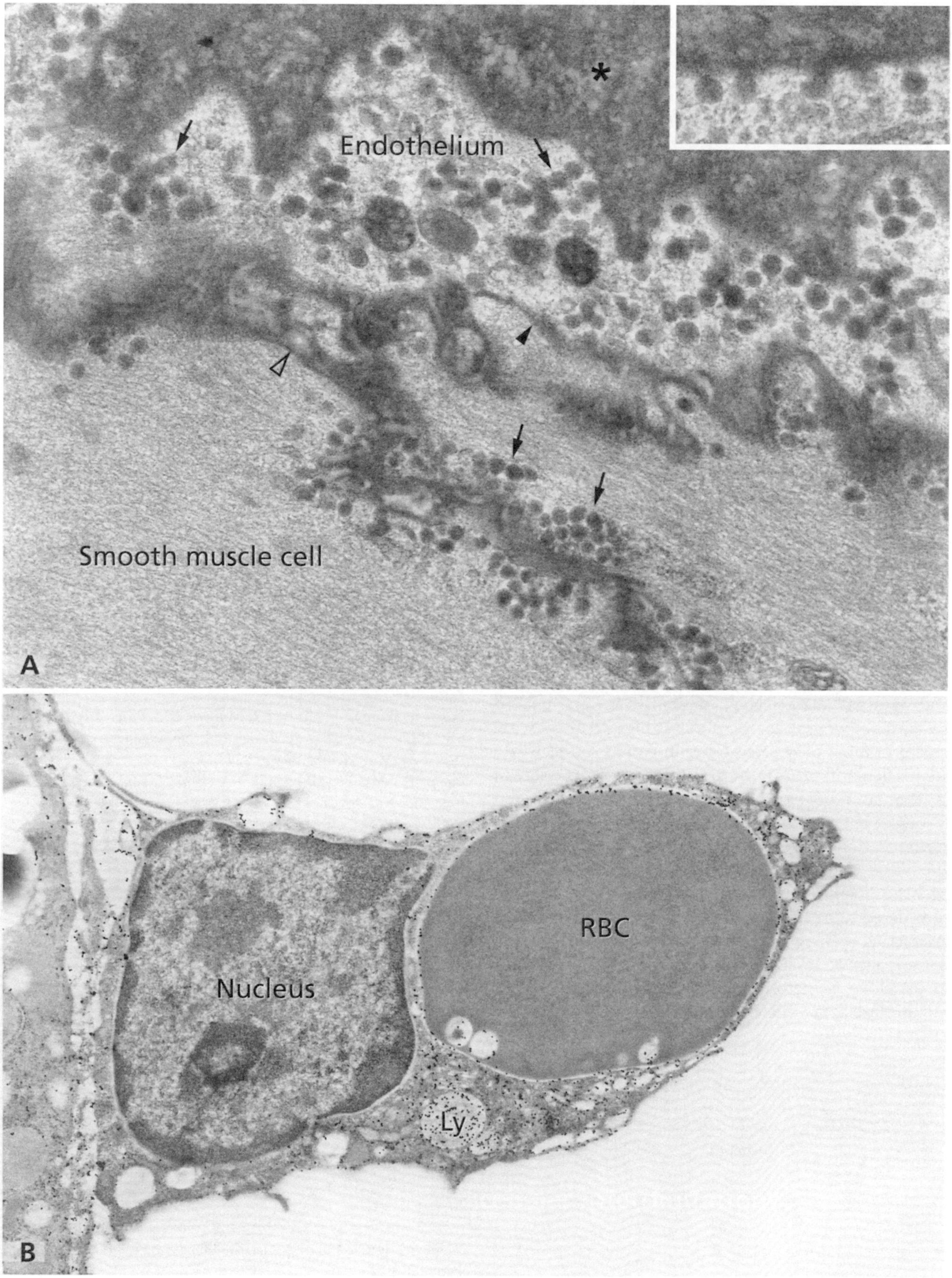
Endothelium
*
Smooth muscle cell
A
Nucleus
RBC
Ly
B

LYSOSOMES

Lysosomes are enriched in hydrolytic enzymes working at low pH that include proteases, lipases, glycosidases, nucleases, phosphatases, and sulfatases, and are responsible for elimination of "unwanted" molecules derived from both the cell itself (autophagy, cf. Fig. 62) or from outside of the cells (heterophagy). However, against the historic view of the lysosome as a terminal degradation compartment and as a unit used in the cells mainly as a disposal for garbage, it is increasingly becoming clear that lysosomes are dynamic organelles, which receive continuous input from three sides, via the biosynthetic, endocytic, and autophagic pathways. Today, lysosomes cannot be seen as terminal or dead-end compartments. They are able to fuse with late endosomes involving "kiss-and-run"-mechanisms and are also capable of fusing with the plasma membrane.

In panels A, B, and C, lysosomes (Ly) in the cytoplasm of rat small intestinal absorptive cells (panels A and B) and in a rat hepatocyte (panel C) are on display. In panel C, the liver cell lysosomes show a particularly distinct and intense contrast after a special uranyl acetate/methyl cellulose adsorption staining, to which the ultrathin sections of the tissue embedded in Lowicryl K4M were exposed. (cf. also Fig. 88). The extreme heterogeneity in the lysosomal sizes, shapes, and luminal materials is evident. Densely packed membrane "whirls", as they occur at high frequency, are visible in most of the lysosomes in panels A and B and are particularly prominent in the large lysosome shown in the inset of panel B. Lysosomes are distributed throughout the cytoplasm of the cells but show higher concentrations in the perinuclear area close to the Golgi apparatus (panels A and C). Lysosome morphologies and the favourite localisations in the Golgi area are considered to be regulated by Rab7 and Rab34 proteins interacting with a particular region of a Rab-interacting lysosomal protein (RILP). In the absorptive cell shown in panel A, numerous lipoprotein particles (LP) are accumulated in dilated Golgi cisternae and large

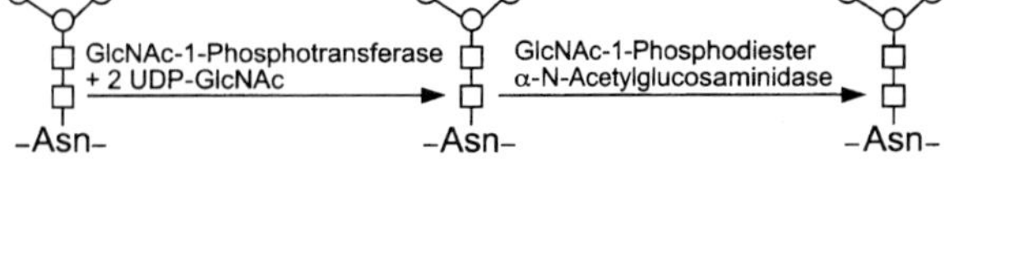

vesicular carrier compartments. The cell segments shown in the micrographs also contain autophagosomes (AV) and multiple mitochondria (M).

The lysosomal enzymes, synthesised and initially glycosylated within the rough endoplasmic reticulum, receive characteristic glycan modifications in the Golgi apparatus. By a two-step process that starts at the *cis*-Golgi side and continues in medial cisternae of the Golgi stacks, mannose-6-phosphate residues are added to the asparagine-linked oligosaccharides of newly synthesised lysosomal enzymes involving two sequentially acting enzymes, *N*-acetyl-glucosaminyl (GlcNAc)-phosphotransferase and GlcNAc-1-phosphodiester alpha-*N*-acetylglucosaminidase (diagram). Presence of the mannose-6-phosphate recognition marker provides the basis for effective sorting through the mannose-6-phosphate receptor of newly synthesised lysosomal enzymes into the lysosomal pathway (cf. Fig. 54).

References

Bucci C, Thomsen P, Nicoziani P, McCarthy J, and van Deurs B (2000) Rab7: A key to lysosome biogenesis. Mol Biol Cell 11: 467

Kornfeld S and Mellman I (1989) The biogenesis of lysosomes. Annu Rev Cell Biol 5: 483

Luzio JP, Poupon V, Lindsay R, Mullock BM, Piper RC, and Pryor PR (2003) Membrane dynamics and the biogenesis of lysosomes (Review). Mol Membr Biol 20: 141

Luzio JP, Pryor PR, and Bright NA (2007) Lysosomes: fusion and function. Nat Rev Mol Cell Biol 8: 622

Roth J, Taatjes DJ, and Tokuyasu KT (1990) Contrasting of Lowicryl K4M thin sections. Histochemistry 95: 123

Storrie B and Desjardins M (1996) The biogenesis of lysosomes: is it a kiss and run continuous fusion and fission process? Bioessays 18: 895

Van Meel E and Klumpermann J (2008) Imaging and imagination: understanding the endo-lysosomal system. Histochem Cell Biol 129: 253

Wang T and Hong W (2002) Interorganellar regulation of lysosome positioning by the Golgi apparatus through Rab34 interaction with Rab-interacting lysosomal protein. Mol Biol Cell 13: 4317

Wang T, Wong KK, and Hong W (2004) A unique region of RILP distinguishes it from its related proteins in its regulation of lysosomal morphology and interaction with Rab7 and Rab34. Mol Biol Cell 15: 815

Magnification: x 26,500 (A); x 26,000 (B); x 17,000 (inset); x 43,000 (C)

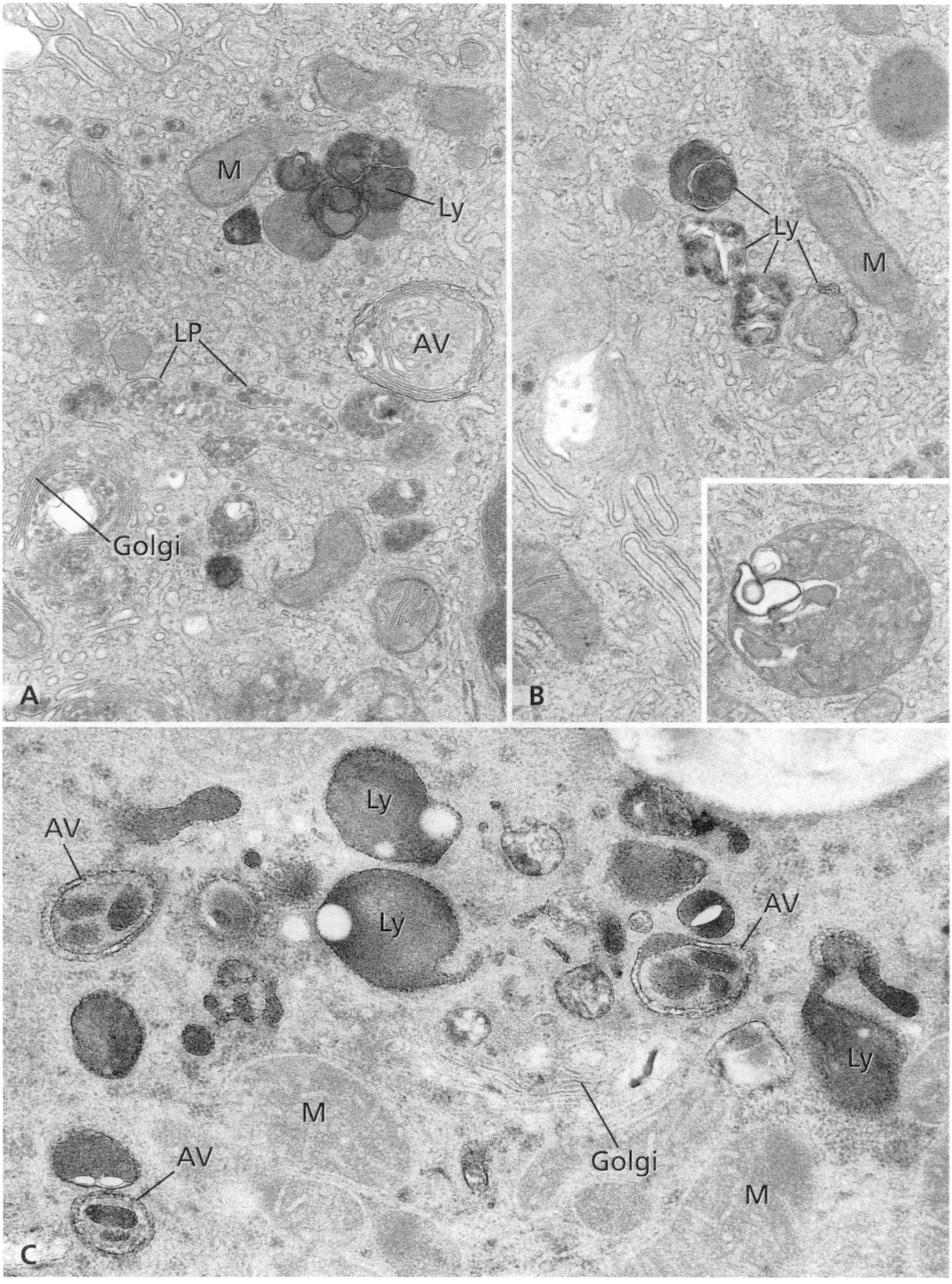
M
Ly
LP
AV
Golgi
A
Ly
M
B
Ly
AV
Ly
AV
Ly
M
Golgi
AV
M
C

LYSOSOMES: LOCALISATION OF ACID PHOSPHATASE, LAMP AND POLYLACTOSAMINE

Lysosomal enzymes and membrane proteins can be located by cytochemical and immunogold methods. Panel A shows enzyme cytochemical staining for acid phosphatase in an acinar cell of the rat pancreas. Electron dense reaction product for acid phosphatase is visible within cisternae at the *trans* Golgi side, coated vesicles budding from those sites (arrow), within condensing vacuoles (CV), and lysosomes (Ly), but not within the mature secretory granules. The enzyme localisations at the *trans* Golgi side reflect the well-studied routes of lysosomal enzymes from the Golgi apparatus to endosomes and lysosomes, although they also are seen in connection with activities of endoproteases or involved in the breakdown of cytidine monophosphate during terminal glycosylation (cf. Fig. 28).

Lysosomal enzymes are sorted from the Golgi apparatus and *trans* Golgi network to the lysosomal pathway by specific binding of the mannose-6-phosphate groups present on the lysosomal enzyme precursors (cf. Fig. 53) by receptors in a divalent cation-dependent or independent manner. The respective receptors, the cation-dependent and independent mannose-6-phosphate receptors (CD-MPR and CI-MPR) are localised in the membranes of the *trans* Golgi network (TGN). By interactions of acidic cluster-dileucine signals present in their cytosolic domains with Golgi-localised, gamma ear-containing ARF-binding proteins (GGAs) and cooperation of GGAs with adapter protein-1 (AP-1), the lysosomal enzyme-MPR complexes are packaged into clathrin-coated carriers that bud from the TGN and function in transporting the lysosomal enzymes to early and late endosomes. The enzymes dissociate from the receptors at the acidic pH of the endosomes and together with the fluid phase are transferred to lysosomes. The receptors are retrieved to the TGN by mechanisms involving AP-1 and a complex of proteins called "retromer".

Panels B and C show constituents of the lysosomal membrane in human cells. The lysosomal membrane, which contains proton (H^+) pumps for transport of H ions into the lumen and proteins that transport the final products of digestion into the cytoplasm, has to resist hydrolysis by its own lytic enzymes. The structural membrane proteins, the lysosome-associated membrane proteins (lamps), lysosomal integral membrane proteins (limps), and lysosomal membrane glycoproteins (lgps) are highly glycosylated at the luminal surfaces. Human lamp-1 and lamp-2 have been characterised as major sialoglycoproteins carrying polylactosamine chains. Panel B shows immunogold labelling for lamp at the membranes of lysosomes (Ly) in HeLa cells. In panel C, polylactosamine is located in lysosomes of HeLa cells by means of gold-labelled *Datura stramonium* lectin.

In some cells, the lysosomal compartment is modified and has also a role as a secretory compartment. Although there are some exceptions, such as the melanocytes, most cells containing secretory lysosomes are derived from the haematopoietic lineage and include granulocytes (cf. Figs. 170, 171), mast cells (cf. Fig. 139), macrophages (cf. Figs. 137, 139) and dendritic cells, B and T lymphocytes (cf. Fig. 174), platelets (cf. Figs. 175, 176), and osteoclasts (cf. Fig. 148).

References

Arighi CN, Hartnell LM, Aguilar RC, Haft CR, and Bonifacino JS (2004) Role of the mammalian retromer in sorting of the cation-independent mannose 6-phosphate receptor. J Cell Biol 165: 123

Blott EJ and Griffiths GM (2002) Secretory lysosomes. Nat Rev Mol Cell Biol 3: 122

Carlsson SR, Roth J, Piller F, and Fukuda M (1988) Isolation and characterisation of human lysosomal membrane glycoproteins, h-lamp-1 and h-lamp-2. Major sialoglycoproteins carrying polylactosaminoglycan. J Biol Chem 263: 18911

Doray B, Ghosh P, Griffith J, Geuze HJ, and Kornfeld S (2002) Cooperation of GGAs and AP-1 in packaging MPRs at the *trans* Golgi network. Science 297: 1700

Eskelinen E-L, Tanaka Y, and Saftig P (2003) At the acidic edge: emerging functions for lysosomal membrane proteins. Trends Cell Biol 13: 137

Geuze HJ, Slot JW, Strous GJAM, Hasilik A, and von Figura K (1985) Possible pathways for lysosomal enzyme delivery. J Cell Biol 101: 2253

Karlsson K and Carlsson SR (1998) Sorting of lysosomel membrane glycoproteins lamp-1 and lamp-2 into vesicles distinct from mannose 6-phosphate receptor/gamma-adaptin vesicles at the *trans* Golgi network. J Biol Chem 273: 18966

Saftig P. and Klumperman J (2009) Lysosome biogenesis and lysosomal membrane proteins: trafficking meets function. Nat Rev Mol Cell Biol 10: 623

Seaman MNJ (2004) Cargo-selective endosomal sorting for retrieval to the Golgi requires retromer. J Cell Biol 165: 111

Magnification: x 21,000 (A); x 42,000 (B); x 55,000 (C)

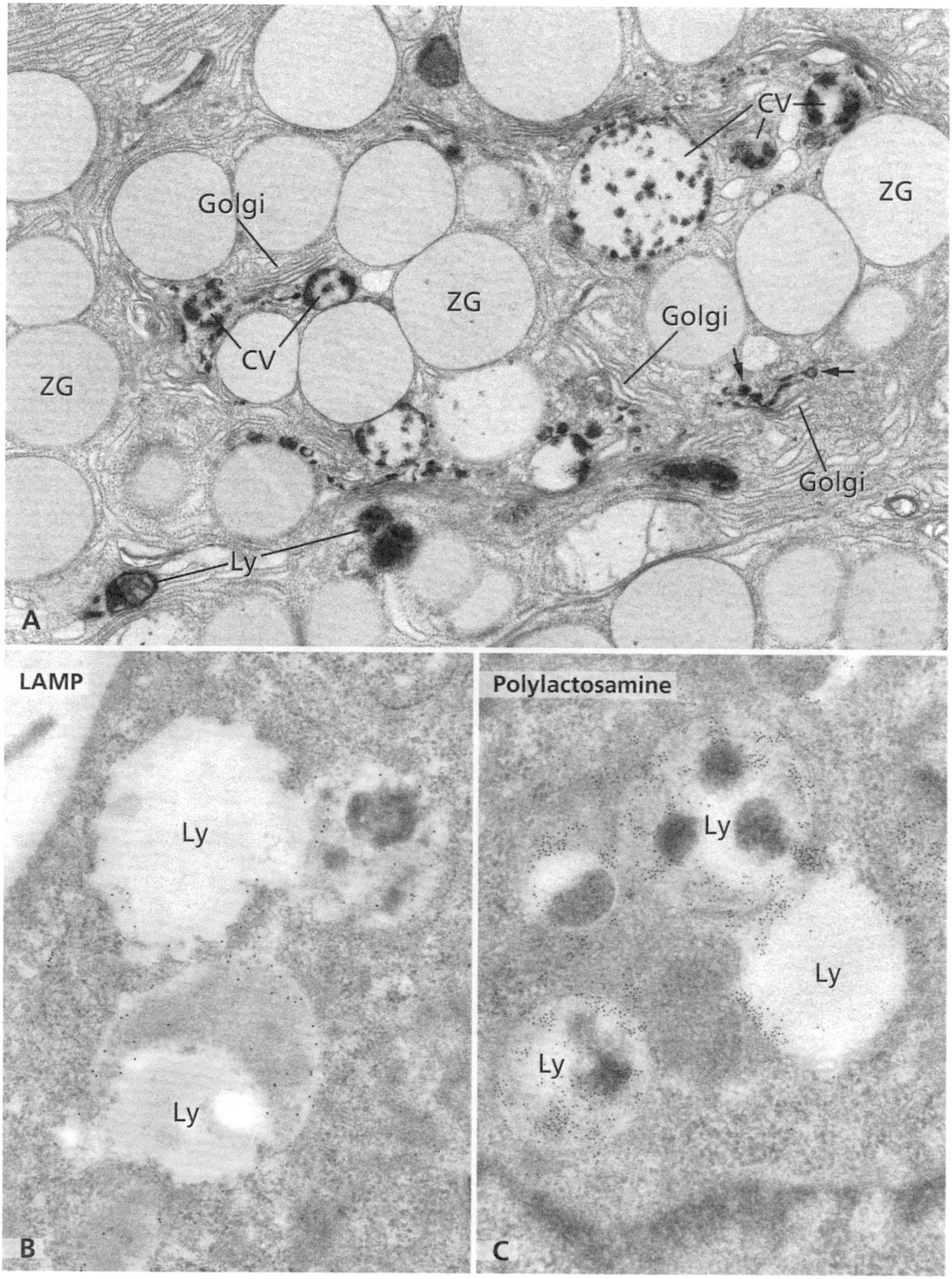
CV
Golgi
ZG
ZG
CV
ZG
Golgi
Golgi
Ly
A
LAMP
Ly
Ly
B
Polylactosamine
Ly
Ly
Ly
C

I-CELL DISEASE

The I-cell disease (mucolipidosis II) is a lysosomal storage disease, inherited in an autosomal recessive manner. The defect lies in the biosynthesis of the mannose 6-phosphate recognition marker for the targeting of lysosomal enzymes into lysosomes. The effect is therefore most far reaching since it affects all lysosomal enzymes. Specifically, the *N*-acetylglucosaminyl-1-phosphotransferase, which catalyses the first step in the synthesis of mannose 6-phosphate (cf. Fig. 53), is defective. The lack of the recognition marker on *de novo* synthesised lysosomal enzymes results in their secretion into the extracellular space. The detection of elevated serum concentrations of multiple lysosomal enzymes is therefore diagnostic. Clinically, I-cell disease is characterised by early onset and progressive severe psychomotoric retardation accompanied by skeletal abnormalities and coarse facial shape. Surprisingly, not all body cells are devoid of lysosomal enzymes, although they are all lacking phosphotransferase. This indicates that certain cells such as hepatocytes, Kupffer cells, and leukocytes are able to endocytose lysosomal enzymes through an unknown alternate, mannose 6-phosphate-independent mechanism.

Morphological investigations have revealed the presence of numerous membrane-limited vacuoles, which have given the disease its name: Inclusion cell disease. In panel A numerous such inclusions are prominently seen in the cytoplasm of dermal fibroblasts present in a skin biopsy. The inset shows the characteristic appearance of the inclusion bodies, which are part empty and part filled with fine granular material. Occasionally, larger electron dense deposits exist as seen in the inset. Panel B is a field from the same skin biopsy showing affected Schwann cells and fibroblasts (arrows). Myelinated nerve fibres (NF) are not affected. The unique ultrastructural features exhibited by the inclusions is highly indicative of I-cell disease.

Phosphotransferase is a complex molecule and composed of three subunits ($\alpha_2\beta_2\gamma_2$). The gene encoding the α and β subunits has been mapped to chromosome 12p and that for the γ subunit to chromosome 16p. The causative genetic defects are not well known except for the lack of transcripts of the α and β subunits, probably due to point mutations or small deletions.

The treatment is symptomatic. Prenatal diagnosis is possible through the detection of elevated lysosomal enzyme activity in amniotic fluid accompanied by decreased lysosomal enzyme activity in cultured amniotic cells.

References

Carey WF, Jaunzems A, Richardson M, Fong BA, Chin SJ, and Nelson PV (1999) Prenatal diagnosis of mucolipidosis II – electron microscopy and biochemical evaluation. Prenat Diagn 19: 252

Endo H, Miyazaki T, Asano S, and Sagami S (1987) Ultrastructural studies of the skin and cultured fibroblasts in I-cell disease. J Cutan Pathol 14: 309

Glickman JN and Kornfeld S (1993) Mannose 6-phosphate-independent targeting of lysosomal enzymes in I-cell disease B lymphoblasts. J Cell Biol 123: 99

Hasilik A, Waheed A, and von Figura K (1981) Enzymatic phosphorylation of lysosomal enzymes in the presence of UDP-*N*-acetylglucosamine. Absence of the activity in I-cell fibroblasts. Biochem Biophys Res Commun 98: 761

Honey NK, Mueller OT, Little LE, Miller AL, and Shows TB (1982) Mucolipidosis III is genetically heterogeneous. Proc Natl Acad Sci USA 79: 7420

Kornfeld S and Sly WS (2001) I-cell disease and pseudo-Hurler polydystrophy: disorders of lysosomal enzyme phosphorylation and localisation. In: The metabolic and molecular bases of inherited disease (Scriver C, Beaudet A, Valle D, and Sly WS, eds). New York: McGraw-Hill, pp 3469

Reitman M, Varki A, and Kornfeld S (1981) Fibroblasts from patients with I-cell disease and pseudo-Hurler polydystrophy are deficient in uridine 5'-diphosphate-*N*-acetylglucosamine: glycoprotein *N*-acetylglucosaminylphosphotransferase activity. J Clin Invest 67: 1574

Vladutiu GD and Rattazzi MC (1979) Excretion-reuptake route of beta-hexosaminidase in normal and I-cell disease cultured fibroblasts. J Clin Invest 63: 595

Magnification: x 7,500 (A); x 21,000 (inset); x 7,000 (B)

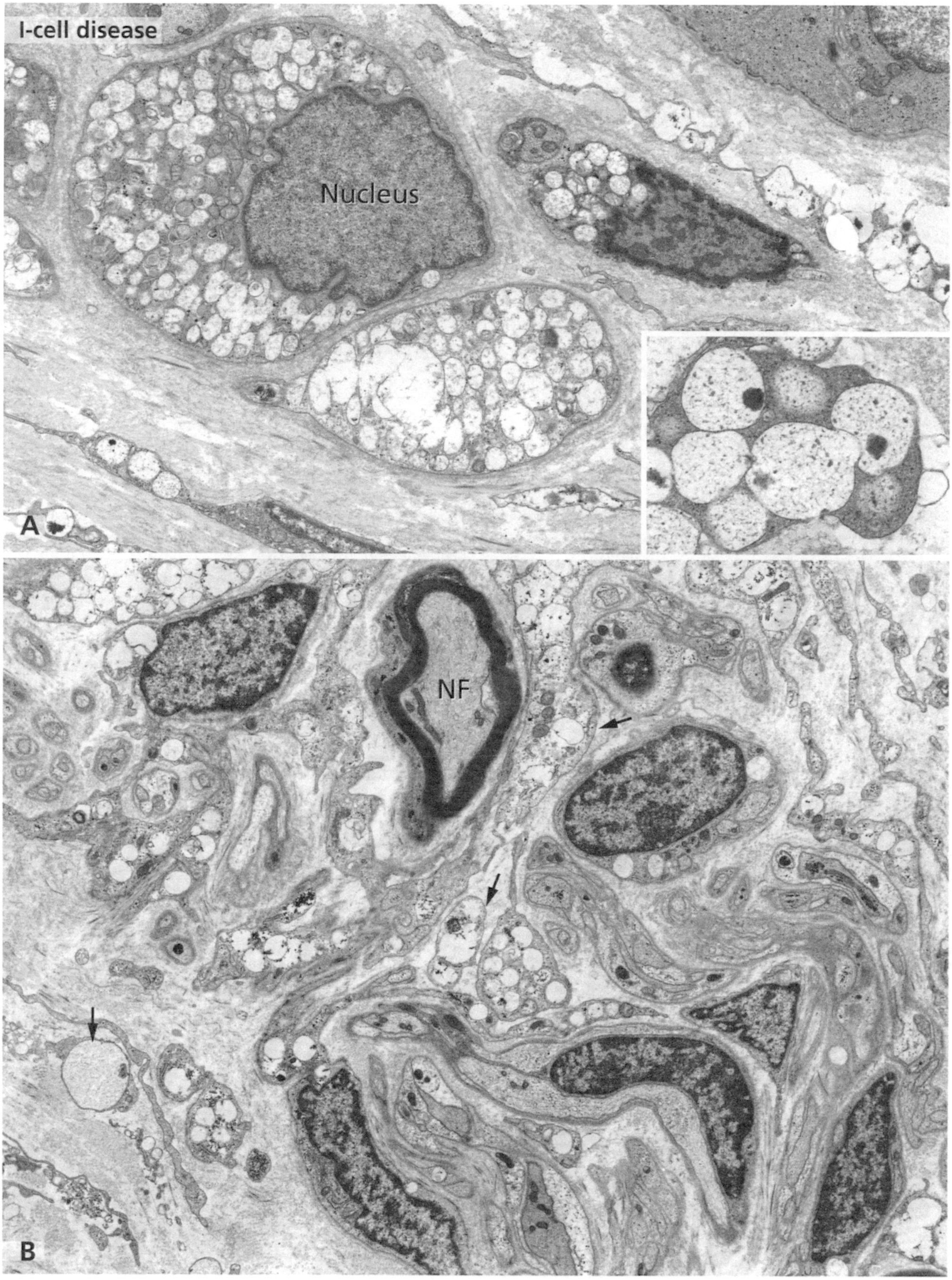
I-cell disease
Nucleus
A
NF
B

GAUCHER'S DISEASE

Gaucher's disease is one of the most common lysosomal storage diseases and comprises three different types. All of them are due to defective hydrolysis of glucosylceramide (glucocerebroside), which results in lysosomal accumulation of this glycolipid mainly in phagocytic cells. The three disease types differ in the clinical course and organ manifestations of the storage.

In type 1 Gaucher's disease, engorged macrophages cause enlargement of liver and spleen and sometimes functional impairment. The bone marrow can be filled with storage cells (macrophages), which affects the normal haematopoiesis. Thus, bleeding due to thrombocytopenia is commonly observed. Bone lesions are often detected as well.

The ultrastructural features observed in cells affected in type 1 Gaucher's disease are of such a characteristic appearance that they permit the diagnosis. The lysosomal accumulation occurs in phagocytic cells, especially in the bone marrow, liver, and spleen. Gaucher's cells become often greatly enlarged (up to 200 μm in diameter) because of the presence of large numbers of elongated and distended lysosomes. Panel A shows an affected Kupffer cell in a liver biopsy of a Gaucher's patient. The pathogonomic appearance of the cytoplasm has been described as wrinkled tissue paper. At higher magnification (panel B), the stored substrate in the lysosomes consists of typical sheaves of twisted tubulus-like structures.

Gaucher's disease is X-linked autosomal recessively inherited and caused by mutations in the gene coding for lysosomal acid β-glucosidase (glucocerebrosidase), which was mapped to chromosome 1q21. Gaucher's disease is most common in the Ashkenazi Jewish population. The disease causing mutations are mainly missense mutations, and the N370S mutation is the most common one. The mutations result in the synthesis of an enzyme with decreased catalytic activity and stability.

As already mentioned, three types of Gaucher's disease exist: type 1 without, type 2 with acute, and type 3 with subacute primary involvement of the central nervous system. Characteristically, the disease is quite heterogeneous genotypically and phenotypically, and type 1 is the most common one.

Type 1 Gaucher's disease represents a classical example of enzyme replacement therapy. The efficacy of the treatment was improved with the use of enzyme genetically tailored to carry oligosaccharides recognised by the cell surface mannose receptor present on macrophages. This resulted in preferential and efficient internalisation of the enzyme by macrophages.

References

Barneveld RA, Keijzer W, Tegelaers FP, Ginns EI, Geurts van Kessel A, Brady RO, Barranger JA, Tager JM, Galjaard H, Westerveld A, and Reuser AJ (1983) Assignment of the gene coding for human beta-glucocerebrosidase to the region q21-q31 of chromosome 1 using monoclonal antibodies. Hum Genet 64: 227

Beutler E and Grabowski G (2001) Gaucher's disease. In: The metabolic and molecular bases of inherited diseases (Scriver C, Beaudet A, Valle D, and Sly WS, eds). New York: McGraw-Hill, pp 3635

Brady R, Kanfer J, and Shapiro D (1965) Metabolism of glucocerebrosides II. Evidence of an enzymatic deficiency in Gaucher's disease. Biochem Biophys Res Commun 18: 221

Desnick RJ and Schuchman EH (2002) Enzyme replacement and enhancement therapies: Lessons from lysosomal disorders. Nat Rev Genet 3: 954

Devine EA, Smith M, Arredondo-Vega FX, Shafit-Zagardo B, and Desnick RJ (1982) Regional assignment of the structural gene for human acid beta-glucosidase to q42 leads to qter on chromosome 1. Cytogenet Cell Genet 33: 340

Dickersin G (2000) Diagnostic electron microscopy. A text/atlas. New York: Springer

Ginns EI, Choudary PV, Tsuji S, Martin B, Stubblefield B, Sawyer J, Hozier J, and Barranger JA (1985) Gene mapping and leader polypeptide sequence of human glucocerebrosidase: implications for Gaucher's disease. Proc Natl Acad Sci USA 82: 7101

Jordan S (1964) Electron microscopy of Gaucher cells. Exp Mol Pathol 3: 76

Magnification: x 7,000 (A); x 29,000 (B)

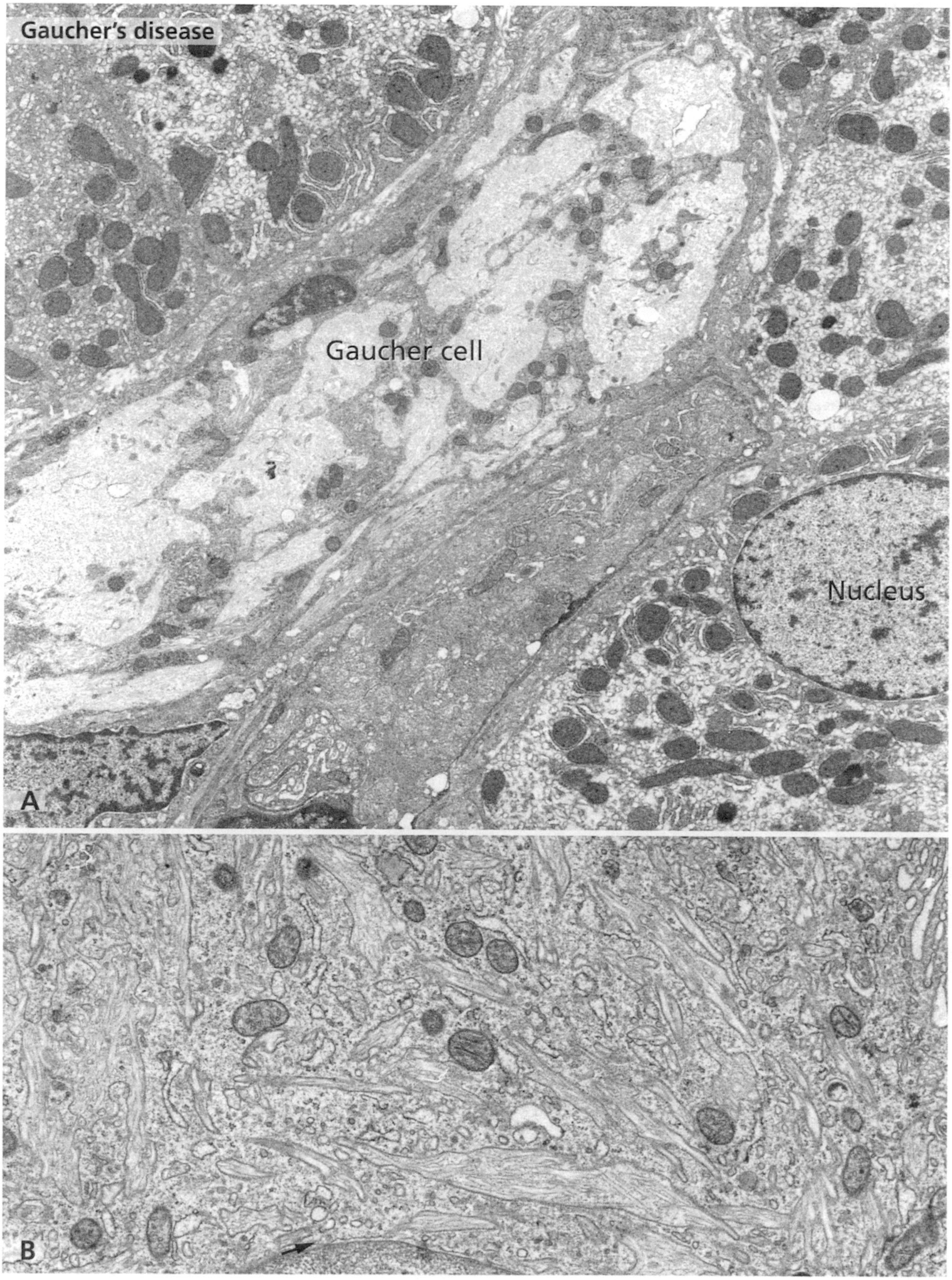
Gaucher's disease
Gaucher cell
Nucleus
A
B

FABRY'S DISEASE

Fabry's disease is an X-linked recessively inherited lysosomal storage disease, caused by a deficiency in catalytic activity or absence of lysosomal α-galactosidase A. This results in the lysosomal accumulation of glycosphingolipids with terminal α-galactosyl residues such as globotriaosylceramide and to a lesser extent of galabiosylceramide and blood group B substance. Clinically, classically affected hemizygotes with undetectable enzyme activity and early disease onset has to be distinguished from atypical hemizygotes with varying levels of residual enzyme activity and either no clinical symptoms or late onset of disease restricted to the heart.

In Fabry's disease, the pattern of glycospingolipid deposition is highly characteristic and classically affects the endothelium and smooth muscle fibres of vasculature as shown in panel A. The endothelia and smooth muscle fibres contain large amounts of crystalline, birefringent lysosomal depositions (asterisks in A), which cause cellular enlargement. Ultrastructurally, the inclusions consist of concentric or parallel lamellar arrangements as shown in panel B. At high magnification alternating electron dense and light bands are obvious, as illustrated in panel C. The same ultrastructural features can be observed in neurons of the central and peripheral nervous system, podocytes and tubular epithelia of kidney, epithelial cells of cornea and heart muscle. In skin, Fabry's disease manifests as angiokeratoma consisting of markedly dilated capillaries of the dermal papillae.

The gene coding for α-galactosidase A has been mapped to chromosome Xq22.1. It has been difficult to establish a genotype/phenotype correlation since neither the type nor the location of a mutation has been of clear-cut predictive value.

The accumulated glycosphingolipids can be detected by the *Bandeiraea simplicifolia* lectin reactive with terminal α-galactosyl residues or a monoclonal anti-globotriasylceramide antibody. The main source for the stored glycosphingolipids is endocytosed globotriasylceramide (secreted by liver hepatocytes) associated with low density lipoprotein and high density lipoprotein and globoside of red blood cells, a precursor for globotriaosylceramide.

The functional consequences of the deposits manifest in characteristic clinical signs. The changes of endothelia and smooth muscle fibres of vasculature in kidney, heart, and brain cause ischemia and may result in thrombosis, conditions known to generate infarcts in these organs. Vascular and tubular lesions together result in end stage renal disease and systemic hypertension. Depositions in nerve cells may be the reason for paresthesias, pain, and hypohydrosis and other general neurological symptoms. In the heart, left ventricular hypertrophy and mitral valve insufficiency are observed, in addition to chronic ischaemic disease and infarcts, since these are the sites of preferential glycosphingolipid deposits. In atypical hemizygotes, cardiac hypertrophy resulting in cardiomyopathy may be the sole clinical manifestation.

References

Cable WJ, Dvorak AM, Osage JE, and Kolodny EH (1982) Fabry's disease: significance of ultrastructural localisation of lipid inclusions in dermal nerves. Neurology 32: 347

Desnick R, Ioannou Y, and Eng C (2001) α-galactosidase A deficiency: Fabry's disease. In: The metabolic and molecular bases of inherited disease (Scriver C, Beaudet A, Sly W, and Valle D, eds). New York: McGraw-Hill, pp 3733

Faraggiana T, Churg J, Grishman E, Strauss L, Prado A, Bishop DF, Schuchman E, and Desnick RJ (1981) Light- and electron-microscopic histochemistry of Fabry's disease. Am J Pathol 103: 247

Roth J, Schulze E, Raabe G, and Waldmann G (1974) Analytische Studie des Morbus Fabry. Virchows Arch A Path Anat 363: 287

Schatzki PF, Kipreos B, and Payne J (1979) Fabry's disease. Primary diagnosis by electron microscopy. Am J Surg Pathol 3: 211

Uchino M, Uyama E, Kawano H, Hokamaki J, Kugiyama K, Murakami Y, Yasue H, and Ando M (1995) A histochemical and electron microscopic study of skeletal and cardiac muscle from a Fabry's disease patient and carrier. Acta Neuropathol (Berlin) 90: 334

Magnification: x 9,800(A); x 25,000 (B); x 220,000 (C)

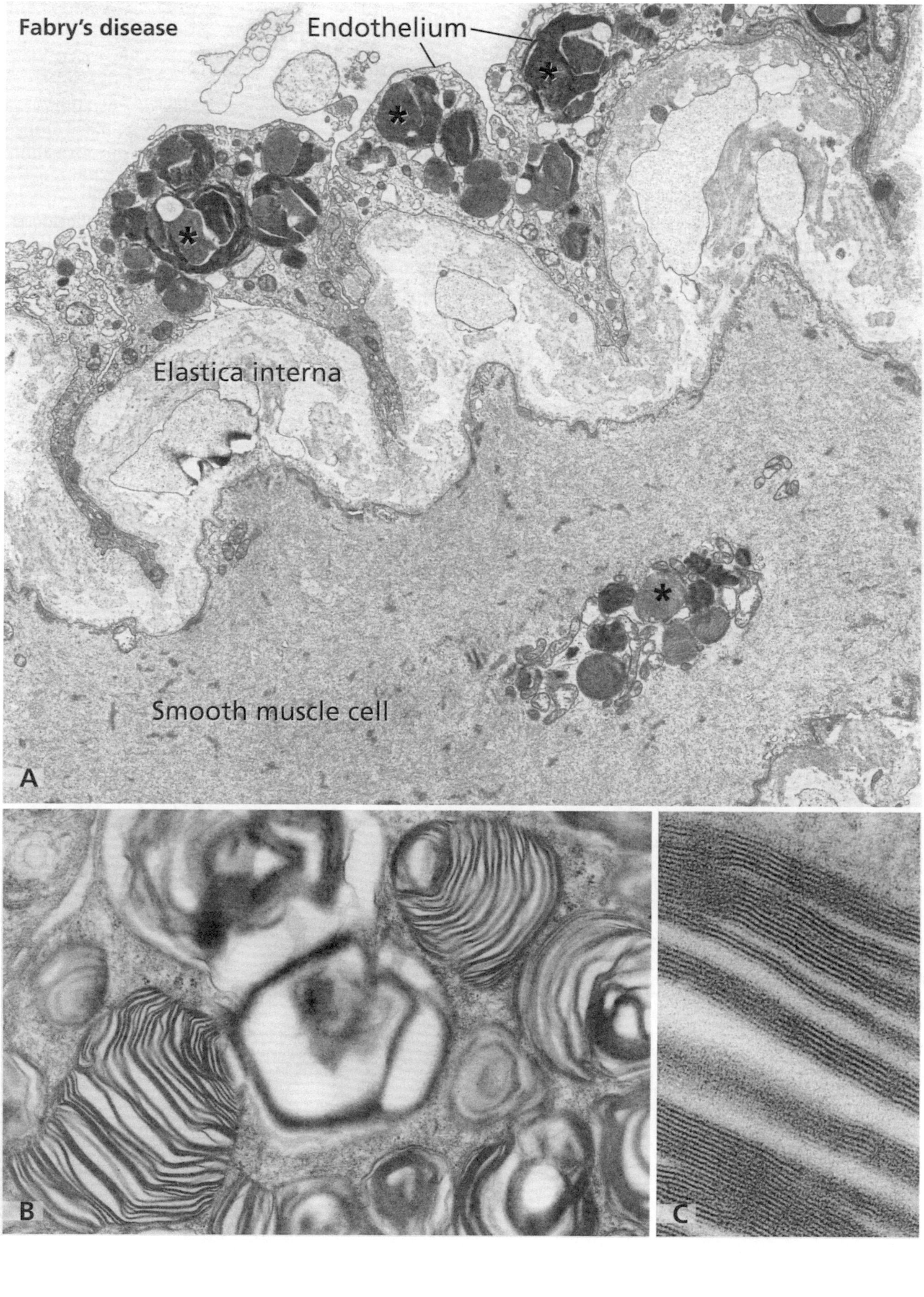
Fabry's disease
Endothelium
*
*
*
*
Elastica interna
Smooth muscle cell
A
B
C

G_{M2} GANGLIOSIDOSES

The G_{M2} gangliosidoses represent a heterogeneous group of lysosomal storage diseases characterised by the deposition of G_{M2} ganglioside and related glycolipids. They are inherited in an autosomal recessive manner. The basis for the various forms of G_{M2} gangliosidoses lies in the multifaceted catabolism of G_{M2} ganglioside, which requires complex formation with the G_{M2} activator before hydrolysis by β-hexosaminidase A and B can occur. Hence, mutations in any of the genes coding for one of these three proteins can result in disease.

In Sandhoff's disease, accumulation of glycosphingolipids occurs not only in neurons but also in visceral organs such as the liver, spleen, lymph nodes, lung, and kidneys. In the liver, storage can be observed in Kupffer cells and in hepatocytes (panels A and B), which occurs in the form of clustered and confluent membranous cytoplasmic bodies (arrows in panel B). However, storage bodies composed of irregular lamellar structures are often found in hepatocytes.

Tay-Sachs disease and its variants are caused by mutations of the *HEXA* gene mapped to chromosome 15, which encodes the α subunit of hexosaminidase A. Sandhoff disease and variants are caused by mutations of the *HEXB* gene mapped to chromosome 5, which encodes the β subunit common to hexosaminidase A and B. G_{M2} activator deficiency is caused by mutations of the *GM2A* gene mapped to chromosome 5.

As complex the genetic causes of the G_{M2} gangliosidoses are as complex are the clinical phenotypes. For all three diseases, infantile and adult as well as late onset forms are known with acute, subacute, or chronic course. All have in common the involvement of the nervous system with storage occurring in all neurons. Characteristically, the perikarya are enlarged and filled with storage bodies. Ultrastructurally they consist of concentrically arranged multilamellar membranous bodies. Their fine structural appearance is therefore different from those observed in Fabry's (cf. Fig. 57), I-cell (cf. Fig. 55) and Farber's disease (cf. Fig. 59) as well as in metachromatic leukodystrophy (cf. Fig. 167) and neuronal ceroid lipofuscinosis (cf. Fig. 168). The presence of membranous cytoplasmic bodies in perikarya causes malfunctioning of neuronal circuits, resulting in dystonia, spinocerebellar degeneration, and motor neuron disease. Abnormal wiring of the neuronal circuits seems to be causative because of the formation of aberrant synaptic contacts. The exact pathogenetic mechanism is unclear however.

Different types of treatment of G_{M2} gangliosidoses have been evaluated such as enzyme replacement therapy, bone marrow transplantation and substrate deprivation. Prenatal diagnosis is possible since activities for hexosaminidase A and B can be readily estimated, for example, in amniotic fluid.

SER: smooth endoplasmic reticulum; M: mitochondria; arrowheads: glycogen particles.

References

Akerman BR, Natowicz MR, Kaback MM, Loyer M, Campeau E, and Gravel RA (1997) Novel mutations and DNA-based screening in non-Jewish carriers of Tay-Sachs disease. Am J Hum Genet 60: 1099

Gravel R, Kaback M, Proia R, Sandhoff K, Suzuki K, and Suzuki K (2001) The G_{M2} gangliosidoses. In: The meatbolic and molecular bases of inherited disease (Scriver C, Beaudet A, Valle D, and Sly WS, eds). New York: McGraw-Hill, p 3827

Myerowitz R and Costigan FC (1988) The major defect in Ashkenazi Jews with Tay-Sachs disease is an insertion in the gene for the alpha-chain of beta-hexosaminidase. J Biol Chem 263: 18587

Purpura DP and Suzuki K (1976) Distortion of neuronal geometry and formation of aberrant synapses in neuronal storage disease. Brain Res 116: 1

Sandhoff K, Andreae U, and Jatzkewitz H (1968) Deficient hexosaminidase activity in an exceptional case of Tay-Sachs disease with additional storage of kidney globoside in visceral organs. Pathol Eur 3: 278

Schroder M, Schnabel D, Hurwitz R, Young E, Suzuki K, and Sandhoff K (1993) Molecular genetics of GM2-gangliosidosis AB variant: a novel mutation and expression in BHK cells. Hum Genet 92: 437

Schroder M, Schnabel D, Suzuki K, and Sandhoff K (1991) A mutation in the gene of a glycolipid-binding protein (GM2 activator) that causes GM2-gangliosidosis variant AB. FEBS Lett 290: 1

Suzuki K (1991) Neuropathology of late onset gangliosidoses. A review. Dev Neurosci 13: 205

Magnification: x 11,000 (A); x 36,000 (B)

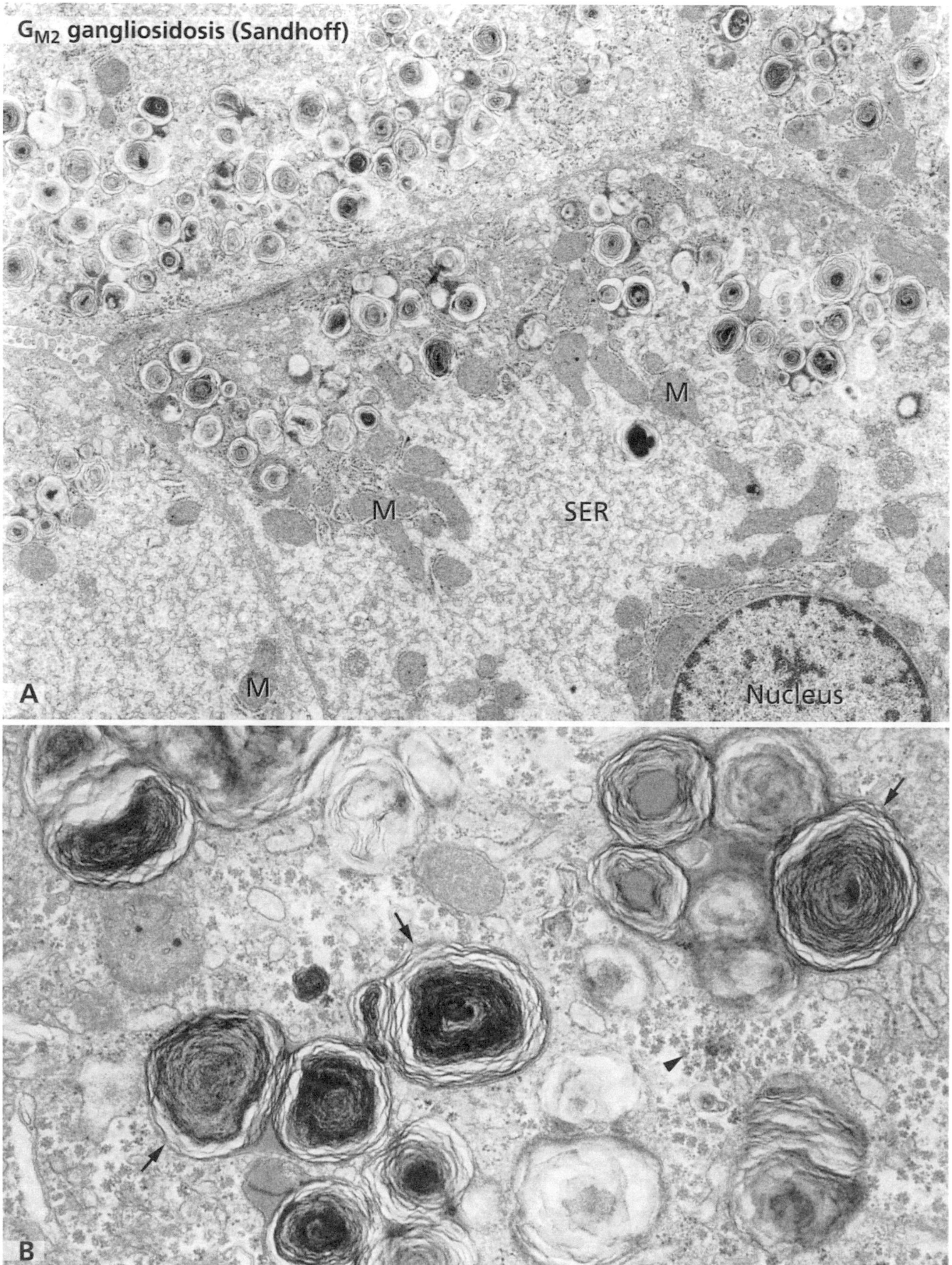
G_{M2} gangliosidosis (Sandhoff)
M
M
SER
M
Nucleus
A
B

FARBER'S DISEASE

Farber's disease, also called acid ceramidase deficiency or Farber lipogranulomatosis, is an autosomal recessively inherited lysosomal disorder in which ceramide accumulates. Although present in large amounts in cells affected by the disease, this ceramide does not function as a signalling molecule since it is enclosed in lysosomes.

The prominent histological lesion are granulomas consisting of foam cells (lipid-loaded macrophages) surrounded by macrophages, lymphocytes, and multinucleated cells or solely lipid-filled macrophages and histiocytes. In panel A, which is from a skin biopsy of a patient with classic Farber's disease, a macrophage is shown which contains numerous, large (2–3 μm) irregularly shaped inclusions that have a limiting membrane (asterisks). The limiting membrane represents the lysosomal membrane. At higher magnification, the highly characteristic ultrastructural appearance of the content of the inclusions is apparent (panel B). It consists of comma-shaped curvilinear tubules referred to as Farber bodies. Because of the particular shape, they are also called banana bodies. These structures are pathognomonic for Farber's disease.

The disease is caused by a deficiency of lysosomal acid ceramidase, which in its mature form is a 53 kDa glycoprotein composed of a 13 kDa α subunit and a 40 kDa β subunit. The rate of ceramide synthesis is normal, but ceramide resulting from the degradation of complex sphingolipids cannot be hydrolysed. The gene coding for both ceramidase subunits has been mapped to chromosome 8p21.3/22. Thus far, some few disease-causing point mutations and two splice site mutations causing deletions of exon 6 and exon 13 have been found, which account for the different types of Farber's disease. Clinically, seven phenotypes can be distinguished, of which five differ solely in severity of illness and organ manifestation. Type 6 is a combination of Farber's disease and Sandhoff's disease, and type 7 corresponds to a combined deficiency of acid ceramidase, glucocerebrosidase, and galactocerebrosidase. The classical symptoms consist of painful swelling of joints, presence of subcutaneous nodules at affected joints and pressure points, and progressive hoarseness due to involvement of the larynx. Additional symptoms are hepatosplenomegaly (type 4) or progressive neurological symptoms (type 5).

The diagnosis can be made on the basis of the typical clinical symptoms mentioned above, measurement of acid ceramidase activity, which is dramatically reduced ($\sim$ 6% of normal), and electron microscopic analysis of skin biopsies. Prenatal diagnosis is possible by measuring acid ceramidase activity in cultured amniotic cells or chorionic villus samples.

References

Burck U, Moser HW, Goebel HH, Gruttner R, and Held KR (1985) A case of lipogranulomatosis Farber: some clinical and ultrastructural aspects. Eur J Pediatr 143: 203

Fusch C, Huenges R, Moser HW, Sewell AC, Roggendorf W, Kustermann-Kuhn B, Poulos A, Carey WF, and Harzer K (1989) A case of combined Farber and Sandhoff's disease. Eur J Pediatr 148: 558

Koch J, Gartner S, Li CM, Quintern LE, Bernardo K, Levran O, Schnabel D, Desnick RJ, Schuchman EH, and Sandhoff K (1996) Molecular cloning and characterisation of a full-length complementary DNA encoding human acid ceramidase. Identification of the first molecular lesion causing Farber's disease. J Biol Chem 271: 33110

Li CM, Park JH, He X, Levy B, Chen F, Arai K, Adler DA, Disteche CM, Koch J, Sandhoff K, and Schuchman EH (1999) The human acid ceramidase gene (ASAH): structure, chromosomal location, mutation analysis, and expression. Genomics 62: 223

Moser H, Linke T, Fensom A, Levadxe T, and Sandhoff K (2001) Acid ceramidase deficiency: Farber lipogranulomatosis. In: The metabolic and molecular bases of inherited disease (Scriver C, Beaudet A, Valle D, and Sly WS, eds). New York: McGraw-Hill, pp 3573

Rauch HJ and Aubock L (1983) "Banana bodies" in disseminated lipogranulomatosis (Farber's disease). Am J Dermatopathol 5: 263.

Schmoeckel C and Hohlfed M (1979) A specific ultrastructural marker for disseminated lipogranulomatosis (Faber). Arch Dermatol Res 266: 187

Sugita M, Dulaney JT, and Moser HW (1972) Ceramidase deficiency in Farber's disease (lipogranulomatosis). Science 178: 1100

Magnification: x 8,500 (A); x 41,000 (B)

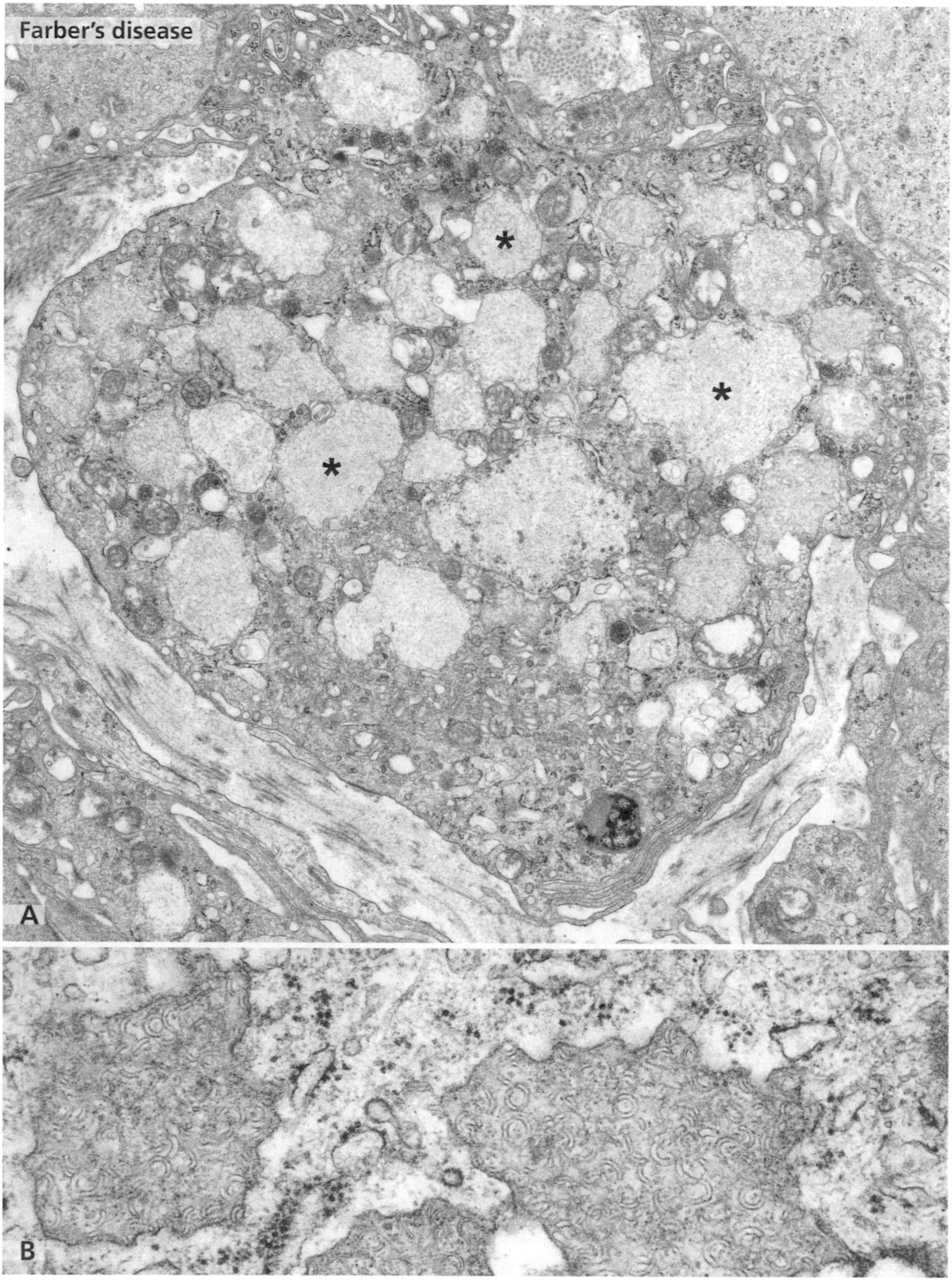
Farber's disease
*
*
*
A
B

WOLMAN'S DISEASE

This lysosomal storage disease is an autosomal recessively inherited disorder and caused by a deficiency of lysosomal acid lipase activity. The structural gene for acid lipase has been located on chromosome 10q23. The mutations in Wolman's disease are diverse and include nonsense mutations, frameshifts, missense mutations, and exon skipping. The deficiency in acid lipase results in massive intralysosomal accumulation of cholesteryl esters and triglycerides, which causes hepatosplenomegaly. Deposits occur also in the adrenals, which cause necrosis and calcification. As a consequence of defective release of free cholesterol from lysosomes, low density lipoprotein receptors and HMG-CoA reductase are upregulated. This results in enhanced *de novo* cholesterol synthesis and enhanced receptor-mediated cholesterol endocytosis. The outcome of this vicious cycle consists in additionally increased lysosomal deposition of lipids.

In hepatocytes of a liver biopsy from an affected individual, droplet-like, enlarged lysosomes (asterisks) that occupy most of the cytoplasm can be easily detected by transmission electron microscopy. These droplet-like structures are limited by a membrane representing the lysosomal membrane (inset). The stored material has no particular structure, which is in contrast to other lysosomal storage diseases such as Gaucher's (cf. Fig. 56), Fabry's (cf. Fig. 57), and Farber's disease (cf. Fig. 59), in which the appearance of the accumulated substrate can be a specific structural hallmark for a single disorder or one group.

The clinical manifestations of Wolman's disease consist of hepatosplenomegaly and gastrointestinal symptoms such as vomiting and diarrhea. Currently – no specific treatment is available, although HMG-CoA reductase inhibitors are used together with cholestyramine to suppress cholesterol synthesis and apolipoprotein B production.

References

Anderson R, Byrum R, Coates P, and Sando G (1994) Mutations at the lysosomal acid cholesteryl ester hydrolase gene locus in Wolman's disease. Proc Natl Acad Sci USA 91: 2718

Assmann G and Seedorf U (2001) Acid lipase deficiency: Wolman's disease and cholesteryl ester storage disease. In: The metabolic and molecular bases of inherited disease (Scriver C, Beaudet A, Valle D, and Sly W, eds). New York: McGraw Hill, p 3551

Lough J, Fawcett J, and Wiegensberg B (1970) Wolman's disease: an electron microscopic, histochemical and biochemical study. Arch Pathol 89: 103

Marshall W, Ockenden B, Fosbrokke A, and Cumings J (1968) Wolman's disease: a rare lipodosis with adrenal calcification. Arch Dis Child 44: 331

Ries S, Büchler C, Schindler G, Aslanidis C, Ameis D, Gasche C, Jung N, Schambach A, Fehringer P, Vanier M, et al. (1998) Different missense mutations in histidine-108 of lysosomal acid lipase cause cholesteryl ester storage disease in unrelated compound heterozygous and hemizygous individuals. Hum Mutat 12: 44

Wolman F, Sterck V, Gatt S, and Frenkel M (1961) Primary familial xanthomatosis with involvement and calcification of the adrenals: report of two more cases in siblings of a previously described infant. Pediatrics 28: 742

Magnification: x 7,900; x 120,000 (inset)

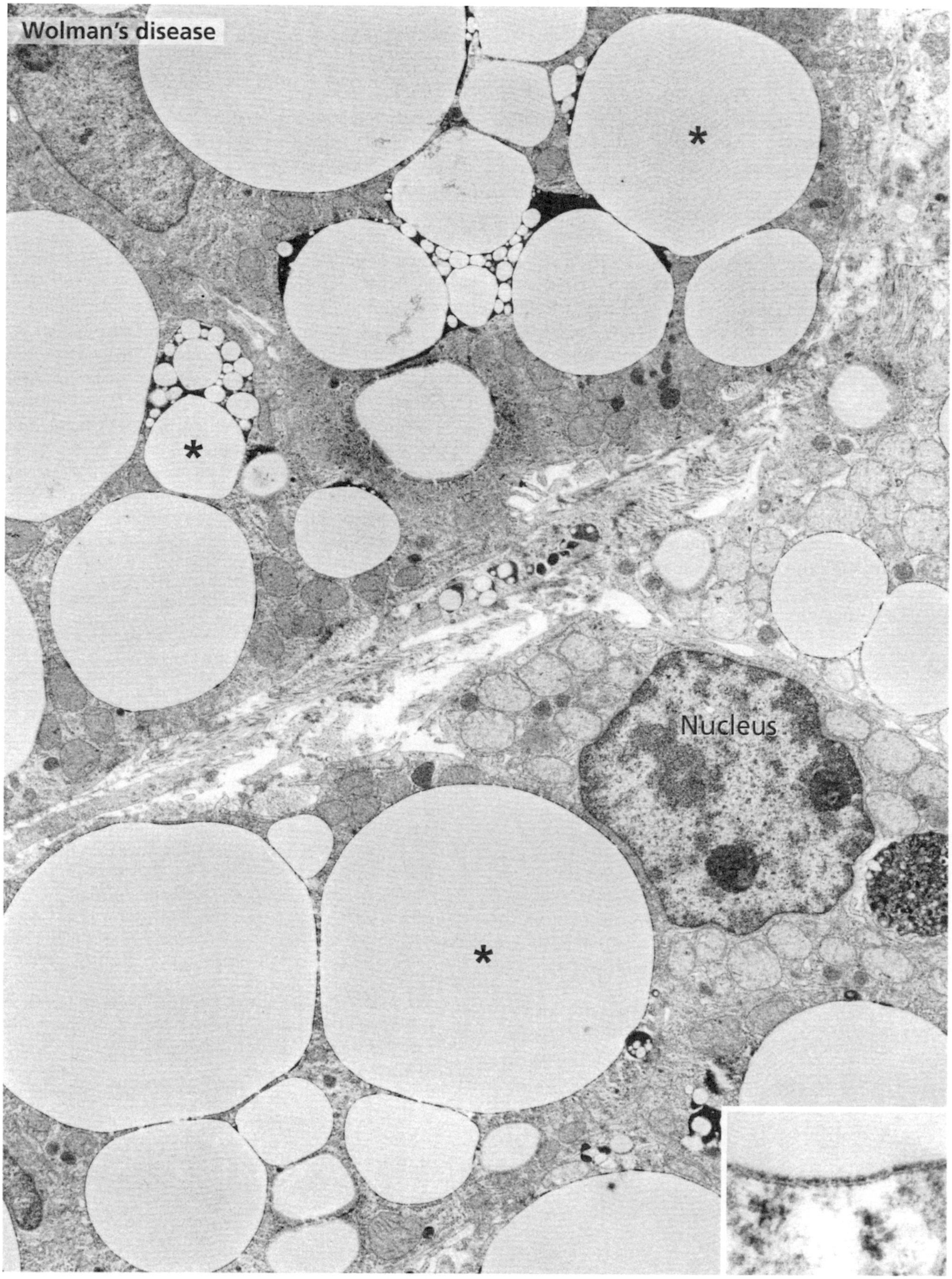
Wolman's disease
*
*
Nucleus
*

GLYCOGENOSIS TYPE II

This lysosomal glycogen storage disease is autosomal recessively inherited and also termed acid maltase deficiency or Pompe's disease. In all other types of glycogenoses, glycogen deposits are cytosolic (cf. Fig. 70 and 153). The intralysosomal glycogen storage results from defective lysosomal acid α-glucosidase activity whose gene has been mapped to chromosome 17q25. A large number of mutations are spread throughout the gene. Clinically, the disease spans a range of phenotypes all of which are associated with myopathy due to glycogen accumulation in cardiac, skeletal and smooth muscle.

The glycogen accumulation in lysosomes of liver hepatocytes, which is also a tissue of major glycogen deposition, is shown in panel A (asterisks). In the inset, the presence of the lysosomal membrane is illustrated (arrows). Structurally, the glycogen appears normal. In addition, cytosolic glycogen is present in the hepatocytes.

Trials of enzyme replacement therapy have been conducted and gene therapy has been attempted experimentally. Thus far, bone marrow transplantation has not given promising results.

References

Hirschhorn R and Reuser A (2001) Glycogen storage disease type II: acid α-glucosidase (acid maltase) deficiency. In: The metabolic and molecular bases of inherited diseases (Scriver C, Beaudet A, Valle D, and Sly W, eds). New York: McGraw-Hill, pp 3389

Huie M, Chen A, Tsujino S, Shanske S, DiMauro S, Engel A, and Hirschhorn R (1994) Aberrant splicing in adult onset glycogen storage disease type II (GSDII): molecular identification of an IVS1(-13T6G) mutation In a majority of patients and a novel IVS0(+1GT6CT) mutation. Hum Mol Genet 3: 1081

Martiniuk F, Mehler M, Pellicer A, Tzall S, LaBadie G, Hobart C, Ellenbogen A, and Hirschhorn R (1986) Isolation of a cDNA for human acid alpha glucosidase and detection of genetic heterogeneity for mRNA in three alpha glucosidase deficient patients. Proc Natl Acad Sci USA 83: 9641

Pompe JC (1932) Over idiopatische hypertrophie van het hart. Ned Tijdschr Geneesk 76: 304

CYSTINOSIS

This is an autosomal recessively inherited, rare disorder of lysosomal membrane transport. The defect lies in the carrier-mediated transport of cystine across the lysosomal membrane. The cystinosis gene, *CNTS*, is located on chromosome 17p13, transcribes a 2.6 kb mRNA that codes for the 367 amino acid long cystinosin, which is a polytope membrane protein. Most cystinosis patients from northern Europe are homozygous for a 57 kb deletion extending from exon 10 upstream. The disease is characterised by the accumulation of free cystine in lysosomes (10 to 100 times the normal lysosomal amount), which then forms crystals (asterisks in panel B). These crystals, which can be observed in the lysosomes of many tissues, lead to a progressive damage of lysosomes. Rectangular cystine crystals are birefringent. The disease manifests mainly in the kidney and causes the renal tubular Fanconi syndrome and glomerular damage requiring dialysis or transplantation. An elevated cystine content in leukocytes of peripheral blood is diagnostic.

References

Gahl W, Thoene J, and Schneider J (2001) Cystinosis: a disorder of lysosomal membrane transport. In: The meatbolic and moledcular bases of inherited diseases (Scriver C, Beaudet A, Valle D, and Sly WS, eds). New York: McGraw-Hill, pp 5085

McDowell G, Town M, van't Hoff W, and Gahl W (2000) Clinical and molecular aspects of nephropatic cystinosis. Pediatr Res 47: 17

Spears G, Slusser R, Tousimis A, Taylor C, and Schulman J (1971) Cystinosis: an ultrastructural and electron-probe study of the kidney with unusual findings. Arch Pathol Lab Med 9: 206

Touchman J, Anikster Y, Dietrich N, Maduro V, McDowell G, Shotelersuk V, and Bouffard G (2000) The genomic region encompassing the nephropathic cystinosis gene (CNTS): complete sequencing of a 200-kb segment and discovery of a novel gene within the common cystinosis-causing deletion. Genome Res 10: 165

Town M, Jean G, Cherqui S, Attard M, Forestier L, Whitmore S, and Callen D (1998) A novel gene encoding an integral membrane protein which is mutated in nephropathic cystinosis. Nat Genet 18: 319

Magnification: x 12,000(A); x 117,000 (inset); x 8,200 (B)

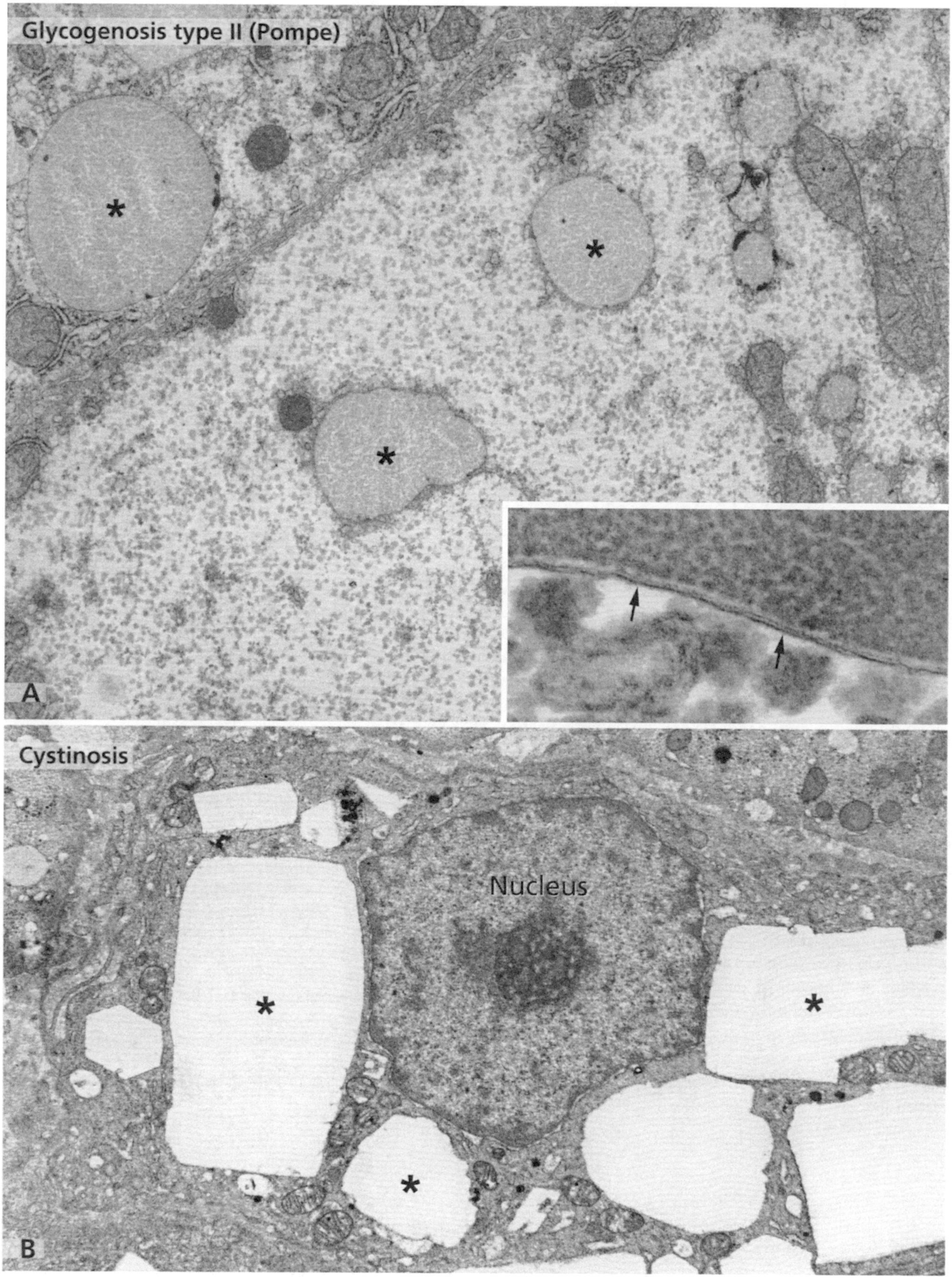
Glycogenosis type II (Pompe)
*
*
*
A
Cystinosis
Nucleus
*
*
*
*
B

AUTOPHAGY: LIMITED SELF-DIGESTION

Autophagy is a basic regulatory process that operates under normal and pathological conditions. It represents a third digestive cellular system besides lysosomes and proteasomes.

Macroautophagy is the digestion of cellular organelles such as endoplasmic reticulum (ER-phagy), mitochondria, endocrine secretory granules (crinophagy) and peroxisomes (pexophagy) and part of the cytoplasm. Microautophagy of small portions of the cytoplasm and hsc73 chaperone-mediated degradation of proteins containing a pentapeptide motif are additional forms of autophagy. Autophagy is enhanced during starvation and after functional overload or cellular injury and results in the remodeling of the cytoplasm and removal of excess organelles. Autophagy is not only an important cellular degradation mechanism, but also plays active roles in cancer, immune defense and apoptosis.

The formation of autophagosomes is a highly coordinated, multistep process and the machinery proteins, encoded by a family of autophagy related genes *(ATG)*, are well characterised. One of them, LC3 (Atg8 in yeast), is useful for the immunocytochemical detection of authophagosomes. Initially, a double membrane is formed, the so-called isolation membrane (also referred to as phagophore). As shown in panels A, B and C, the isolation membrane (arrowheads) curves around part of the cytoplasm and upon elongation engulfs it (panel C). The closure of the isolation membrane by the sealing of its edges results in the formation of a unique double membrane-limited vesicle, the autophagosome. Such a double membrane-limited autophagosome, which contains cytoplasm, is shown in panel D. The formation of the autophagosomes is a fast process, which takes about 10 minutes for completion. Autophagosomes have no digestive capacity. Only following fusion with lysosomes, they obtain both lysosomal membrane proteins and lysosomal enzymes and are termed now autolysosomes. An autolysosome is shown in panel E. It contains cytoplasm and part of endoplasmic reticulum as well as electron dense lamellar material. Autophagy also functions in the removal of cytosolic aggregates of misfolded proteins in chronic neurodegeneration. Cytosolic protein aggregates also occur following experimental inhibition of proteasomes as demonstrated in panel F by the presence of electron dense material in the cytoplasm (arrows). Panel E exemplifies the autophagic removal of such cytosolic protein aggregates, which can be seen inside a double membrane autophagosome (asterisk).

In addition, fusion between autophagosomes and endosomes may occur, which results in the formation of amphisomes. Amphisomes contain not only cellular but also endocytosed foreign material.

The origin of the isolation membrane is still a matter of debate, but most probably there is not a single source. In the maturation model, the isolation membrane is supposed to be derived from the endoplasmic reticulum upon induction of autophagy. In the assembly model, it is postulated that the isolation membrane is assembled *de novo* by non-vesicular transport or local synthesis.

References

Axe EL, Walker SA, Manifava M, Chandra P, Roderick HL, Habermann A, Griffiths G, and Ktistakis NT (2008) Autophagosome formation from membrane compartments enriched in phosphatidylinositol 3-phosphate and dynamically connected to the endoplasmic reticulum. J Cell Biol 182: 685

Dunn W (1990) Studies on the mechanism of autophagy: formation of the autophagic vacuole. J Cell Biol 110: 1923

Dunn W (1990) Studies on the mechanism of autophagy: maturation of the autophagic vacuole. J Cell Biol 110: 1935

Kabeya Y, Mizushima N, Ueno T, Yamamoto A, Kirisako T, Noda T, Kominami E, Ohsumi Y, and Yoshimori T (2000) LC3, a mammalian homologue of yeast Apg8p, is localized in autophagosome membranes after processing. EMBO J 19: 5720

Klionsky D (2004) Autophagy: Landes Bioscience, Georgetwon, TX

Mizushima N, Klionsky DJ (2007) Protein turnover via autophagy: implications for metabolism. Annu Rev Nutr 27: 19

Xie Z and Klionsky DJ (2007) Autophagosome formation: core machinery and adaptations. Nat Cell Biol 9: 1102

Yokota S, Himeno M, Roth J, Brada D, and Kato K (1993) Formation of autophagosomes during degradation of excess peroxisomes induced by di-(2-ethylhexyl)phthalate treatment. II. Immunocytochemical analysis of early and late autophagosomes. Eur J Cell Biol 62: 372

Magnification: x 41,000 (A); x 60,000 (B); x 74,000 (C–E); x 50,000 (F); x 70,000 (G)

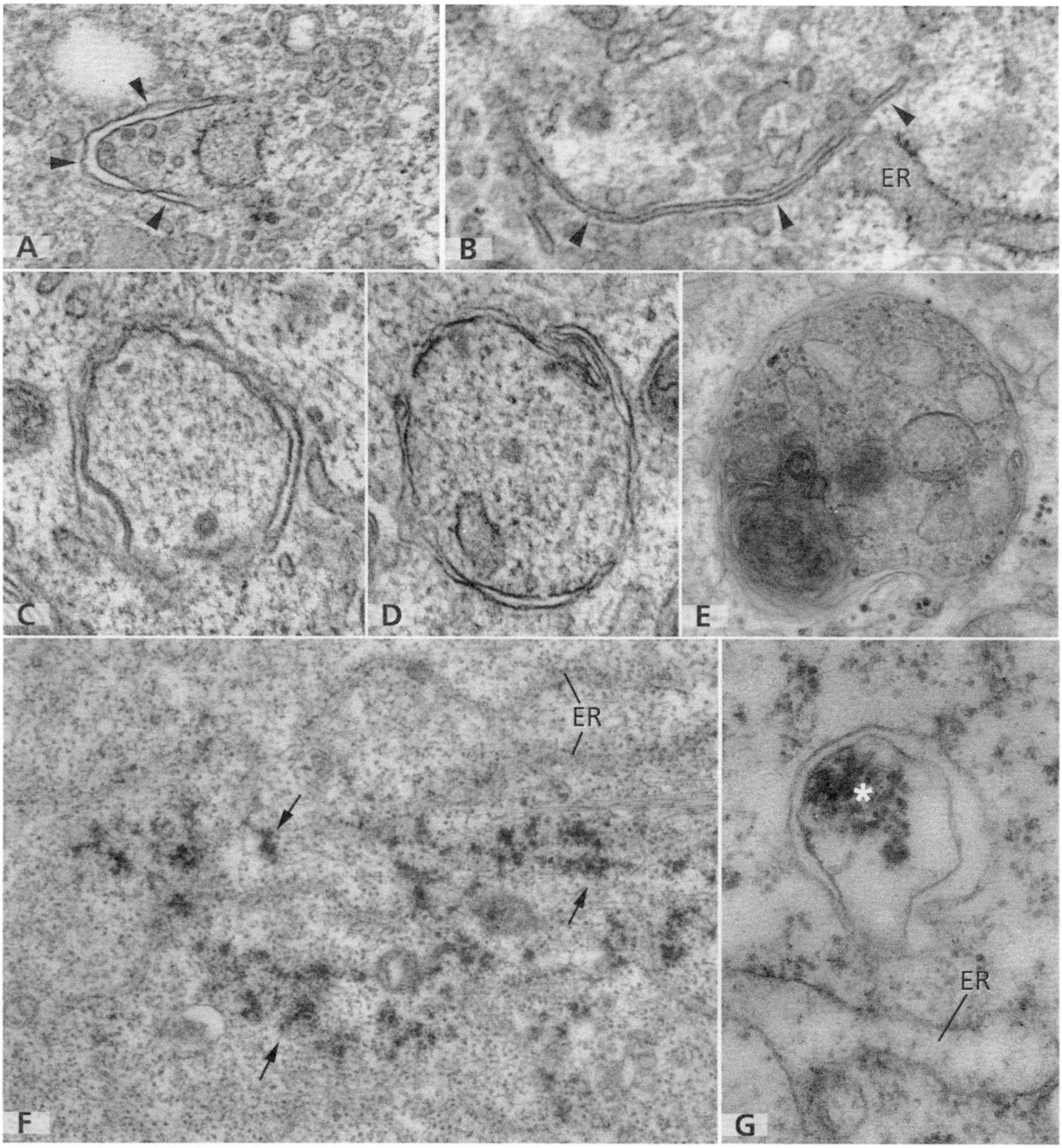
A
B
ER
C
D
E
F
ER
G
*
ER

PEXOPHAGY: AUTOPHAGY OF PEROXISOMES

The autophagic removal of peroxisomes is called pexophagy. Pexophagy goes on at a basal level. It also results in the removal of excess peroxisomes induced by experimental manipulation of rats as illustrated and discussed here.

Panel A represents the initial stage of pexophagy. Excess peroxisomes (P), induced in rat liver by administration of di-(2-ethylhexyl)phthalate, are surrounded and sequestered by a double membrane, the isolation membrane (arrowheads in panel A). This isolation membrane is most probably derived from the endoplasmic reticulum because at this stage it exhibits the endoplasmic reticulum markers glucose-6-phosphatase, carboxyesterase 1, and the endoplasmic reticulum located trimming glucosidase II (cf. Fig. 18). Autophagosomes are formed after elongation and closure of the isolation membrane. Early autophagosomes lack lysosomal hydrolases, and the isolation membrane lacks lysosomal membrane glycoproteins. In the next step, fusion between autophagosomes and lysosomes occurs, which results in the presence of lysosomal hydrolases. Panel B shows the fusion between an autophagosome (arrowheads), which contains a peroxisome and a mitochondrium, and a lysosome (site of membrane fusion is marked by arrows). Panel C also shows fusion between autophagosomes and lysosomes (arrows). The immunogold labelling for a lysosomal membrane glycoprotein is restricted to the fusing lysosomes (arrows in panel C) and not yet detectable along the autophagosome membrane (arrowheads in panel C). The fusion between autophagosomes and lysosomes, as illustrated in panels B and C, results in the formation of autolysosomes. The presence of lysosomal hydrolases results in the degradation of the inner membrane of the original autophagosomes and their content. In panel D, intense immunogold labelling for the lysosomal enzyme cathepsin B is present in several autolysosomes (asterisks). An early autolysosome (arrowheads) is only weakly labelled for cathepsin B. On completion of the digestion, the autolysosomes may contain dense, lamellar material composed of peroxidized lipids and other undigestable substances, and are referred to as residual bodies.

References

Dunn W (1990) Studies on the mechanism of autophagy: formation of the autophagic vacuole. J Cell Biol 110: 1923

Dunn W (1990) Studies on the mechanism of autophagy: maturation of the autophagic vacuole. J Cell Biol 110: 1935

Klionsky DJ (2004) Autophagy. Georgetwon, TX: Landes Bioscience

Xie Z and Klionsky DJ (2007) Autophagosome formation: core machinery and adaptations. Nat Cell Biol 9: 1102

Yokota S (2003) Degradation of normal and proliferated peroxisomes in rat hepatocytes: regulation of peroxisomes quantity in cells. Microsc Res Tech 61: 151

Yokota S, Himeno M, Roth J, Brada D, and Kato K (1993) Formation of autophagosomes during degradation of excess peroxisomes induced by di-(2-ethylhexyl)phthalate treatment. II. Immunocytochemical analysis of early and late autophagosomes. Eur J Cell Biol 62: 372

Yokota S and Fahimi HD (2009) Degradation of excess peroxisomes in mammalian liver cells by autophagy and other mechanisms. Histochem Cell Biol 131: 455

Magnification: x 32,000 (A, C); x 45,000 (B); x 23,000 (D)

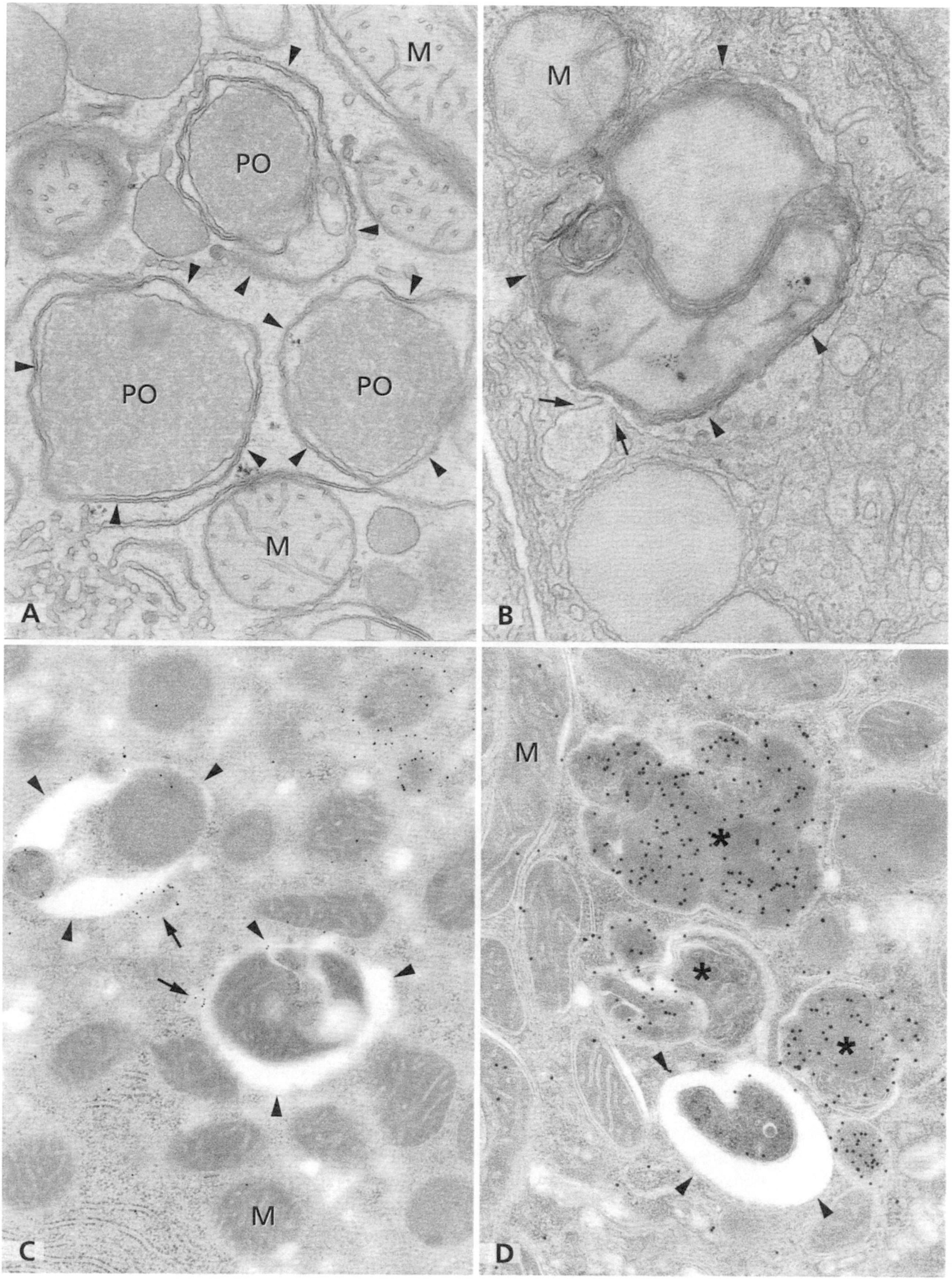
M
PO
PO
PO
M
A
M
B
C
M
M
D

MITOCHONDRIA: CRISTA AND TUBULUS TYPES

Although known since the early days of microscopy, mitochondria attract continuous interest because of their unique functions in both cell life and death. Dimensions, shapes, and locations of mitochondria are strikingly different in diverse types of cells in relation to the specific cellular functions. Because of the obligatory double membranes and the characteristic inner compartmentalisation, it is easy to discriminate mitochondria from other membrane-bound organelles, such as peroxisomes (cf. Fig. 66), under the electron microscope.

Panels A and B show mitochondria of the crista type in a rat pancreatic acinar cell, surrounded by cytoplasm with densely packed cisternae of the rough endoplasmic reticulum. Mitochondria are located mainly at sites where energy is needed and according to the requirements may change their locations and also undergo temporary alterations in shape. In both panels, outer and inner mitochondrial membranes are visible enclosing the intermembrane space. The outer membrane contains voltage-dependent anion channels, the mitochondrial "porins", which allow ions and small molecules to enter the intermembrane space, creating a milieu resembling that of the cytoplasm. The inner membrane is impermeable to ions because of its enrichment in the phospholipid cardiolipin. It surrounds the innermost compartment, the mitochondrial matrix, and contains the proteins for the oxidation reactions of the respiratory electron-transport chain, for adenosine triphosphate (ATP) synthesis, and the regulation of the metabolite transport into and out of the matrix. The inner membrane is enlarged by folds projecting into the matrix. In the majority of cells, the inner membrane forms cristae (arrowheads in panel B) but other shapes may occur as well. In cell producing steroids, the inner membrane typically forms tubular projections. Mitochondria of the tubule type in an endocrine cell of the ovary are shown in panel C (T-tubular projections of the inner membrane).

The matrix contains the enzymes of the citric acid cycle and enzymes engaged in fatty acid β-oxidation. In panels A and B, dense matrix granules are visible, which are important for the storage of Ca^{2+} and other divalent cations. Furthermore, the mitochondrial DNA and the machineries for protein synthesis, ribosomes and tRNAs are contained in the matrix. Only some of the mitochondrial proteins are encoded by the mitochondrial genome and synthesised in the matrix. Mitochondrial proteins are mostly synthesised on free ribosomes in the cytoplasm and are post-translationally translocated across the mitochondrial membranes to reach their functional destinations inside the mitochondria. Three membrane protein complexes, the translocase of the outer membrane (TOM) that includes the general import pore (GIP), the presequence translocase of the inner membrane (TIM 23), and the protein insertion complex of the inner membrane (TIM 22) build up the central machineries for recognition and translocation of mitochondrial precursor proteins.

Mitochondria are dynamic organelles. They increase in number by division throughout the interphase, and the divisions are independent of the cell cycle. Ongoing fission and fusion events are required for the maintenance of their regular structure, which is necessary for efficient function and impacts mitochondrial DNA inheritance.

Mitochondria are sensitive to cellular stress and have a pivotal role in the initiation of programmed cell death. As a major event, cytochrome c is released from the intermembrane space into the cytoplasm, initiating the cascade of proteolytic reactions that result in the apoptotic changes of the cell (cf. Fig. 11).

References

Debatin KM, Poncet D, and Kroemer G (2002) Chemotherapy: targeting the mitochondrial cell death pathway. Oncogene 21: 8786

Perkins G, Bossy-Wetzel E, and Ellisman MH (2009) New insights into mitochondrial structure during cell death. Exp Neurol 218: 183

Pfanner N and Wedermann N (2002) Mitochondrial protein import: two membranes, three translocases. Curr Opin Biol 14: 400

Scott SV, Cassidy-Stone A, Meeusen SL, and Nunnari J (2003) Staying in aerobic shape: how the structural integrity of mitochondria and mitochondrial DNA is maintained. Curr Opin Cell Biol 15: 482

Susin SA, Zamzami N, and Kroemer G (1998) Mitochondria as regulators of apoptosis: doupt no more. Biochim Biophys Acta 1366: 151

Wang B, Li N, Sui L, Wu Y, Wang X, Wang Q, Xia D, Wan T, and Cao X (2004) HuBMX-mcP, a novel member of mitochondrial carrier superfamily, enhances dendritic cell endocytosis. Biochim Biophys Res Commun 314: 292

Magnification: x 40,000 (A); x 49, 000 (B); x 49,500 (C)

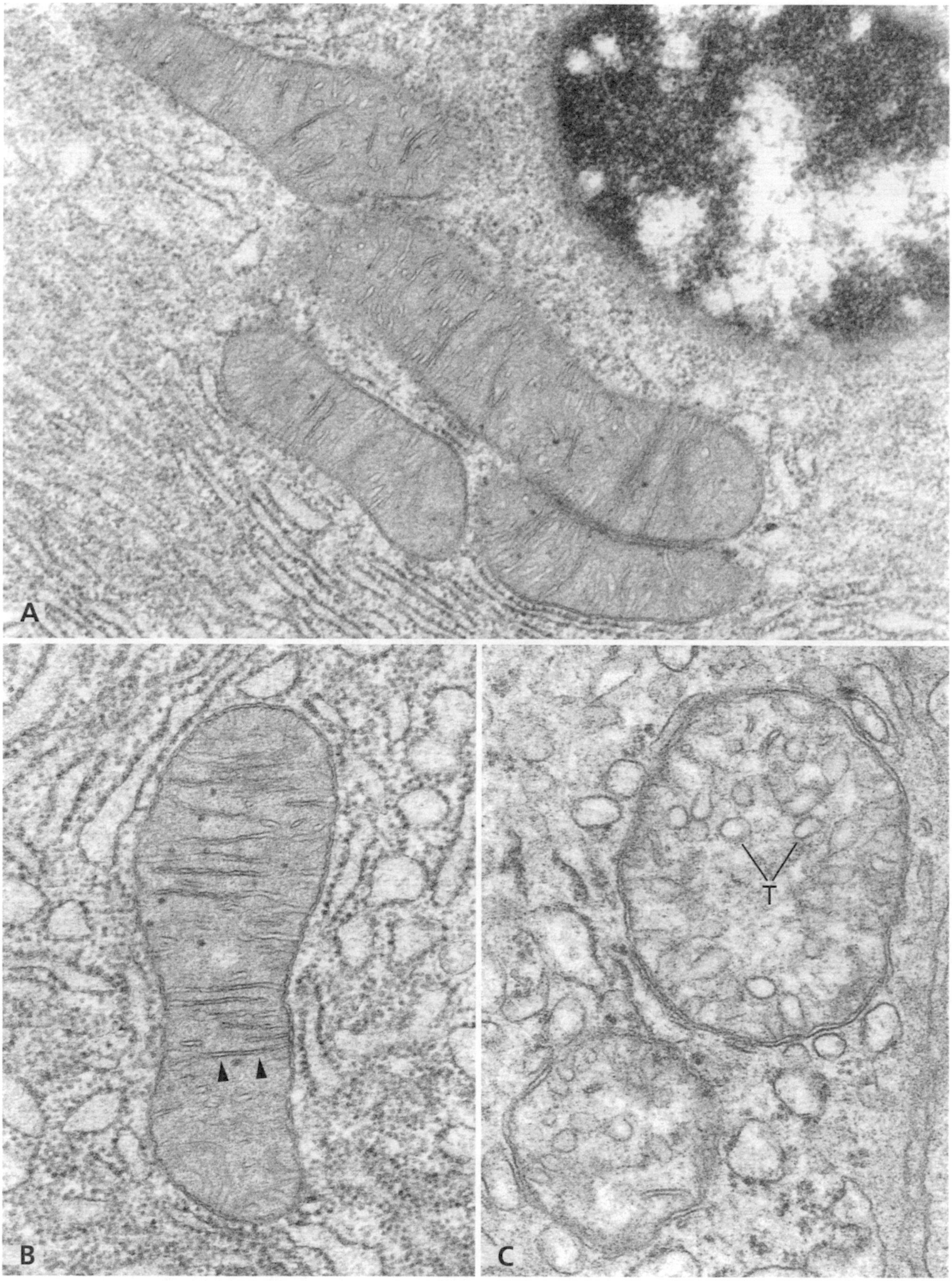
A
B
C
T

ABNORMALITIES OF MITOCHONDRIA

Structural abnormalities of mitochondria are found most often in inherited disorders affecting the skeletal muscle (myopathies) and central nervous system and in addition can be caused by drug toxicity (alcohol, hydrazine, some antiretroviral drugs). Although these are quite different diseases involving different pathogenetic mechanisms, the observed structural changes of the mitochondria are alike. These abnormalities consist not only of an increase of the number of mitochondria but also of an enlargement and abnormal shape, variation in the number of cristae and particular patterns of cristae, and in abnormal inclusions. The functional consequences of these mitochondrial abnormalities can be far reaching and systemic due to the common underlying impairment of oxidative phosphorylation.

In panel A, a detail from a skeletal muscle fibre (cross-sectioned myofibrils marked by an asterisk) is shown with numerous mitochondria that all contain several paracrystalline inclusions. Such paracrystalline inclusions can be observed in mitochondrial encephalomyopathies, which are a heterogeneous group of disorders. Furthermore, such type of inclusions occur in specific mitochondrial disorders.

In panel B, a group of mitochondria of various size and shape is shown. Some of them contain cristae arranged in parallel order, which is normal, as is their size (arrows). However, other mitochondria are greatly enlarged and are filled with concentric cristae, which is abnormal (arrowheads). The mitochondrial matrix appears to be inexistent.

In panel C, a mitochondrial abnormality due to drug toxicity is exemplified. In mitochondria of hepatocytes from a liver biopsy of a patient receiving an antiretroviral drug, cristae can be observed only rarely (arrow) and the matrical substance is increased. However, the outer and inner mitochondrial membrane is unmistakably seen, excluding the possibility that these could be peroxisomes (cf. Fig. 66). Paracrystalline mitochondrial inclusions as shown in panel A can also occur during the course of treatment with antiretroviral drugs. Antiretroviral treatment with nucleoside analog reverse transcriptase inhibitors results in impaired oxidative phosphorylation through inhibition of DNA polymerase γ. This causes myopathy, neuropathy, hepatic steatosis (lipid accumulation in hepatocytes), and lactic acidosis. Treatment with AZT (zidovudine) results in oxidative damage of mitochondrial DNA through production of peroxide. As a consequence this seems to affect mitochondrial DNA replication and to lessen mitochondrial renewal. However, free radical scavengers can be applied to protect against the AZT-induced oxidative damage of mitochondrial DNA.

References

Barile M, Valenti D, Quagliariello E, and Passarella S (1998) Mitochondria as cell targets of AZT (zidovudine). Gen Pharmacol 31: 531

Dickersin G (2000) Diagnostic electron microscopy. A text/atlas. New York: Springer

Garcia de la Asuncion J, del Olmo M, Sastre J, Millan A, Pellin A, Pallardo F, and Vina J (1998) AZT treatment induces molecular and ultrastructural oxidative damage to muscle mitochondria. Prevention by antioxidant vitamins. J Clin Invest 102: 4

Lewis W and Dalakas M (1995) Mitochondrial toxicity of antiviral drugs. Nat Med 1: 417

Li V, Hostein J, Romero NB, Marsac C, Mezin P, Bost R, Degoul F, Fardeau M, and Fournet J (1992) Chronic intestinal pseudoobstruction with myopathy and ophthalmoplegia. A muscular biochemical study of a mitochondrial disorder. Dig Dis Sci 37: 456

Lindal S, Lund I, Torbergsen T, Aasly J, Mellgren SI, Borud O, and Monstad P (1992) Mitochondrial diseases and myopathies: a series of muscle biopsy specimens with ultrastructural changes in the mitochondria. Ultrastruct Pathol 16: 263

Magnification: x 56,000 (A); x 9,500 (B); x 62,500 (C)

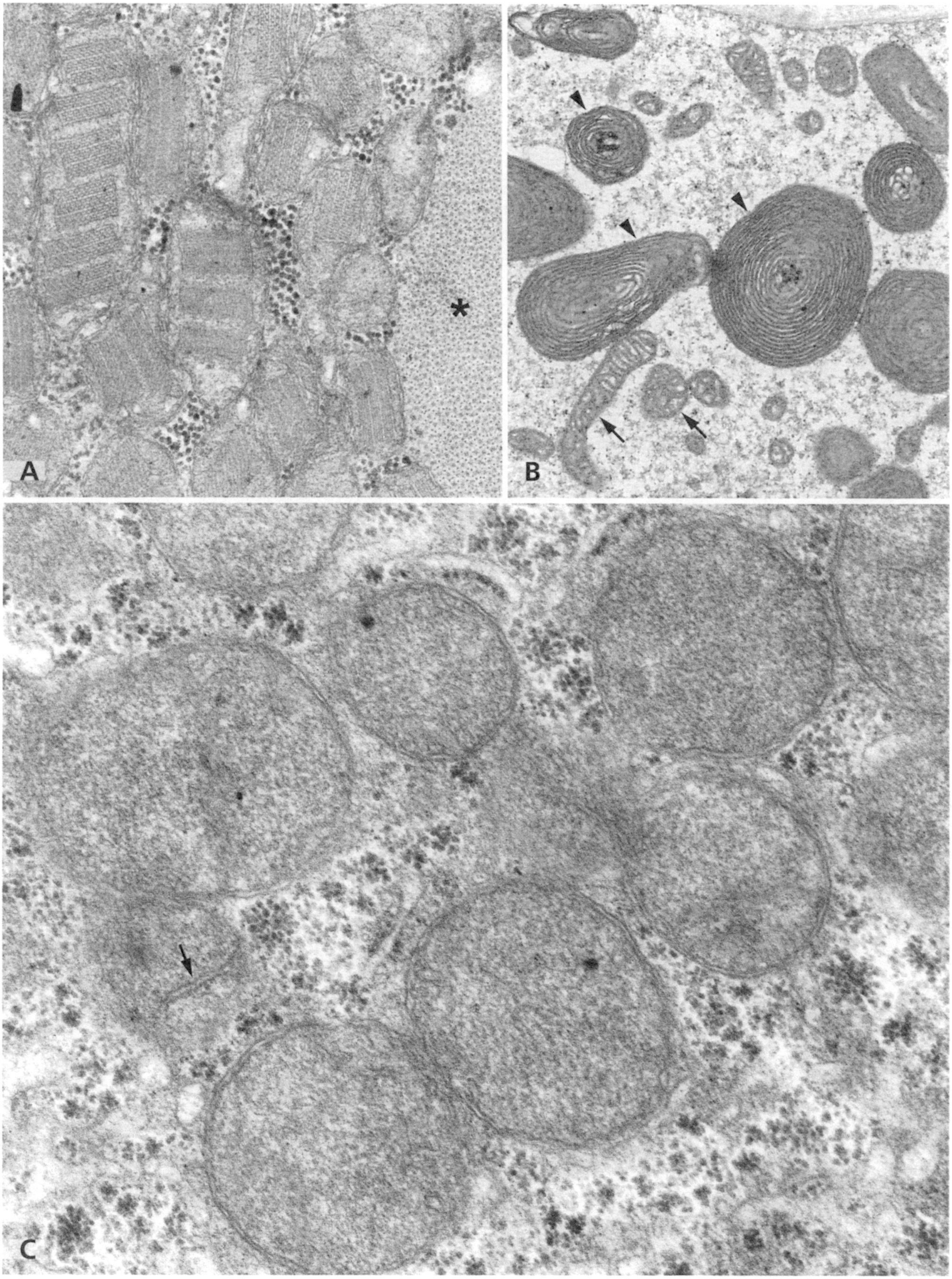
A
B
C

PEROXISOMES: MULTITALENTED ORGANELLES

Peroxisomes (formerly called microbodies) are ubiquitous organelles that contain catalase and oxidative enzymes producing H_2O_2. Depending on cell type their number, shape, and size vary. By electron microscopy, peroxisomes have a single membrane that encloses a dense matrix, which contains a crystalloid core in some species (e.g., rat hepatocytes) but not in others (e.g., human hepatocytes). The typical fine structure of peroxisomes (Po) in rat liver hepatocytes with a dense crystalloid core (asterisk) is shown in panel A, which also illustrates how different they are from mitochondria (M). Cisternae of the endoplasmic reticulum (ER) can be closely associated with peroxisomes, alike as with mitochondria. Usually, peroxisomes are spherical, with a diameter as large as 1 μm in hepatocytes or as small as 0.1 μm in fibroblasts. However, in kidney tubular cells they may be angular. In specialized mammalian cells, such as proliferating hepatocytes after partial hepatectomy and some yeasts, peroxisomes may form an interconnected network of tubules and cup shaped structures, which have been referred to as peroxisomal networks.

A main function of peroxisomes is in lipid metabolism. Oxidases catabolize long chain unsaturated fatty acids by β-oxidation to acetyl CoA, and β-oxidize bile acid intermediates, leukotrienes, and prostaglandins. The oxidative enzymes use molecular oxygen to carry out oxidative reactions that result in the formation of hydrogen peroxide. Hydrogen peroxide is used by peroxisomal catalase to oxidize substrates such as alcohol, phenol, formaldehyde, and formic acid. In hepatocyte and kidney epithelia, this represents an important detoxification reaction. In the liver, peroxisomes function in cholesterol metabolism and gluconeogenesis. In the central nervous system, peroxisomes catalyze the first biosynthetic reaction in the formation of plasmalogens, the most abundant class of myelin phospholipids. In the sebaceous glands of skin, they are involved in the synthesis of complex lipids in sebum.

Immunoelectron microscopy has shown a remarkable degree of compartmentation in peroxisomes. Catalase was detectable in the matrix but not in the crystalloid core of rat hepatocyte peroxisomes (panel B). On the contrary, urate oxidase (panel C), α-hydroxy acid oxidase A and xanthine oxidase were confined to the crystalloid core. These cores are composed of parallel bundles of hollow tubules. Ten primary tubules (5 nm inner diameter) are arranged in a circular fashion and form a centrally located secondary tubule (20 nm outer diameter). Remarkably, both urate and xanthine oxidase were restricted to the lumen of primary tubules. An extreme example of compartmentation was observed for angular peroxisomes of beef kidney. The crystalline core contained urate oxidase, the non-crystalline central region of the matrix D-amino acid oxidase, the peripheral matrix region catalase and α-hydroxy acid oxidase A, and the marginal dense plates close but separated from the peroxisomal membrane α-hydroxy acid oxidase B. Panel D shows the presence of the 70 kDa peroxisomal membrane protein (PMP-70). Under conditions of peroxisome proliferation, the PMP-70 was found not only along the peroxisomal membrane (panel D) but additionally in membranous loops in continuity with it. PMP-70 was also identified in membranes of peroxisomes of a lower density, probably pre-peroxisomes (cf. Fig. 67).

References

Angermüller S, Bruder G, Völkl A, Wesch H, and Fahimi HD (1987) Localization of xanthine oxidase in crystalline cores of peroxisomes. A cytochemical and biochemical study. Eur J Cell Biol 45: 137

Baumgart E (1994) Morphology of peroxisomes in light- and electron microscopy. In: Peroxisomes (Latruffe N and Bugaut M, eds). Heidelberg: Springer, pp 37

Fahimi DH, Reich D, Völkl A, and Baumgart E (1996) Contributions of the immunogold technique to the elucidation of the biology of peroxisomes. Histochem Cell Biol 106: 105.

Mannaerts G and Van Veldhoven P (1996) Functions and organization of peroxisomal β-oxidation. Ann NY Acad Sci 804: 99

Tsukada H, Mochizuki Y, and Fujiwara S (1966) The nucleoids of rat liver cell microbodies. Fine structure and enzymes. J Cell Biol 28: 449

Wanders R and Waterham H (2006) Biochemistry of mammalian peroxisomes revisited. Annu Rev Biochem 75: 295

Zaar K (1992) Structure and function of peroxisomes in the mammalian kidney. Eur J Cell Biol 59: 233

Magnification: x 50,000 (A); x 63,000 (C–D)

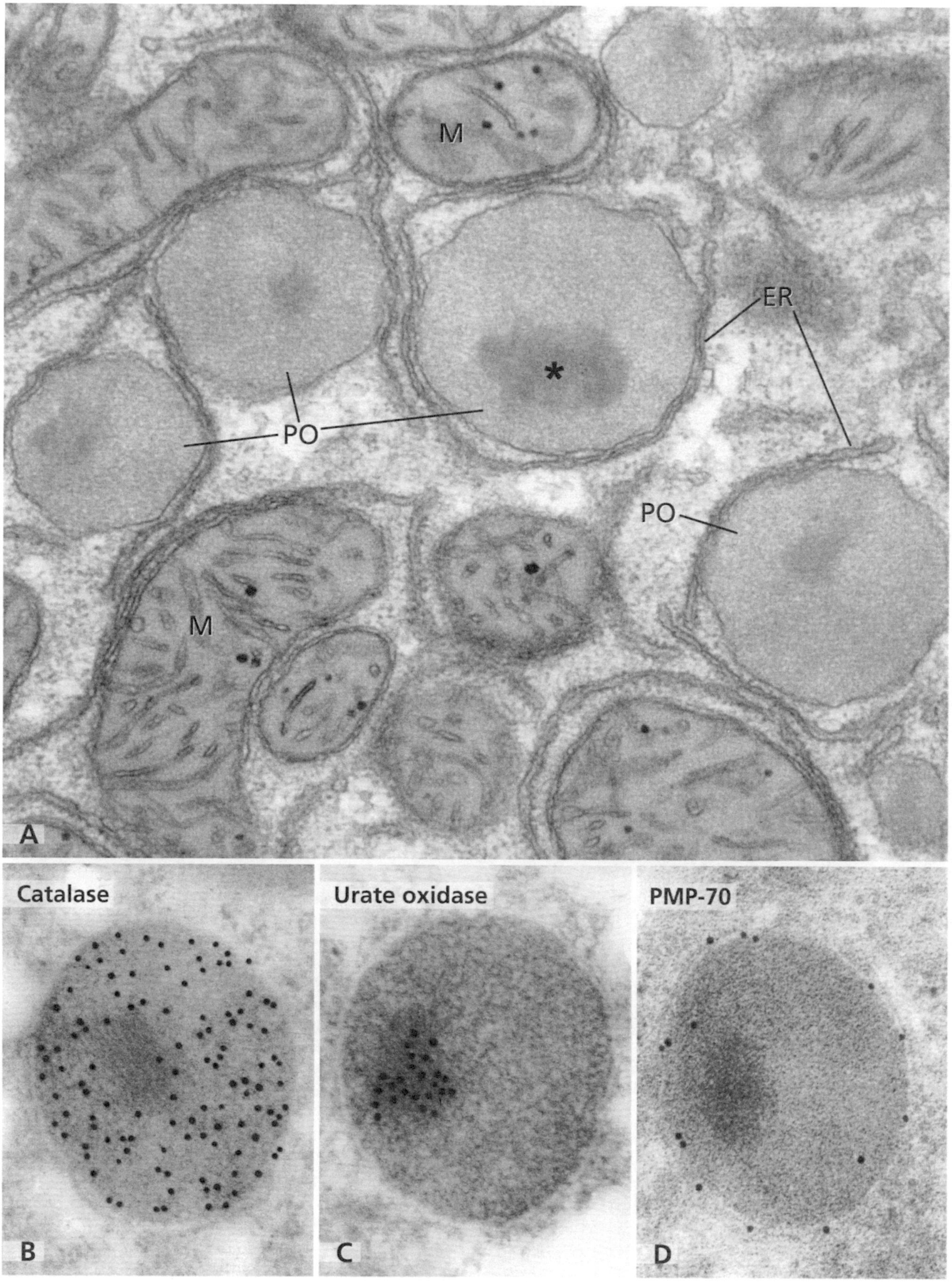
M
ER
PO
PO
PO
M
A
Catalase
B
Urate oxidase
C
PMP-70
D

PEROXISOME BIOGENESIS

Peroxisomes have no DNA. Their matrix and membrane proteins are synthesized on cytosolic polysomes. They are recognized by peroxins (Pex) through peroxisomal targeting signals (PTS) and targeted to the peroxisomal membrane and matrix. PTS1 is a conserved sequence of three amino acids (serine-lysine-leucine) at the extreme C-terminus of most matrix proteins that are targeted by the Pex5p receptor. PTS2 is a broad consensus nonapeptide sequence found on few matrix proteins at or near their N-terminus, and Pex7p is the main receptor. Matrix proteins do not need to unfold for import and can cross the peroxisomal membrane as oligomeric protein, which is in strong contrast to the import of mitochondrial proteins. Pex8 is necessary for peroxisomal matrix protein import via the PST1 and PST2 pathways. The targeting signals for peroxisomal membrane proteins are less well known although it is clear that their pathway is separate from that of matrix proteins. For it, Pex3p, Pex17p and Pex19p are important receptor proteins. The peroxisome membrane contains phosphatidyl choline and phosphatidyl ethanolamine as principal phospholipids that are synthesized in the endoplasmic reticulum. The mechanism of their transport to peroxisomes is uncertain. Transport by phospholipid exchange proteins, through sites of close apposition with endoplasmic reticulum and by specialized vesicles has been proposed.

Several pathways of peroxisome biogenesis and division have been proposed. One is the growth and division model, which implies that peroxisomes are autonomous organelles. This model is based on experimental evidence of peroxisomal protein targeting and electron microscopic analysis Peroxisome division includes three processes. The first is peroxisome elongation, which is mediated by Pex11 proteins; the second is constriction; and the third peroxisome fission by dynamin-related proteins. The 3-D reconstruction of proliferating peroxisomes in regenerating liver has also supported this model. In addition to spherical peroxisomes, interconnected peroxisomes forming a network were found. In panels A and B, catalase positive peroxisomes (Po) with a dumbbell shape (arrowheads in A) or with tail-like extensions (arrows in B) connecting peroxisomes are shown.

Initially, the endoplasmic reticulum was proposed as the source of peroxisomes, which were assumed to bud from the endoplasmic reticulum. Recently, this concept has been revived, mainly based on data obtained in yeast, and peroxisome formation has been proposed to be a multistep maturation pathway. Pex3, an integral membrane protein, which is essential for peroxisome proliferation in yeast, has been shown to assemble in the endoplasmic reticulum and to give raise to new peroxisomes through a budding process. During this process, Pex19, a mostly cytosolic protein, becomes enriched initially at the Pex3-containing endoplasmic reticulum domains and then on the maturing peroxisomes. Probably, mature peroxisomes are formed by fusion of the small peroxisomes under the influence of Pex1 and Pex6.

References

Baumgart E, Völkl A, Hashimoto T, and Fahimi HD (1989) Biogenesis of peroxisomes: immunocytochemical investigation of peroxisomal membrane proteins in proliferating rat liver peroxisomes and in catalse-negative membrane loops. J Cell Biol 108: 2221

Delille HK, Alves R, and Schrader M (2009) Biogenesis of peroxisomes and mitochondria: linked by division. Histochem Cell Biol 131: 441

Geuze HJ, Murk JL, Stroobants AK, Griffith JM, Kleijmeer MJ, Koster AJ, Verkleij AJ, Distel B, and Tabak HF (2003) Involvement of the endoplasmic reticulum in peroxisome formation. Mol Biol Cell 14: 2900

Hoepfner D, Schildknegt D, Braakman I, Philippsen P, and Tabak HF (2005) Contribution of the endoplasmic reticulum to peroxisome formation. Cell 122: 85

Koch A, Yoon Y, Bonekamp N, McNiven M, and Schrader M (2005) A role for Fis1 in both mitochondrial and peroxisomal fission in mammalian cells. Mol Biol Cell 16: 507

Lazarow PB (2003) Peroxisome biogenesis: advances and conundrums. Curr Opin Cell Biol 15: 489

Novikoff A and Shin W (1964) The endoplasmic reticulum in the Golgi zone and its relations to microbodies, Golgi apparatus, and autophagic vacuoles in rat liver cells. J Microsc (Paris) 3: 187

Subramani S, Koller A, and Snyder W B (2000) Import of peroxisomal matrix and membrane proteins. Annu Rev Biochem 69: 399

Tabak H, van der Zand A, and Braakman I (2008) Peroxisomes: minted by the ER. Curr Opin Cell Biol 20: 393

Tam Y, Fagarasanu A, Fagarasanu M, and Rachubinski R (2005) Pex3p initiates the formation of a preperoxisomal compartment from a subdomain of the endoplasmic reticulum in Saccharomyces cerevisiae. J Biol Chem 280: 34933

Yamamoto K and Fahimi HD (1987) Three-dimensional reconstruction of a peroxisomal reticulum in regenerating rat liver: evidence of interconnections between heterogeneous segments. J Cell Biol 105: 713

Magnification: x 49,000 (A); x 83,000 (B)

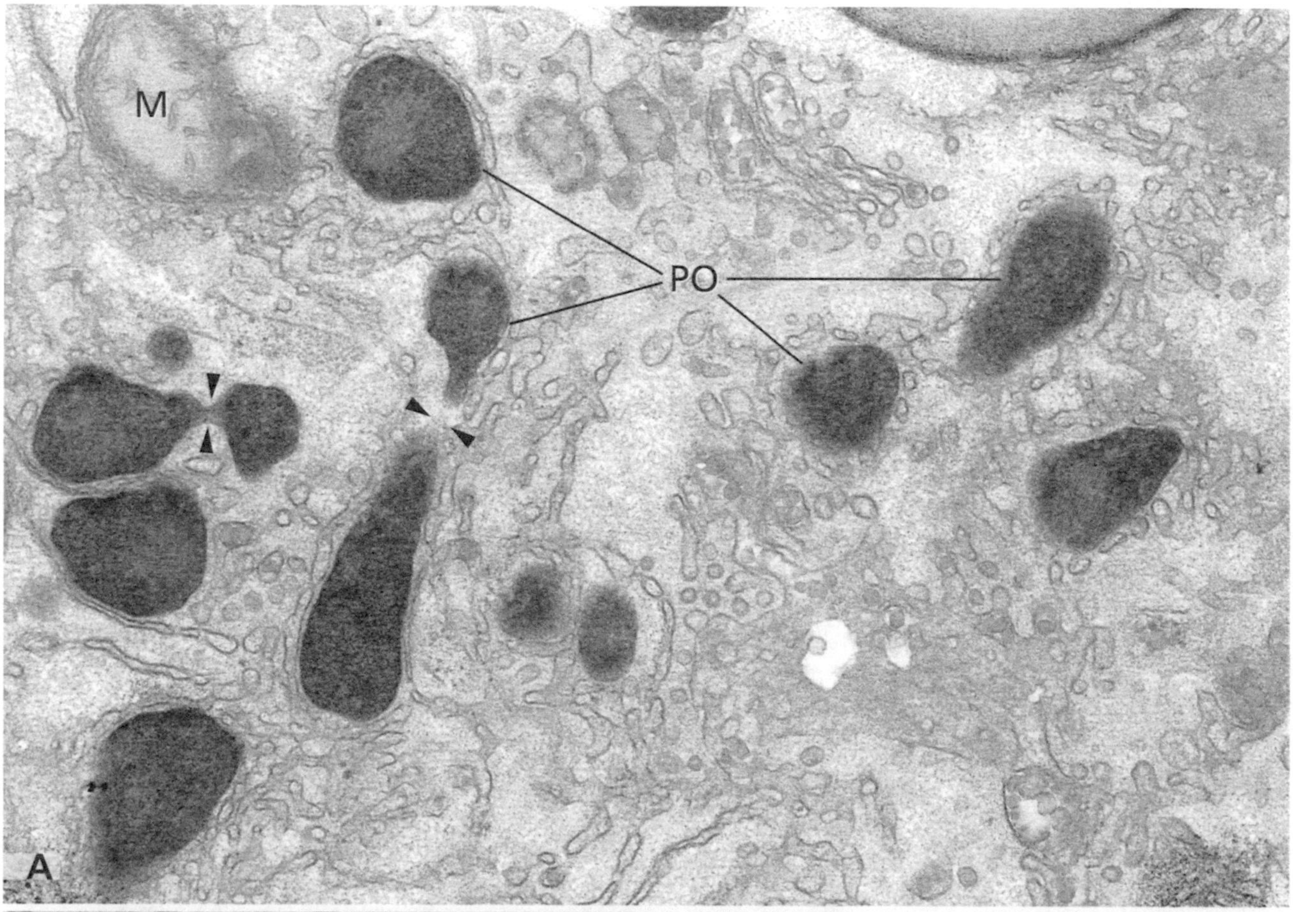
M
PO
A

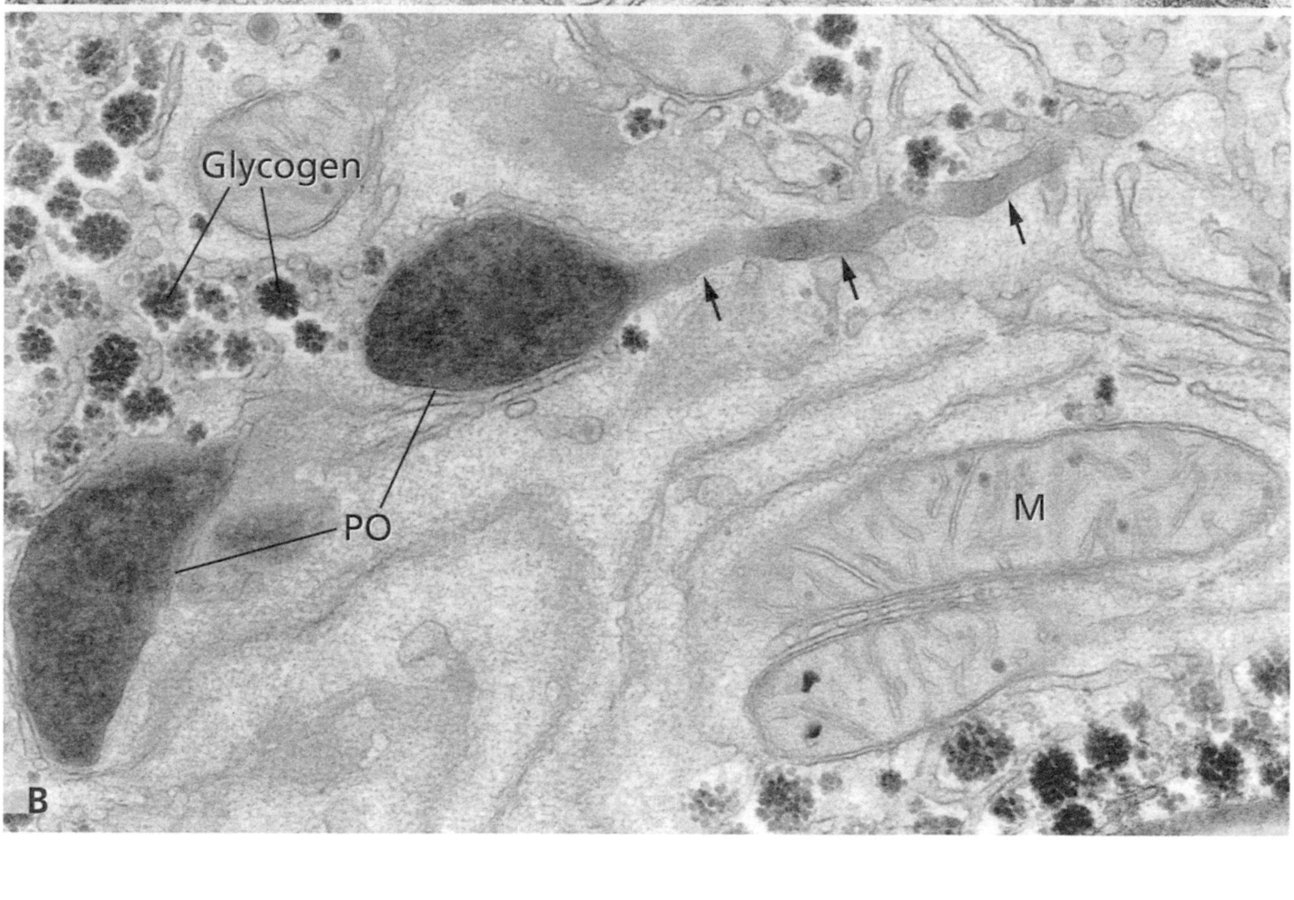
Glycogen
PO
M
B

PEROXISOMES: ADAPTIVE CHANGES

Peroxisomes are remarkably flexible organelles that adapt quickly to changing requirements. It is therefore not surprising that changes in the size and number of peroxisomes and in the composition and amount of oxidative enzymes occur in response to treatment with certain drugs and after exposure to environmental pollutants. The enzyme induction is not only dose-dependent but usually stronger in male compared with female experimental animals and is species related since rat and mice respond in a more pronounced way than guinea pigs and humans. The adaptive response is mediated by proteins belonging to the superfamily of steroid hormone nuclear receptors referred to as peroxisome proliferator activated receptors (PPARs). It has been shown that the peroxin Pex11p is involved in the regulation of peroxisome proliferation, whereas Pex14p has an essential role in selective peroxisome degradation.

An example of the peroxisomal changes observed in the liver after treatment of rats with a hypocholesterolemic drug is given in panels A and B. These animals received BM 15766, a compound that inhibits the 7-dehydrocholesterol-Δ^7-reductase. Peroxisomes were detected by histochemical demonstration of catalase. Panel A shows a control with catalase-positive, black stained peroxisomes (PO), mitochondria (M), and rough as well as smooth endoplasmic reticulum in a hepatocyte. As shown in panel B, treatment with BM 15766 resulted in a marked proliferation of peroxisomes (PO), which tended to form clusters and showed high variability in shape and size. Aggregates of peroxisomes were more often observed in female rats. These changes were more pronounced in the perivenous zone of the hepatic lobules. A similar regionally different response of peroxisomes in liver occurred in response to other xenobiotics. The proliferation of peroxisomes was accompanied by a proliferation of the smooth endoplasmic reticulum. Peroxisomes were often surrounded by multiple layers of smooth endoplasmic reticulum. This change was more pronounced in female animals.

Panels C–D provide examples of immunoelectron microscopic evaluation of a specific peroxisomal enzyme in response to different treatments. The changes in immunogold labelling can be assessed quantitatively by counting the number of gold particles and relating them to surface unit of peroxisomal matrix. Treatment of rats with recombinant human tumour necrosis factor-α (TNF-α) resulted in a marked reduction in immunogold labelling for the multifunctional peroxisomal enzyme hydratase-dehydrogenase-epimerase, as shown in panel C. The higher labelling intensity for this peroxisomal enzyme in peroxisomes of untreated control rats can be appreciated easily in panel E. Treatment of rats with the cholesterol synthesis inhibitor bezafibrate, however, resulted in induction of the enzyme, and this was reflected in a marked increase of immunogold labelling (panel D).

References

Baumgart E, Stegmeier K, Schmidt F, and Fahimi HD (1987) Proliferation of peroxisomes in pericentral hepatocytes of rat liver after administration of a new hypocholesterolemic agent (BM 15766). Lab Invest 56: 55

Beier K and Fahimi HD (1992) Environmental pollution by common chemicals and peroxisome proliferation: efficient detection by cytochemistry and automatic image analysis. Progr Histochem Cytochem 25: 150

Bellu A, Komori M, Van der Klei I, Kiel J, and Veenhuis M (2001) Peroxisome biogenesis and selective degradation converge at Pex14p. J Biol Chem 276: 44570

Chang C, South S, Warren D, Jones J, and Moser A (1999) Metabolic control of peroxisome abundance. J Cell Sci 112: 1579

Erdmann R and Blobel G (1995) Giant peroxisomes in oleic acid-induced Saccharomyces cerevisiae lacking the peroxisomal membrane protein Pmp27p. J Cell Biol 135: 111

Green S and Wahli W (1994) Peroxisome proliferator-activated receptors – finding the orphan a home. Mol Cell Endocrinol 100: 149

Lindauer M, Beier K, Völkl A, and Fahimi HD (1994) Zonal heterogeneity of peroxisomal enzymes in rat liver: differential induction by three divergent hypolipidemic drugs. Hepatology 20: 475

Moody D and Reddy J (1976) Morphometric analysis of the ultrastructural changes in rat liver induced by the peroxisome proliferator SaH 42-348. J Cell Biol 71: 768

Magnification: x 35,000 (A, B); x 62,000 (C–E)

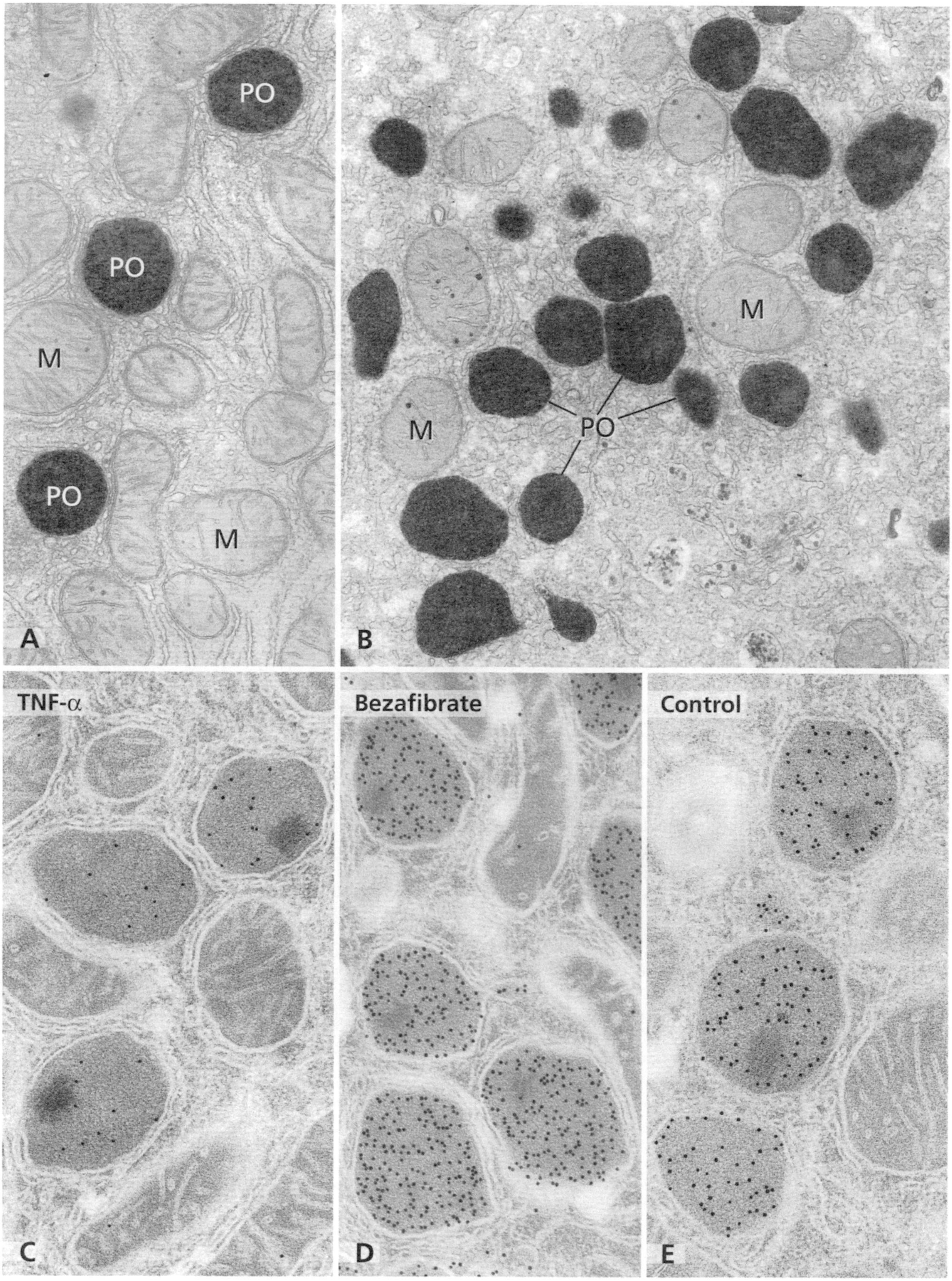
PO
PO
M
PO
M
M
PO
M
A
B
TNF-α
Bezafibrate
Control
C
D
E

PEROXISOMAL DISORDERS

Peroxisomal disorders are inherited recessively and are due either to deficiencies in single peroxisomal enzymes or in peroxisome biogenesis.

Single peroxisomal enzyme deficiencies mainly affect the lipid metabolism and are represented by β-oxidation disorders, defects in plasmalogen (etherphospholipid) biosynthesis, isoprenoid biosynthesis, and detoxification. Peroxisome biogenesis disorders (PBDs) are as complex as the biogenesis of peroxisomes. For example, the Zellweger spectrum comprises the Zellweger syndrome, neonatal adrenoleukodystrophy and infantile Refsum's disease. All PBDs are characterized by defective peroxisomes. In patients with Zellweger syndrome, absence of peroxisomes or structurally altered mitochondria are a classic finding. Depending on the severity of a PBD, a noticeable number of peroxisomes can be observed by electron microscopy. The explanation for these findings is that PBDs are caused by defects in peroxisomal matrix protein import. However, the targeting of peroxisomal membrane proteins is not affected. Therefore, varying numbers of peroxisome ghosts can be observed by electron microscopy. Depending on the severity of the import defect, some peroxisomes may contain variable amounts of matrix proteins.

The molecular defect in PBDs lies in mutations in various peroxin genes, and most of the human *PEX* genes can be affected. The peroxin Pex5p is the receptor for PST1 and involved in the initial steps of peroxisomal matrix transport. Severe Pex5p deficiency causes Zellweger syndrome and less severe deficiency infantile adrenoleukodystrophy. By electron microscopy, all studied patients deficient in Pex5p had cells with numerous peroxisomes that were lacking matrix proteins to a different degree.

Animal models of Zellweger syndrome were created by disrupting the *PEX5* gene, which permitted detailed morphological and biochemical analyses. In contrast to normal mice, the $PEX5^{-/-}$ animals had a severe peroxisomal matrix import defect and no detectable peroxisomes. Panel A shows several catalase-positive peroxisomes (Po), abundant glycogen and mitochondria (M) in a hepatocyte from control mouse ($PEX^{+/+}$). Panels B and C illustrate the fine structural changes in hepatocytes of Zellweger syndrome ($PEX5^{-/-}$) mice. They lacked typical peroxisomes, but a few hepatocytes contained peroxisome ghosts as identified by immunolabelling for PMP70. However, hepatocytes had an increased number of aggregates of pleomorphic mitochondria, which showed abnormalities of their outer and inner membrane and most prominently of cristae. The latter consisted mainly of rarification of cristae, and curvilinear and circular alterations of the cristae (arrows). These ultrastructural changes were associated with lower expression levels and reduced activity of mitochondrial respiratory chain complexes. It has been proposed that the mitochondrial alterations were caused by defective rescue of reactive oxygen species and defective detoxification function of peroxisomes.

References

Baumgart E, Vanhoorebeek I, Grabenbauer M, Borgers M, Declerq P, Fahimi H D, and Baes M (2001) Mitochondrial alterations caused by defective peroxisomal biogenesis in a mouse model for Zellweger syndrome (PEX5 knockout mouse). Am J Pathol 159: 1477

Chang C, South S, Warren D, Jones J, and Moser A (1999) Metabolic control of peroxisome abundance. J Cell Sci 112: 1579

Gould SJ, Raymond GV, and Valle D (2001) The peroxisome biogenesis disorders. In: The metabolic and molecular bases of inherited disease (Scriver C, Beaudet A, Valle D, Sly WS, Childs B, Kinzler K, and Vogelstein B, eds). New York: McGraw-Hill, pp 3181

Wanders R, Barth P, and Heymans H (2001) Single peroxisomal enzyme deficiencies. In: The metabolic and molecular bases of inherited disease (Scriver C, Beaudet A, Valle D, Sly WS, Childs B, Kinzler K, and Vogelstein B, eds). New York: McGraw-Hill, pp 3219

Weller S, Gould S, and Valle D (2003) Peroxisome biogenesis disorders. Annu Rev Genomics Hum Genet 4: 165

Wiemer E, Nuttley W, Bertolaet B, Li X, Franke U, Wheelock M, Anné U, Johnson K, and Subramani S (1995) Human peroxisomal targeting signal-1 receptor restores peroxisomal protein import in cells from patients with fatal peroxisomal disorders. J Cell Biol 130: 51

Magnification: x 54,000 (A); x 36,000 (B, C)

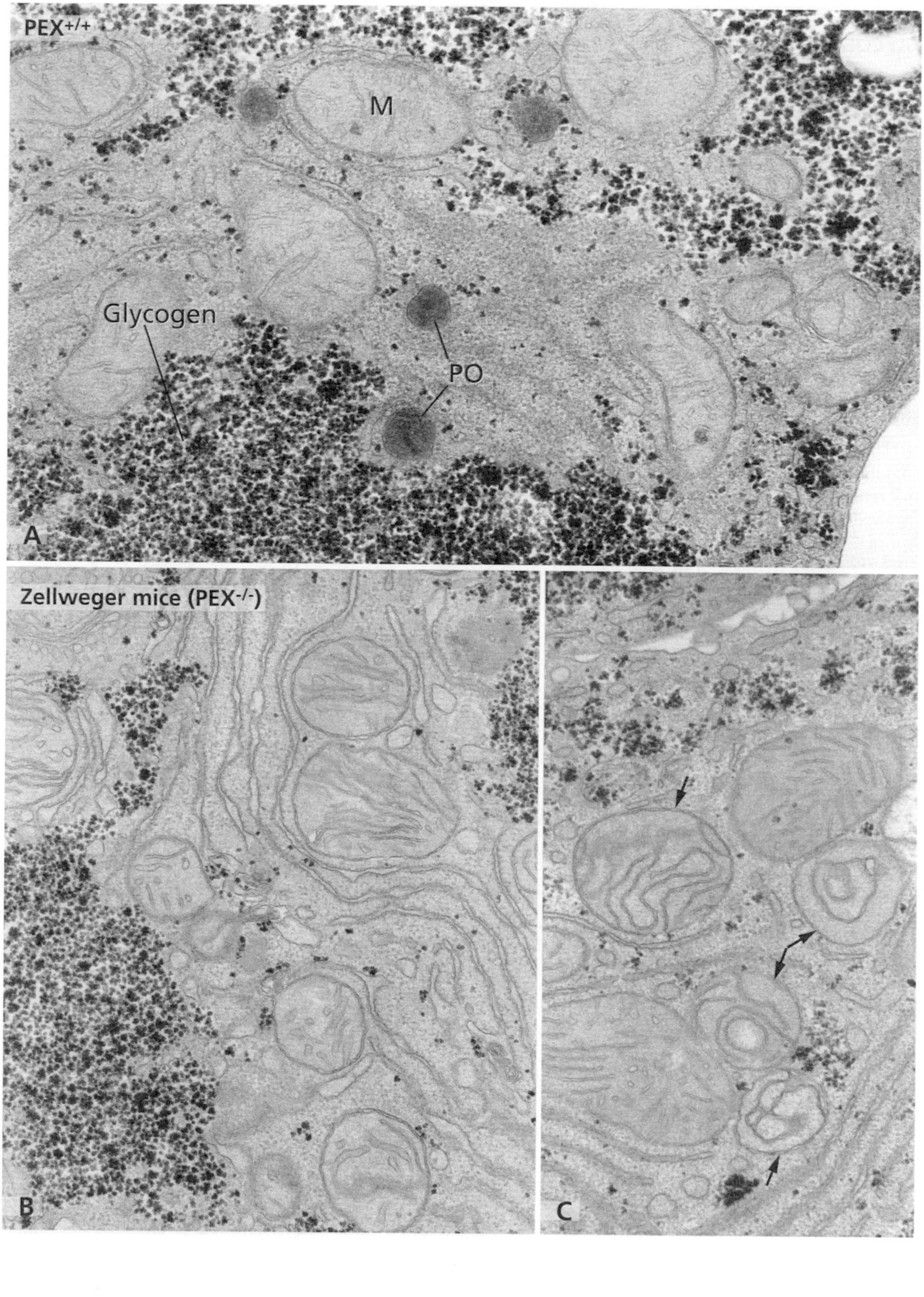
PEX+/+
M
Glycogen
PO
A
Zellweger mice (PEX-/-)
B
C

GLYCOGEN

Glucose is an important source for energy and glycogen is its cellular storage form, which is most abundant in liver and muscle. Glycogen is found in the cytoplasm in the form of granules ranging from 10 to 40 nm in diameter, the so-called β particles, which are typical for muscle cells. In hepatocytes, the β particles assemble to form characteristic rosettes of glycogen, the α particles (arrows). The α particles do not consist solely of glycogen but additionally contain various enzymatic proteins involved in the synthesis of glycogen, hence the name glycosomes. During glycogen synthesis, glycogenin, which initiates the synthesis, and glycogen synthase, which elongates the glucose chain, form a complex with glucose.

The glycosomes are often closely related to the smooth endoplasmic reticulum. The smooth endoplasmic reticulum contains glucose-6-phosphatase, which is involved in the final step of breakdown of glycogen and hydrolyses glucose-6-phosphate to glucose and phosphate. Remarkably, this enzyme is present in high levels only in liver, kidney, and the insulin-producing pancreatic beta cells. Inherited deficiency of this enzyme results in a glycogen storage disease (see below).

References

Burchell A and Waddell ID (1991) The molecular basis of the hepatic microsomal glucose-6-phosphatase system. Biochim Biophys Acta 1092: 129

Rybicka KK (1996) Glycosomes – the organelles of glycogen metabolism. Tissue Cell 28: 253

GLYCOGENOSIS TYPE I

As complex as the biosynthesis and breakdown of glycogen are as multifaceted are the inherited disorders of glycogen metabolism of which currently over 12 disease types are known.

The type-I glycogen storage disease (glucose-6-phosphatase deficiency, von Gierke's disease) is an autosomal recessive trait and caused by deficiency of glucose-6-phosphatase activity. The disease may be caused by a partial or complete deficiency of the catalytic enzyme subunit or the entire enzyme. The gene coding for the enzyme has been mapped to chromosome 17q21 and that for the translocase to chromosome 11q23. Mutations that affect the transmembrane domain of glucose-6-phosphatase cause more severe reduction in enzyme activity than mutations in one of the two luminal loops. Interestingly, ethnic-specific mutations have been found.

By electron microscopy, the glycogen deposits are observed as glycogen particles in the cytosol of the hepatocytes (panel B). These deposits are massive and fill most of the cytoplasm of the hepatocytes. Structurally similar cytosolic glycogen depositions occur in the other types of glycogen storage diseases, with the exception of one lysosomal glycogen storage disease (cf. Fig. 61). The accumulation of glycogen occurs in liver, kidney and intestinal mucosa and causes hypoglycemia and lactic acidosis. The treatment is directed to establish and maintain normal blood glucose concentrations by special nutritional regimen.

References

Annabi B, Hiraiwa H, Mansfield BC, Lei KJ, Ubagai T, Polymeropoulos MH, Moses SW, Parvari R, Hershkovitz E, Mandel H, et al. (1998) The gene for glycogen-storage disease type 1b maps to chromosome 11q23. Am J Hum Genet 62: 400

Burchell A (1992) The molecular basis of the type-1 glycogen storage diseases. Bioessays 14: 395

Chen Y (2001) Glycogen storage diseases. In: The metabolic and molecular bases of inherited diseases (Scriver C, Beaudet A, Valle D, and Sly WS, eds). New York: McGraw-Hill, pp 1521

Chou J and Mansfield B (1999) Molecular genetics of type-1 glycogen storage diseases. Trends Endocrinol Metab 10: 104

Cori G and Cori C (1952) Glucose-6-phosphatase of the liver in glycogen storage disease. J Biol Chem 199: 661

Lei KJ, Shelly LL, Pan CJ, Sidbury JB, and Chou JY (1993) Mutations in the glucose-6-phosphatase gene that cause glycogen storage disease type 1a. Science 262: 580

Veiga-da-Cunha M, Gerin I, Chen YT, de Barsy T, de Lonlay P, Dionisi-Vici C, Fenske CD, Lee PJ, Leonard JV, Maire I, et al. (1998) A gene on chromosome 11q23 coding for a putative glucose- 6-phosphate translocase is mutated in glycogen-storage disease types Ib and Ic. Am J Hum Genet 63: 976

von Gierke E (1929) Hepato-nephro-megalia glycogenica (Glykogenspeicherkrankheit der Leber und Nieren). Beitr Pathol Anat 82: 497

Magnification: x 23,500 (A); x 7,200 (B)

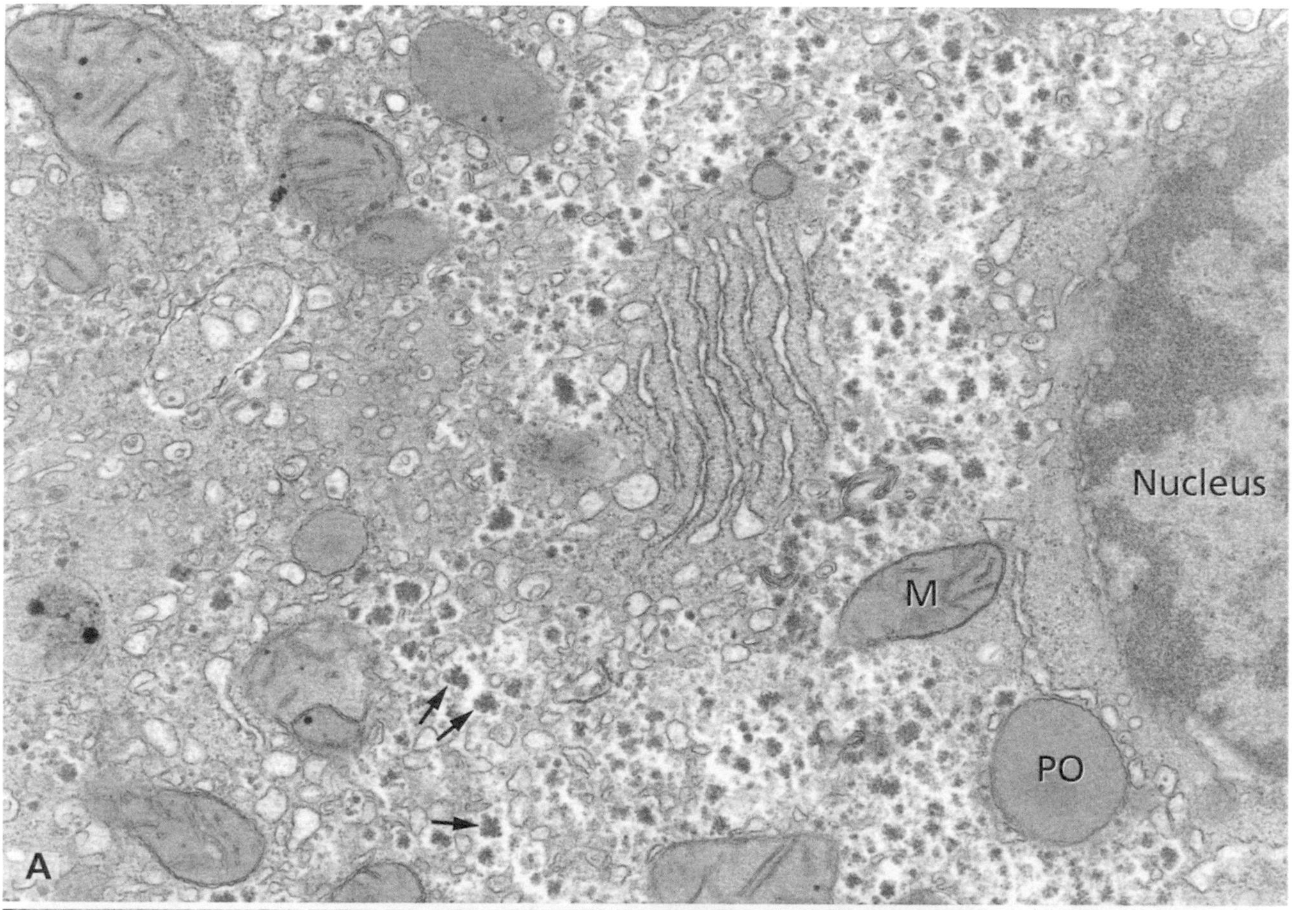
Nucleus
M
PO
A

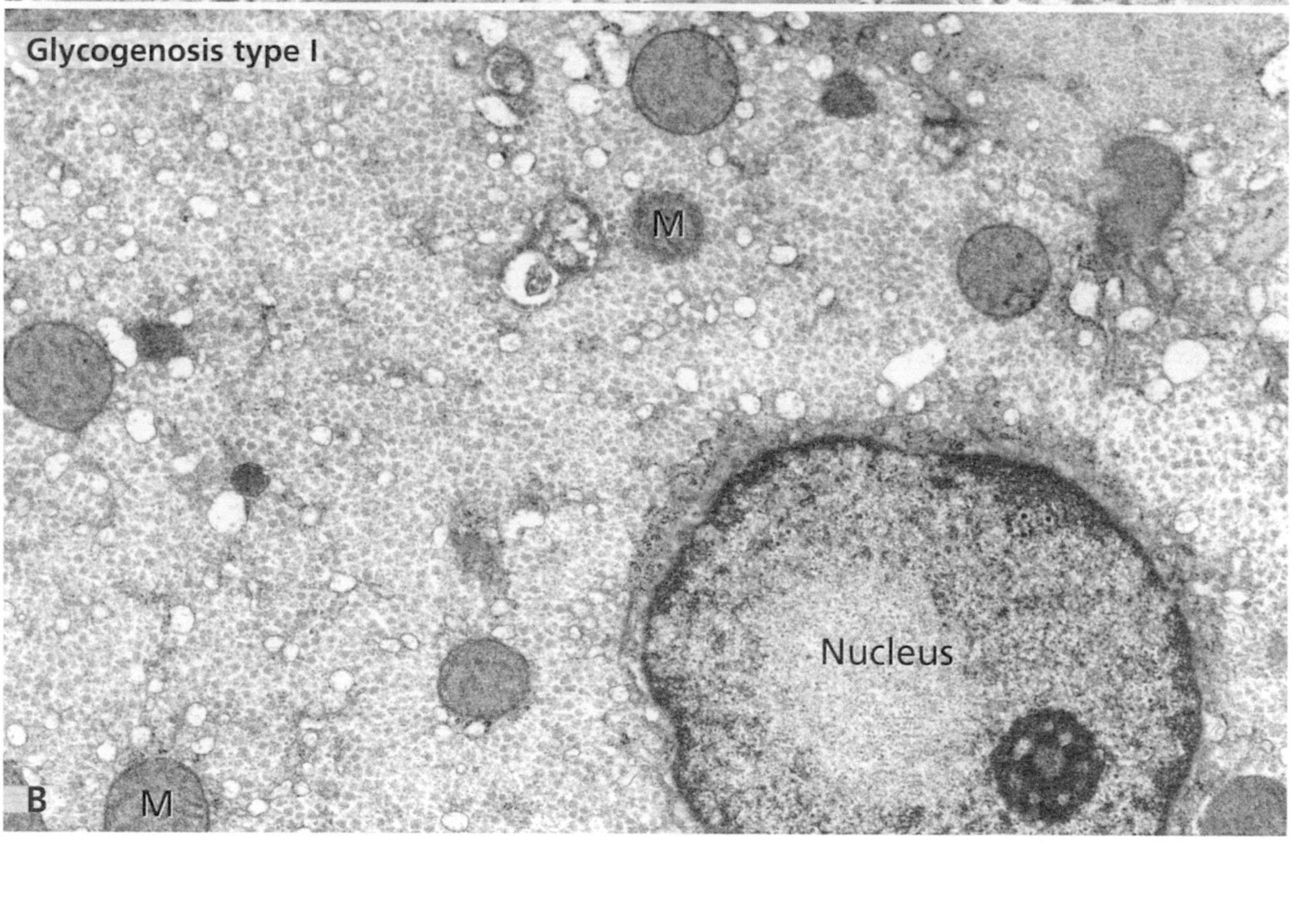
Glycogenosis type I
M
Nucleus
M
B

ERYTHROPOIETIC PROTOPORPHYRIA

This is an autosomal dominant disease in which a partial deficiency of the enzyme ferrochelatase is the causative defect. Ferrochelatase is present in the inner mitochondrial membrane and is the last acting enzyme in the haem biosynthesis. It inserts the iron into protoporphyrin IX to yield the haem. The gene locus coding for the enzyme is at chromosome 18q23.1 and a whole spectrum of disease-causing mutations has been detected in all 11 exons. In patients, protoporphyrin accumulates in the erythroid cells of bone marrow and circulating erythrocytes and is elevated in the plasma, bile and feces and can cause cutaneous photosensitivity.

The liver is affected only in a minority of the patients giving rise to characteristic hepatobiliary complications. By electron microscopy, numerous star-burst, crystalline inclusions of varying sizes (arrows) are found in the cytoplasm of liver hepatocytes. These characteristic inclusions are composed of a filamentous crystalline material. They are also present in Kupffer cells, bile ductal epithelia and ductal lumen as well as bile canalicular lumens.

References

Gouya L, Deybach JC, Lamoril J, Da Silva V, Beaumont C, Grandchamp B, and Nordmann Y (1996) Modulation of the phenotype in dominant erythropoietic protoporphyria by a low expression of the normal ferrochelatase allele. Am J Hum Genet 58: 292

Gouya L, Puy H, Lamoril J, Da Silva V, Grandchamp B, Nordmann Y, and Deybach JC (1999) Inheritance in erythropoietic protoporphyria: a common wild-type ferrochelatase allelic variant with low expression accounts for clinical manifestation. Blood 93: 2105

Magnus: I, Jarrett A, Prankerd T, and Rimington C (1961) Erythropoietic protoporphyria. A new porphyria syndrome with solar urticaria to protoporphyrinaemia. Lancet 2: 448

Rademakers LH, Cleton MI, Kooijman C, Baart de la Faille H, and van Hattum J (1990) Early involvement of hepatic parenchymal cells in erythrohepatic protoporphyria? An ultrastructural study of patients with and without overt liver disease and the effect of chenodeoxycholic acid treatment. Hepatology 11: 449

Rademakers LH, Cleton MI, Kooijman C, Baart de la Faille H, and van Hattum J (1991) Ultrastructural aspects of the liver in erythrohepatic protoporphyria. Curr Probl Dermatol 20: 154

Rufenacht UB, Gouya L, Schneider-Yin X, Puy H, Schafer BW, Aquaron R, Nordmann Y, Minder EI, and Deybach JC (1998) Systematic analysis of molecular defects in the ferrochelatase gene from patients with erythropoietic protoporphyria. Am J Hum Genet 62: 1341

Taketani S, Inazawa J, Nakahashi Y, Abe T, and Tokunaga R (1992). Structure of the human ferrochelatase gene. Exon/intron gene organisation and location of the gene to chromosome 18. Eur J Biochem 205: 217

Magnification: x 18,000

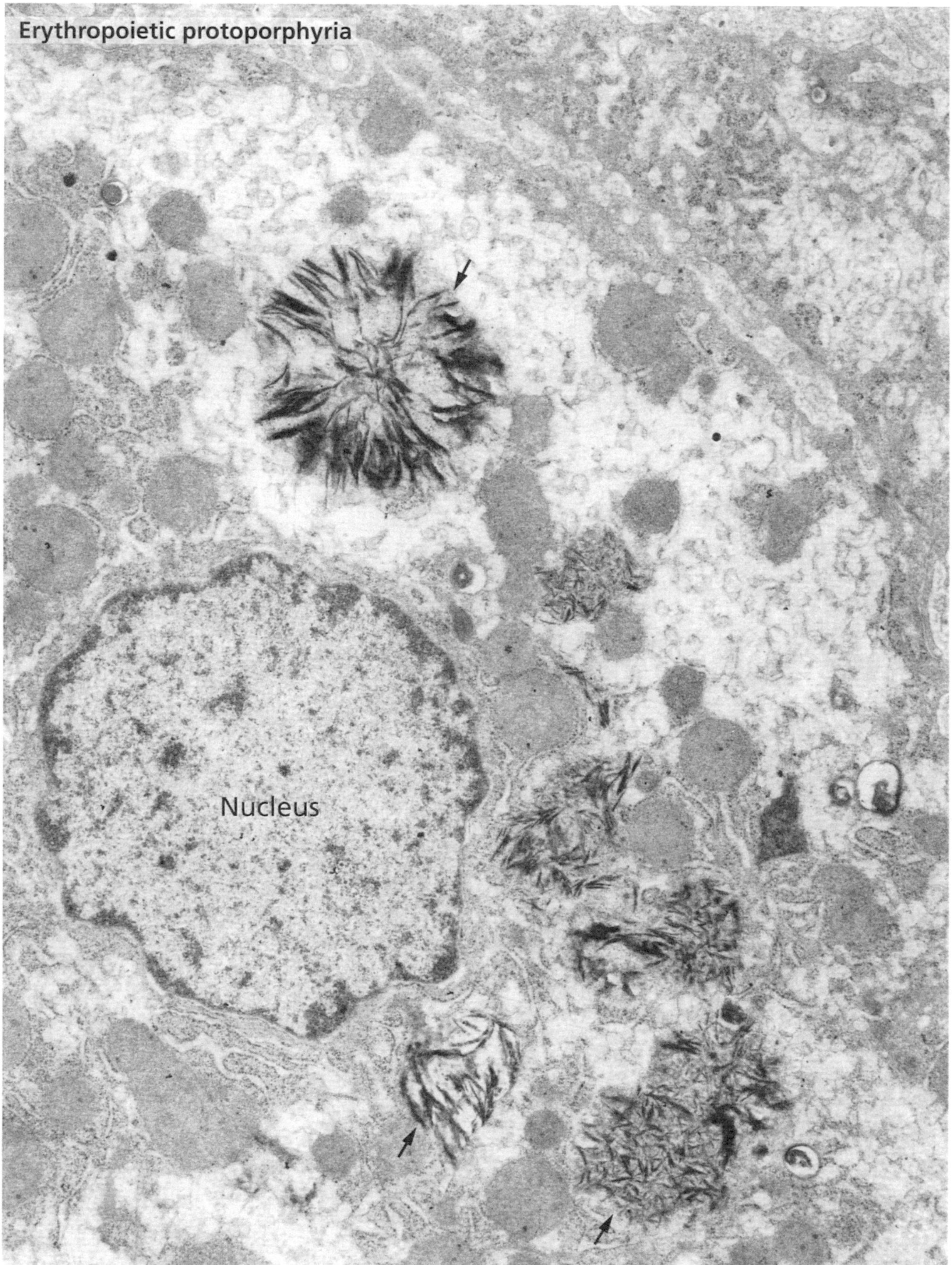
Erythropoietic protoporphyria
Nucleus

CYTOCENTRE, CENTROSOME, AND MICROTUBULES

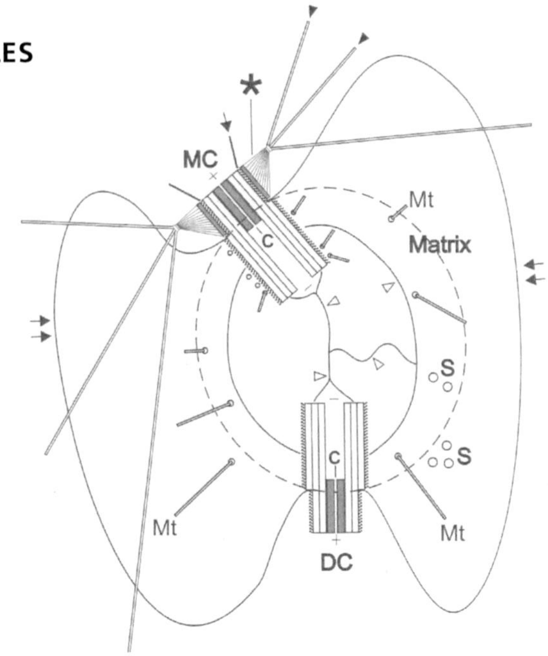

The electron micrograph shows the cytocentre of a bone marrow cell of the granulopoietic lineage. It consists of the centrosome with one of the pair of centrioles in the centre, abundant radiating microtubules (arrows), and stacks of the Golgi apparatus organised in a circle around the centrosome. In interphase cells, the cytocentre is most commonly located close to the nucleus, which can be seen in the uppermost part of the micrograph. The centrosome is dynamic and thus, it is important to distinguish between growing and quiescent cells. It contains the centrioles embedded in an amorphous protein matrix and functions as the microtubule-organising centre with gamma-tubulin rings in the matrix, serving as nucleation sites for the growth of microtubules.

Microtubules are essential components of the cytoskeleton and involved in multiple functions such as intracellular transport of organelles, cell migration, and correct segregation of chromosomes during mitosis. Microtubules measure 24 nm in diameter; the wall consists of 13 protofilaments of α- and β-tubulin dimers in a circular array, which polymerise in an end-to-end fashion. The stable minus ends of the microtubules are embedded in the centrosome, and the dynamic plus ends elongate towards the cell periphery forming kinds of tracks within the cells along which organelles are transported with the help of motor proteins. This network is constantly remodelled by growth and sudden shrinkage of the microtubules, a process known as dynamic instability. The centrioles, to which microtubules are anchored, are not only the most prominent structures in the cytocentre but also have a pivotal role for the structural organisation of the centrosome. A centriole has the shape of a cylinder of about 0.2 μm length, the walls consist of nine triplets of microtubules oriented parallel to the longitudinal axis. The microtubules are fused and have a common wall. Only the innermost A-tubule consists of a complete ring of 13 protofilaments, the B- and C-microtubules are crescent-shaped, each consisting of 10 protofilaments. The microtubule triplets are visible in the cross-section of the centriole in the centre of the micrograph, which also shows cartwheel-shaped structures in its lumen, where centrin (C in the diagram) is known to be localised, and a ring of appendages, where microtubule asters are anchored.

The diagram (drawn according to Bornens, 2002) shows the main components of the centrosome of a cell in the G1 phase containing a differentiated mother centriole (MC) and a daughter centriole (DC). The centrioles are linked by a matrix (dotted line) and embedded in a larger pericentriolar area (outer line, double arrows). Assembly of the matrix is assumed to be triggered by the centrioles through various microtubule-binding proteins (open arrowheads). Juxtacentriolar structures, known as satellite (S), are seen as precursor complexes occurring during centriole duplication. Microtubules are nucleated in the vicinity of both centrioles (Mt), but only the fully differentiated mother centrioles possess appendages, distal appendages (arrow) and subdistal appendages (asterisk), where microtubule asters (filled arrowheads) are anchored. The centriole shown in the micrograph is cross-sectioned through its distal part, thus both the centrin core and a corona of subdistal appendages with prominent tips, are visible. Part of a microtubule aster is to be seen associated with the right lower appendage.

References

Barr F and Egerer J (2005) Golgi positioning: are we looking at the right MAP? J Cell Biol 168: 993

Bornens M (2002) Centrosome composition and microtubule anchoring mechanisms. Curr Opin Cell Biol 14: 25

Galjart N and Perez F (2003) A plus-end raft to control microtubule dynamics and function. Curr Opin Cell Biol 15: 48

Janson ME, de Dood ME, and Dogterom M (2003) Dynamic instability of microtubules is regulated by force. J Cell Biol 161: 1029

Rios RM and Bornens M (2003) The Golgi apparatus at the cell centre. Curr Opin Cell Biol 15: 60.

Stearn T (2004) The centrosome yields its secrets. Nature Cell Biol 6: 14

Strnad P and Gönczy P (2008) Mechanisms of procentriole formation. Trends Cell Biol 18: 389

Magnification: x 78,500

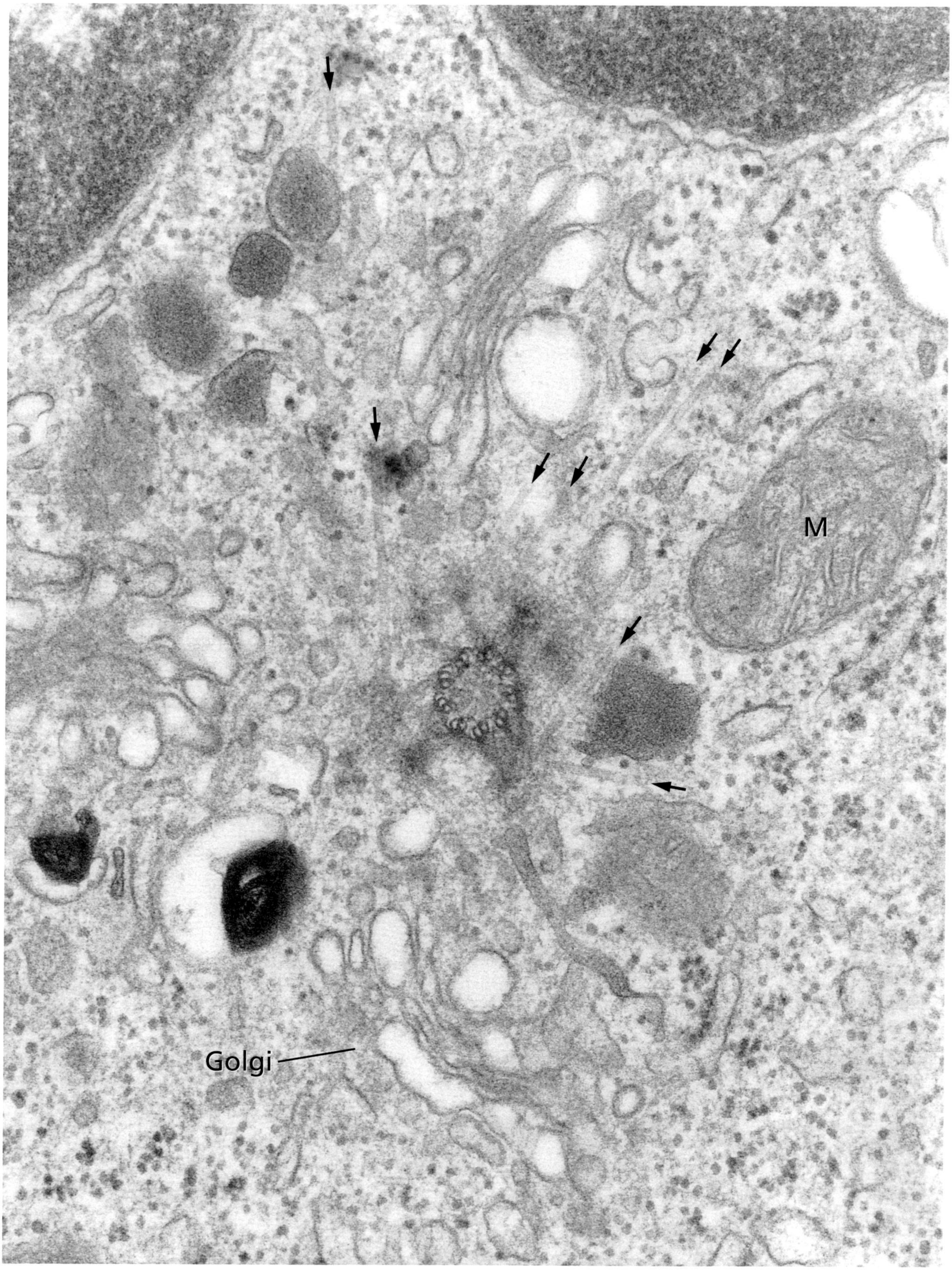
M
Golgi

EFFECTS OF MICROTUBULE DISRUPTION

The involvement of microtubules in main cellular tasks can indirectly be studied by impairment of microtubule function either by disruption of the microtubule network, e.g. by treatment with colchicine or nocodacole, or by administration of substances that stabilise microtubules, e.g. taxol. Because of the important role of microtubules in the intracellular traffic, application of antimicrotubular agents leads to an impairment or block of multiple cellular pathways, including biosynthetic and secretory routes, endocytosis pathways, and transcytosis. Microtubules are required for regular organisation and localisation of the Golgi apparatus and also have a crucial role in the polarisation of cells. Disruption of the microtubule system leads to characteristic cell changes.

In many types of cells, the Golgi apparatus is localised around the centrosome (cf. Fig. 72) and several non-motor microtubule-binding proteins have been shown to be associated with the Golgi apparatus. The transport of pre-Golgi intermediates from endoplasmic reticulum exit sites in peripheral regions of cells to the cytocentre and finally the Golgi apparatus also occurs along microtubules and involves minus end-directed motor proteins. After disruption of the cytoplasmic microtubules, the Golgi apparatus loses its characteristic position, becomes vacuolised, and dispersed throughout the cytoplasm. This is shown in small intestinal absorptive cells of rats 6 hours after treatment with colchicine (panels A and B). The Golgi apparatus, which is localised normally in the supranuclear cytoplasm of the absorptive cells (cf. Fig. 111), has disappeared from this site and vacuolised Golgi components are present in uncommon positions close to the basal cell surface (arrowheads). The former Golgi stacks, although redistributed and vacuolised, still contain lipoprotein particles (asterisks), alike Golgi cisternae of untreated cells. Treatment with colchicine also leads to changes of the typical polarity of the absorptive cells. A brush border of densely packed microvilli, which is usually restricted to the apical cell surfaces, occurs at the basolateral cell surfaces (Bb in panels A and C). It occupies more than 3% of the basolateral surface at 6 hours after colchicine administration. The diagram summarises the cellular changes occurring after microtubule disruption 6 hours after administration of colchicine.

C: cytocentre, GA: Golgi apparatus, G: dispersed Golgi elements, aBb: apical brush border, bBb: basolateral brush border.

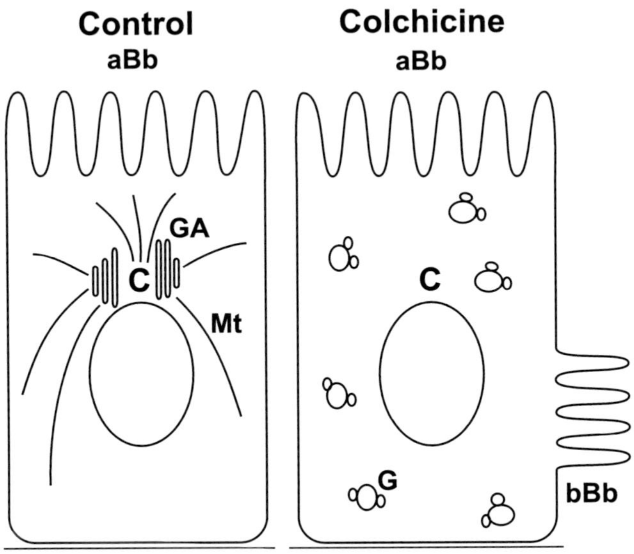

References

Brownhill K, Wood L, and Allan V (2009) Molecular motors and the Golgi complex: Staying put and moving through. Seminars Cell Dev Biol 20: 784

Cole NB, Sciaky N, Marotta A, Song J, and Lippincott-Schwartz J (1996) Golgi dispersal during microtubule disruption: regeneration of Golgi stacks of peripheral endoplasmic reticulum exit sites. Mol Biol Cell 7: 631

Ellinger A, Pavelka M, and Gangl A (1983) Effect of colchicine on rat small intestinal absorptive cells. II. Distribution of label after incorporation of (^{3}H)fucose into plasma membrane glycoproteins. J Ultrastr Res 85: 260

Pavelka M and Gangl A (1983) Effects of colchicine on the intestinal transport of endogenous lipid. Ultrastructural, biochemical, and radiochemical studies in fasting rats. Gastroenterology 84: 544

Reaven EP and Reaven GM (1977) Distribution and content of microtubules in relation to the transport of lipid. An ultrastructural quantitative study of the absorptive cell of the small intestine. J Cell Biol 75: 559

Rios RM, Sanchis A, Tassin AM, Fedriani C, and Bornens M (2004) GMAP-210 recruits γ-tubulin complexes to *cis*-Golgi membranes and is required for Golgi ribbon formation. Cell 118: 323

Sandoval IV, Bonifacino JS, Klausner RD, Henkart M, and Wehland J (1984) Role of microtubules in the organisation and localisation of the Golgi apparatus. J Cell Biol 99: 113s

Thyberg J and Moskalewski S (1985) Microtubules and the organisation of the Golgi complex. Exp Cell Res 159: 1

Magnification: x 14,000 (A); x32,000 (B); x 25,000 (C)

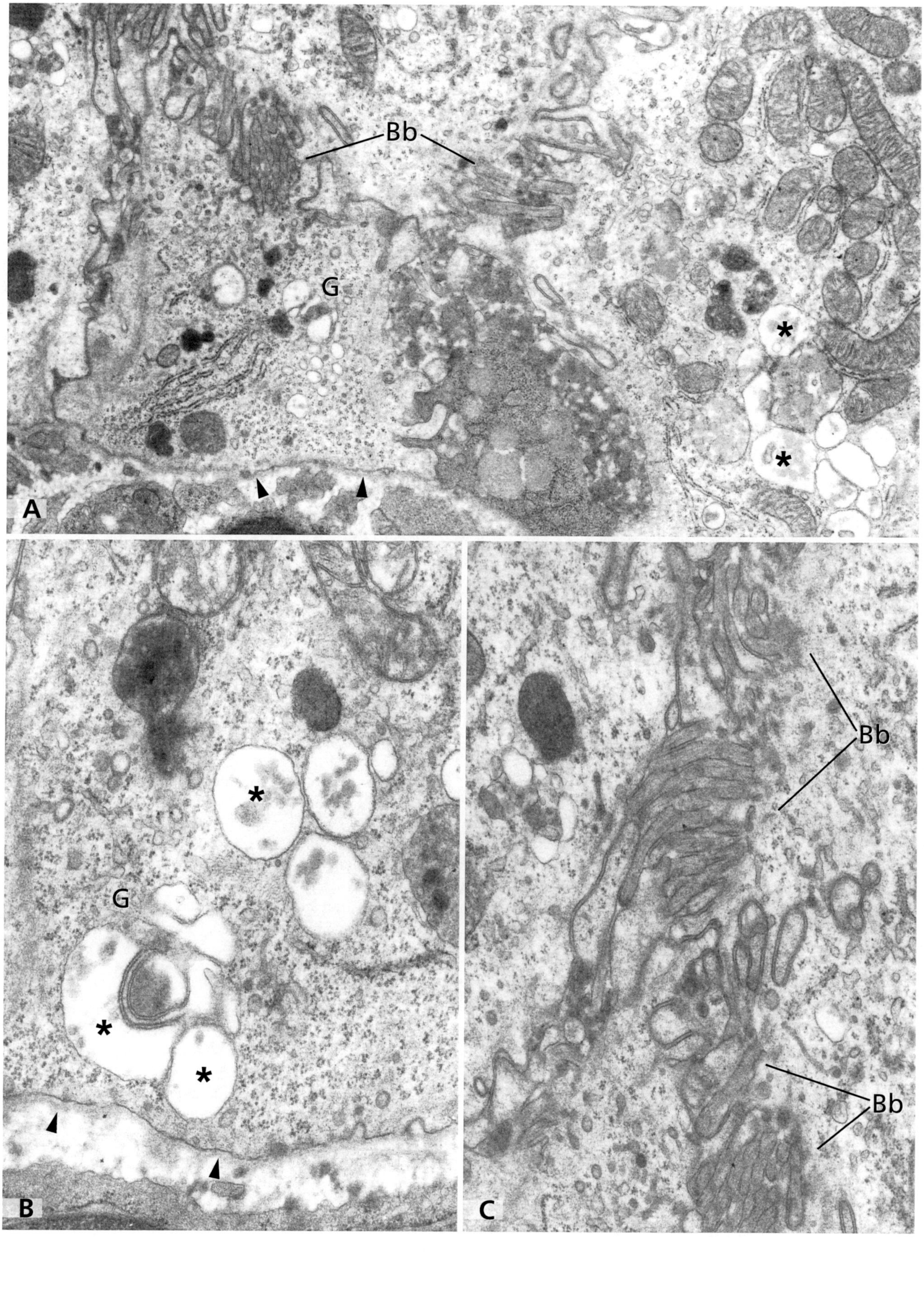
Bb
G
A
B
G
C
Bb
Bb

ACTIN FILAMENTS

Actin filaments form a dynamic skeletal and motility system in all eukaryotic cells. Interacting with the other main components of the cytoskeleton, microtubules (cf. Fig. 72) and intermediate filaments (cf. Fig. 75), they are involved in all kinds of cellular motion, including cell migration and division, changes of cell shapes during exocytosis and endocytosis, formation of filopodia and lamellipodia during cell crawling, and transport of organelles and particles, including internalised microorganisms. Free actin molecules (G-actin) in the cytoplasm assemble by polymerisation into a linear double helical array to form filaments, measuring 6–8 nm in diameter. Similar to microtubules, actin filaments are polarised structures possessing a dynamic, fast growing plus end and a slowly growing minus end. Multiple actin-associated proteins are involved in the regulation of actin polymerisation, and the arrangements and remodelling of higher architectures, such as networks and filament bundles. Actin is recognised to be responsible for the viscoelastic properties of cells. According to the requirements, actin filaments build up stabile or dynamic zones, cortical networks beneath the cell surfaces reinforcing the plasma membrane and excluding organelles, stress fibres forming cables between adhesive junctions, or tracks for movement of myosin motor proteins. Actin assembly also is reported to power the movement of intracellular organelles. By actin filament comets, internalised pathogens such as Listeria monocytogenes propel themselves through the cytoplasm.

Panels A–E show examples of actin filament arrangements. Within the microvilli of the brush border of resorptive cells, actin filaments, crosslinked and stabilised by associated proteins, such as fimbrin, fascin, and villin, and connected with the plasma membrane by myosins and calmodulin, form regular bundles shown longitudinally sectioned in panel A and cross-sectioned in panel C. The filamentous core of the microvilli continues into rootlets that protrude into the cytoplasm beneath the brush border and, by being interconnected by spectrins and attached to intermediate cytokeratin filaments, contribute to the arrangement of the terminal web. The filamentous system of the brush border and the terminal web has a dual stabilising and motility function. It is responsible for the upright positions of the microvilli and the allover organisation of the brush border. Furthermore, by being in connection with the belt desmosome of the junctional complex, it builds up a cell-to-cell spanning motion system that changes the diameters of the apical parts of the resorptive cells. This leads to a tilting of the brush border microvilli, facilitating contact with nutrients to be ingested. Panel D shows cross-sectioned actin filament rootlets. Profiles of plasma membrane invaginations are visible close to the actin filament bundles. They are assumed to represent a membrane reservoir for adaptations of the apical cell surface. Panels B and E show actin filament bundles and tuft-like filament arrangements (arrows) in the cytoplasm of high-pressure-frozen cultured hepatoma cells. Close associations with vesicular organelles (panel B) and cisternae of the endoplasmic reticulum (panel E) are visible, recalling the prominent function of actin filaments as a traffic system for cell constituents including transport of messenger-RNAs.

References

Denker SP and Barber DL (2002) Ion transport proteins anchor and regulate the cytoskeleton. Curr Opin Cell Biol 14: 214

De Heredia ML and Jansen RP (2004) mRNA localisation and the cytoskeleton. Curr Opin Cell Biol 16: 80

Fehrenbacher KL, Boldogh IR, and Pon LA (2003) Taking the A-train: actin-based force generators and organelle targeting. Trends Cell Biol 13: 472

Lee E and de Camilli P (2001) Dynamin at actin tails. Proc Natl Acad Sci USA 99: 161

Mejillano MR, Kojima S, Applewhite DA, Gertler FB, Svitkina TM, and Borisy GG (2004) Lamellipodial versus filopodial mode of the actin nanomachinery: Pivotal role of the filament barbed end. Cell 118: 363

Pollard TD, Blanchoin L, and Mullins RD (2000) Molecular mechanisms controlling actin filament dynamics in nonmuscle cells. Annu Rev Biophys Biomolec Struct 29: 545

Rodriguez OC, Schaefer AW, Mandato CA, Forscher P, Bement WM, and Waterman-Storer CM (2003) Conserved microtubule-actin interactions in cell movement and morphogenesis. Nat Cell Biol 5: 599

Schleicher M and Jockusch BM (2008) Actin: its cumbersome pilgrimage through cellular compartments. Histochem Cell Biol 129: 695

Small JV and Kaverina I (2003) Microtubules meet substrate adhesions to arrange cell polarity. Curr Opin Cell Biol 15: 40

Stamnes M (2002) Regulating the actin cytoskeleton during vesicular transport. Curr Opin Cell Biol 14: 428

Steven AC and Aebi U (2003) The next ice age: Cryo-electron tomography of intact cells. Trends Cell Biol 13: 107

Taunton J (2001) Actin filament nucleation by endosomes, lysosomes and secretory vesicles. Curr Opin Cell Biol 13: 85

Wallar BJ and Alberts AS (2003) The formins: active scaffolds that remodel the cytoskeleton. Trends Cell Biol 13: 435

Magnification: x 45,000 (A); x 44,000 (B); x 134,000 (C); x 77,500 (D); x 41,000 (E)

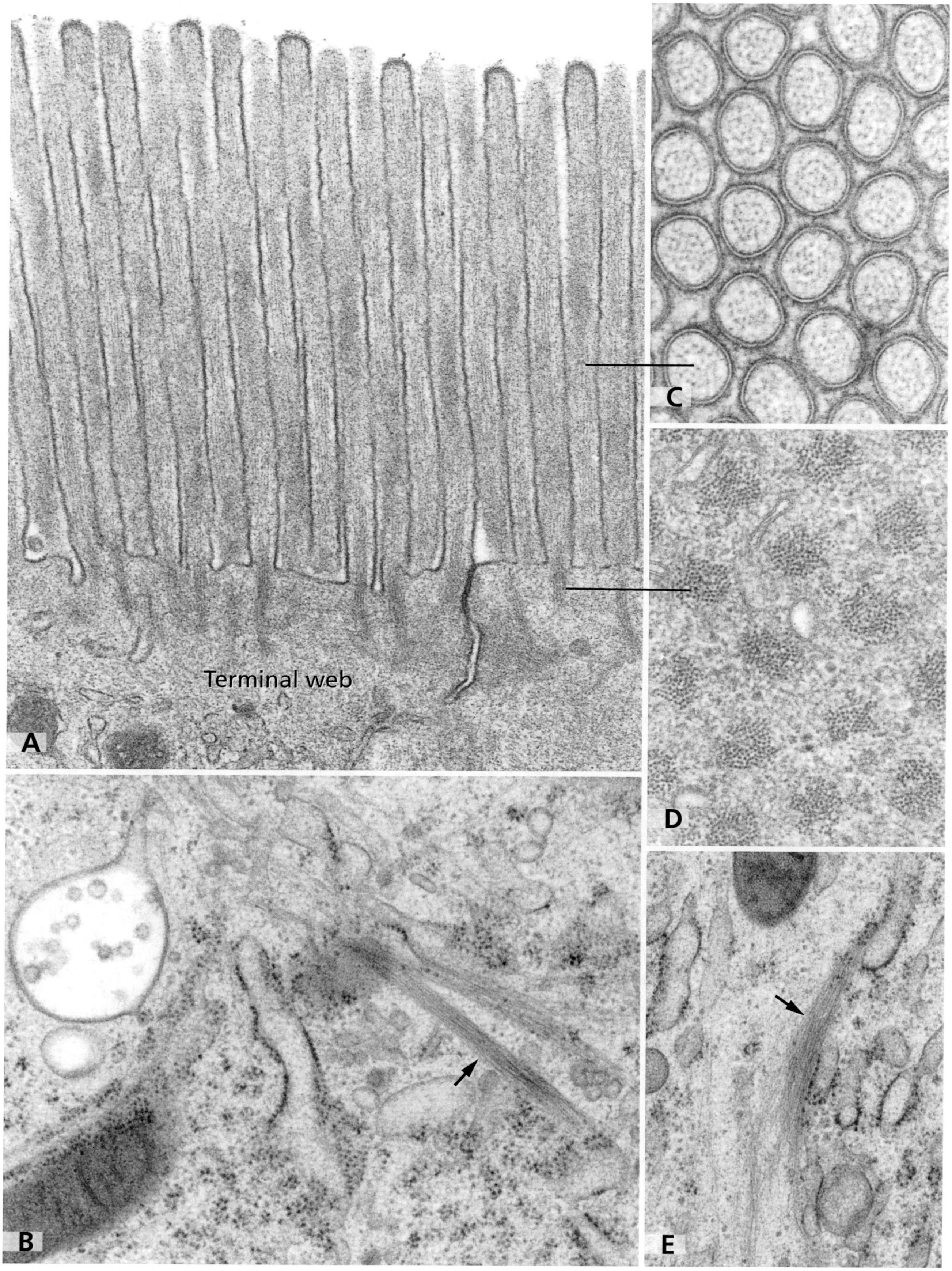
Terminal web
A
B
C
D
E

INTERMEDIATE FILAMENTS

The network of intermediate filaments, in addition to those of microtubuli (cf. Fig. 72) and actin filaments (cf. Fig. 74), constitutes the third part of the distinct, yet interconnected, cytoskeletal systems required for cellular stability and dynamics. Intermediate filaments, which are often organised in thick fibrils, provide mechanical support for the cell. However, the intermediate filament scaffolds within cells are not rigid but are dynamic, motile elements of the cytoskeleton. Intermediate filaments proved to be even more dynamic than other parts of the cytoskeleton. They are interconnected and in crosstalk with the microtubuli and actin filament motion systems. With the aid of molecular motors, such as kinesins and dyneins, filament precursors and short filaments are delivered with high speed to specific cellular regions, where they assemble to form long filaments.

The architecture of cytokeratin intermediate filaments is illustrated in panels A and B. Cytokeratins (keratins), grouped in acidic cytokeratins (type I) and basic ones (type II), are the proteins of intermediate filaments of epithelial cells. The electron micrographs show keratinocytes of the epidermal spinous layer. Cytokeratin filaments are bundled to form thick fibrils, the tonofilaments (arrows), which form a dense network within the cytoplasm excluding only the perinuclear areas. Tonofilaments radiate into the intercellular bridges and associate with the desmosomes, thus connecting adjacent cells and forming a cell spanning mechanical support of the tissue. In panel C, immunogold labelling for keratin is shown which resulted in an intense labelling of the network of cytokeratin filaments in a keratinocyte.

The term "intermediate filament" refers to the 10 nm diameter of the filaments being in between the 24 nm and 7 nm diameters of microtubules and actin filaments, respectively. Intermediate filaments are assembled from fibrous proteins encoded by a multigene family of more than fifty members. The proteins possess a mostly alpha-helical central rod domain that is flanked by globular domains on either end. Apolar protofilaments of 3 nm diameter are formed from pairs of helical monomers that twist around each other to generate coiled-coil dimers, which again twist around each other in antiparallel fashion to form staggered tetramers. By lateral associations, two protofilaments form protofibrils, and 4 protofibrils associate to build up a 10 nm intermediate filament. Intermediate filament proteins, including the lamins of the nuclear lamina (cf. Fig. 3), are classified on the basis of sequence similarities in the rod domain into six types. The patterns of intermediate filament protein expression are cell type-specific, and several distinct proteins are often co-expressed in a given cell type for a certain time during cell differentiation (cf. Figs. 119 and 120) and embryonic development.

References

Chou Y-H, Helfand BT, and Goldman RD (2001) New horizons in cytoskeletal dynamics: transport of intermediate filaments along microtubule tracks. Curr Opin Cell Biol 13: 106

Dinsdale D, Lee JC, Dewson G, Cohen GM, and Peter ME (2004) Intermediate filaments control the intracellular distribution of caspases during apoptosis. Am J Pathol 164: 395

Franke WW, Schiller DL, Moll R, Winter S, Schmid E, Engelbrecht I, Denk H, Krepler R, and Platzer B (1981) Diversity of cytokeratins. Differentiation specific expression of cytokeratin polypeptides in epithelial cells and tissues. J Mol Biol 153: 933

Fuchs E and Cleveland DW (1998) A structural scaffolding of intermediate filaments in health and disease. Science 279: 514

Helfand BT, Chang L, and Goldman RD (2004) Intermediate filaments are dynamic and motile elements of cellular architecture. J Cell Sci 117: 133

Herrmann H and Aebi U (2000) Intermediate filaments and their associates: multi-talented structural elements specifying cytoarchitecture and cytodynamics. Curr Opin Cell Biol 12: 79

Kumemura H, Harada M, Omary MB, Sakisaka S, Suganuma T, Namba M and Sata M (2004) Aggregation and loss of cytokeratin filament networks inhibit Golgi organisation in liver-derived epithelial cell lines. Cell Motility Cytoskeleton 57: 37

Lariviere RC and Julien JP (2004) Functions of intermediate filaments in neuronal development and disease. J Neurobiol 58:131

Moll R, Divo M, and Langbein L (2008) The human keratins: biology and pathology. Histochem Cell Biol 129: 705

Mücke N, Kreplak L, Kirmse R, Wedig T, Herrmann H, Aebi U, and Langowski J (2004) Assessing the flexibility of intermediate filaments by atomic force microscopy. J Molec Biol 335: 1241

Strnad P, Stumpner C, Zatloukal K, and Denk H (2008) Intermediate filament cytoskeleton of the liver in health and disease. Histochem Cell Biol 129: 735

Magnification: x 8,500 (A); x 27,000 (B); x 20,000 (C)

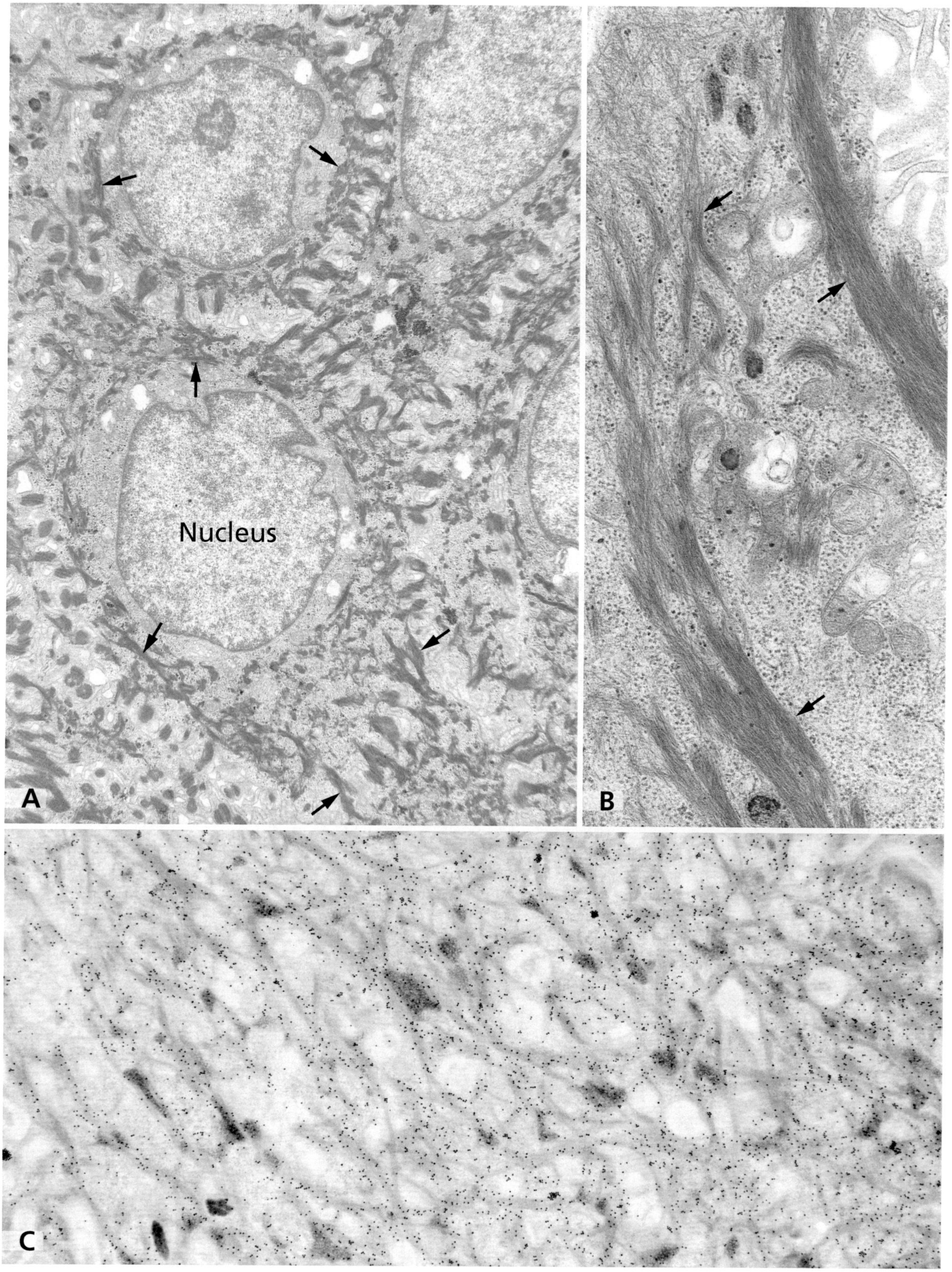
Nucleus
A
B
C

MALLORY BODIES

Mallory bodies represent cytoplasmic inclusion bodies that are different from aggresomes (cf. Fig. 20). They are characteristically observed in hepatocytes in chronic alcoholic liver disease. However, they are not specific for alcohol induced hepatocyte damage since they can be experimentally induced in mice by prolonged feeding of griseofulvin or dicarboxy-diethoxydihydro-collidine and were observed under various other diseased states and experimental conditions. Furthermore, Mallory bodies are not hepatocyte specific inclusion bodies since they were observed in other cell types such as alveolar pneumocytes and muscle fibres. Mallory bodies share compositional similarities with Lewy bodies found in Parkinson's disease, neurofibrillar tangles in Alzheimer's disease, and neuronal inclusions in motor neuron disease.

Mallory bodies (MB) are of a complex filamentous nature, composed of hyperphosphorylated and ubiquitinated keratins 8 and 18 and several non-keratin proteins such as proteasomes, p62 and chaperones. Their size can be variable, approaching that of the nucleus as seen in panel A. By transmission electron microscopy, Mallory bodies consist of a meshwork of randomly oriented intermediate filaments that are free of organelles such as the endoplasmic reticulum, mitochondria (M) and Golgi apparatus (panels A and B). The dashed line in panel B marks the border between the cytoplasm and a Mallory body. The intermediate filaments can be identified by immunoelectron microscopy as keratins (panel C). The Mallory bodies shown in panels A–C were induced in cultured rat hepatocytes stably expressing a misfolded polytope membrane protein.

References

Denk H, Franke WW, Eckerstorfer R, Schmid E, and Kerjaschki D (1979) Formation and involution of Mallory bodies ("alcoholic hyalin") in murine and human liver revealed by immunofluorescence microscopy with antibodies to prekeratin. Proc Natl Acad Sci USA 76: 4112

Denk H, Gschnait F, and Wolff K (1975) Hepatocellular hyalin (Mallory bodies) in long term griseofulvin-treated mice: a new experimental model for the study of hyalin formation. Lab Invest 32: 773

Denk H, Stumptner C, and Zatloukal K (2000) Mallory bodies revisited. J Hepatol 32: 689

Hirano K, Roth J, Zuber C, Ziak M (2002) Expression of a mutant ER-retained polytope membrane protein in cultured rat hepatocytes results in Mallory body formation. Histochem Cell Biol 117: 41

Jensen K and Gluud C (1994) The Mallory body: theories on development and pathological significance. Hepatology 20: 1330

Kawaguchi Y, Kovacs JJ, McLaurin A, Vance JM, Ito A, and Yao TP (2003) The deacetylase HDAC6 regulates aggresome formation and cell viability in response to misfolded protein stress. Cell 115: 727

Komatsu M, Waguri S, Koike M, Sou YS, Ueno T, Hara T, Mizushima N, Iwata J, Ezaki J, Murata S, et al. (2007) Homeostatic levels of p62 control cytoplasmic inclusion body formation in autophagy-deficient mice. Cell 131: 1149

Nakanuma Y and Ohta G (1986) Expression of Mallory bodies in hepatocellular carcinoma in man and its significance. Cancer 57: 8

Seibenhener ML, Babu JR, Geetha T, Wong HC, Krishna NR, and Wooten MW (2004) Sequestosome 1/p62 is a polyubiquitin chain binding protein involved in ubiquitin proteasome degradation. Mol Cell Biol 24: 8055

Stumptner C, Fuchsbichler A, Lehner M, Zatloukal K, and Denk H (2001) Sequence of events in the assembly of Mallory body components in mouse liver: clues to the pathogenesis and significance of Mallory body formation. J Hepatol 34: 665

Stumptner C, Omary MB, Fickert P, Denk H, and Zatloukal K (2000) Hepatocyte cytokeratins are hyperphosphorylated at multiple sites in human alcoholic hepatitis and in a Mallory body mouse model. Am J Pathol 156: 77

Yokoo H, Harwood TR, Racker D, and Arak S (1982) Experimental production of Mallory bodies in mice by diet containing 3,5-diethoxycarbonyl-1,4-dihydrocollidine. Gastroenterology 83: 109

Yuan QX, Marceau N, French BA, Fu P, and French SW (1995) Heat shock in vivo induces Mallory body formation in drug primed mouse liver. Exp Mol Pathol 63: 63

Zatloukal K, Stumptner C, Fuchsbichler A, Heid H, Schnoelzer M, Kenner L, Kleinert R, Prinz M, Aguzzi A, and Denk H (2002) p62 is a common component of cytoplasmic inclusions in protein aggregation diseases. Am J Pathol 160: 255

Magnification: x 14,600 (A); x 50,000 (B); x 22,000 (C)

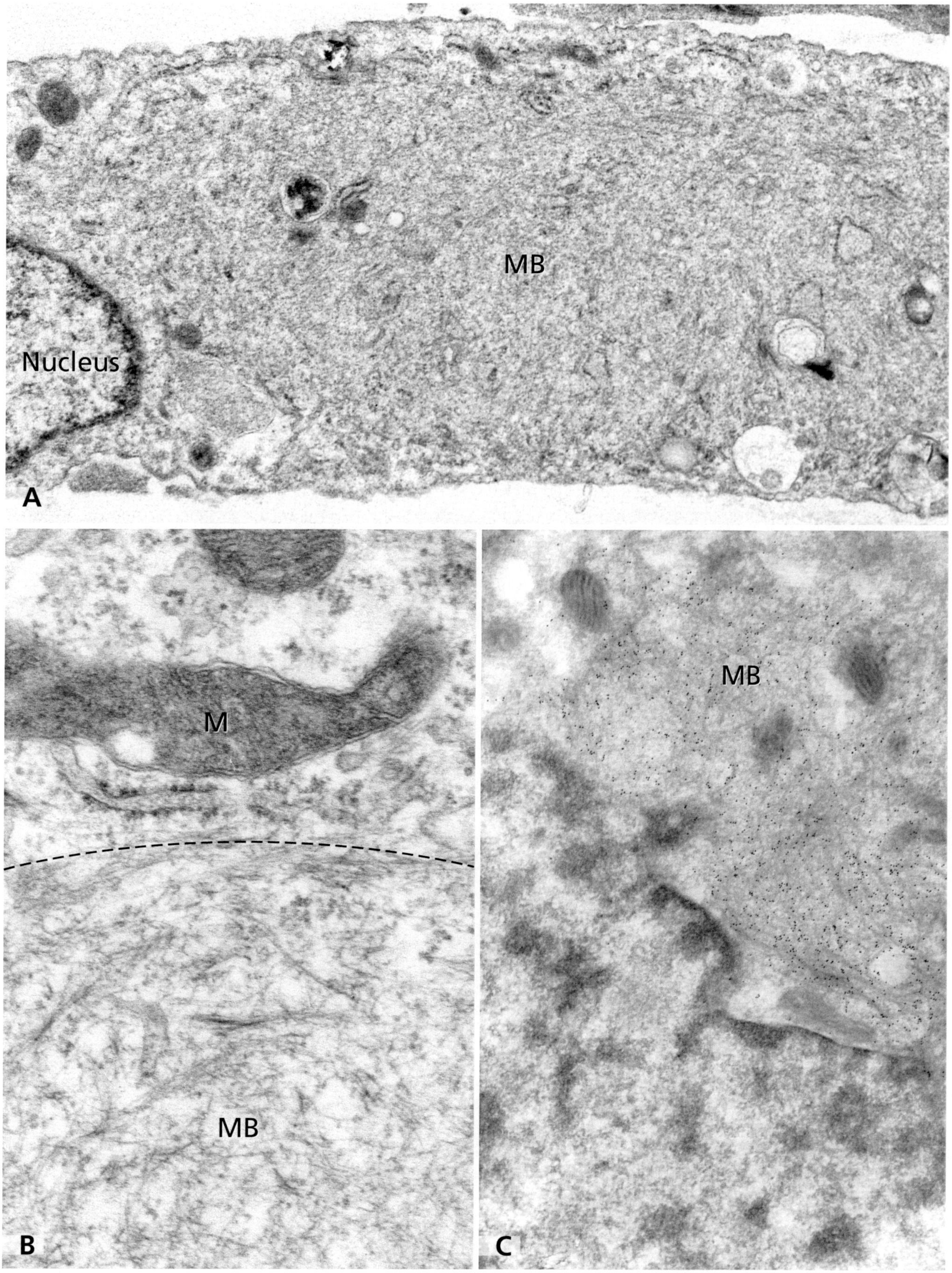
MB
Nucleus
A
M
MB
B
MB
C

THE PLASMA MEMBRANE

Cells are surrounded by a plasma membrane, which forms the boundary between their cytoplasm and the environment. The principal components of the plasma membrane and of all other cellular membranes are (glyco)lipids and (glyco)proteins.

In ultrathin sections, the plasma membrane appears quite simple in structure (panel A). It consists of two electron dense leaflets and a lucent space in between, together reaching a thickness of about 75 nm. In the electron micrograph shown, the trilamellar plasma membranes of two adjacent enterocytes and the narrow intercellular space are visible. This fine structural monotony does not reflect the asymmetric and complex composition and the dynamic nature of the plasma membrane, which additionally differ between cell types.

The freeze-fracture electron microscopy is highly suitable for the study of membranes and provided proof for the presence of membrane spanning proteins. The fracture plane preferentially passes through the hydrophobic membrane interior and produces two membrane halves: the P-face which is the cytosolic membrane half and the E-face which corresponds to the external membrane half. In panel B, the E- and the P-face of two neighbouring red blood cells are shown. Both membrane faces are studded with intramembraneous particles, which are related to fractured transmembrane proteins. The smooth parts of the fracture faces principally correspond to membrane lipids. As seen in panel B, the P-face usually contains a higher density of intramembraneous particles. Variants of the freeze-fracture technique applied to cell cultures permitted the preparation of plasma membrane fracture faces of enormous size as shown in panel C. In contrast to the uniform distribution of intramembraneous particles in the erythrocyte plasma membrane, those of cultured hepatocytes are irregularly arranged. The clusters of intramembraneous particles correspond to coated pits involved in receptor-mediated endocytosis (cf. Fig. 46). The numerous elevations correspond to plasma membrane processes (cf. Fig. 78).

The plasma membrane performs two basic functions. On the one side, the lipid bilayer constitutes an impermeable barrier for most water-soluble molecules. On the other, its membrane-spanning proteins make it porous for bidirectional transmembrane transport and diffusion, communication and signaling as well as cell-cell and cell-matrix interactions. The lipid bilayer represents a two-dimensional fluid in which both lipids and proteins are relatively mobile in the plane of the membrane. However, lipids and proteins may be confined to specific membrane regions, the microdomains. Hence, the name fluid mosaic model of membranes.

The lipid bilayer consists of phospholipids, cholesterol and glycolipids which are differentially distributed in the two membrane leaflets. The oligosaccharide side chains of glycolipids are found on the outer plasma membrane surface together with the oligosaccharides of glycoproteins and form the glycocalyx (cf. Figs. 80 and 81). The membrane proteins can extend through the lipid bilayer, these are the transmembrane proteins. Others are only partially embedded in the cytosolic lipid layer by an amphipathic α helix or a covalently linked lipid chain. Other proteins are inserted in the outer lipid layer through a glycosylphosphatidlyinositol anchor. The peripheral membrane proteins are not inserted in the lipid bilayer at all and noncovalently bound to other membrane proteins.

The plasma membrane asymmetry is not only confined to the two lipid layers. The apical and the basolateral plasma membrane in polarized cells can differ in their protein and lipid composition in relation to their specific functions.

References

Daley D (2008) The assembly of membrane proteins into complexes. Curr Opin Struct Biol 18: 420

Frye L and Edidin M (1970) The rapid intermixing of cell surface antigens after formation of mouse-human heterokaryons. J Cell Sci 7: 319

Lajoie P, Goetz JG, Dennis JW, and Nabi IR (2009) Lattices, rafts, and scaffolds: domain regulation of receptor signaling at the plasma membrane. J Cell Biol 185:381

Maxfield FR (2002) Plasma membrane microdomains. Curr Opin Cell Biol 14: 483

Mukherjee S and Maxfiled F (2004) Membrane domains. Annu Rev Cell Dev Biol 20:839

Simson R, Yang B, Moore SE, Doherty P, Walsh FS, and Jacobson KA (1998) Structural mosaicism on the submicron scale in the plasma membrane. Biophys J 74: 297

Singer S and Nicolson G (1972) The fluid mosiac model of the structure of biological membranes. Science 175: 720

van Meer G (1993) Transport and sorting of membrane lipids. Curr Opin Cell Biol 5: 661

Magnification: x 75,000 (A); x 85,000 (B); x 39,000 (C)

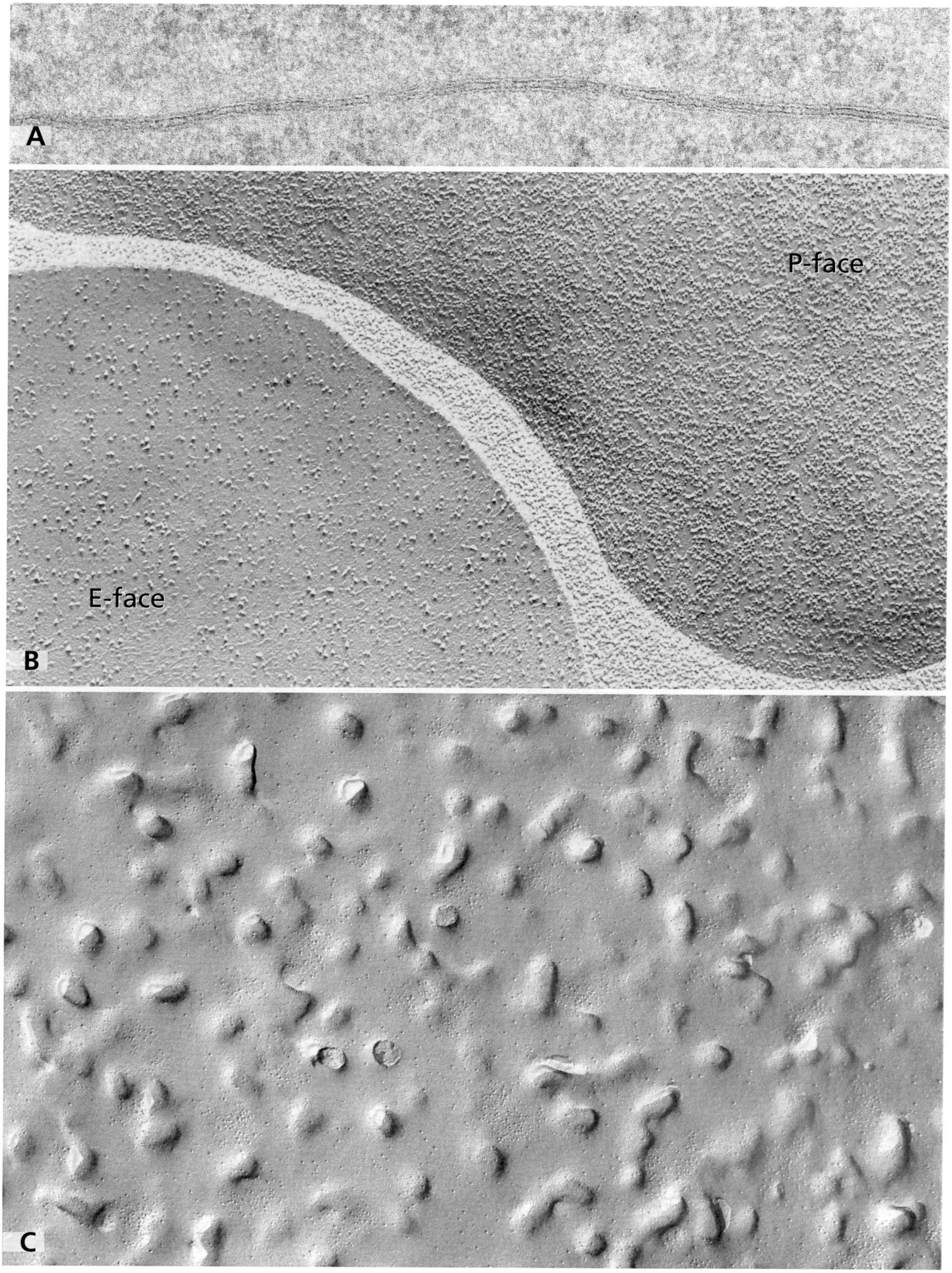
A
P-face
E-face
B
C

CELLS IN CULTURE

Various cell types including stem cells, epithelial, neuronal and mesenchymal cells as well as different tumour cell types can be grown *in vitro* in tissue culture dishes or bottles. Routinely, plastic tissue culture dishes are used and cells can survive and multiply when supplied with appropriate culture medium, temperature and atmosphere. Epithelial cells in tissues are polarised and monolayers of polarised epithelial cells can be obtained when grown on porous membranes. Cell cultures provide excellent experimental tools since they can be studied microscopically or analysed biochemically and can be used as host to synthesise and secrete foreign proteins such as monoclonal antibodies and recombinant proteins.

Observation of living cells by light microscopy and of fixed cells by scanning electron microscopy has provided a wealth of information on cell spreading and locomotion under culture conditions. Spreading of fibroblasts from a cell suspension occurs through thin thread-like protrusions, the filopodia, which establish initial contacts with the substrate (arrowhead in panel A) that finally results in a well attached, flattened cell. Attached fibroblasts and other cell types can crawl over the substratum. This represents a directional movement associated with the formation of lamellipodia, which are flat, two-dimensional protrusions formed at the leading edge of the cells. Thus, moving cells are distinctly polarised (arrows in panel A). Both types of cell protrusions contain actin: filopodia have long bundled filaments and lamellipodia orthogonally, cross-linked meshworks essentially parallel arranged to the substratum. The actin filaments of the lamellipodia in concert with myosin and microtubules and accessory cytoskeletal proteins are the active principle for the cell movement. The cytoskeleton is actively reorganised during cellular locomotion, which includes the formation, contraction and disassembly of actin networks in lamellipodia.

Panel B shows a group of rat hepatocytes attached to a plastic support, which form a coherent monolayer. Their surface is covered by microvilli-like membrane extensions which can be well seen in an ultrathin section cut perpendicularly to the plane of the cell monolayer (panel C). The ultrathin section shown in panel C reveals that the microvilli-like extensions are restricted to the free cell surface and that the basal cell surface is rather flat and focally attached to the plastic support. Epithelial cells grown on a solid plastic or glass support have a discoid shape as seen in panels B and C and form adherens junctions and desmosomes at sites of lateral cell-cell contacts. As mentioned, when grown on permeable, porous membranes, epithelial cells such as kidney epithelial cells form a highly polarised cell monolayer. This represents a most useful system to analyse aspects of polarity of cellular traffic and cytoarchitecture.

Cell crawling is a basic phenomenon in living organisms during embryogenesis and in adult organs. It is important for the function of cells involved in inflammation and immune defense, wound healing and tissue remodelling as well as the spread of malignant cells.

References

Balcarova-Ständer J, Pfeiffer S, Fuller S, and Simons K (1984) Development of cell surface polarity in the epithelial Madin-Darby canine kidney (MDCK) cell line. EMBO J 3: 2687

Bershadsky A and Vasiliev J (1988) Cytoskeleton. New York: Plenum Press

Bray D (2001) Cell movements: From molecules to motility, 2nd edn. New York: Garland Publishing

Cereijido M, Robbins E, Dolan W, Rotunna C, and Sabatini D (1978) Polarized monolayers formed by epithelial cells on a permeable and translucent support. J Cell Biol 77: 853

Condeelis J (1993) Life at the leading edge: the formation of cell protrusions. Annu Rev Cell Biol 9: 411

Pantaloni D, Le Clainche C, and Carlier M (2001) Mechanism of actin-based motility. Science 292: 1502

Rodriguez-Boulan E and Powell SK (1992) Polarity of epithelial and neuronal cells. Annu Rev Cell Biol 8: 395

Simons K and Fuller S (1985). Cell surface polarity in epithelia. Annu Rev Cell Biol 1: 243

Magnification: x 500 (A); x 2,900 (B); x 3,500 (C)

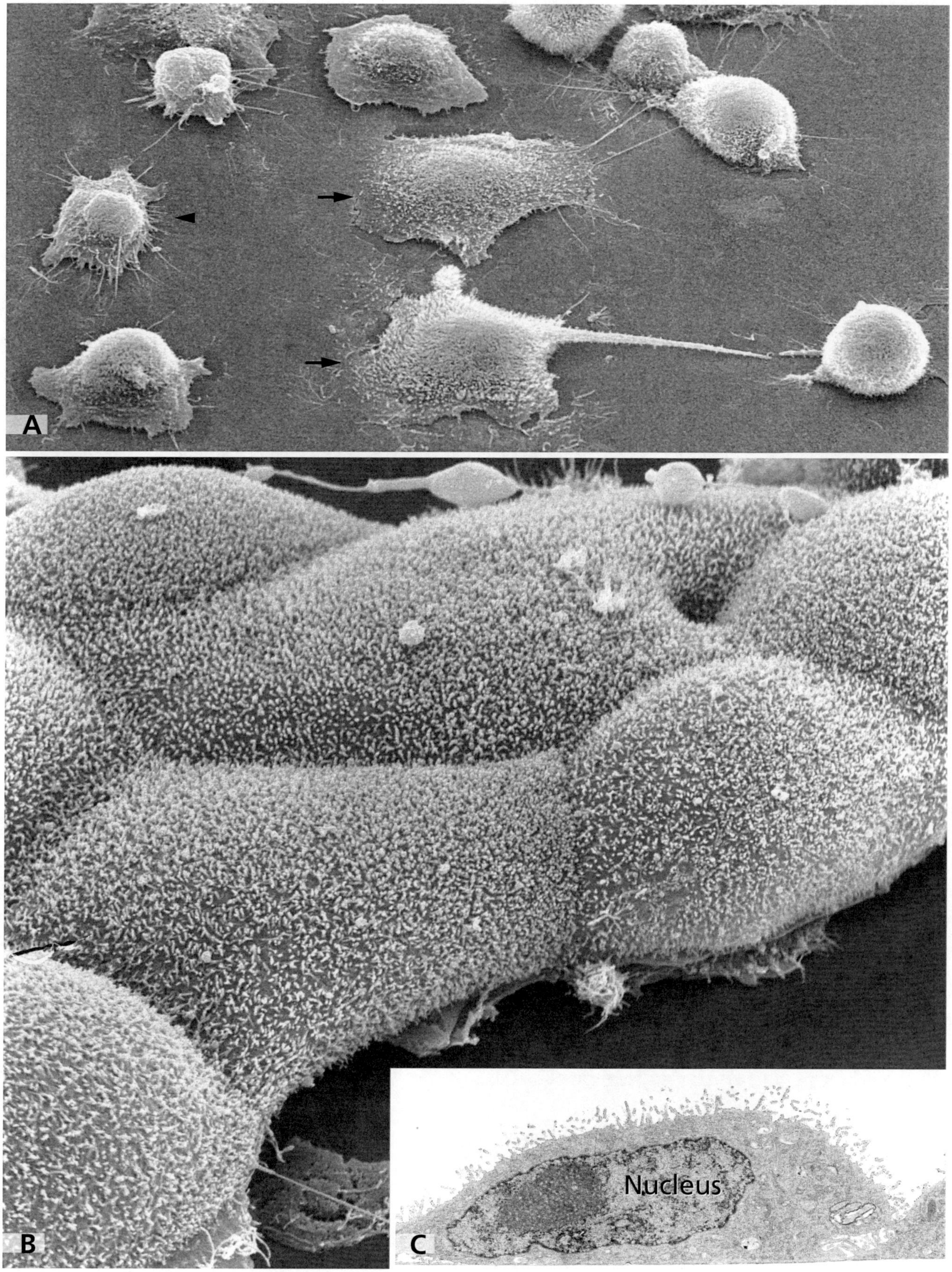
A
B
C
Nucleus

BRUSH CELL

Specialised surface differentiations exist in brush cells, alternatively called tuft cells or caveolated cells. Brush cells belong to a cell population that is widely distributed in the epithelial organs of the gastrointestinal and respiratory tracts, such as the stomach, the small and large intestine, the bile duct, the gallbladder and pancreatic duct, the tracheal epithelium, and the lung. The cells are called after their brush-like or tuft-like apical surface specialisations formed by long and thick microvilli, which differ from the microvilli of the brush borders of absorptive cells (cf. Figs.74 and 111). A brush cell of the rat colon, which shows cytochemical labeling for demonstration of sialic acid residues, is on display. The apical microvilli forming the tuft (arrow) protrude into the gut's lumen and tower above the brush border microvilli of the neighbouring absorptive cells (arrowhead). The differences in dimensions are clearly visible. Densely packed actin filaments build up the core of the tuft microvilli and extend into deep regions of the apical cytoplasm, where they still are bundled (open arrow) and accompanied by microtubules, intermediate filaments, and membrane vesicles.

Although the definite function of brush cells and the particular role of the apical cell specialisations are still obscure, several structural and functional characteristics led us to consider a connection with chemoreceptive tasks and with the regulation of electrolyte concentrations in the secretory fluids of hollow organs. The apical ultrastructure resembles receptor cells in sensory epithelia, such as the sensory cells in taste buds. Brush cells of the stomach, intestine, and pancreatic duct system express a taste cell-specific GTP-binding protein, alpha-gustducin, which is particularly concentrated in the apical pole of the cells, similar to the taste cells, where it is associated with sweet and bitter gustatory functions. Brush cells also are particularly rich in enzymes involved in the production of nitric oxide (NO), such as NO synthase I. Brush cells may have a role as chemoreceptive cells and use NO as a paracrine gaseous messenger.

The microvilli plasma membrane contains a specialised composition of glycoconjugates and seems to be turned over rapidly. Both features also are discussed in favour of a receptive cell function of brush cells, a concept that is supported further by the presence of intermediate filaments that are characteristic for mature neurons. Brush cells express two types of intermediate filaments, cytokeratin 18 filaments and neurofilaments, a combination that is not known to occur in other healthy cells. Cytokeratin 18 is densely concentrated in a network of intermediate filament bundles extending from the cell periphery to the perinuclear cytoplasm, whereas is absent from the apical cytoplasmic regions, where actin filaments and microtubules are assembled with neurofilaments.

References

Gebert A, Al-Samir K, Werner K, Fassbender S, and Gebhard A (2000) The apical membrane of intestinal brush cells possesses a specialized, but species-specific, composition of glycoconjugates – on-section and in vivo lectin labelling in rats, guinea-pigs and mice. Histochem Cell Biol 113: 389

Höfer D and Drenckhahn D (1996) Cytoskeleton marker allowing discrimination between brush cells and other epithelial cells of the gut including enteroendocrine cells. Histochem Cell Biol 105: 405

Höfer D and Drenckhahn D (1998) Identification of the taste cell G-protein, alpha-gustducin, in brush cells of the rat pancreatic duct system. Histochem Cell Biol 110: 303

Luciano L, Groos S, and Reale E (2003) Brush cells of rodent gallbladder and stomach epithelia express neurofilaments. J Hictochem Cytochem 51: 187

Ogata T (2000) Mammalian tuft (brush) cells and chloride cells of other vertebrates share a similar structure and cytochemical reactivities. Acta Histochem Cytochem 33: 439

Roth J, Lucocq JM, and Charest PM (1984) Light and electron microscopic demonstration of sialic acid residues with the lectin from Limax flavus: A cytochemical affinity technique with the use of fetuin-gold complexes. J Histochem Cytochem 32: 1167

Magnification: x 29,500

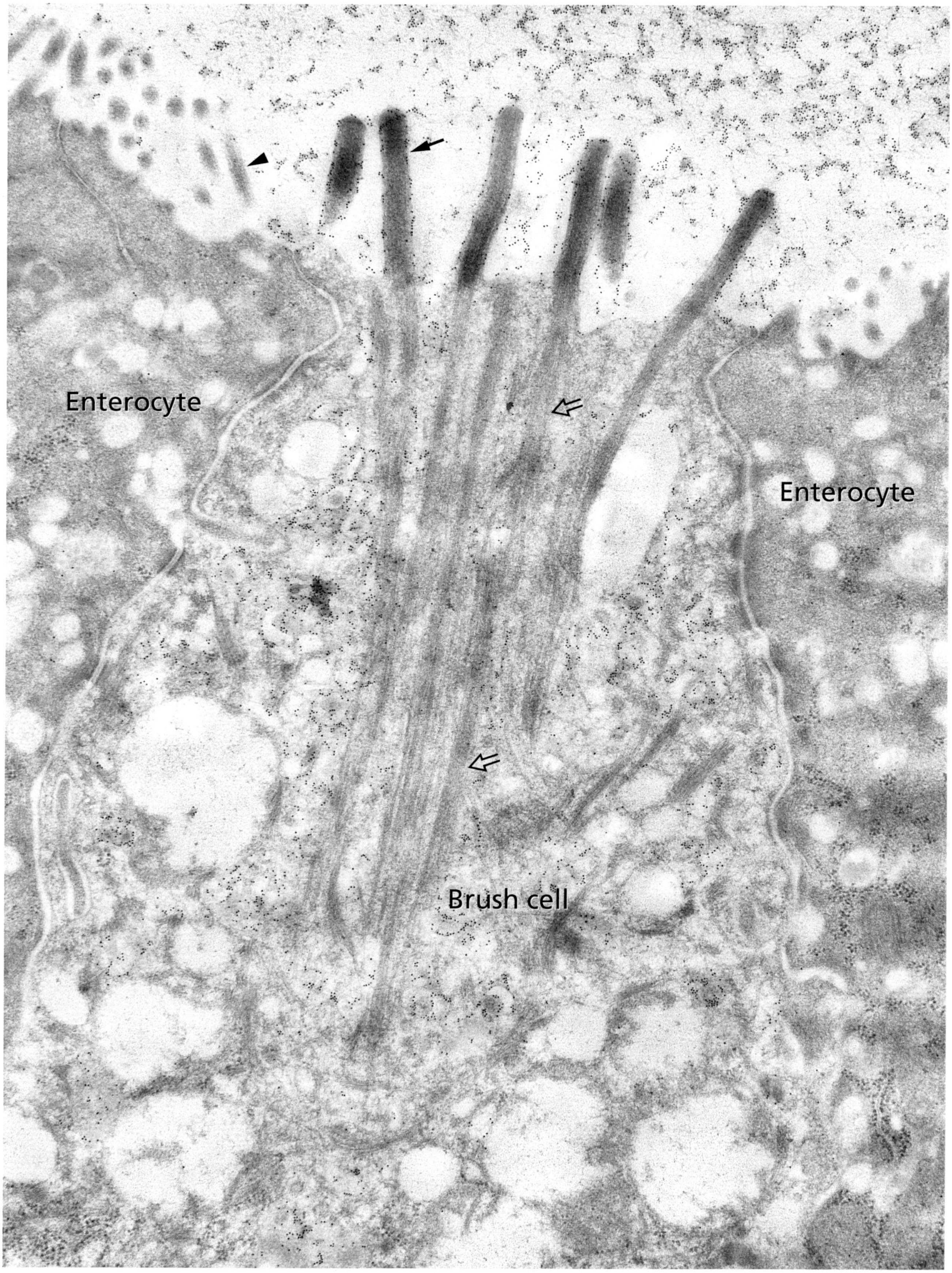
Enterocyte
Enterocyte
Brush cell

GLYCOCALYX (CELL COAT)

The outer surface of all animal cells is covered by a glycocalyx composed of oligosaccharides (glycans) of glycoproteins and glycolipids and a layer of secreted mucus particularly in the gastrointestinal, respiratory, and urogenital tracts. The biological roles of the glycocalyx are diverse. In general terms, it exerts stabilising and protective functions. Specific functions are related to the glycan structure and cell type and to specific recognition and interaction of glycans with other molecules. Certain glycans are important for development and differentiation of organs through modulation of cell-cell and cell-matrix interactions and signalling. They also can function as receptors for certain pathogens or may represent ligands for various receptors and can be involved in turnover and trafficking of molecules.

The glycocalyx may be so well developed that it can be observed by ordinary transmission electron microscopy. One example are the intestinal absorptive cells. In panel A, the brush border of an absorptive enterocyte is shown. At the tips of the microvilli, the glycocalyx appears as prominently visible antennulae microvillares (arrows). Panel B represents cross-sectioned microvilli of an enterocyte and staining with ruthenium red, a cationic dye that binds electrostatically to ionised carboxylic acid groups of acid mucopolysaccharides. This staining results in a highly increased contrast of the glycocalyx. The drawbacks of this staining are the non-discriminatory reaction with polyanions and the membrane impermeability of the dye, which limit its use for compact tissues and staining of intracellular glycans. These major limitations have been overcome with the use of lectins and monoclonal anti-carbohydrate antibodies of defined specificity and the use of tissue sections from Lowicryl K4M-embedded tissue or ultrathin frozen tissue sections. In panel C, the sialic acid specific lectin from *Limax flavus* has been used to detect such residues in the glycocalyx of absorptive enterocytes. This resulted in gold particle labelling of the antennullae microvillares in the Lowicryl thin sections. In panel D, a monoclonal antibody against the blood group A substance has been applied to thin sections from Lowicryl-embedded human duodenum of a blood group A subject. The blood group A determinant, which is terminal non-reducing *N*-acetylgalactosamine, is present in the glycocalyx, and the adhering mucus as indicated by the gold particle labelling.

References

Blanquet P (1976) Ultrahistochemical study on the ruthenium red surface staining. I. Processes which give rise to electron dense marker. Histochemistry 47: 63

Crocker PR, Paulson JC, and Varki A (2007) Siglecs and their roles in the immune system. Nat Rev Immunol 7: 255

Goldstein IJ and Poretz RD (1986) Isolation, physicochemical characterisation, and carbohydrate-binding specificity of lectins. In: The lectins. Properties, functions and applications in biology and medicine (Liener IE, Sharon N, and Goldstein IJ, eds). Orlando: Academic Press, pp 35

Hayat M (1993) Stains and cytochemical methods. New York London: Plenum Press

Luft JH (1971) Ruthenium red and violet. I. Chemistry, purification methods of use for electron microscopy, and mechanism of action. Anat Rec 171: 347

Rambourg A (1971) Morphological and histochemical aspects of glycoproteins at the surface of animal cells. Int Rev Cytol 31: 57

Roth J (1983) Application of lectin-gold complexes for electron microscopic localisation of glycoconjugates on thin sections. J Histochem Cytochem 31: 987

Roth J (1996) Glycosylation in the endoplasmic reticulum and the Golgi apparatus and cell type-specificity of cell surface glycoconjugate expression: analysis by the protein A-gold and lectin-gold techniques. Histochem Cell Biol 106: 79

Varki A (1997) Sialic acids as ligands in recognition phenomena. FASEB J 11: 248

Magnification: x 87,500 (A); x 120,000 (B); x 55,500 (C); x 52,000 (D)

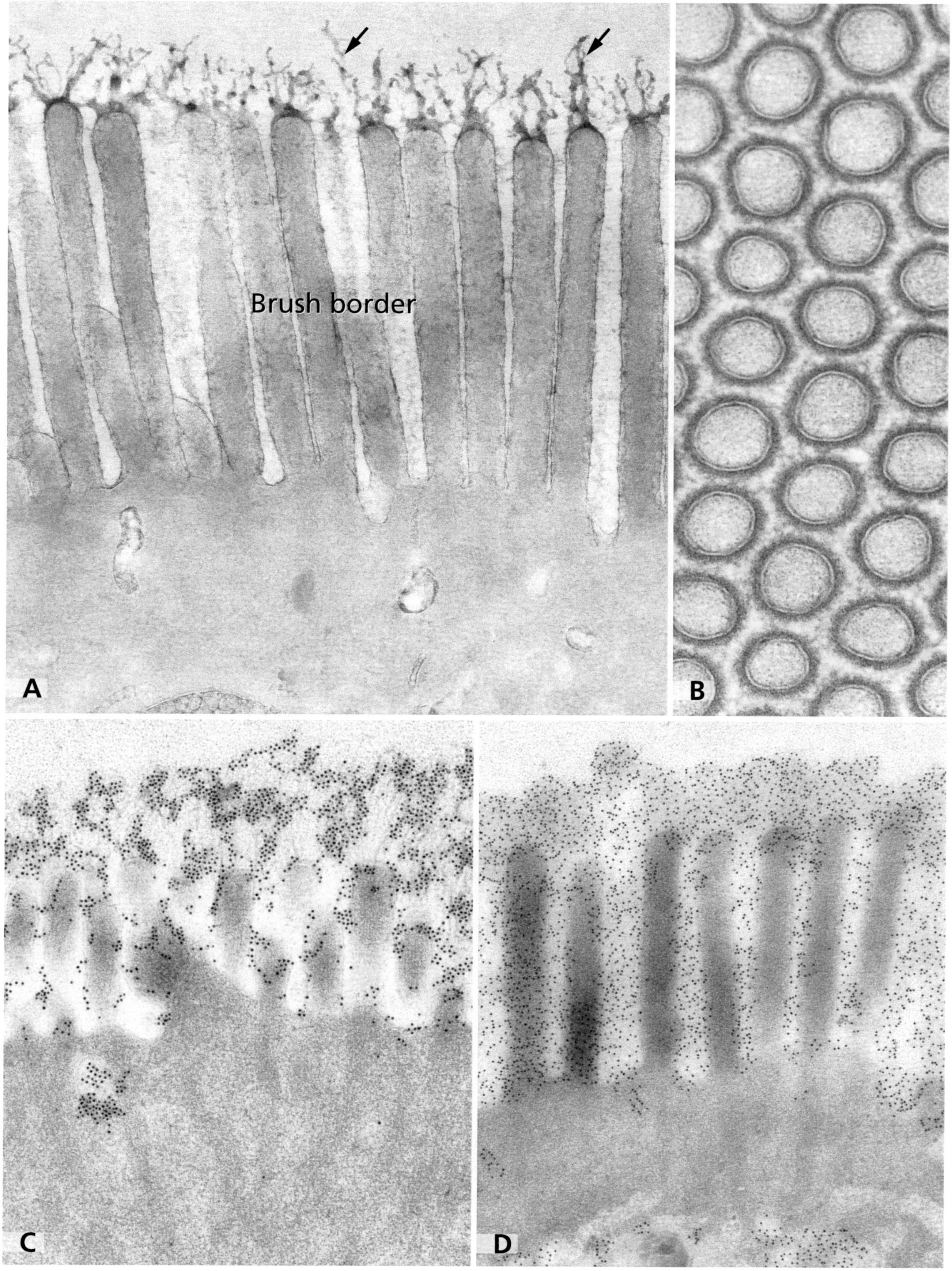
Brush border
A
B
C
D

GLYCOCALYX: CELL TYPE SPECIFICITY AND DOMAINS

The analysis of the expression of various glycosyltransferases by Northern blot analysis and specific enzyme assays has shown their differential tissue expression. On the basis of this the view is held that specific glycan structures in general reflect the expression of the respective glycosyltransferases. Cytochemical *in situ* studies provide distinct advantages over Northern blot analyses and specific enzyme assays since particular glycans can be localised to specific cell types in organs with complex cellular composition. With the application of lectins and monoclonal antibodies, histochemistry therefore represents an important tool in studies of cell type-specific glycosylation and may provide a clue about their possible functions.

Panel A shows the apical portions of a dark and a light cell of normal human breast duct epithelium and provides an example of cell type-specific and plasma membrane domain-related expression of a glycan. This Lowicryl thin section was incubated with a monoclonal antibody reactive with an *O*-glycan in normal human breast and breast carcinoma. The plasma membrane of the dark cell type is labelled (filled arrowheads), whereas the one of the adjacent light cells is unlabelled (open arrowheads). A further detail is that the labelling of the dark cell is restricted to the apical plasma membrane domain, which is separated from the lateral plasma membrane by junctions (arrows).

In panel B part of a capillary loop from rat renal glomerulum is seen and exemplifies another grade of glycocalyx domain formation in the podocyte plasma membrane. The labelling in the base of the podocyte foot processes (arrowheads) is due to binding of gold-labelled *Helix pomatia* lectin, which has a nominal specificity for terminal, non-reducing *N*-acetylgalactosamine residues. Lectin binding did not occur to the rest of the podocyte plasma membrane as no labelling was detectable in the capillary endothelial cells (asterisk) and glomerular basement membrane. This showed that the podocyte foot process plasma membrane is highly specialised in terms of the composition of their glycocalyx. In contrast, the wheat germ lectin, which has a nominal specificity for *N*-acetylglucosamine residues, did label all regions of the podocyte plasma membrane, the glomerular basement membrane, and capillary endothelia (panel C).

References

Brown D, Roth J, and Orci L (1985) Lectin-gold cytochemistry reveals intercalated cell heterogeneity along rat kidney collecting ducts. Am J Physiol 248: C348

Gersten KM, Natsuka S, Trinchera M, Petryniak B, Kelly RJ, Hiraiwa N, Jenkins NA, Gilbert DJ, Copeland NG, and Lowe JB (1995) Molecular cloning, expression, chromosomal assignment, and tissue-specific expression of a murine alpha-(1,3)-fucosyltransferase locus corresponding to the human ELAM-1 ligand fucosyl transferase. J Biol Chem 270: 25047

Kerjaschki D, Noronha-Blob L, Sacktor B, and Farquhar M (1984) Microdomains of distinctive glycoprotein composition in the kidney proximal tubule brush border. J Cell Biol 98: 1505

Kitagawa H and Paulson JC (1994) Differential expression of five sialyltransferase genes in human tissues. J Biol Chem 269: 17872

Le Hir, M, Kaissling B, Koeppen BM, and Wade JB (1982) Binding of peanut lectin to specific epithelial cell types in the kidney. Am J Physiol 242: C117

Roth J, Brown D, and Orci L (1983) Regional distribution of N-acetyl-D-galactosamine residues in the glycocalyx of glomerular podocytes. J Cell Biol 96: 1189

Yoshida Y, Kurosawa N, Kanematsu T, Taguchi A, Arita M, Kojima N, and Tsuji S (1996) Unique genomic structure and expression of the mouse alpha 2,8-sialyltransferase (ST8Sia III) gene. Glycobiology 6: 573

Magnification: x 20,000 (A); x 35,500 (B, C)

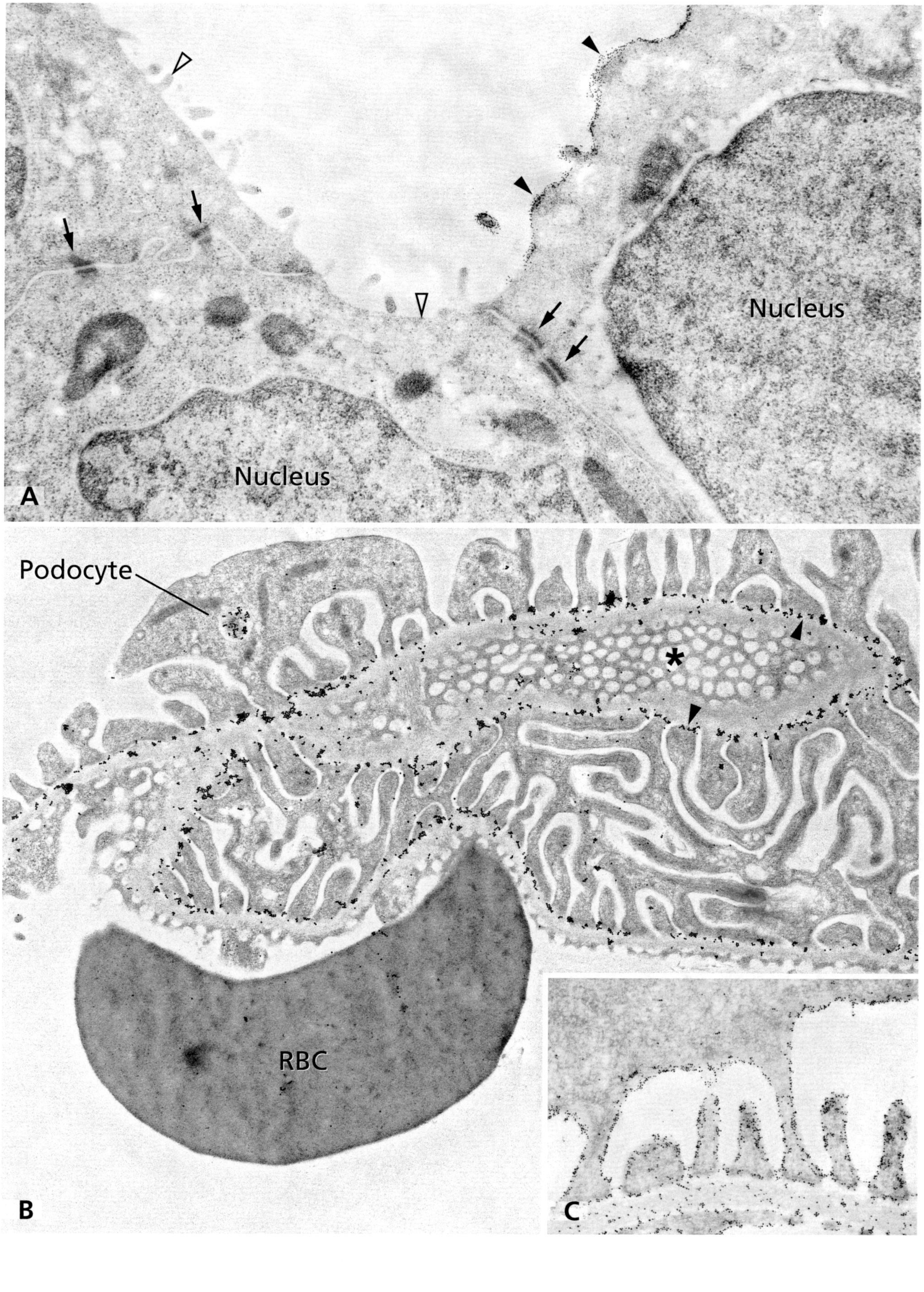
Nucleus
Nucleus
A
Podocyte
RBC
B
C

GLYCOCALYX CHANGES IN TUMOURS

Particular glycans exhibit spatiotemporal expression patterns during embryonic development and may become re-expressed in malignant human tumours. Carcinoma-associated cell surface glycans can be involved in invasive and metastatic growth or of clinical importance as predictive markers. A commonly observed change is the increased synthesis of β1,6-branched tri- and tetra-antennary glycans. This correlates with the metastatic potential of certain tumours and is an independent predictive marker in colon carcinoma. Sialylated glycans terminated in α2,6-linked sialic acid or the sialosyl Tn antigen are associated with colon carcinoma progression and of predictive value. These findings, however, cannot be generalised.

Sialic acids also exist as homopolymers in α2,8-ketosidic linkages and such a polysialic acid is present on the neural cell adhesion molecule NCAM. This unique glycan modulates cell-cell and cell-substratum interactions during brain development and neuronal functions in the adult. Unexpectedly, polysialylated NCAM was detected in embryonic kidney and found to be re-expressed in the Wilms tumour, a highly malignant kidney tumour. In panel A, the presence of an electron dense surface coat of variable thickness (arrowheads) in a Wilms tumour is demonstrated. Small lumina (asterisks in panels A and C) are formed at sites of high surface coat thickness because of reduced cell adhesion. This surface coat has been interpreted as a basement membrane, but as shown in panels B and C consists of polysialic acid as revealed by immunogold labelling with a monoclonal antibody. Polysialic acid exists in malignant neuroendocrine tumours and is of diagnostic importance. Experimental studies with small cell lung carcinoma cells have directly demonstrated the role of polysialic acid for invasive and metastatic growth properties. Clinical studies revealed the importance of measuring serum levels of polysialic acid in patients with neuroendocrine tumours as an indicator of tumour stage and progression.

References

André S, Kopitz J, Kaltner H, Villalobo A, and Gabius H-J (2009) Glycans as functional markers in malignancy? In: The Sugar Code. Fundamentals of Glycosciences (Gabius H-J, ed). Weinheim: Wiley/VCH, pp 419

Brockhausen I (1999) Pathways of O-glycan biosynthesis in cancer cells. Biochim Biophys Acta 1473: 67

Dennis JW, Granovsky M, and Warren CE (1999) Glycoprotein glycosylation and cancer progression. Biochim Biophys Acta 1473: 21

Gluer S, Schelp C, Madry N, von Schweinitz D, Eckhardt M, and Gerardy-Schahn R (1998) Serum polysialylated neural cell adhesion molecule in childhood neuroblastoma. Br J Cancer 78: 106

Hakomori SI (1996) Tumour-associated carbohydrate antigens and modified blood group antigens. In: Glycoproteins and disease (Montreuil J, Vliegenthart J, and Schachter H, eds). Amsterdam Lausanne New York Oxford Shannon Tokyo: Elsevier, pp 243

Kobata A (1996) Cancer cells and metastasis. The Warren-Glick phenomenon – a molecular basis of tumourigenesis and metastasis. In: Glycoproteins and disease (Montreuil J, Vliegenthart JFG, and Schachter H, eds). Amsterdam Lausanne New York Oxford Shannon Tokyo: Elsevier, pp 211

Roth J, Rutishauser U, and Troy FA (1993) Polysialic acid. From microbes to man. Basel Boston Berlin: Birkhäuser

Roth J, Taatjes DJ, Wagner P, Weisgerber C, Heitz PU, Goridis C, and Bitter-Suermann D (1988) Reexpression of poly(sialic acid) units of the neural cell adhesion molecule in Wilms tumour. Proc Natl Acad Sci USA 85: 2999

Seelentag WKF, Li WP, Schmitz SFH, Metzger U, Aeberhard P, Heitz PU, and Roth J (1998) Prognostic value of beta 1,6-branched oligosaccharides in human colorectal carcinoma. Cancer Res 58: 5559

Tanaka F, Otake Y, Nakagawa T, Kawano Y, Miyahara R, Li M, Yanagihara K, Inui K, Oyanagi H, and Yamada T (2001) Prognostic significance of polysialic acid expression in resected non-small cell lung cancer. Cancer Res 61: 1666

Magnification: x 8,500 (A); x 58,000 (B); x 55,000 (C)

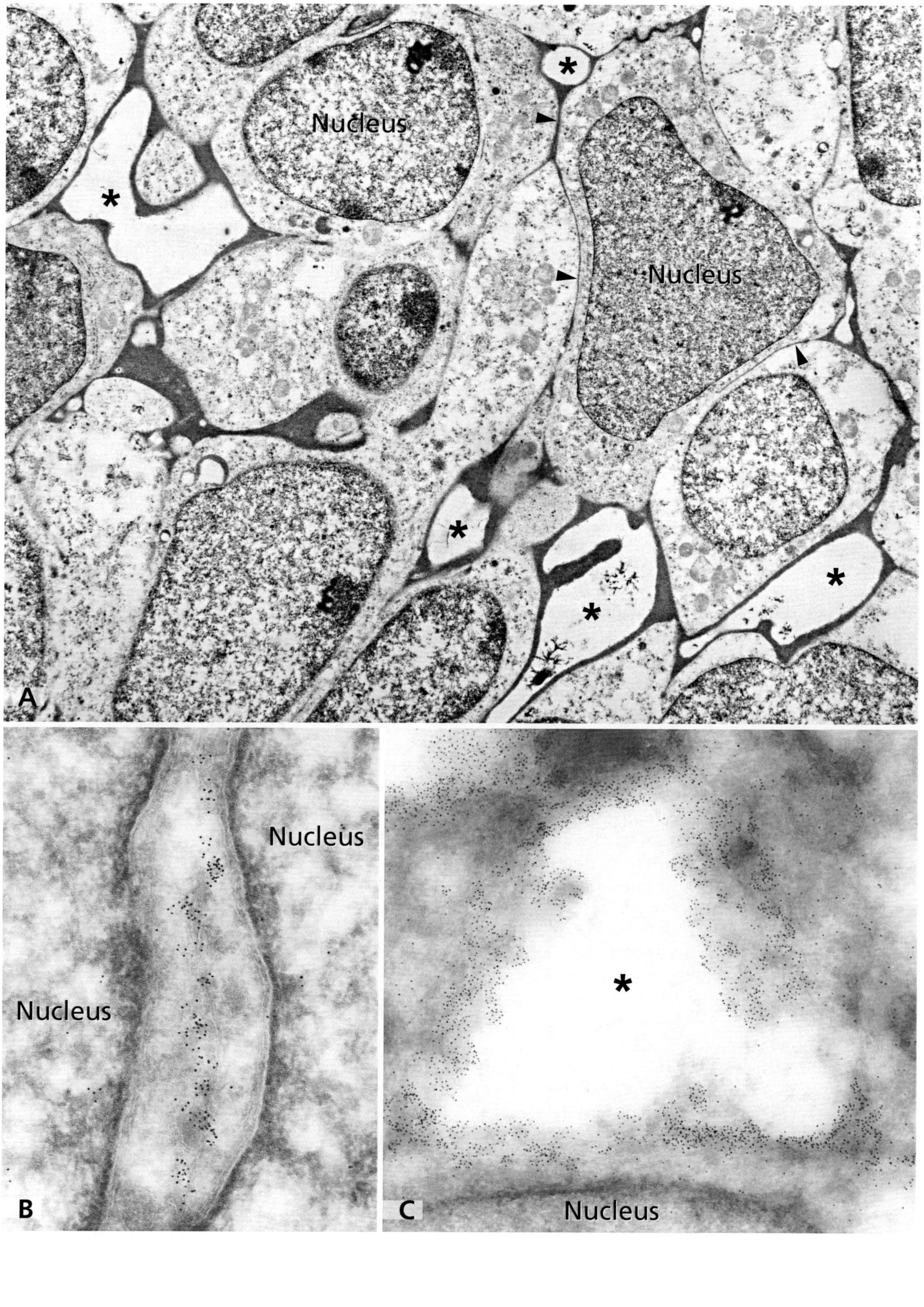
Nucleus
Nucleus
A
Nucleus
Nucleus
B
C
Nucleus

JUNCTIONAL COMPLEX

Cell to cell contacts are necessary for any higher organisation of cells and provide the basis for the formation of tissues and organs. In epithelia, sheets of cells are formed by close attachment of cells to each other, which is facilitated by cell adhesion molecules and further stabilised and differentiated by formation of specialised cell junctions. In multiple epithelia, cells are attached to each other by junctional complexes, composed of a characteristic combination of occluding and adhering junctions that regulate paracellular traffic and stabilise the tissue. Junctional complexes are symmetrical structures formed between adjacent cells and consist of three components: Firstly, a band of tight junctions, forming an occluding zone in the top position (zonula occludens, ZO, cf. also Fig. 84), secondly, a band of anchoring junctions in the middle position (belt desmosome, zonula adhaerens, ZA), and thirdly, a circle of spot desmosomes in the bottom position (maculae adhaerentes, MA, cf. also Fig. 86). The complexes are localised mainly in the apical epithelial regions, as shown in panel A, where all three parts of a junctional complex of the small intestinal epithelium are on display (bracket), together with adhering interdigitated regions of the lateral cell surfaces (arrow).

The uppermost occluding zone (ZO) occupies cell surface areas close to the apical microvilli and determines the border between apical and basolateral regions of the plasma membrane. The plasma membranes of the neighbouring cells are completely linked at regular intervals, the sealing sites forming interconnected ridges and networks that surround the cells completely. In the thin section of the occluding zone shown in panel A, the sealing ridges appear as connection points ("kissing points"). The anastomosing ridges are visible in freeze fracture replica electron micrographs shown on the next page (cf. Fig. 84).

The zonula adhaerens (ZA) is usually localised below the occluding zone. It also forms a belt surrounding the cells completely. In panel A, the adherent zone is cross-sectioned. A flat section through the adherent belt at a connection site of three neighbouring cells is on display in panel B. The adherent belt is associated with densely packed actin filaments. In the cytoplasm nearby, multiple cross-sections of the actin filament rootlets of the brush border microvilli extending into the terminal web are visible (arrows; cf. Fig. 74). Formation of belt desmosomes requires interactions of membrane-spanning cadherins (desmocollins, desmogleins), plaque proteins (desmoplakin, plakoglobin), and catenins. E-cadherin molecules bind to partners localised in the apposing plasma membrane of neighbouring cells. At the cytoplasmic side, interactions of the E-cadherin tails with alpha- and beta-catenins and formin-1 induce actin polymerisation and the assembly of the adherent belt-associated actin cytoskeleton, which is responsible for the cell-to-cell spreading skeleton and motion systems (cf. also Fig. 74).

Spot desmosomes are shown in panel A (MA) and, at higher magnification, in panel C. In a circle-like arrangement below the belt desmosome, they represent the third part of junctional complexes, but they also exist independently of other cell contacts and, like buttons, attach cells to each other at multiple sites of the lateral cell surfaces. They are associated with intermediate filaments, bands of which build up interconnections between the individual spot desmosomes at the lateral walls of cells and the basal hemidesmosomes (cf. Fig. 93). The intercellular space is wider than it is in the adhering belt and is occupied by the extracellular glycosylated portions of cadherins appearing as dense zipper-like midline (arrow in panel C). Cytoplasmic plaques (arrowhead) attached to the plasma membrane function as anchors for the cadherins and for the intermediate filaments, which insert into the plaques in a hairpin-like fashion (cf. Fig. 86).

References

Bazzoni G (2003) The JAM family of junctional adhesion molecules. Curr Opin Cell Biol 15: 525

Ebnet K (2008) Organization of multiprotein complexes at cell-cell junctions. Histochem Cell Biol 130: 1

Green KJ and Gaudry CA (2000) Are desmosomes more than thethers for intermediate filaments? Nat Rev Mol Cell Biol 1: 208

Kobielak A, Pasolli HA, and Fuchs E (2004) Mammalian forming-1 participates in adherens junctions and polymerisation of linear actin cables. Nat Cell Biol 6: 21

Tsukita S, Furuse M, and Itoh M (2001) Multifunctional strands in tight junctions. Nat Rev Mol Cell Biol 2: 285

Zigmond S (2004) Forming adherens junctions. Nat Cell Biol 6: 12

Magnification: x 88,000 (A), x 20,300 (B), x 120,000 (C)

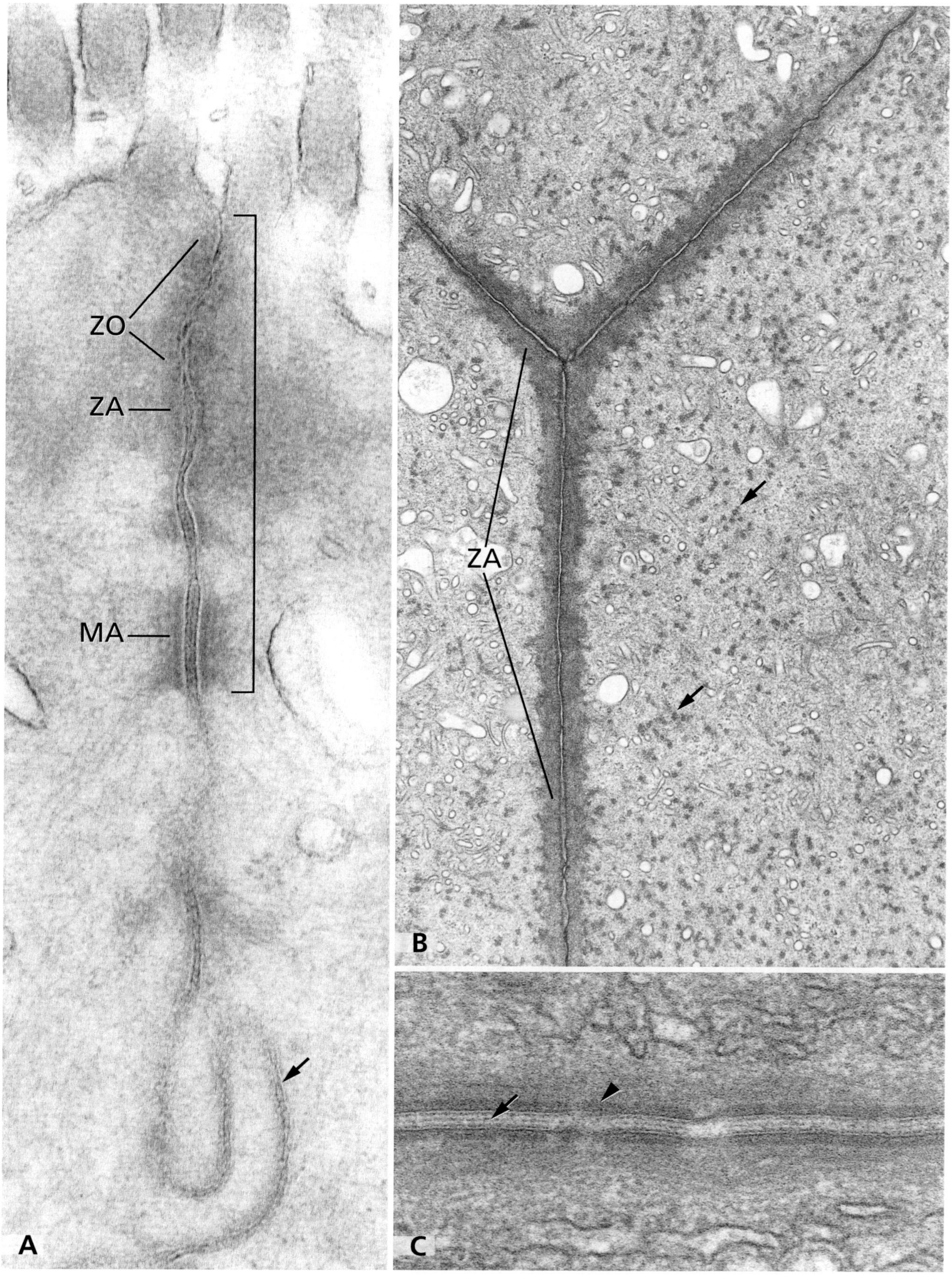
ZO
ZA
MA
A
ZA
B
C

TIGHT JUNCTIONS AND GAP JUNCTIONS

Tight junctions have several major functions. They seal the intercellular space in epithelial and endothelial cell layers and prevent free paracellular passage of substances. They determine the polarity of epithelial cells by creating a boundary between the apical domain of the plasma membrane and the basolateral domain and prevent diffusion of lipids and proteins between them. Tight junctions recruit multiple cytoskeletal and signalling molecules at their cytoplasmic surfaces, which is seen in connection with regulatory processes involving the actinomyosin cytoskeleton and with intercellular adhesion signalling within epithelia and endothelia. Tight junctions may occur independently of other cell contacts but are more often part of junctional complexes forming an occluding belt in the top position (cf. Fig. 83).

In freeze-fracture replica electron micrographs, tight junctions are particularly well visible as continuous, anastomosing strands of particles appearing on the protoplasmic face (P-face; TJ in panels A and B), forming a band or complex network. The number of tight junction strands is a crucial factor in determining the barrier properties of tight junctions. Transepithelial resistance increases with the number of parallel tight junction strands. The strands consist mainly of aggregations of the proteins claudin and occludin, and associated zonula proteins (ZO-1, ZO-2, and ZO-3). Individual strands associate with those of the apposing membrane of an adjacent cell to form paired strands and to establish the "kissing points", visible in thin sections under the electron microscope (cf. Fig. 83). There is evidence that claudins form the backbone of tight junction strands. Not only proteins but also lipids are assumed to contribute to the formation of tight junction strands by forming inverted cylindrical micelles.

Most types of epithelia and endothelia contain occluding junctions. They have a crucial role in the formation of distinct barriers, such as the blood-brain barrier (cf. Fig. 158) and the blood-thymus barrier in endothelial cells, the blood-testis barrier built by the Sertoli cells, and the barrier sealing the bile canaliculi in the liver epithelium. The latter is shown in an *in vitro* culture system, using hepatoma cells that organise themselves three-dimensionally and form anastomosing canaliculi lined by an apical dense microvilli border (panel A, inset). Peroxidase-conjugated wheat germ agglutinin, visualised by electron dense oxidized diaminobenzidine, labels both the apical microvilli membane of the canaliculi and the basolateral plasma membrane, shown at the lefthand side, but is prevented from entering the tight junction area (TJ).

Gap junctions, or communicating junctions, or nexus, are cell contacts providing cell to cell communication by transport of ions and small molecules up to approximately 1 kDa. They are formed by integral membrane proteins, the connexins. Six connexins assemble to form a hollow cylindrical structure called connexon. Connexons align with connexon partners present in the apposing membrane of neighbouring cells, forming hydrophilic channels of communication between the cytoplasms of adjacent cells. High Ca^{2+}-concentrations lead to a closing of the connexon channels. Gap junctions allow chemical and electrical coupling of adjacent cells that are particularly critical for heart and smooth muscle cell action and regular embryogenesis.

The connexons are visible in freeze fracture replicas under the electron microscope appearing as aggregated particles organised in spots or large areas (GJ in panel B). They can be distinguished clearly from the tight junction strands (TJ). Electron micrographs of thin sections show that the gap junction plasma membranes are closely apposed. The intercellular space is extremely regular and narrow but is visible as a gap none the less (panel B, inset), which is how the term "gap junction" was coined.

References

Förster C (2008) Tight junctions and the modulation of barrier function in disease. Histochem Cell Biol 130: 55

Kachar B and Reese TS (1982) Evidence for the lipid nature of tight junction strands. Nature 296: 464

Kumar NM and Gilula NB (1996) The gap junction communication channel. Cell 84: 381

Prochnow N and Dermietzel R (2008) Connexons and cell adhesion: a romantic phase. Histochem Cell Biol 130: 71

Tsukita S, Furuse M, and Itoh M (2001) Multifunctional strands in tight junctions. Nat Rev Mol Cell Biol 2: 285

Magnification: x 90,500 (A), x 34,000 (inset), x 71,000 (B), x 81,000 (inset)

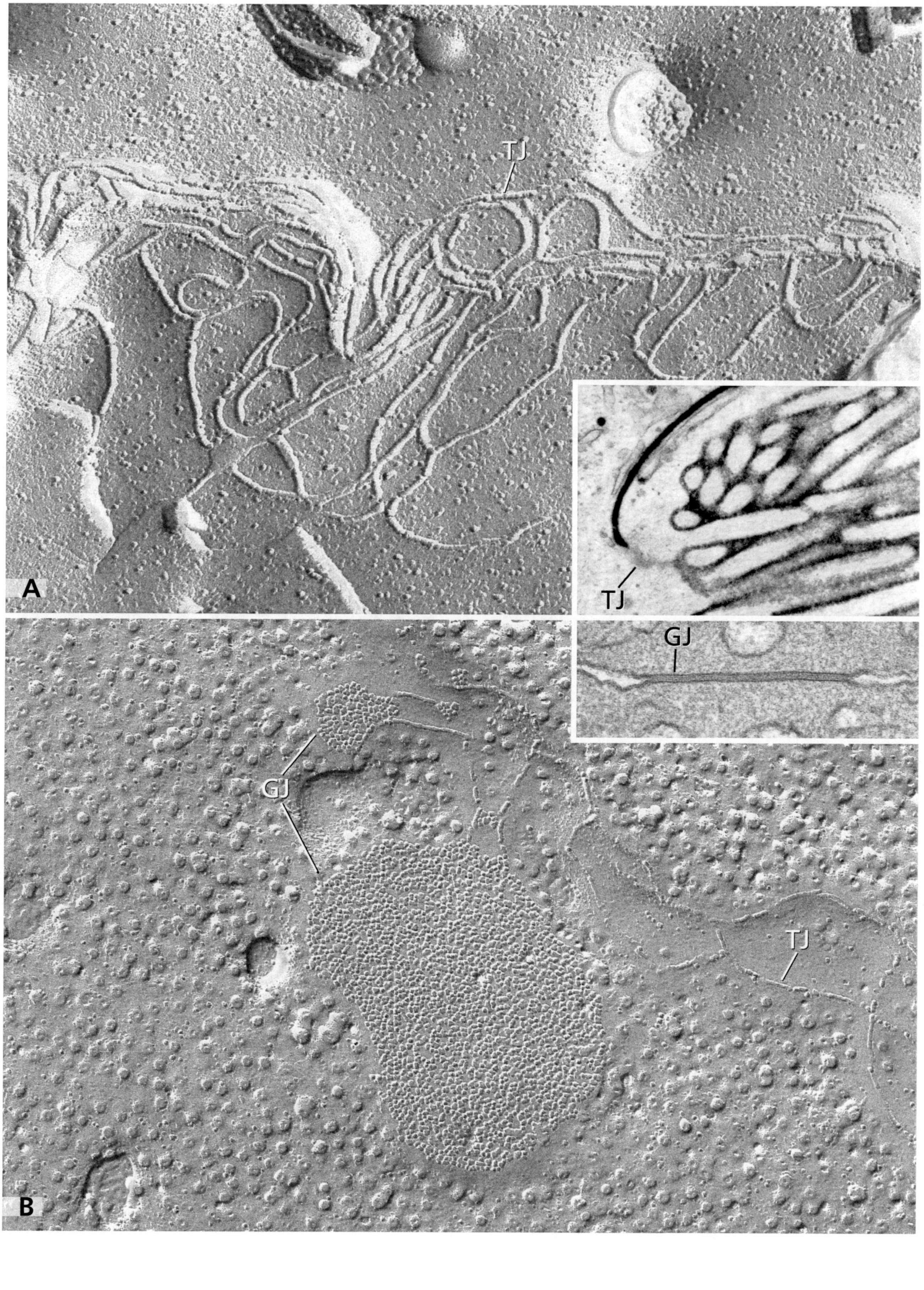
TJ
TJ
A
GJ
GJ
TJ
B

TUNNELING NANOTUBES

Cells in the diverse organs and when grown in culture can communicate with each other by several means such as gap junctions (cf. Fig. 84) and chemical synapses (cf. Fig. 158). A novel type of cell-cell communication was discovered, which is represented by tunneling nanotubes (TNT).

TNT are membraneous nanotubes with a diameter varying depending on the cell type. In the pheochromocytoma cell line PC12, their diameter is 50–200 nm and their length up to several cell diameters as shown in panel A. This is a scanning electron micrograph showing a TNT bridging two PC12 cells, the cell type in which TNT were discovered. In contrast to other cellular protrusions, TNT of cultured cells do not contact the substratum and bridge two cells at their nearest distance. In panel B, a transmission electron micrograph of an TNT connecting two PC12 cells is shown. Panels C and D are higher resolution micrographs from consecutive serial sections of this TNT, which demonstrate the continuity between the TNT membrane and the plasma membrane of the two connected cells. In all cell types analysed, TNT contain F-actin. The seamless membrane contact provided by TNT permits the lateral diffusion of certain plasma membrane proteins and viruses between the connected cells. Probably equally important, TNT permit the intercellular transfer of organelles and cytoplasmic proteins as well as ions. Organelles delivered between cells by TNT are represented by endosomal/lysosomal elements and mitochondria.

TNT are formed *de novo* and have a dynamic nature with a variable lifetime, thereby providing only transient cell-cell contacts. For their formation, actin polymerization appears to be an important aspect.

There appears to exist structural diversity of TNT. In PC12 cells and many other cell types, TNT are open-ended structures as described above and illustrated in the panels B–D. In T-cells, however, the TNT are of closed-ended type. Here, the TNT formed by one cell only protrudes into an invagination of the other cell and establishes a junctional contact.

Although TNT were observed in many types of cultured cells, they appear to exist also in tissues where they may participate in communicative processes in embryonic and adult tissues.

References

Gousset K, Schiff E, Langevin C, Marijanovic Z, Caputo A, Browman DT, Chenouard N, de Chaumont F, Martino A, Enninga J, et al. (2009) Prions hijack tunnelling nanotubes for intercellular spread. Nat Cell Biol 11: 328

Davis DM and Sowinski F (2008) Membrane nanotubes: dynamic long-distance connections between animal cells. Nat Rev Molec Cell Biol 9: 431

Gurke S, Barroso JF, and Gerdes HH (2008) The art of cellular communication: tunneling nanotubes bridge the divide. Histochem Cell Biol 129: 539

Hodneland E, Lundervold A, Gurke S, Tai XC, Rustom A, and Gerdes HH (2006) Automated detection of tunneling nanotubes in 3D images. Cytometry A 69: 96

Rustom A, Saffrich R, Markovic I, Walther P, and Gerdes HH (2004) Nanotubular highways for intercellular organelle transport. Science 303: 1007

Sowinski S, Jolly C, Berninghausen O, Purbhoo MA, Chauveau A, Kohler K, Oddos S, Eissmann P, Brodsky FM, Hopkins C, et al. (2008) Membrane nanotubes physically connect T-cells over long distances presenting a novel route for HIV-1 transmission. Nat Cell Biol 10: 21

Magnification: x 5,800 (A); x 7,500 (B); x 40,000 (C, D)

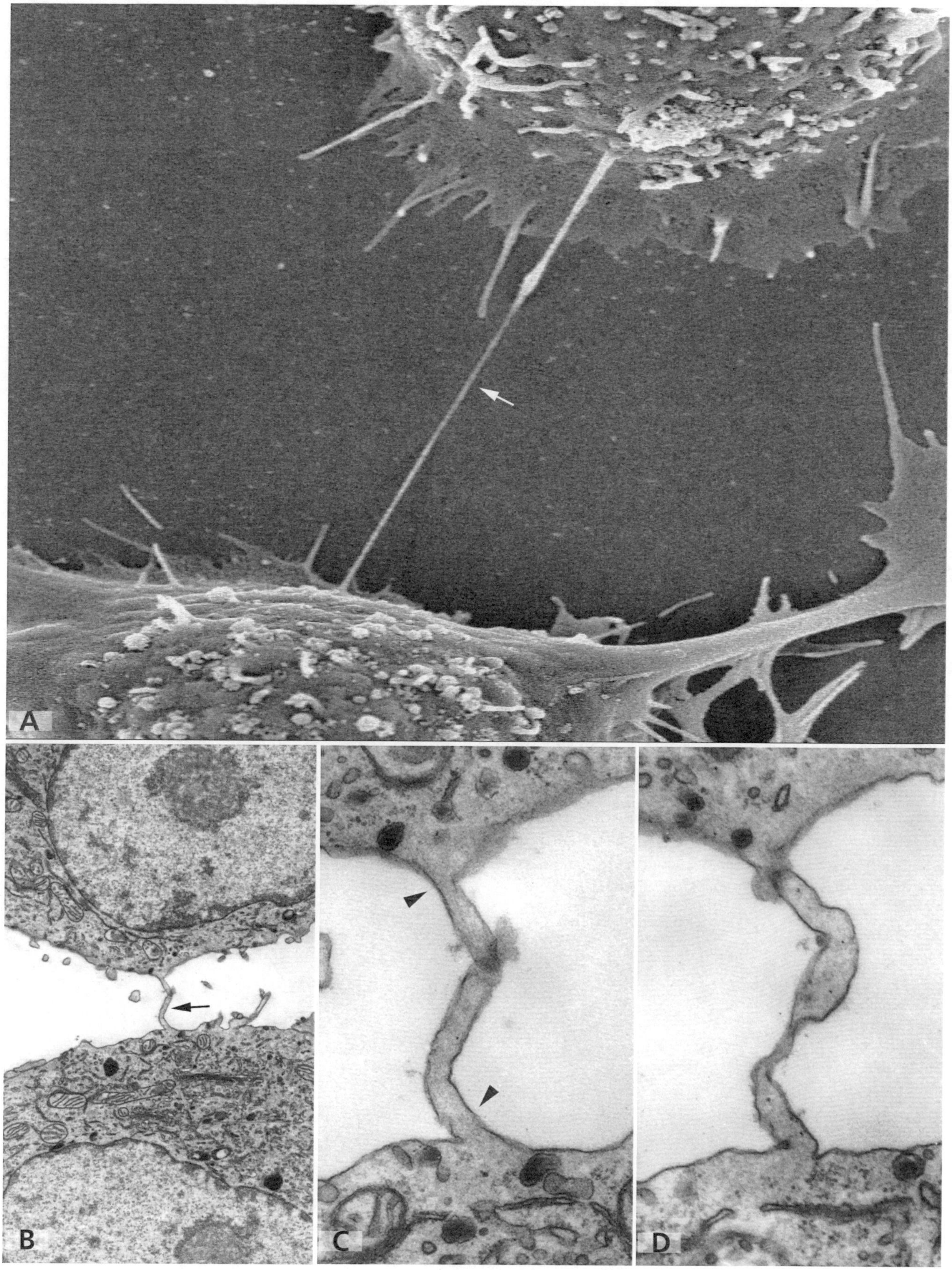
A
B
C
D

SPOT DESMOSOMES

Spot desmosomes (maculae adhaerentes) are anchoring junctions associated with intermediate filaments. They are arranged in circles in the lowermost position of junctional complexes (cf. Fig. 83), but are common also outside complex junctions, occurring independently of other cell contacts. Spot desmosomes are particularly abundant in all tissues that are exposed to mechanical stress. Like buttons, they link neighbouring cells to each other and help to stabilise cells and tissues and make them resistant to mechanical injuries. However, it is increasingly becoming clear that spot desmosomes not merely have a mechanical role by welding cells together. They also hold signalling functions and are considered to have a role as sensors that respond to cellular and environmental signals by changing their organisation and modulating their assembly state. Intracellular calcium homeostasis is crucial for desmosomal adhesion.

Panel A shows a segment of the stratified epithelium of the spinous layer of the epidermis (cf. Figs. 119 and 120). Within the extended interdigitations, the neighbouring keratinocytes are attached to each other by numerous spot desmosomes (arrows). Intermediate (cytokeratin) filaments, called tonofilaments, form thick bundles of tonofibrils (TF). They build up stabilising cables within the cytoplasm, and are anchored in the plaques of the spot desmosomes (arrows). At higher magnification (inset), the dense midline and the inner and outer zones of the plaques, where the tonofilaments attach, are visible. The arrowheads in panel A point to melanin granules.

In panel B, cytokeratin in the epidermis is localized by immunogold labelling by the use of an anti-cytokeratin monoclonal antibody recognising a conserved epitope localised on the surface of cytokeratin filaments. Gold particles label in the outer parts of the dense desmosome plaques (arrows), where cytokeratin filaments are anchored by interacting with desmoplakin, which is one of the plaque proteins. Desmoplakin, via other plaque proteins, plakoglobin and plakophyllin, in turn is connected with membrane-spanning proteins of the cadherin family, desmocollins and desmogleins. Heterophilic, rather than homophilic, interactions of desmocollins and desmogleins are required for desmosomal adhesion and are responsible for the electron dense midline visible under the electron microscope (inset and Fig. 83 C).

References

Chiraev NA and Troyanovsky SM (1997) Direct Ca^{2+}-dependent heterophilic interaction between desmosomal cadherins, desmoglein and desmocollin, contributes to cell-cell adhesion. J Cell Biol 138: 193

Franke WW, Winter S, von Overbeck J, Gudat F, Heitz PU, and Stahli C (1987) Identification of the conserved, conformation-dependent cytokeratin epitope recognised by monoclonal antibody (lu-5). Virch Arch A Pathol Anat Histopathol 411: 137

Green KJ and Gaudry CA (2000) Are desmosomes more than thethers for intermediate filaments? Nat Rev Mol Cell Biol 1: 208

Von Overbeck J, Stahli C, Gudat F, Carmann H, Lautenschlager C, Durmuller U, Takacs B, Miggiano V, Staehelin T, and Heitz PU (1985) Immunohistochemical characterisation of an anti-eipthelial monoclonal antibody (mAB lu-5). Virch Arch A Pathol Anat Histopathol 407: 1

Studer D, Humbel BM, and Chiquet M (2008) Electron microscopy of high pressure frozen samples: bridging the gap between cellular ultrastructure and atomic resolution. Histochem Cell Biol 130: 877

Waschke J (2008) The desmosome and pemphigus. Histochem Cell Biol 130: 21

Magnification: x 38,000 (A), x 60,000 (inset), x 72,000 (B)

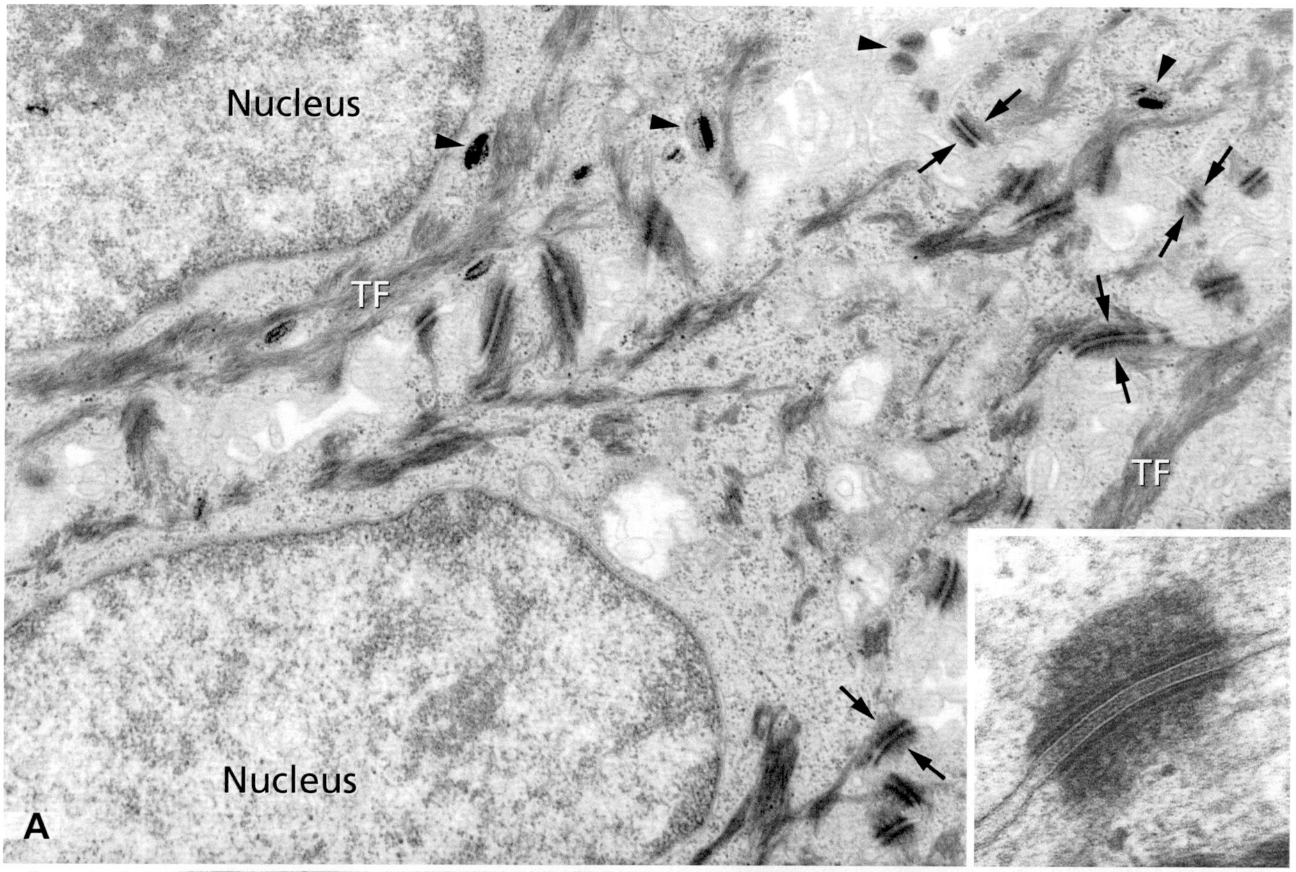
Nucleus
TF
TF
Nucleus
A

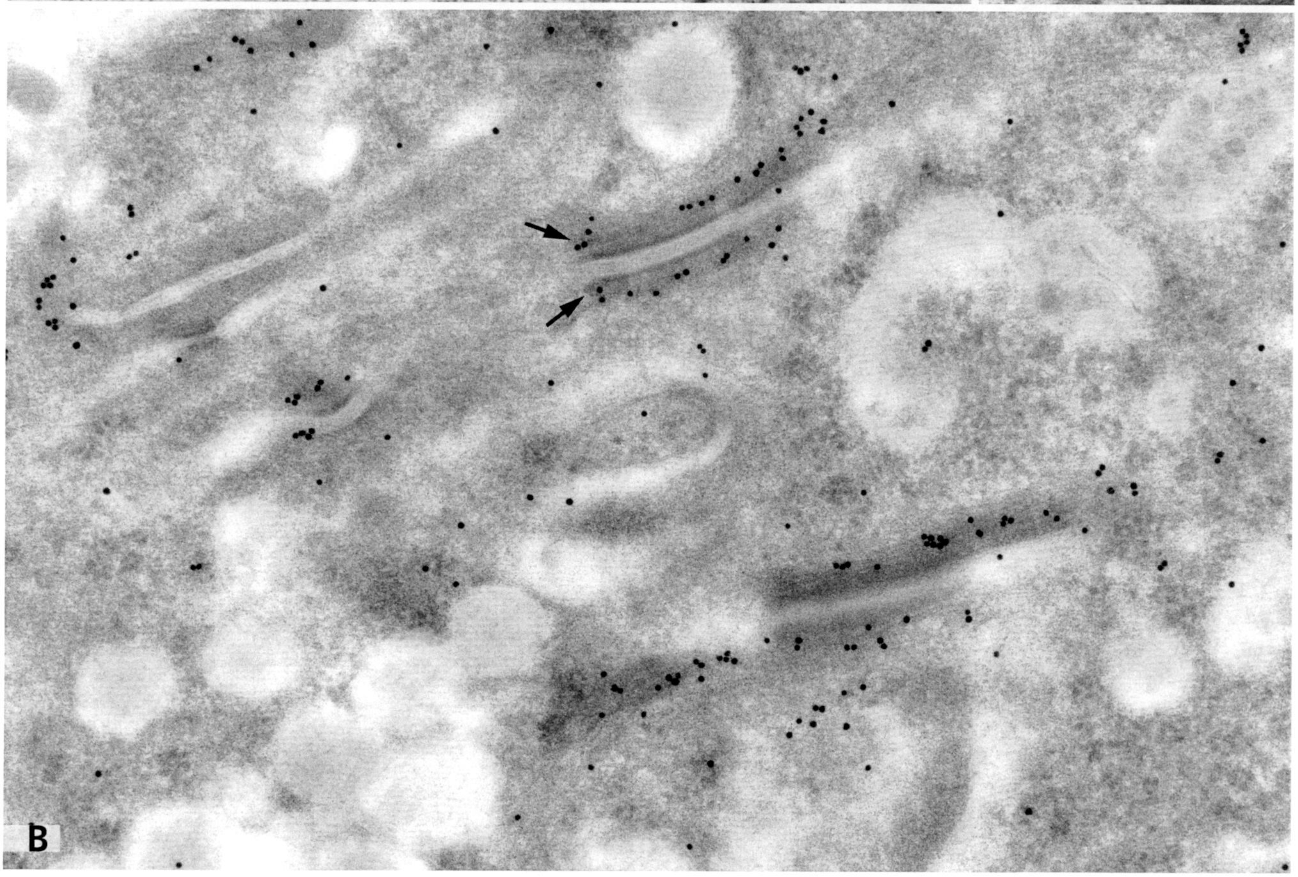
B

SELECTIN – LIGAND-MEDIATED CELL-CELL INTERACTION

Neutrophilic leukocytes and naïve lymphocytes may interact in a specific manner with the endothelial cells lining the capillaries in various organs. This occurs during physiological recirculation (or homing) of lymphoid cells and diapedesis (extravasation) of leukocytes, and during infection or injury as well as inflammation by interaction of these cells with the activated endothelium. For this cell-cell interaction to occur, selectins and cell-surface glycans (as ligands) are important.

In panel A, a neutrophilic leukocyte is shown, which has established contact with an endothelial cell (End) of the sinus in the liver. In panel B, part of a lymphoid cell is shown which appears firmly attached to the endothelium (End) of a liver sinus. In both instances, this cell-cell interaction was strong enough to withstand the mechanical sheering forces during vascular perfusion with the fixative through which all other circulating blood cells were washed out. The labelling by gold particles seen in panels A and B is for lectin binding sites and not related to the particular cell-cell interaction discussed. The interaction between neutrophilic leukocytes and lymphoid cells on one side and normal endothelium on the other is mediated by L-selectin molecules expressed in their plasma membrane microvilli. L-selectins interact with ligands present in the plasma membrane of the endothelia. Such ligands are represented by a spectrum of different fucosylated and sialylated *O*-glycans.

The process of this interaction can be divided into distinct, well-coordinated steps. The selectin-ligand-mediated interactions enable leukocytes or lymphoid cells to tether to and role along the endothelium. This interaction is weak and reversible. This is followed by adhesion to the endothelium, which is a strong interaction. The adhesion is eventually followed by transendothelial migration. Although the initial phase of this cell-cell interaction is mediated only by selectins, integrins become involved during adhesion and lateron.

References

Denucci CC, Mitchell JS, and Shimizu Y (2009) Integrin function in T-cell homing to lymphoid and non-lymphoid sites: getting there and staying there. Crit Rev Immunol 29: 87

Ley K, Laudanna C, Cybulsky MI, and Nourshargh S (2007) Getting to the site of inflammation: the leukocyte adhesion cascade updated. Nat Rev Immunol 7: 678

Mulligan MS, Polley MJ, Bayer RJ, Nunn MF, Paulson JC, and Ward PA (1992) Neutrophil-dependent acute lung injury – requirement for P-selectin (GMP-140). J Clin Invest 90: 1600

Polley MJ, Phillips ML, Wayner E, Nudelman E, Singhal AK, Hakomori S, and Paulson JC (1991) CD62 and endothelial cell-leukocyte adhesion molecule 1 (ELAM-1) recognize the same carbohydrate ligand, sialyl-Lewis x. Proc Natl Acad Sci USA 88: 6224

Rosen SD (2004) Ligands for L-selectin: Homing, inflammation, and beyond. Annu Rev Immunol 22: 129

Springer TA (1994) Traffic signals for lymphocyte recirculation and leukocyte emigration – the multistep paradigm. Cell 76: 301

True DD, Singer MS, Lasky LA, and Rosen SD (1990) Requirement for sialic acid on the endothelial ligand of a lymphocyte homing receptor. J Cell Biol 111: 2757

Magnification: x 20,000 (A, B)

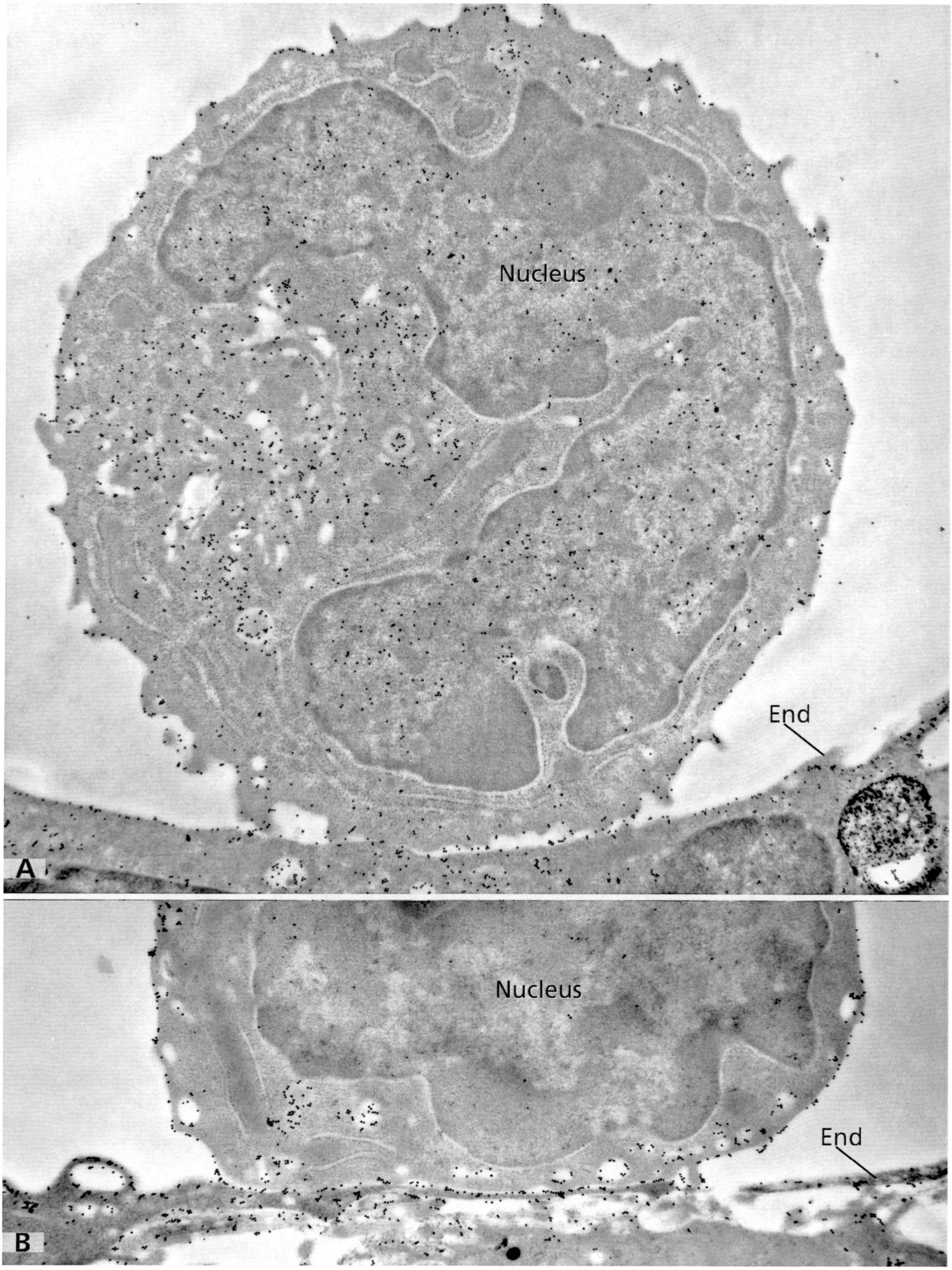
Nucleus
End
A
Nucleus
End
B

CELLULAR INTERDIGITATIONS

Within tissues and organs, the individual cells are organised according to specific patterns, which require particular cell contacts and attachments. Cell junctions of different types are responsible for mechanical, chemical, and electrical coupling of cells and for formation of particular barrier functions in epithelia and endothelia (cf. Figs. 83–86). However, cell contacts also exist outside the specific junctions. For example, the formation of sheets of epithelia building up the lining of internal hollow organs requires tight attachment of cells to each other and to the basal lamina. This is shown in micrographs of the small intestinal epithelium cross-sectioned and sectioned perpendicularly to the surface, presented in panels A and B, respectively. Throughout extended areas, neighbouring cells are tightly attached to each other, and the epithelium is stabilised additionally by multiple interdigitations of the adjacent cells (arrows in panels A and B). Flat ridges of neighbouring cells are intensely interlocked, and the intercellular spaces are closed. However, the cell to cell attachments are variable and intercellular spaces open and close, depending on the functional state of cells (cf. Fig. 112). Panel B of Fig. 112 shows a segment of the small intestinal epithelium with multiple interdigitating ridges of adjacent absorptive cells, but dilated intercellular spaces containing lipoprotein particles, products of the cells secreted at the lateral surfaces.

The electron micrographs shown on the opposite page in panels A and B show thin sections of tissue embedded in Lowicryl K4M and exposed to a special uranyl acetate/methyl cellulose adsorption staining, which yields intense contrast of the plasma membrane and cell junctions (cf. Fig. 83). Intracellularly, membranes of organelles involved in the synthesis and transport of complex glycans show particularly intense staining. Such intense membrane contrast is visible in the Golgi apparatus of the absorptive cells (Golgi in panels A and B).

References

Gumbiner BM (1996) Cell adhesion: The molecular basis of tissue architecture and morphogenesis. Cell 84: 345

Roth J, Taatjes DJ, and Tokuyasu KT (1990) Contrasting of Lowicryl K4M thin sections. Histochemistry 95: 123

Magnification: x 34,500 (A), x 34,500 (B)

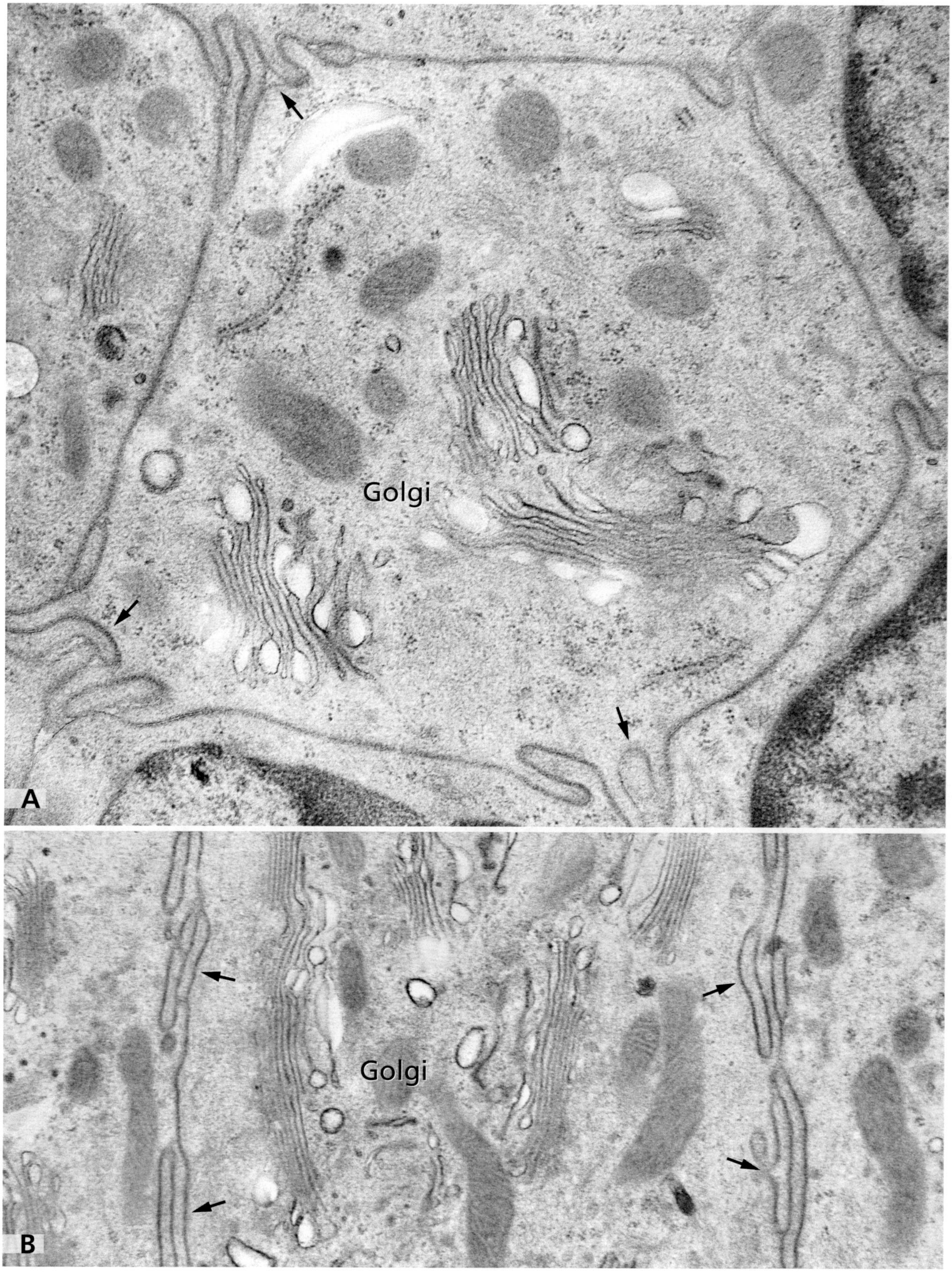
Golgi
A
Golgi
B

BASAL LABYRINTH

Epithelia involved in extensive transcellular transport not only have characteristic apical differentiations (cf. Figs. 113 and 114) but also show basal characteristics, summarised with the term basal "membrane labyrinth". The basal labyrinth is a huge membrane convolute at the basal cell domain, consisting of infoldings of the basal plasma membrane and interdigitating basal ridges of neighbouring cells, which both lead to a tremendous increase of the basolateral surfaces of the epithelial cells. Close to the membranes, mitochondria are accumulated and possibly aligned in a vertical basal to apical orientation. Basal labyrinths are particularly prominent in the epithelia of the renal proximal and distal tubules, in the renal cortical collecting ducts, and in the striated ducts of salivary glands. The "striations" visible in the light microscope result from the vertically aligned mitochondria.

Panels A, B, and C show segments of renal proximal and distal tubules. According to their transport functions, both of those parts of the nephron are equipped with a basal labyrinth, which is shown in survey in panel A, where a proximal tubule (PT) is visible in the upper part of the micrograph and a distal convoluted tubule (DCT) is on display in the lower micrograph segment. Adjacent epithelial cells of a proximal tubule are shown at higher magnification in panel B and the basal domains of distal tubular cells are featured at higher magnification in panel C. Both proximal and distal tubular cells in panels B and C show lectino-gold labeling for demonstration of sialic acid residues. In the polarised epithelial cells of the proximal tubules, the apical surface is the site of reabsorption of a high percentage of the filtered glucose, water, Na^+, Cl^-, K^+, and small proteins from the primary urine (cf. Fig. 113). The basal cell pole is equipped for excretion into the extracellular space. Membranes of the basal labyrinth contain the enzymes for active transport of ions and the long mitochondria localised nearby provide the adenosine triphosphate necessary. Distal tubular cells possess short apical microvilli, but basal mitochondria are more numerous and the basal infoldings of the plasma membrane and interdigitating cell ridges are more extended and more regular in comparison with the proximal tubule cells (panels A and C).

References

Fukudome H (2001) A combined SEM and TEM study on the basal labyrinth of the collecting duct in the rat kidney. Arch Histol Cytol 64: 339

Roth J, Lucocq JM, and Charest PM (1984) Light and electron microscopic demonstration of sialic acid residues with the lectin from Limax flavus: A cytochemical affinity technique with the use of fetuin-gold complexes. J Histochem Cytochem 32: 1167

Tsuchiya K, Wang W, Giebisch G, and Welling PA (1992) ATP is a coupling modulator of parallel Na, K-ATPase-K-channel activity in renal proximal tubule. Proc Natl Acad Sci USA 89: 6418

Magnification: x 73,000 (A), x 15,200 (B), x 18,000 (C)

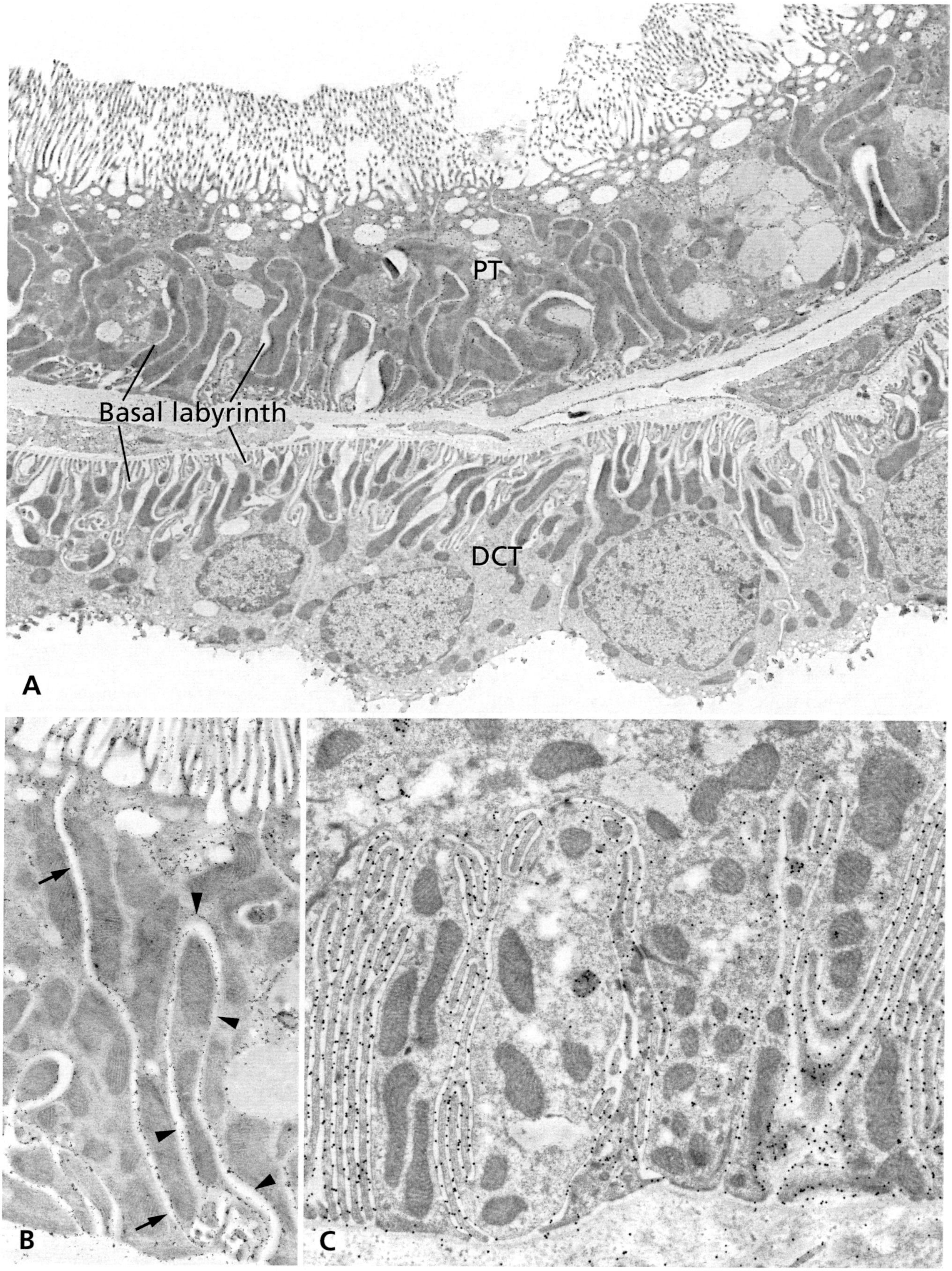
PT
Basal labyrinth
DCT
A
B
C

BASEMENT MEMBRANE

Basement membranes are thin sheets of specialised extracellular matrix, supporting epithelial cell layers and covering muscle cells and nerve fibres. Basement membranes not only provide a particular support for tissues, but also have essential roles in cell differentiation and movement, in morphogenesis and new formation of organs and in multiple pathological processes, such as tumour growth and migration. Basement membranes consist of two different layers (laminae), which can easily be distinguished under the electron microscope: the basal lamina (Lb, lamina basalis) and the reticular lamina (Lf, lamina fibroreticularis).

The basal lamina is in direct contact with the plasma membrane of those cells that synthesise and secrete its components, epithelial cells, glial cells and muscle cells The basement membrane of a smooth muscle cell is shown on the opposite page. With few exceptions, the basal lamina is well visible as a continuous, fine, electron dense layer close to the surface of the respective cells (lamina densa, Ld), but frequently separated from the plasma membrane by a narrow space, designated as lamina rara (Lr, cf. also Fig. 93). The basal lamina mainly contains type IV collagen, fibronectin, laminin, entactin (also known as nidogen), and heparan sulfate proteoglycans (also called perlacan). The characteristic structure results from a self assembly process of laminin molecules with type IV collagen, entactin, and proteoglycans. Integrins localised in the plasma membrane are the major adhesion receptors connecting cells with components of the extracellular matrix. Integrins interact directly with laminin and fibronectin present in the basal lamina and intracellularly contact actin through intermediate proteins, such as alpha-actinin, vinculin, and talin. Integrin-mediated cell adhesion modulates different signal transduction cascades, influences the expression of genes related to cell differentiation, and is involved in cell migration, proliferation, and survival. Other proteins of the lamina basalis participating in diverse supramolecular structures belong to the family of fibulins. ADAMs, named after their adhesive and metalloproteinase domains, are considered to be modulators of cell-matrix interactions and have a crucial role in the remodelling of components of the extracellular matrix.

The reticular lamina is a product of fibroblasts of the adjacent connective tissue and mainly contains type III collagen. It supports the lamina basalis. As shown in the inset, the reticular lamina may form a sheet of regularly organised reticular fibrils beneath and close to the dense part of the lamina basalis.

References

Brown E and Dejana E (2003) Cell-to-cell contact and extracellular matrix. Editorial overview: Cell-cell and cell-matrix interactions – running, jumping, standing still. Curr Opin Cell Biol 15: 505

Danen EHJ and Sonnenberg A (2003) Integrins in regulation of tissue development and function. J Pathol 200: 471

Sakai T, Larsen M, and Yamada KM (2003) Fibronectin requirement in branching morphogenesis. Nature 423: 876

Schwarzbauer JE and Sechler JL (1999) Fibronectin fibrillogenesis: a paradigm for extracellular matrix assembly. Curr Opin Cell Biol 11: 622

Timpl R, Sasaki T, Kostka G, and Chu M-L (2003) Fibulins: A versatile family of extracellular matrix proteins. Nat Rev Mol Cell Biol 4: 479

White JM (2003) ADAMs: modulators of cell-cell and cell-matrix interaction. Curr Opin Cell Biol 15: 598

Magnification: x 55,600, x 59,000 (inset)

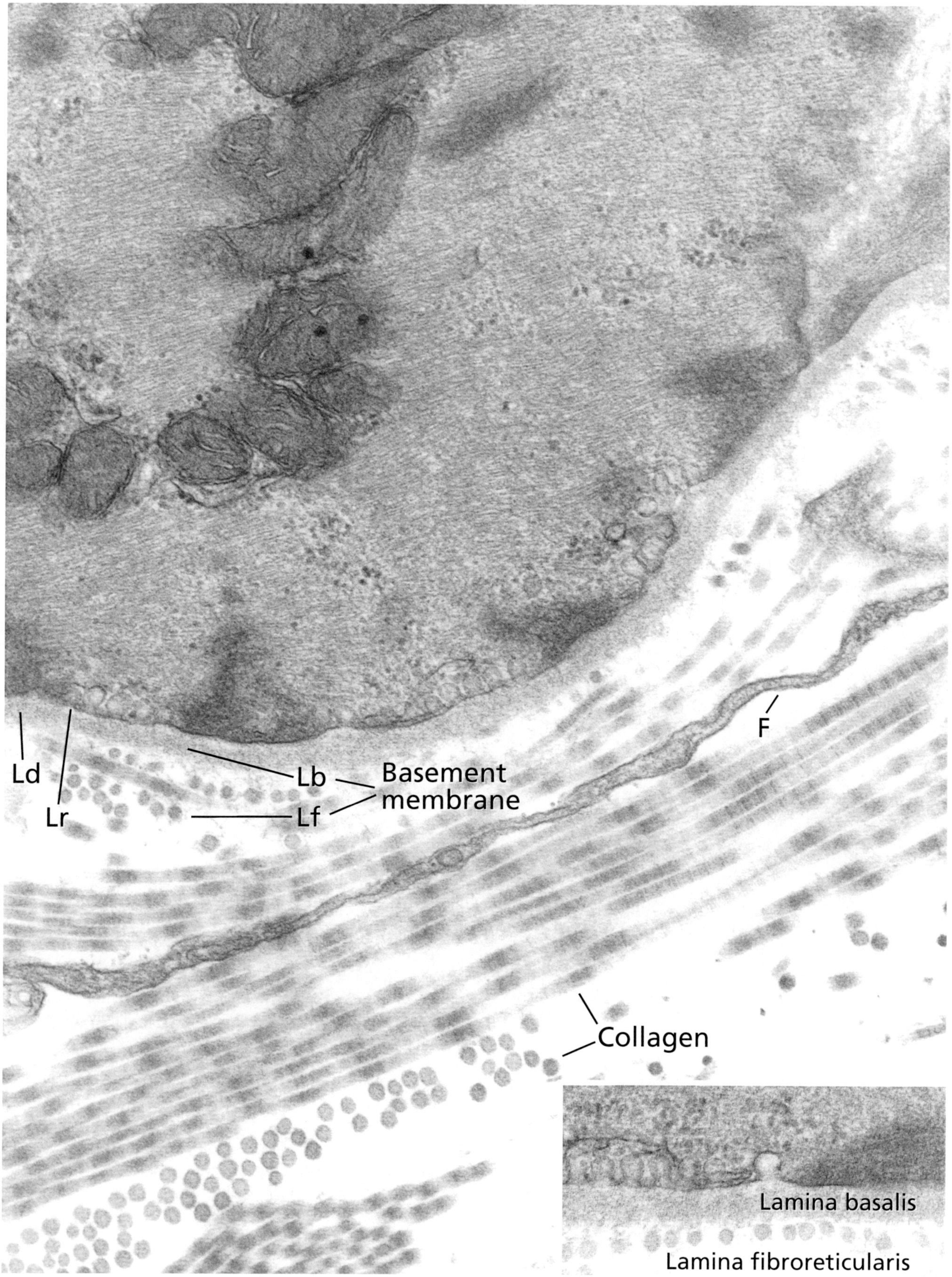
F
Ld
Lr
Lb
Lf
Basement
membrane
Collagen
Lamina basalis
Lamina fibroreticularis

GLOMERULAR BASEMENT MEMBRANE

The glomerular basement membrane shown in panel A differs from basement membranes in other locations in that it is faced by a cell layer on either side, namely the endothelia and the podocytes and that it is thicker, because it is the product of fusion of the basement membrane of endothelial cells and podocytes. It is composed of three layers: a central electron dense lamina densa and a layer of lower electron density on either side, the lamina rara interna toward the endothelia and the lamina rara interna toward the podocytes. The lamina densa is composed of a compact meshwork of 3 nm filaments. The lamina rara interna contains 10 nm fibrils and both laminae rarae are crossed by filamentous material, which reaches the plasma membrane of endothelia and podocytes. The glomerular basement membrane is extensively cross-linked by disulfide bonds and collagen-type bonds. It is made up of various components also found in other basement membranes: type IV collagen, heparan sulfate and chondroitin sulfate proteoglycans, laminin, nidogen and BM-40/osteonectin/SPARC. Collagen type IV does not form fibrils and represents the scaffold of the lamina densa. The proteoglycans form a quasi-regular, lattice-like array in both laminae rarae, whereas laminin and nidogen are present in all three layers of the glomerular basement membrane. It is disputed whether or not fibronectin is a glomerular basement membrane component.

Arrows: slit diaphragms between podocyte foot processes; arrowheads: tangentially sectioned pores of endothelia.

References

Farquhar M (1991) The glomerular basement membrane. A selective macromolecular filter. In: Cell biology of extracellular matrix (Hay E, eds). New York: Plenum Press, pp 365

Kriz A and Kaissling B (2000) Structural organisation of the mammalian kidney. In: The Kidney: Physiology and pathophysiology (Seldin D, and Giebisch G, eds). New York: Lippincott, Williams and Wilkins), pp 587

ALPORT'S SYNDROME (HEREDITARY NEPHRITIS)

This is mainly an X-linked dominant inherited trait and the disease manifests in childhood and is progressive. The mutations are located in the *COL4A5* gene encoding the α5 chain of type IV collagen. Autosomal Alport syndrome is caused by mutations in the *COL4A3* and *COL4A4* genes encoding the α3 and α4 chain of type IV collagen.

Some of the characteristic electron microscopic findings are illustrated in panel B. They typically consist of a thickening of the glomerular basement membrane (GBM) and longitudinally lamination, splitting, fragmentation and formation a net-like pattern of the lamina densa. Other segments of the glomerular basement membrane are extremely thin (not seen in the micrograph). Furthermore, these changes are accompanied by endothelial hypertrophy, podocyte edema and podocyte foot process effacement or fusion. The serrated outer and inner contours of the glomerula basement membrane are also evident in panel B. The presence of all these changes permits the diagnosis.

Clinical signs of the structural alteration and malfunction of the GBM include hematuria, proteinuria and edema. Sensori-neural deafness is usually associated with Alport's syndrome. Therapy consists in renal transplantation.

References

Barker DF, Hostikka SL, Zhou J, Chow LT, Oliphant AR, Gerken SC, Gregory MC, Skolnick MH, Atkin CL, and Tryggvason K (1990) Identification of mutations in the COL4A5 collagen gene in Alport syndrome. Science 248: 1224

Heidet L and Gubler MC (2009) The renal lesions of Alport syndrome. J Am Soc Nephrol 20: 1210

Jefferson JA, Lemmink HH, Hughes AE, Hill CM, Smeets HJ, Doherty CC, and Maxwell AP (1997) Autosomal dominant Alport syndrome linked to the type IV collage alpha 3 and alpha 4 genes (COL4A3 and COL4A4). Nephrol Dial Transplant 12: 1595

Kwoh C, Shannon MB, Miner JH, and Shaw A (2006) Pathogenesis of nonimmune glomerulopathies. Annu Rev Pathol 1: 349

Kuroki A, Ito J, Yokochi A, Kato N, Sugisaki T, Sueki H, Akizawa T (2008) Diagnosing Alport syndrome using electron microscopy of the skin. Kidney Int 73: 364

Tryggvason K and Martin P T (2001) Alport syndrome and basement membrane collagen. In: The metabolic and molecular bases of inherited disease (Scriver C, Beaudet A, Valle D, and Sly W, eds). New York: McGraw-Hill, pp 5453

Wang XP, Fogo AB, Colon S, Giannico G, Abul-Ezz SR, Miner JH, Borza DB (2005) Distinct epitopes for anti-glomerular basement membrane alport alloantibodies and goodpasture autoantibodies within the noncollagenous domain of alpha3(IV) collagen: a janus-faced antigen. J Am Soc Nephrol 16: 3563

Zollinger H and Mihatsch M (1978) Renal pathology in biopsy. Berlin Heidelberg New York: Springer

Magnification: x 65,000 (A); x 28,000 (B)

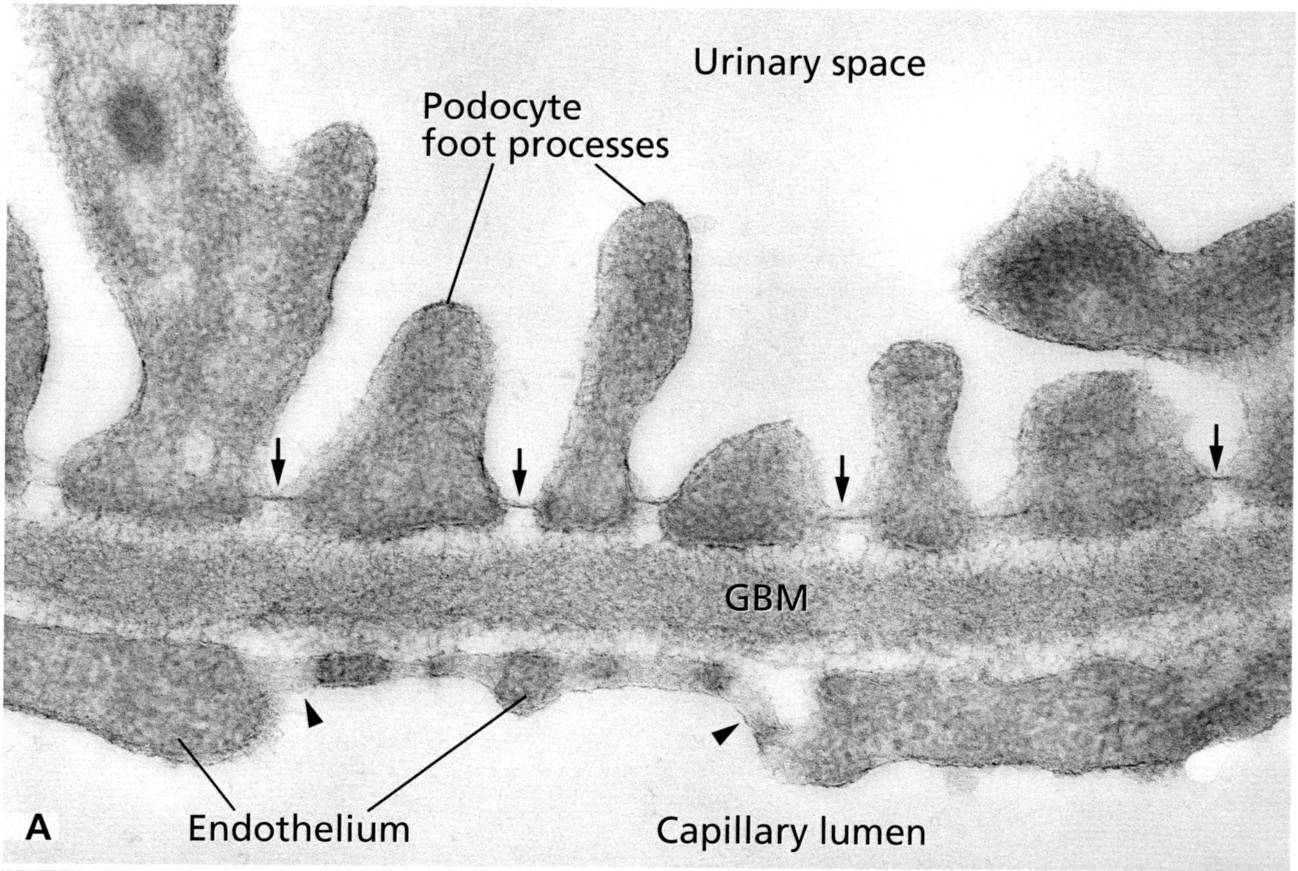
Urinary space
Podocyte
foot processes
GBM
Endothelium
Capillary lumen
A

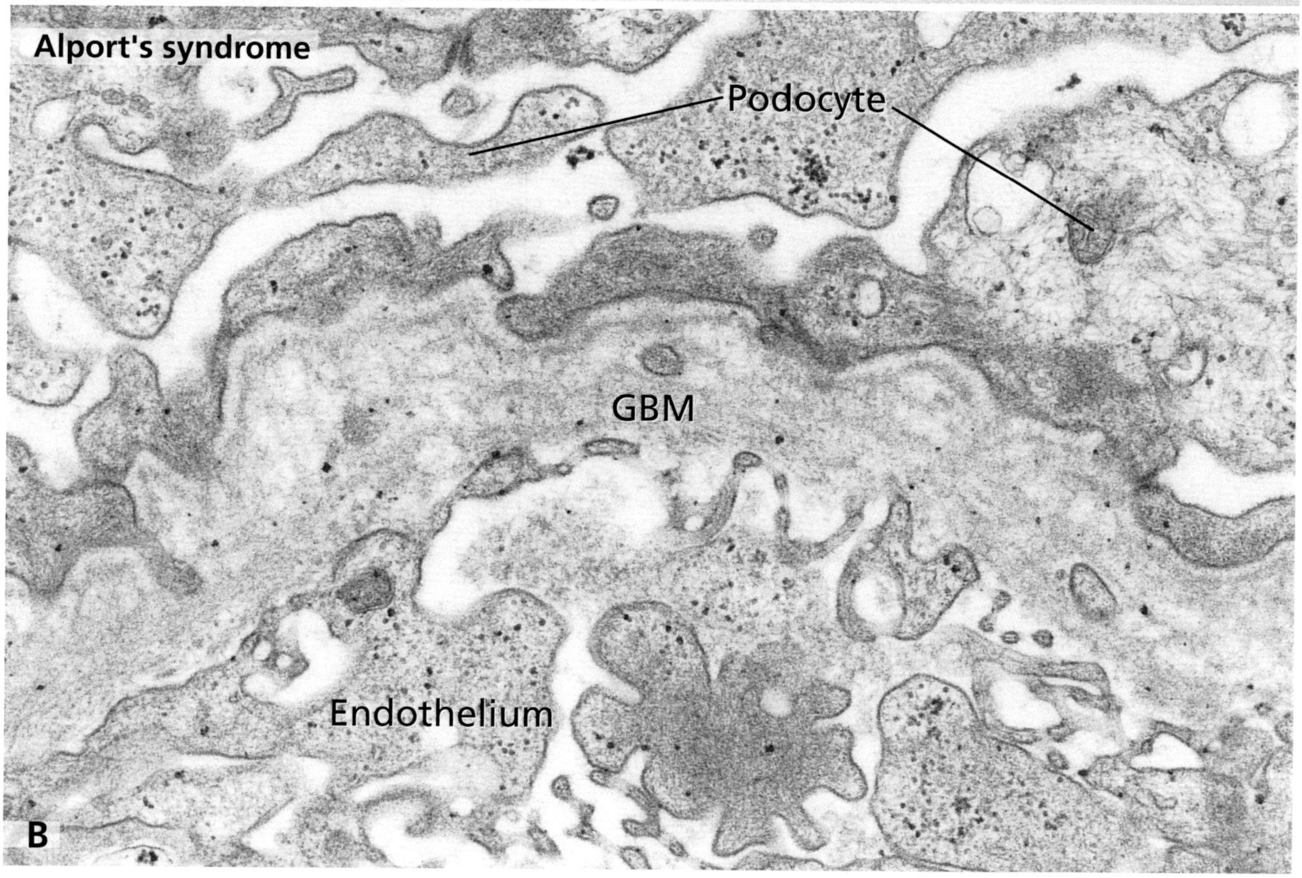
Alport's syndrome
Podocyte
GBM
Endothelium
B

DESCEMET'S MEMBRANE

Descemet's membrane is an extraordinary thick basement membrane, which is unique in the body with respect to both its dimension and composition. It is built by the cells of the flat squamous epithelium that lines the posterior surface of the cornea and is designated as corneal endothelium. Measuring 5–10 μm in thickness, the membrane of Descemet, like other basement membranes, consists of two distinct layers, a posterior layer adjacent to the endothelium and produced by the endothelial cells and an anterior layer formed by collagen lamellae and proteoglycans. Both layers are on display in panel A, which also shows the endothelium in the upper segment and the corneal stroma in the lower segment. The different organisation of the Descemet's membrane in comparison to the collagen fibril organisation in the stroma is clearly visible. Panel B shows the posterior part of Descemet's membrane at higher magnification.

The membrane of Descemet is composed of a range of proteins, including laminin and fibronectin, and proteoglycans that contain keratan sulfate, heparan sulfate, and dermatan sulfate. Comparable with the situation in the corneal stroma, content and qualities of proteoglycans are essential for hydration of the tissue, which in turn is responsible for its transparency. Descemet's membrane also contains collagen with types IV and VIII being predominant. Type IV collagen is a usual component of the basal lamina part of basement membranes (cf. Fig. 90). In contrast, type VIII collagen is rare in other parts of the body, but is contained in Descemet's membrane at high concentration. It forms characteristic lattices and networks, which appear as ladder-like structures under the electron microscope. Such structures are particularly prominent and well visible in the micrograph shown in panel A.

The thickness of Descemet's membane increases throughout life. At birth, the membrane of Descemet in the human is about 3 μm thick and consists mainly of the anterior part, whereas the posterior layer is produced by the endothelial cells after birth. It slowly increases in thickness measuring about 10 μm in older adults.

The corneal endothelium visible in panels A and B in the upper segments of the micrographs holds pivotal functions in regulating corneal hydration and maintaining the critical state of hydration required for transparency of the corneal tissues. The endothelium permits passage of nutrients from the aqueous humor contained in the posterior and anterior chambers of the eye into the cornea. On the other hand, it counteracts swelling of the corneal stroma by removing excess stromal fluid through the activity of "ionic pumps" localised in the basolateral plasma membranes. Aquaporins forming water selective channels also have an important role in the traffic of fluids across the endothelium.

References

Abrams GA, Bentley E, Nealey PF, and Murphy CJ (2002) Electron microscopy of the canine corneal basement membranes. Cells Tiss Org 170: 251

Benya PD and Padilla SR (1986) Isolation and characterisation of type-VIII collagen synthesised by cultured rabbit corneal endothelial cells. J Biol Chem 261: 4160

Joyce NC (2003) Proliferative capacity of the corneal endothelium. Progr Ret Eye Res 22: 359

Kapoor R, Sakai LY, Funk S, Roux E, Bornstein P, and Sage EH (1988) Type-VIII collagen has a restricted distribution in specialised extracellular matrices. J Cell Biol 107: 721

Verkman AS (2003) Role of aquaporin water channels in eye function. Exp Eye Res 76: 137

Magnification: x 15,200 (A); x 60,000 (B)

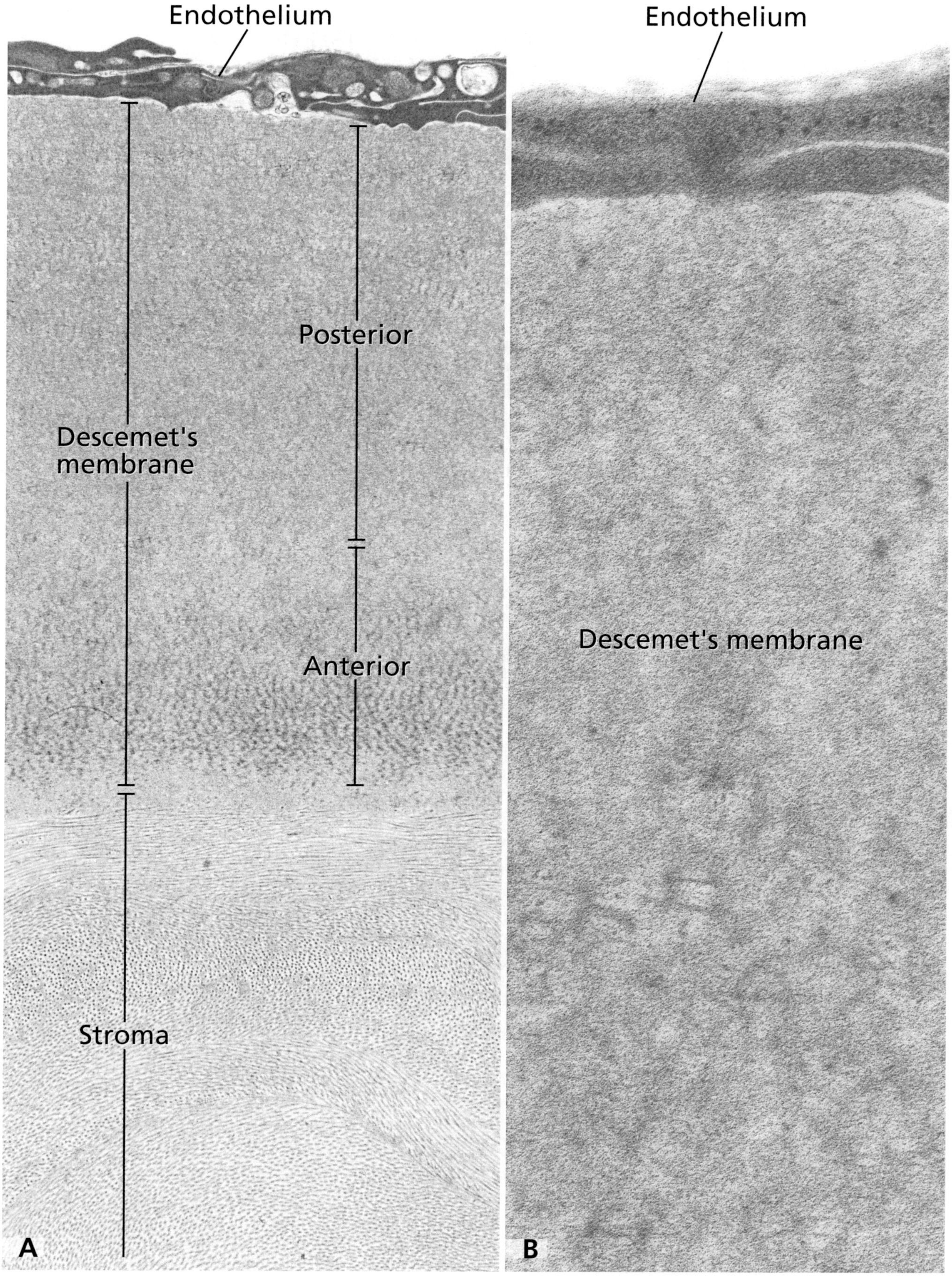
Endothelium
Posterior
Descemet's membrane
Anterior
Stroma
A
Endothelium
Descemet's membrane
B

SKIN BASEMENT MEMBRANE AND KERATINOCYTE HEMIDESMOSOMES: AN EPITHEL-CONNECTIVE TISSUE JUNCTIONAL COMPLEX

In the skin, the squamous epithelial layer of the epidermis and the connective tissue of the underlying dermis are firmly connected to each other as illustrated in panels A and B. The structural arrangement of their interface, the dermal-epidermal junction, provides an illuminating example of how these two tissue components are linked together.

The main structures involved are the basement membrane with the lamina densa derived from the basal cell layer of the epidermis, the anchoring fibrils (asterisk in panel B) of the dermis and the hemidesmosomes present in the basal plasma membrane of the basal cells (arrows in panel A and B).

The basement membrane is composed of a lamina densa with a thin lamina rara toward the basal cell layer of the epidermis as seen in panels A and B. The dermis is composed of fibroblasts, bundles of collagen fibres and other fibres. Hemidesmosomes, in contrast to desmosomes, are asymmetrical, highly specialised integrin-mediated junctions with a complex molecular composition strikingly different from that of desmosomes (cf. Fig. 86). By electron microscopy, each hemidesmosome is composed of a dense cytoplasmic plaque, which sits on a plate and thin anchoring filaments extending from the plate across the lamina rara to the lamina densa of the basement membrane (arrows in panels A and B). The major components of the hemidesmosome are the $\alpha_6\beta_4$ integrin, type VII collagen (identical to bullous pemphigoid antigen 2) and the bullous pemphigoid antigen 1. Intermediate filaments of the keratin-type associate with the cytoplasmic plaque establishing the cytoskeleton-hemidesmosome association. They are linked through plectin and the bullous pemphigoid antigen 1 to the unusually long tail of β_4 integrin. The basal cells of the epidermis contain keratins K5 and K14. The hemidesmosome-basement membrane association occurs through complexes formed by β_4 integrin and laminin 5. The latter, which forms thin thread-like structures of 2–4 nm in diameter, acts as anchoring filaments. This represents another major difference to desmosomes which link neighboring epithelia by the glycosylated extracellular domain of interlocking cadherins such as desmoglein1 and desmocollin 3. The dense lamina of the basement membrane is attached through fibrils to the dermis. The involvement of type VII collagen – laminin 5 interactions in linking the lamina densa with the underlying dermis has been demonstrated. The type VII collagen itself entraps banded collagen fibres, elastic microfibrils and beaded microfilaments present in the stroma of the dermis.

Although it has been established since long that hemidesmosomes are anchors for epithelial cells to the basement membrane and that perturbation of their integrity results in separation and blisters (cf. Fig. 94), there is evidence accumulating for their involvement in signalling phenomena.

Asterisk in panel A: melanocyte dendrites between keratinocytes; arrowheads in panel A: melanosomes in keratinocytes.

References

Borradori L and Sonnenberg A (1996) Hemidesmosomes: roles in adhesion, signaling and human diseases. Curr Opin Cell Biol 8: 647

Burgeson R and Christiano A (1997) The dermal-epidermal junction. Curr Opin Cell Biol 9: 651

Dowling J, Yu QC, and Fuchs E (1996) Beta 4 integrin is required for hemidesmosome formation, cell adhesion and cell survival. J Cell Biol 134: 559

Green K and Gaudry C (2000) Are desmosomes more than tethers for intermediate filaments? Nat Rev Mol Cell Biol 1: 208

Jones J, Hopkinson S, and Goldfinger L (1998) Structure and assembly of hemidesmosomes. Bioessays 20: 488

Kowalczyk A, Bornslaeger E, Norvell S, Palka H, and Green KJ (1999) Desmosomes: intercellular adhesive junctions specialised for attachment of intermediate filaments. Int Rev Cytol 185: 237

Mainiero F, Pepe A, Wary K, Spinardi L, Mohammadi M, Schlessinger J, and Giancotti F (1995) Signal transduction by the $\alpha_6\beta_4$ integrin: Distinct β_4 subunit sites mediate recruitment of Shc/Grb2 and association with the cytoskeleton of hemidesmosomes. EMBO J 14: 4470

Magnification: x 15,000 (A); x 60,000 (B)

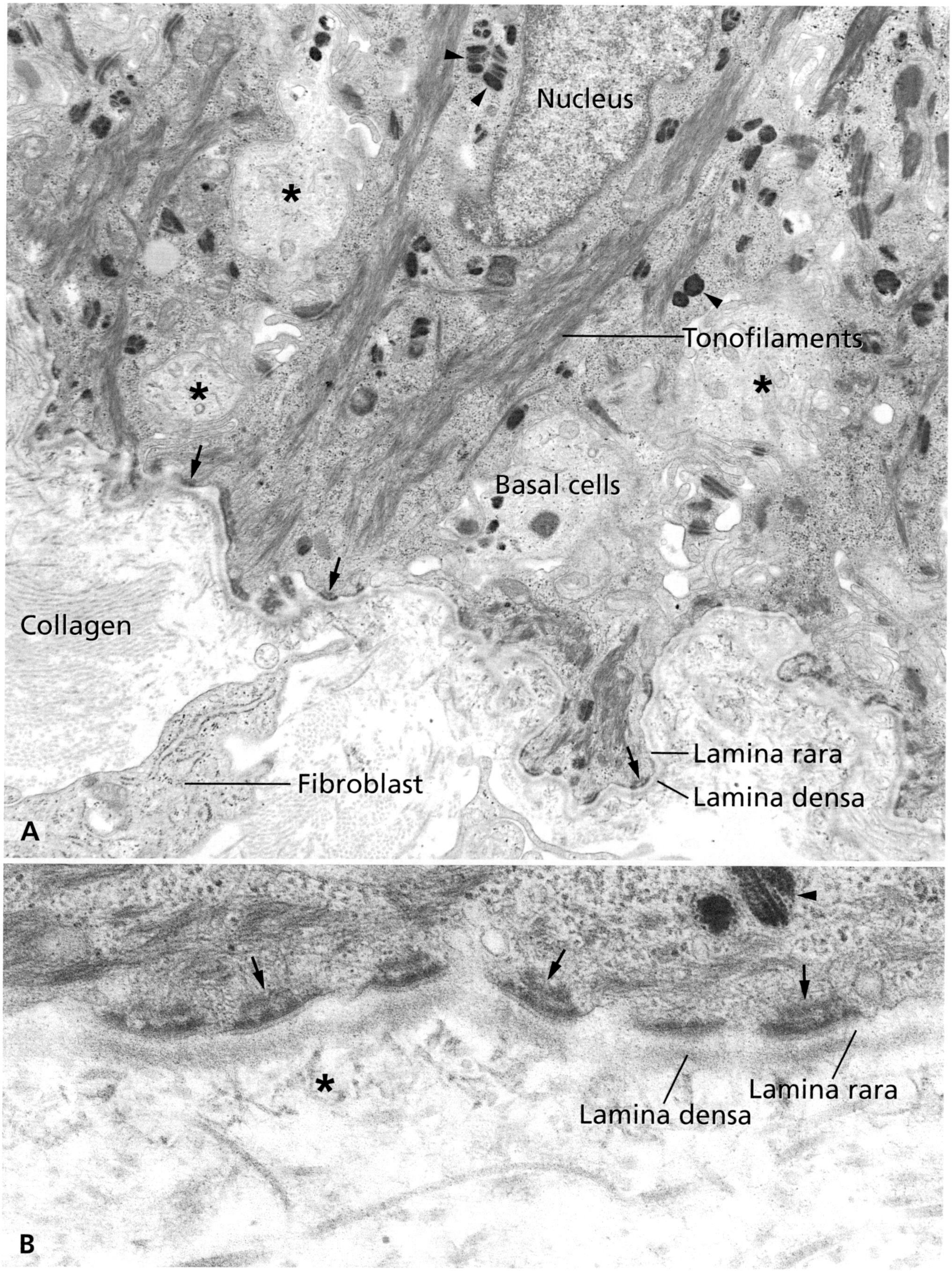
Nucleus
Tonofilaments
Basal cells
Collagen
Lamina rara
Fibroblast
Lamina densa
A
Lamina rara
Lamina densa
B

EPIDERMOLYSIS BULLOSA SIMPLEX

Disruption of the dermal-epidermal junction is associated with diseases of the skin resulting in skin fragility and formation of blisters. These changes are most apparent in regions susceptible to mechanical trauma such as palms and sole of foot.

In panels A and B, the epidermolysis bullosa simplex Weber-Cockayne, a mild form, is shown. By electron microscopy, lysis of basal cells is obvious and rupturing of cells can occur between the nucleus and the hemidesmosomes. The separation of epidermis from dermis resulting in blister formation can be unequivocally seen in panel A. In this mild form, keratin filaments and filament bundles may appear structurally normal (panel B) as are the hemidesmosomes (arrows in panel B). In severe epidermolysis bullosa simplex Dowling-Meara, clumps or aggregates of keratin filaments in the basal cell cytoplasm are the fine structural hallmark of this form.

Epidermolysis bullosa simplex is a disorder of keratins K5 and K14 present in the basal cell layer and occurs in different clinical forms. Different point mutations in the keratin-encoding genes cause defects in the assembly of the keratin filaments. Depending on the location of the mutation in the gene, clinically severe forms with defects in the elongation of keratin filaments or mild forms with impaired lateral interactions within the filaments can be observed. Severe forms are always associated with mutations either in the N-terminal end of helix 1A or the C-terminal end of helix 2B.

In addition to the disorders of keratins K5 and K14, disorders of keratins present in the suprabasal cell layers are known. They involve the keratins K1 and K10 and cause epidermolytic hyperkeratosis with excessive keratinisation and breakdown of the epidermis. Although the suprabasal cells are degenerating, the basal cells are not affected because they lack of keratins K1 and K10. Furthermore, disorders related to intermediate linker proteins such as plectin and bullous pemphigoid antigen 1 have been detected. Impaired anchoring of keratin filaments to the hemidesmosomes results. In humans, an epidermolysis bullosa simplex with muscular dystrophy due to mutant bullous pemphigoid antigen 1 has been described.

Epidermolysis bullosa is a group of disorders in which laminin 5 and type VII collagen, the anchoring fibrils between hemidesmosomes, lamina densa and dermis are involved. Junctional epidermolysis bullosa is due to a lack of laminin 5. By electron microscopy, blisters occur as the result of separation in the lamina rara with the lamina densa and anchoring fibrils being unaffected. Dystrophic epidermolysis bullosa results of mutations in the type VII collagen gene with morphological altered anchoring fibrils which may be reduced in numbers or completely missing.

References

Chan Y, Yu Q, Fine J, and Fuchs E (1993) The genetic basis of Weber-Cockayne epidermolysis bullosa simplex. Proc Natl Acad Sci USA 90: 7414

Eady R, McGrath J, and McMillan J (1994) Ultrastructural clues to genetic disorders of the skin: the dermal-epidermal junction. J Invest Dermatol 103: 13S

Fuchs E (2001) Disorders of intermediate filaments and their associated proteins. In: The metabolic and metabolic bases of inherited disease (Scriver C, Beaudet A, Valle D, and Sly WS, eds). New York: McGraw-Hill, pp 5629

Ishida-Yamamota A, Mc Grath J, Chapman S, Leigh I, Lane E, and Eady R (1991) Epidermolysis bullosa simplex (Dowling-Meara type) is a genetic disease characterised by abnormal keratin-filament network involving keratins K5 and K14. J Invest Dermatol 97: 959

Letai A, Coulombe P, McCormick M, Yu Q, Hutton E, and Fuchs E (1993) Disease severity correlates with position of keratin point mutations in epidermolysis bullosa simplex. Proc Natl Acad Sci USA 90: 3197

Pearson R and Spargo B (1961) Electron microscope studies of dermal-epidermal separation in human skin. J Invest Dermatol 36: 213

Uitto J and Pulkinnen L (2001) Epidermolysis bullosa: the disease of the cutaneous basement membrane zone. In: The metabolic and metabolic bases of inherited disease (Scriver C, Beaudet A, Valle D, and Sly W, eds). New York: McGraw-Hill, pp 5655

Wilgram G and Caulfield J (1966) An electron microscopic study of epidermolytic hyperkeratosis. Arch Dermatol 94: 127

Magnification: x 3,000 (A); x 45,000 (B)

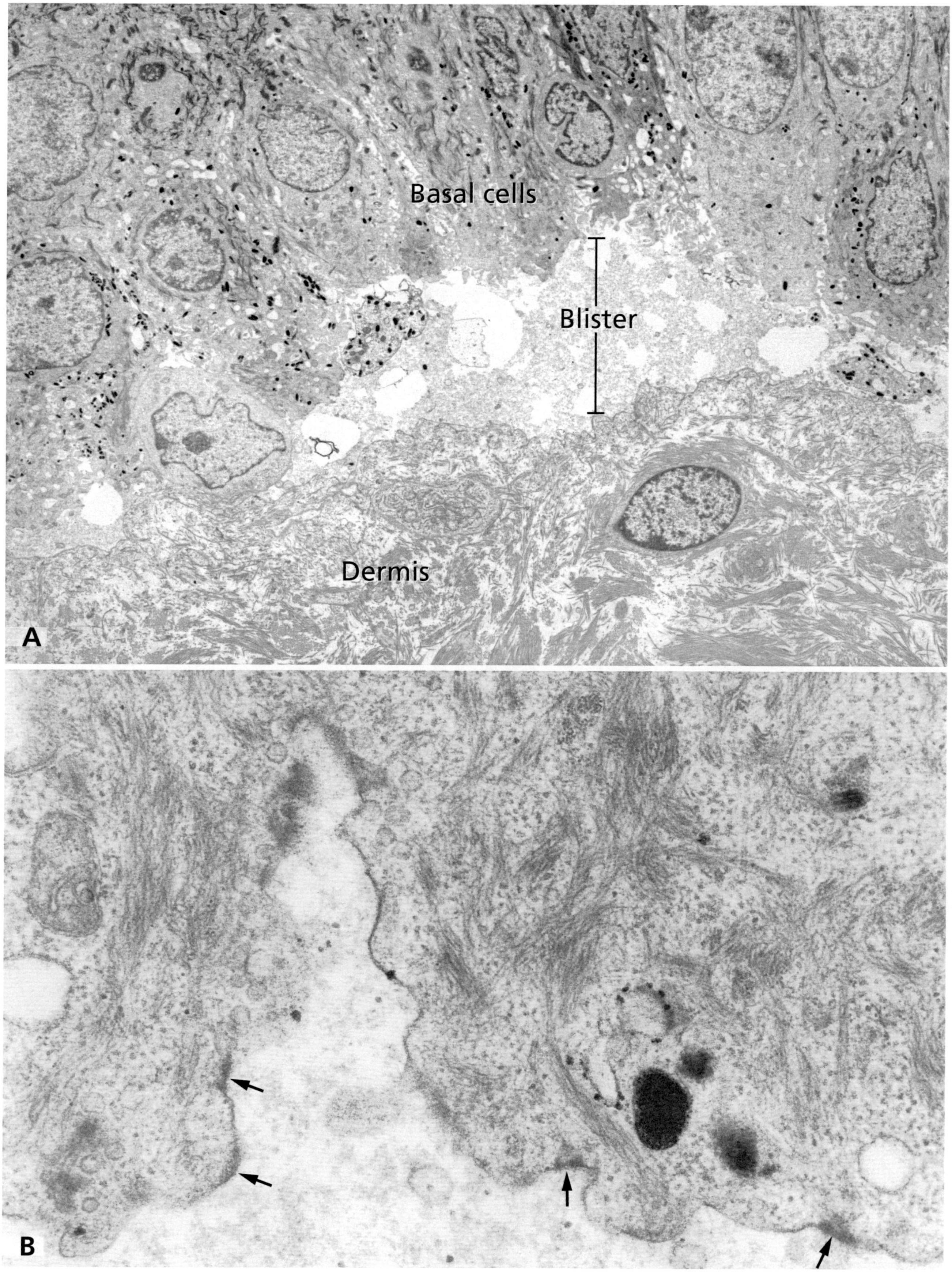
Basal cells
Blister
Dermis
A
B

Principles of Tissue Organisation

PANCREATIC ACINUS

The pancreas is a combined exocrine and endocrine gland, localised close to the posterior abdominal wall and composed of four anatomic components, a head, a neck, a body and a tail.

The acinus, as shown cross-sectioned in the micrograph on the opposite page, represents the structural and functional secretory unit of the exocrine part of the pancreas (for the endocrine pancreas cf. Figs.102, 103).

The exocrine pancreas is the major digestive gland in the human body, draining directly into the duodenum at the ampulla of Vater. It is a tubuloacinar gland, which means that the tubular excretory ducts end up in grape-like bodies, the pancreatic acini, being composed of groups of secretory cells, which produce the precursors of the pancreatic enzymes. The pancreatic fluid (pancreatic juice) contains of a mixture of 22 digestive enzymes. Inactive precursor enzymes are synthesised, processed, and stored in zymogen granules. They are secreted into the acinar lumen in a regulated manner via the vegetative nerve and the endocrine systems, transported along the excretory ducts and finally activated in the lumen of the duodenum, then being able to digest all classes of nutrients.

The grape-like acini are built up by multiple pyramidal secretory cells, which are joined to each other by apical junctional complexes and with their apical surfaces line the lumen in the centre of the acinus (AL). The acinar cells' ultrastructures clearly mirror their functions and remind us that these secretory cells were used as model cells in basic studies of the intracellular pathways of protein synthesis.

Flattened cisternae of the rough endoplasmic reticulum are densely packed in the basal and perinuclear parts of the cells (RER). They define the "ergastoplasm"-domain of the cells (cf. Fig. 14) and are the sites of synthesis of the pancreatic enzymes. After co-translational folding, release into the RER-lumen and passing quality controls, the newly synthesised enzymes are exported out of the ER and taken up into the Golgi apparatus. The Golgi apparatus stacks (Golgi) are in typical supranuclear position. In the Golgi stacks, the newly synthesized enzymes are modified and, at the *trans* Golgi side, packed into secretory vesicles (for details cf. Fig. 25). The immature secretory vesicles or condensing vacuoles, are sites of further processing and condensation of the enzymes; they are intermediates that develop to mature zymogen granules (ZG), containing the densely packed, mostly still inactive precursors of the digestive enzymes. The numerous zymogen granules accumulated in the apical cytoplasm of the cells are the intracellular storage compartments of the proenzymes. The proenzymes are released into the acinar lumen upon specific stimulation after dietary intake.

Situated closely side by side, the acini comprise the major portion of the pancreatic parenchyma. Acini are surrounded by fine connective tissue leading blood capillaries, lymphatics, and unmyelinated nerve fibres (cf. Fig. 128).

References

Palade G (1975) Intracellular aspects of the process of protein synthesis. Science 189: 347

Schmidt K, Dartsch H, Linder D, Kern H-F, and Kleene R (2000) A submembranous matrix of proteoglycans on zymogen granule membranes is involved in granule formation in rat pancreatic acinar cells. J Cell Sci 113: 2233

Magnification: x 6,200

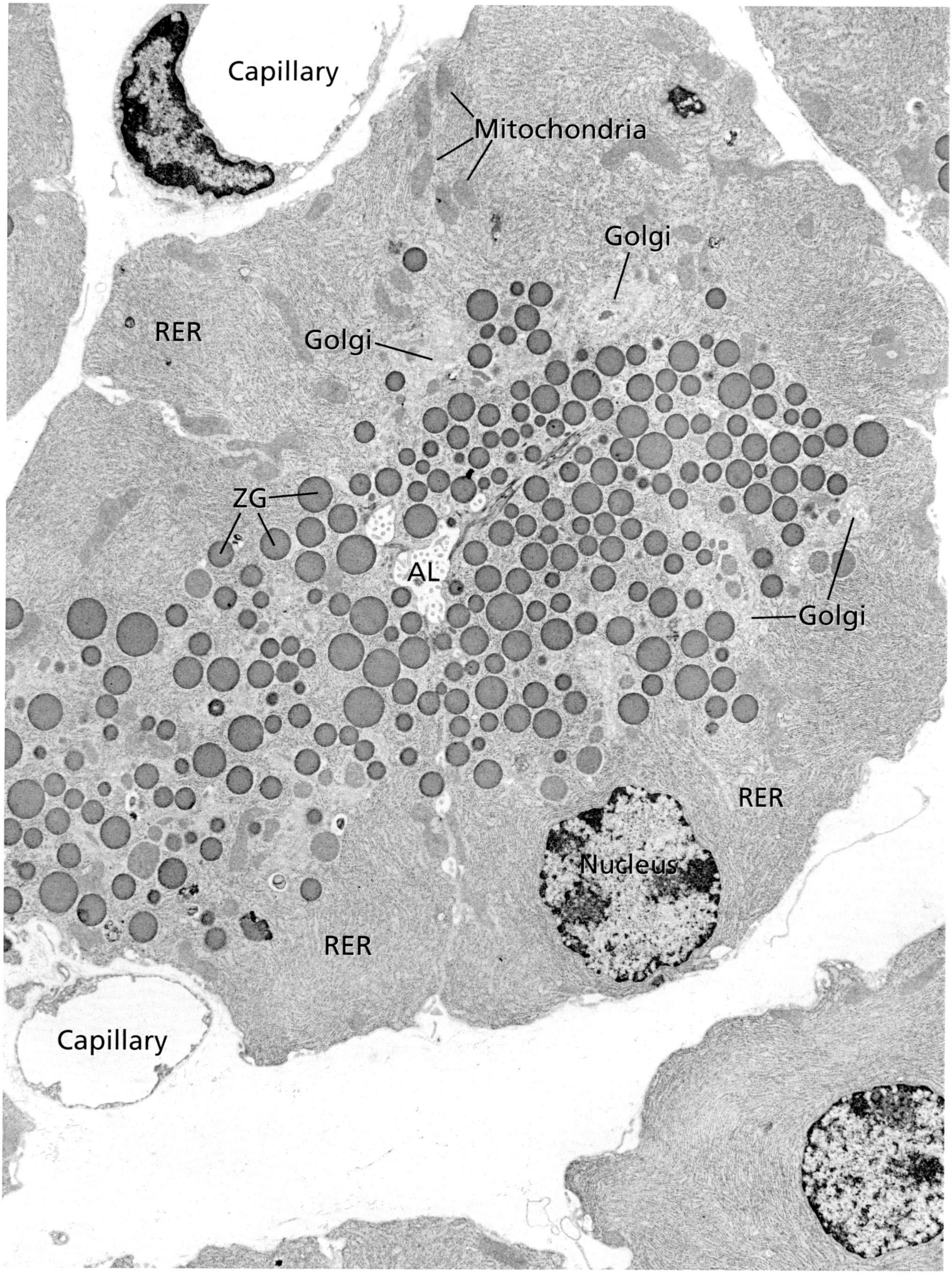
Capillary
Mitochondria
Golgi
RER
Golgi
ZG
AL
Golgi
RER
Nucleus
RER
Capillary

ACINAR CENTRE: ACINAR AND CENTROACINAR CELLS

The narrow ductuli in the centre of the acini build up the initiation of the secretory duct system. Small intercalated canaliculi (intercalated ducts) export the secretion out of the acini, lead to the intralobular excretory ducts, which converge to form interlobular ducts. These, by anastomoses, further build up the main large pancreatic duct.

The micrograph shows details of the centre of an acinus. The secretory acinar cells are joined both to each other and to adjacent centroacinar cells (CAC) by junctional complexes (asterisk), composed of a zone of tight junctions, a zone of adhering junctions and a circle of spot desmosomes (cf. Fig. 83). The apical tight junctions prevent leakage of pancreatic proenzymes from the acinar lumen (AL) into the intercellular spaces.

All parts of the secretory system are displayed in the secretory acinar cells shown in the micrograph: rough endoplasmic reticulum in the basal and lateral parts of the cells, Golgi apparatus (Golgi), condensing vacuoles (CV), and zymogen granules (ZG) dominating in the apical cytoplasm. The increased electron densities of the zymogen granules compared with the condensing vacuoles reflect the condensation processes occurring at this level of the secretory pathway (cf. Figs. 25 and 43). Condensing vacuoles originate from *trans* Golgi cisternae and therefore are found mostly close to the *trans* Golgi side. By contrast, the multiple mature zymogen granules are abundant in the entire apical cytoplasm and often take position close to the apical plasma membrane. The contents of the zymogen granules, consisting of mostly still inactive proenzymes, are released into the acinar lumen by regulated exocytosis, which is connected with the fusion of the granule membrane with the apical plasma membrane and formation of a pore (for details cf. Figs. 43 and 44). Proenzyme discharge occurs after dietary intake and is mainly induced by binding of cholecystokinin to specific receptors localised in the basolateral plasma membrane of the acinar cells.

Autophagosomes and lysosomes multiply apparent in the apical cytoplasm of the acinar cells mirror lysosomal degradation of excessively produced secretory enzymes by a mechanism, common to many secretory cells of the regulated type and called crinophagy. A highly flexible secretory system with new production of enzymes, storage, release into the acinar lumen, and degradation is necessary since the amount of pancreatic fluid produced and the concentration of enzymes in the zymogen granules vary with the dietary intake. A diet rich in carbohydrates induces a selective production of amylases and a decreased protease synthesis. Insulin produced by the pancreatic islets' beta cells (cf. Fig. 102) regulates amylase gene expression, an event that stresses the significance of an insulo-acinar portal system, built up by capillaries that leave the endocrine islets and supply blood to the surrounding acini.

Two acinar lumina are on display in the micrograph. They are shown at different functional states and demonstrate the high plasticity of the duct system: The lower acinar ductulus lacks secretory contents, and its wall shows occupation by dense microvilli. By contrast, the upper one apparently is in a state following zymogen discharge. The lumen is filled, the lining plasma membranes are flat, and only few microvilli protrude into the luminal space. It is evident that the acinar lumina are lined by the apical plasma membranes of both secretory acinar cells and centroacinar cells. Centrocinar cells are the most proximal epithelial cells of the small intercalated ducts and reside in the centre of the acini. Centroacinar cells are unique to the pancreas. They lack zymogen granules, contain abundant free polyribosomes and sparse endoplasmic reticulum and, with their apical domains, together with those of the secretory acinar cells, form the walls of the acinar ductuli.

References

Jahn R and Südhof TC (1999) Membrane fusion and exocytosis. Annu Rev Biochem 68: 863

Schmidt K, Dartsch H, Linder D, Kern H-F, and Kleene R (2000) A submembranous matrix of proteoglycans on zymogen granule membranes is involved in granule formation in rat pancreatic acinar cells. J Cell Sci 113: 2233

Magnification: x 20,000

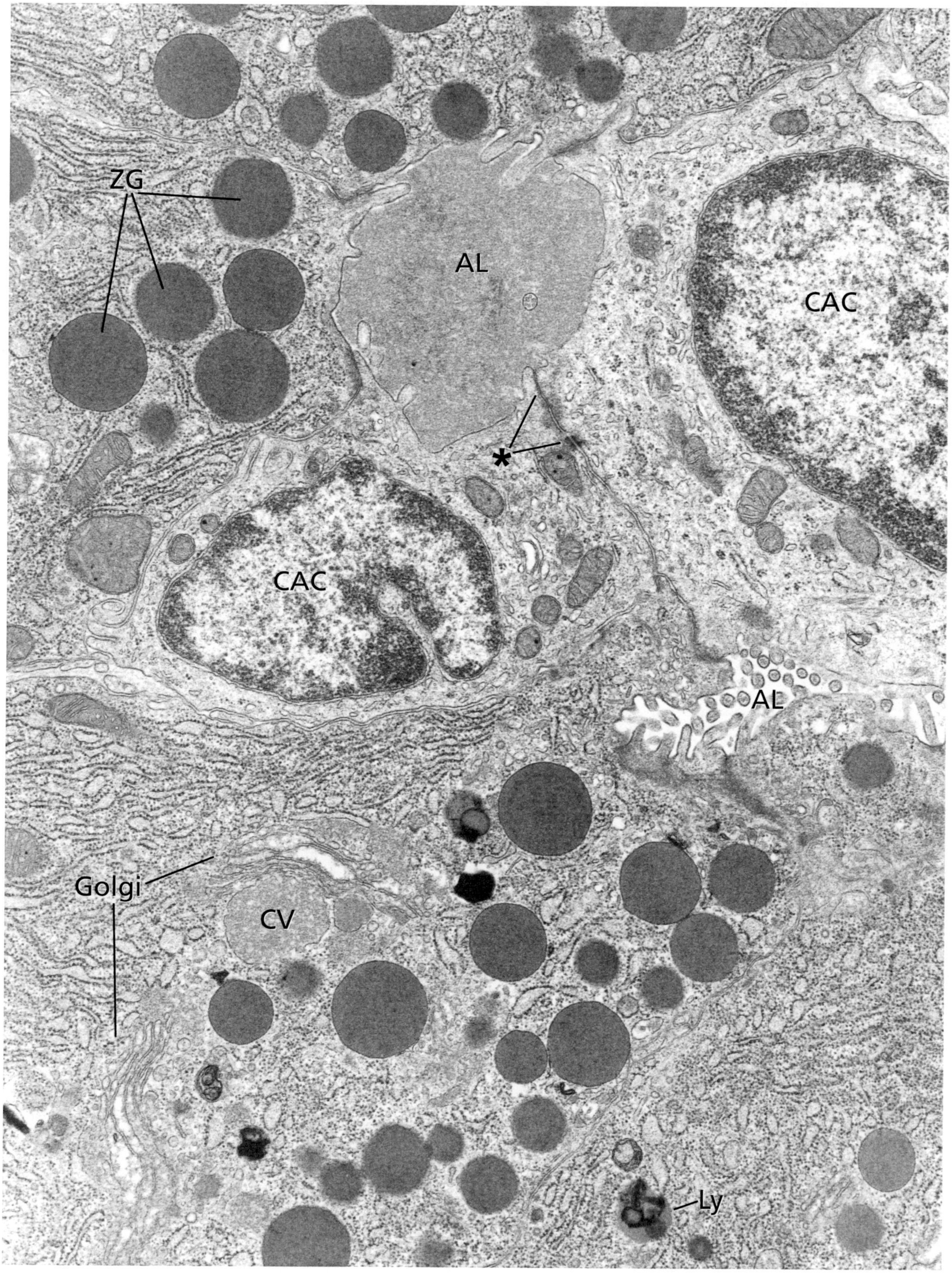
ZG
AL
CAC
*
CAC
AL
Golgi
CV
Ly

PANCREATIC INTERCALATED DUCT

The narrow secretory ductuli in the acinar centres (cf. Figs. 95 and 96) continue into small intercalated ducts that lead out of the acini into the connective tissue stroma and further converge to form the intralobular and interlobular excretory ducts. Striated ducts, such as those present in other salivary glands (cf. Fig. 98) and myoepithelial cells are lacking in the pancreas. Intercalated ducts are surrounded by an own basal lamina and are localised within small roads of connective tissue.

Panel A shows a cross section, panel B a longitudinal section through a small pancreatic intercalated duct. In both pictures, pancreatic fluid is present in the lumen and appears as a dense or flocculent content. In panel B, amylase contained in the pancreatic fluid is shown by immunogold labelling.

The pancreatic fluid contains mostly still inactive enzymes, such as trypsin, chymotrypsin, and carboxylpeptidases, which are going to be activated in the duodenal lumen by HCO_3^- ions and the alkaline secretion of Brunner's glands in the submucosa of the duodenum. Trypsin and other proteases are prevented from being activated earlier within the ducts by an inhibitor, which is produced and released into the lumen by the secretory acinar cells together with the enzymes.

As occurs in most exocrine glands, the primary secretion is modified in the duct system. Bicarbonate and water are added to the secretion by the intercalated duct epithelium, triggered by specific binding of the small intestinal hormone secretin to plasma membrane receptors.

Multiple basolateral folds, as are visible in the left hand part of panel A, indicate an involvement of the intercalated duct cells in ion transport. The cells interact with each other by cell to cell contacts that build up extensive apical junctional complexes (arrows). In the apical parts of the complexes by a belt of tight junctions, the epithelium is sealed and pancreatic fluid prevented from entering the intercellular spaces.

In the duct lumen shown in panel A, several microvilli and fine cilia appear cross-sectioned. Some of them contain microtubuli, resembling the end pieces of kinocilia (cf. Fig. 121).

Magnification: x 28,000 (A), x 30,000 (B)

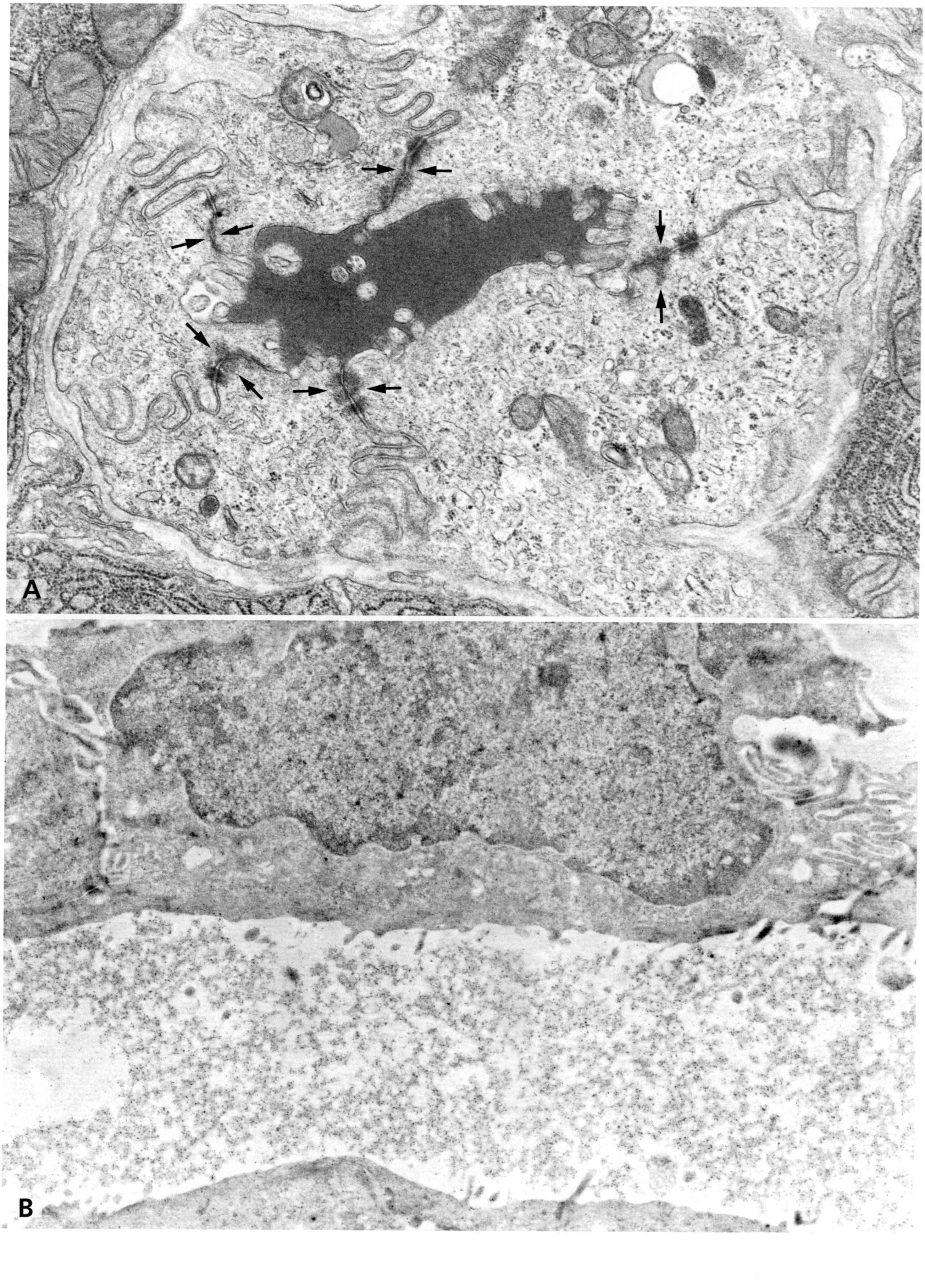
A
B

SUBMANDIBULAR GLAND

This low magnification electron micrograph of a pig submandibular gland shows main functional parts of a compound salivary gland, a secretory end piece with mixed acini (1 and 2) and the proximal portions of the excretory duct system, which include intercalated (3) and striated (4) ducts.

The submandibular gland is a compound tubulo-acinar gland that produces a mixed mucous and serous secretion that makes up to 70% of the saliva. Together with the products of other major and minor salivary glands, the salivary fluid of the submandibular gland has a role that is important in multiple ways in lubricating the surfaces of the oral cavity and forming a thin protective film, dissolving and moistening food and protecting against microorganisms, due to the contents in lysozyme, lactoferrin, and immunoglobulin A.

The saliva-producing end-pieces of the gland either are serous acini or form mixed pieces, in which mucous and serous secretory cells coexist in the same acinus. The micrograph shows mixed sero-mucous acini in its right hand part. Typically, the mucous cells (1), containing abundant densely packed secretory mucous droplets, reside closer to the lumen of the acinus, compared with the serous cells, and their apical domains build up the luminal walls. The serous cells (2) are localised at the acinus base and form a crescent-like region embracing the mucous cells (serous demilune). Fine intercellular canaliculi connecting the serous cells with the acinar lumen serve to transport the serous secretion. The serous cells are in contact with the basal lamina or are surrounded by myoepithelial cells, which in this very low magnification electron micrograph are hardly discernible. The contractile myoepithelial cells form a kind of basket around the acini and support secretion of the salivary fluid, which is drained sequentially by intercalated ducts, striated ducts, and large interlobular excretory ducts.

Two intercalated ducts (3) are visible in the left upper corner of the micrograph showing a cuboidal epithelium and flat myoepithelial cells located at the epithelial base. At the left lower corner, a segment of a striated duct (4) is on display. This duct consists of columnar epithelial cells exhibiting multiple basal infoldings that indicate its involvement in ion and water transport. In the striated ducts, kallikrein is secreted that processes several proteins in the primary saliva. Furthermore, Na^{+} and Cl^{-} are reabsorbed and the saliva becomes hypo-osmotic. Via transcytosis, IgA secreted by plasma cells in the connective tissue surrounding the acini is transported into both the acinus and striated duct lumina.

The complex duct system of salivary glands is formed during embryonic development by epithelial branching connected with repetitive cleft and bud formation. Growth factors, actin microfilaments, and components of the basement membrane have key roles in the formation of clefts and buds. In particular, fibronectin is important controlling type III collagen accumulations at clefts. It is assumed that local, developmentally programmed expression of epithelial cell fibronectin might regulate branching morphogenesis associated with the conversion of cell-cell adhesions to cell-matrix adhesions.

References

Sakai T, Larsen M, and Yamada KM (2003) Fibronectin requirement in branching morphogenesis. Nature 423: 876

Magnification: x 1,500

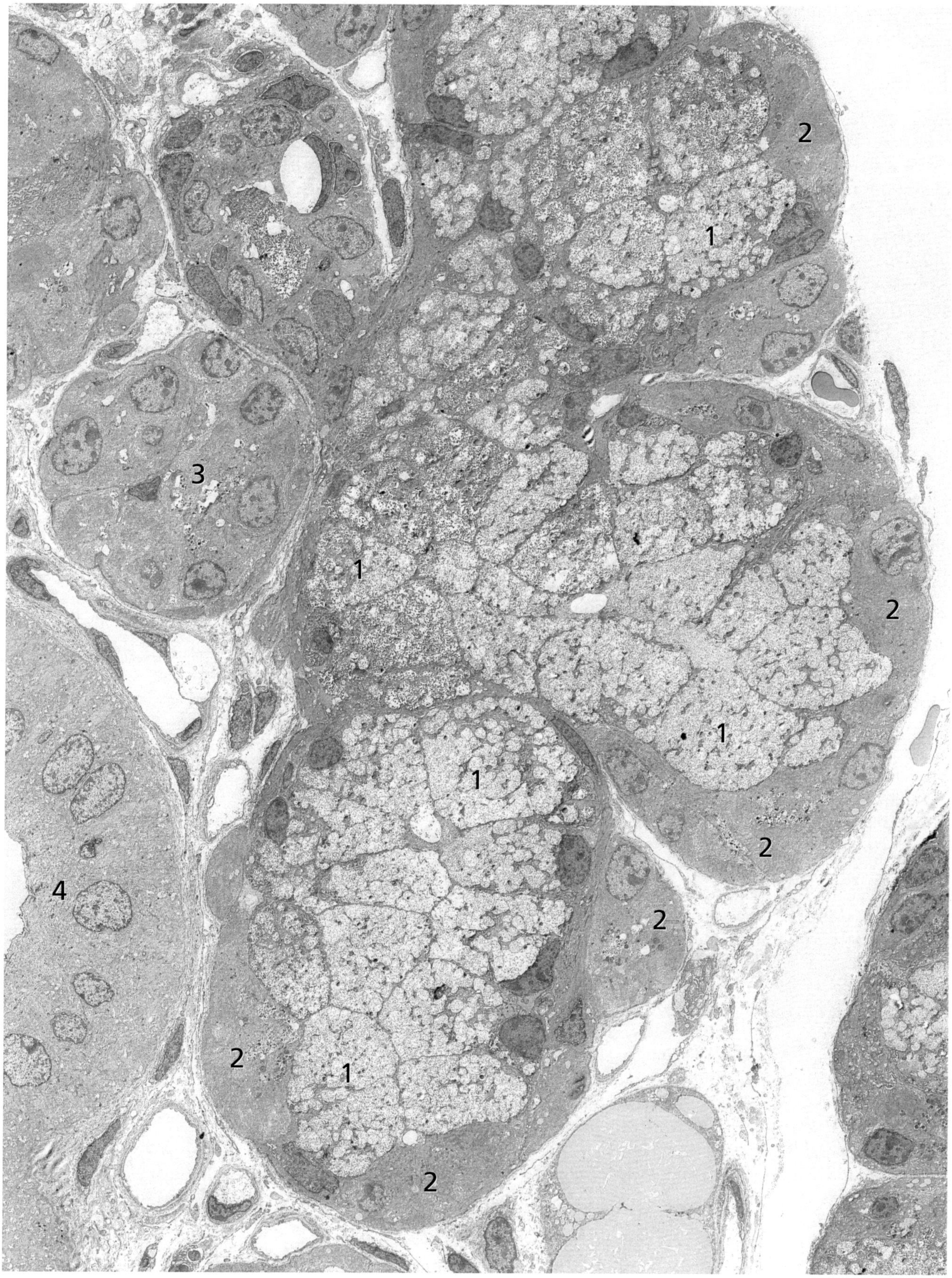
2
1
3
1
2
1
1
2
4
2
2
1
2

GOBLET CELLS – UNICELLULAR GLANDS

Exocrine glands usually are highly complex organs composed of a single or different secretory cell types, which form secretory units, so-called acini. However, there exists secretory cells forming unicellular glands in the mucosa of the respiratory, digestive and urogenital tract as well as in the conjunctiva of the eye and in the gall bladder. The archetype of such a unicellular gland is represented by the flask-shaped, mucus-producing and storing intestinal goblet cells shown in the electron micrograph. Mucus-producing single secretory cells exist dispersed in the mucosa of various other organs. Mucus, their major secretory product together with water, ions and other glycoproteins forms a highly hydrated, visco-elastic blanket at the surface of the mucosa. This mucus layer not only protects against mechanical and chemical irritants as well as bacterial infection but also prevents dehydration of the mucosa and is a regulator of the specific organ flora.

As seen in the electron micrograph taken from a human duodenal biopsy, a single goblet cell is typically flanked by absorptive enterocytes exhibiting their characteristic apical brush border. The arrows point to intercellular digitations (cf. Fig. 88) formed by the plasma membrane of the neighbouring cells. The apical cytoplasm of the goblet cell is filled with mucus droplets. The nucleus as well as the Golgi apparatus and endoplasmic reticulum cisternae, which has a characteristic wide lumen, are located in the lower half of the cell. The mucus droplets, like other secretory granules, are formed in distended parts of *trans* Golgi cisternae and mature to form membrane-limited mucin-containing granules.

The content of the secretory granules of the goblet cells consists of highly glycosylated proteins and is rich in calcium. Mucin proteins consist of a heterogenous group of secretory and membrane proteins. They all have in common a high content of serine/threonine-linked glycans which can make up more than 50% of their molecular mass. Depending on the tissue and cell type, the apomucin and their *O*-glycans can differ significantly in structure. Despite this diversity, all mucin proteins have the same kind of domain structure. A central domain consists of repeats of serine and threonine-rich sequences, all of which represent potential glycosylation sites. Both the N- and C-terminal domains are rich in cysteine. *De novo* synthesized apomucin in the endoplasmic reticulum is modified by *N*-glycans and becomes dimerized by disulfide bridges at the C-terminal part. Upon transport to the *cis* Golgi apparatus, apomucin is further modified by *O*-glycans (cf. Fig. 31). In the *trans* Golgi apparatus, further dimerization at the C-terminal region takes place, which results in the formation of the mature multimeric mucins.

The content of the mature mucin droplets is released by exocytosis. In addition to a basal secretion rate, compound exocytosis can be observed (cf. Fig. 45). Such compound exocytosis can occur after various stimuli such as non-specific mechanical and chemical irritation. Specific stimuli involve activators of phospholipase C and of adenylate and guanylate cyclase. Details of the mechanism of exocytosis are described for Figs. 43 and 44.

References

Burgoyne RD and Morgan A (2003) Secretory granule exocytosis. Physiol Rev 83: 581

Davis CW and Dickey BF (2008) Regulated airway goblet cell mucin secretion. Annu Rev Physiol 70: 487

Dekker J, Rossen JW, Buller HA, and Einerhand AW (2002) The MUC family: an obituary. Trends Biochem Sci 27: 126

Freeman JA (1966) Goblet cell fine structure. Anat Rec 154: 121

Gundelfinger E, Kessels M, and Qualmann B (2003) Temporal and spatial coordination of exocytosis and endocytosis. Nat Rev Molec Cell Biol 4: 127

Hattrup CL and Gendler SJ (2008) Structure and function of the cell surface (tethered) mucins. Annu Rev Physiol 70: 431

Lagow E, DeSouza MM, and Carson DD (1999) Mammalian reproductive tract mucins. Hum Reprod Update 5: 280

Perez-Vilar J and Hill RL (1999) The structure and assembly of secreted mucins. J Biol Chem 274: 31751

Pickett J and Edwardson J (2006) Compound exocytosis: mechanisms and functional significance. Traffic 7: 109

Specian RD and Oliver MG (1991) Functional biology of intestinal goblet cells. Am J Physiol 260: C183

Tabak LA (1995) In defense of the oral cavity: structure, biosynthesis and function of salivary mucins. Annu Rev Physiol 57: 547

Thornton D, Rousseau K, and McGuckin M (2008) Structure and function of the polymeric mucins in airways mucus. Annu Rev Physiol 70: 459

Verhage M and Toonen R (2007) Regulated exocytosis: merging ideas on fusing membranes. Curr Opin Cell Biol 19: 402

Magnification: x 10,500

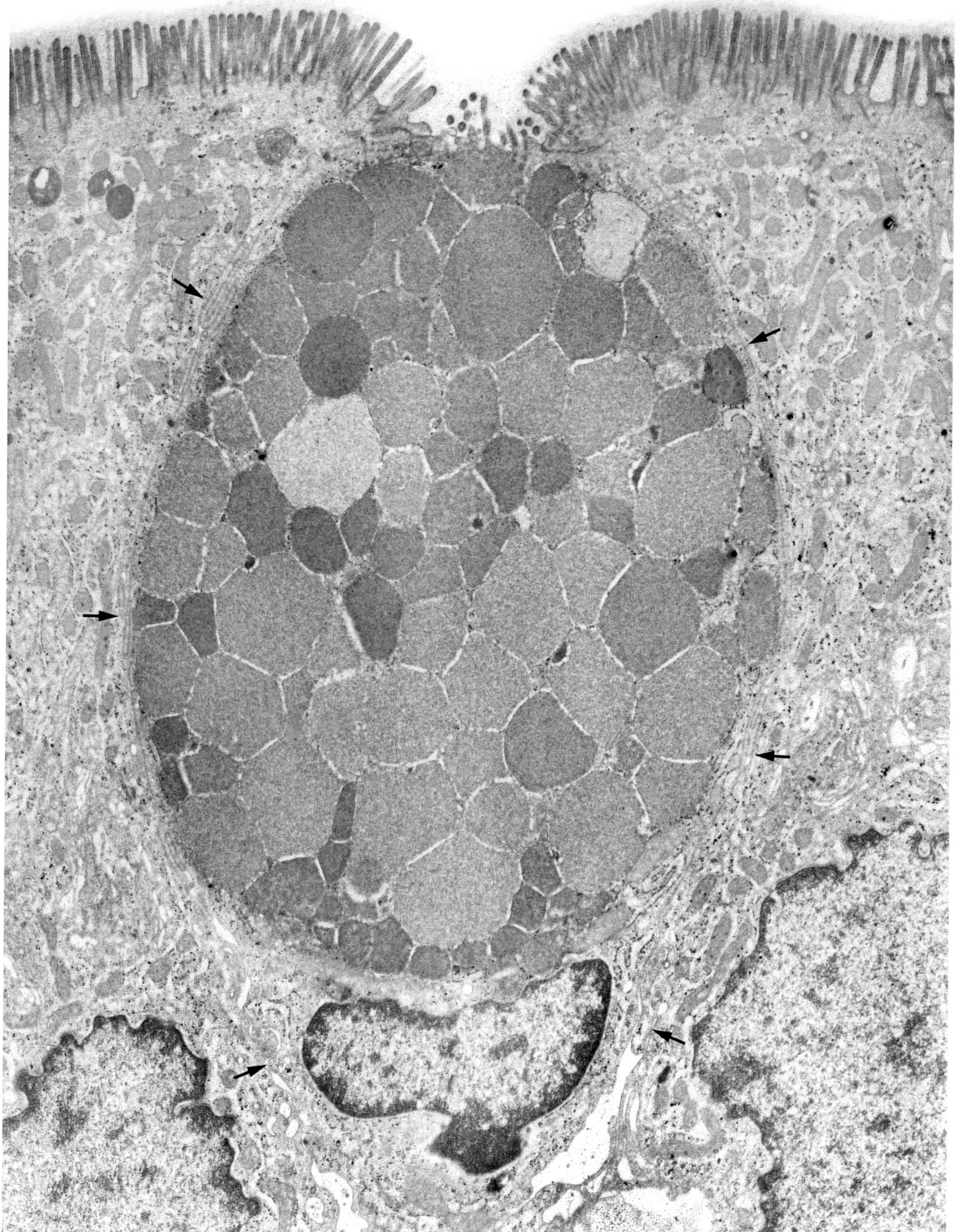

PARIETAL CELLS OF STOMACH: SECRETION OF ACID

The parietal cells are located in the middle part of the glands of the fundus-body region of the stomach. They secrete gastric acid, i.e. hydrochloric acid (approximately 0.16 M, pH $\geq$ 0.8) and secrete the intrinsic factor, which is a vitamin B_{12}-binding protein.

The highly specialised function of the gastric parietal cells is reflected in a particular fine structure (panel A) and its changes dependent on different functional states. Their apical plasma membrane forms numerous channels, the so-called canaliculi, which reach deep into the cytoplasm as can be easily recognized in longitudinal (asterisks in panel A) and cross sections (asterisk in panel B). The canaliculi are rich in microvilli. Beneath the apical plasma membrane and around the canaliculi is a complex membrane system located, which consists of numerous tubulo-vesicles and cisternae (tv in panel C). The tubulo-vesicles are rich in H^+, K^+-ATPase. Another characteristic feature of parietal cells is their abundance of mitochondria (M). They are required since one ATP molecule is consumed for the coupled export of a single H^+ and the import of a single K^+. The H^+, K^+-ATPase of the parietal cells is a P-type ATPase, which consist of a catalytic α-subunit and a β-subunit. The α-subunit contains a recognition sequence for apical sorting and the β-subunit is not only important for the stabilization of the α-subunit but also for recycling since it contains a recognition sequence for endocytosis. The H^+, K^+-ATPase is the primary gastric proton pump. Under resting conditions, it is preferentially located in the intracellular tubulo-vesicles.

Since its discovery, the three-dimensional structure of the membrane system of the parietal cells is a matter of debate. It was and continues to be intensely studied not only by conventional techniques of tissue preparation, but more importantly by high-pressure, rapid-freezing and subsequent freeze-substitution of tissue. Three-dimensional reconstruction is then achieved by serial section analysis. In resting parietal cells, the membrane system was found to be mainly composed of cisternae, that were not connected to canaliculi, and of a network of winding tubuli. Following stimulation of acid secretion, translocation of H^+, K^+-ATPase-containing tubulo-vesicles in the apical plasma membrane occured, which resulted in an about 10-fold enlargement of its surface. In parallel, the number of microvilli of the canaliculi increased. In the expanded apical plasma membrane, the presence of H^+, K^+-ATPase resulted in K^+ and Cl^- conductance. This led to active proton pumping with secretion of H^+ and Cl^- and water. Therefore, the H^+, K^+-ATPase catalyzes the electroneutral exchange of intracellular protons for extracellular potassium ions, thus generating the enormous proton gradients associated with gastric acid secretion. Upon termination of the stimulus, membrane recycling results in the re-establishment of the elaborate intracellular membrane system. Currently, the recruitment-recycling model of parietal cell activation for gastric acid secretion is most widely accepted.

The secretion of gastric acid is tightly regulated. In parietal cells, ligands such as acetylcholine, gastrin, and histamine bind to their receptors present at basolateral plasma membrane. Stimulation by histamine released from neighboring entero-endocrine cells is most important.

The autoimmune gastritis is due to autoantibodies directed against the gastric H^+, K^+-ATPase and results in lack of gastric acid and intrinsic factor. The resulting vitamin B_{12} deficiency causes pernicious anaemia.

N: nucleus.

References

Duman JG, Pathak NJ, Ladinsky MS, McDonald KL, and Forte JG (2002) Three-dimensional reconstruction of cytoplasmic membrane networks in parietal cells. J Cell Sci 115: 1251
Forte J and Yao X (1996) The membrane-recruitment-and-recycling hypothesis of gastric HCl secretion. Trends Cell Biol 6: 45
Lindstrom E, Chen D, Norlen P, Andersson K, and Hakanson R (2001) Control of gastric acid secretion:the gastrin-ECL cell-parietal cell axis. Comp Biochem Physiol A Mol Integr Physiol 128: 505
Ogata T and Yamasaki Y (2000) Morphological studies on the translocation of tubulovesicular system toward the intracellular canaliculus during stimulation of the gastric parietal cell. Microsc Res Tech 48: 282
Samuelson LC and Hinkle KL (2003) Insights into the regulation of gastric acid secretion through analysis of genetically engineered mice. Annu Rev Physiol 65: 383
Schubert ML and Peura DA (2008) Control of gastric acid secretion in health and disease. Gastroenterology 134: 1842
Tamura A, Kikuchi S, Hata M, Katsuno T, Matsui T, Hayashi H, Suzuki Y, Noda T, Tsukita S, and Tsukita S (2005) Achlorhydria by ezrin knockdown: defects in the formation/expansion of apical canaliculi in gastric parietal cells. J Cell Biol 169: 21
Yao X and Forte JG (2003) Cell biology of acid secretion by the parietal cell. Annu Rev Physiol 65: 103

Magnification: x 6,000 (A); x 7,500 (B); x 46,000 (C)

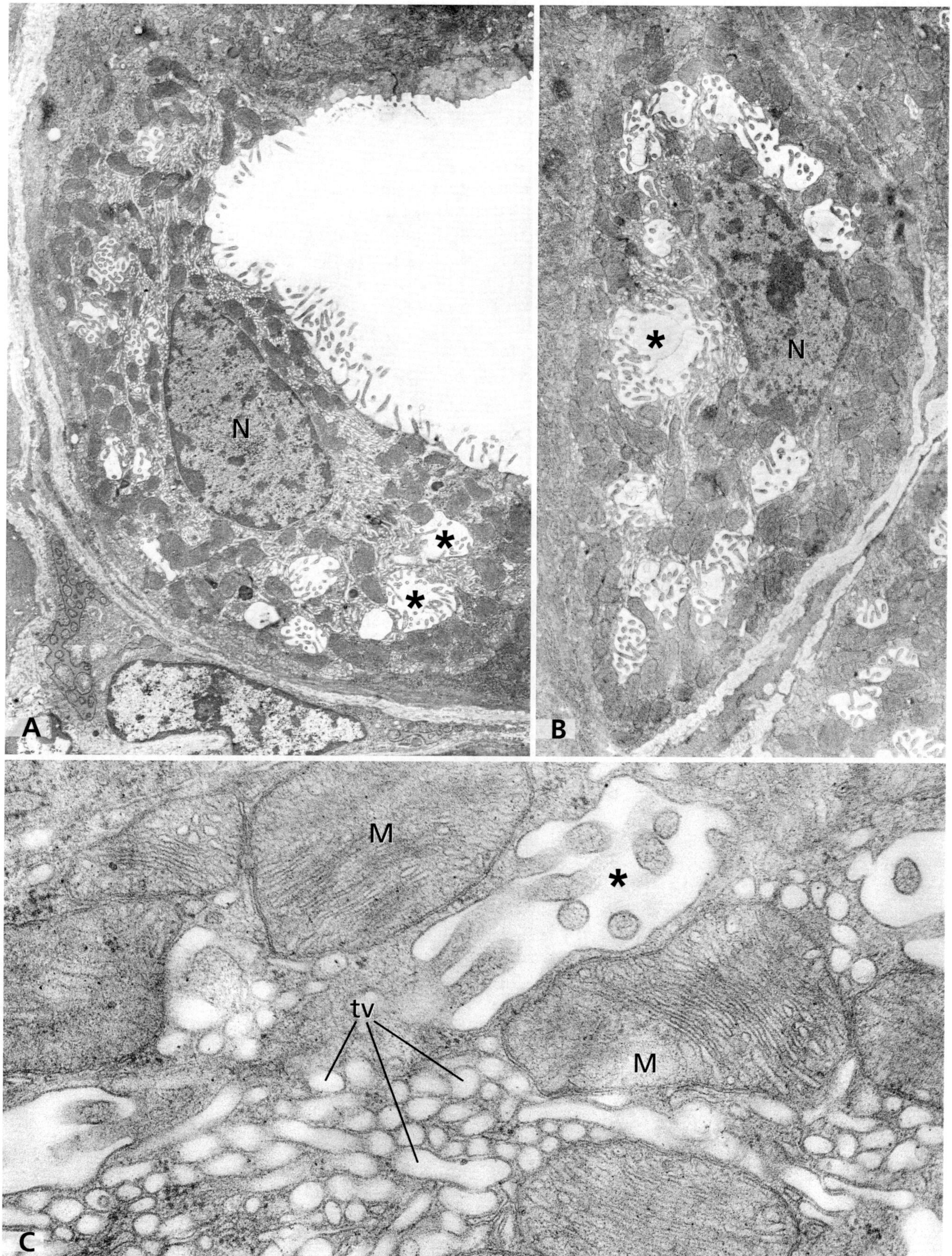
A
N
*
*
B
N
*
C
M
*
tv
M

INTERCALATED CELLS OF KIDNEY: IMPORTANT REGULATORS OF ACID-BASE BALANCE

Intercalated cells (ICs) mediate H^+ and HCO_3^- secretion, and Cl^- and K^+ reabsorption. They are primarily present in the cortical and outer medullary collecting ducts of kidney as well as connecting tubules and, depending on animal species, may extend into distal convoluted tubules and more distal parts of collecting ducts. ICs are interspersed among the segment specific principal cells (panels A and B, arrowheads point to lateral interdigitations) and express specific proteins such as carbonic anhydrase II, H^+-ATPase and band 3 through which they can be identified immunocytochemically. Actually, the expression of vacuolar H^+-ATPase is a most characteristic feature of ICs.

By transmission electron microscopy, ICs are distinguished by a dark cytoplasm and for that reason often called dark cells (panel A), a high density of mitochondria, tubulovesicular structures in the apical cytoplasm (recognisable even at low magnification in panel B), and abundant apical microvilli and microfolds. In contrast to the principal cells, a basal membrane labyrinth is less extensively developed (panels A and B). Fine structurally, ICs show considerable variability. The cytoplasm may be less dark and the number of mitochondria lower in ICs of the outer medulla (panel B) as compared to ICs of cortex (panel A). Furthermore, the apical pole of ICs can be wide and protruding and be covered by many microvilli, but with only few cytoplasmic tubulovesiclar profiles (panel A). Conversely, the apical pole may be extremely narrow and constricted at the level of the tight junction belt but with abundant flat profile vesicles in the apical cytoplasm (panel B). These structural differences of the apical part of ICs are related to different functional states and manifestation of an extensive membrane recycling between the apical plasma membrane and intracellular vesicles.

Functionally and immunocytochemically, at least two major subtypes of ICs can be distinguished: the α and β ICs. Both subtypes exist in cortical collecting ducts, whereas outer medullary collecting ducts have only α ICs. The α ICs have an apical H^+-ATPase and a basolateral $Cl^-:HCO_3^-$ anion exchanger and excrete HCO_3^- basolaterally through apical proton (H^+) secretion. The β ICs have an opposite membrane polarity of these transporters and secrete HCO_3- apically, in the tubular lumen. There is evidence for plasticity of this functional polarity resulting in the conversion of β ICs in an α-like IC phenotype. Both, α and β ICs have chloride channels and a H^+K^+-ATPase, which are present in the basolateral plasma membrane.

References

Al-Awqati Q (2003) Terminal differentiation of intercalated cells: the role of Hensin. Annu Rev Physiol 65: 567

Brown D, Hirsch S, and Bluck S (1988) An H^+-ATPase in opposite plasma membrane domains in kidney epithelial cell subpopulations. Nature 331: 622

Brown D and Stow JL (1996) Protein trafficking and polarity in kidney epithelium: From cell biology to physiology. Physiol Rev 76: 245

Kaissling B and Kriz W (1992) Morphology of the loop of Henle, distal tubule, and collecting duct. In: Handbook of physiology, (Windhager EE, ed). New York: American Physological Society, pp 109

Kriz A and Kaissling B (2000) Structural organisation of the mammalian kidney. In: The kidney: Physiology and pathophysiology (Seldin D, and Giebisch G, eds). New York: Lippincott, Williams and Wilkins, pp 587

Schuster VL (1993) Function and regulation of collecting duct intercalated cells. Annu Rev Physiol 55: 267

Schwartz GJ, Tsuruoka S, Vijayakumar S, Petrovic S, Mian A, and Al-Awqati Q (2002) Acid incubation reverses the polarity of intercalated cell transporters, an effect mediated by hensin. J Clin Invest 109: 89

Tsuganezawa H, Kobayashi K, Iyori M, Araki T, Koizumi A, Watanabe S, Kaneko A, Fukao T, Monkawa T, Yoshida T, et al. (2001) A new member of the $HCO_3^{(-)}$ transporter superfamily is an apical anion exchanger of beta-intercalated cells in the kidney. J Biol Chem 276: 8180

Van Adelsberg J, Edwards JC, Takito J, Kiss B, and Al-Awqati Q (1994) An induced extracellular matrix protein reverses the polarity of band 3 in intercalated epithelial cells. Cell 76: 1053

Magnification: x 8,500 (A); x 8,000 (B). From rabbit kidney cortex.

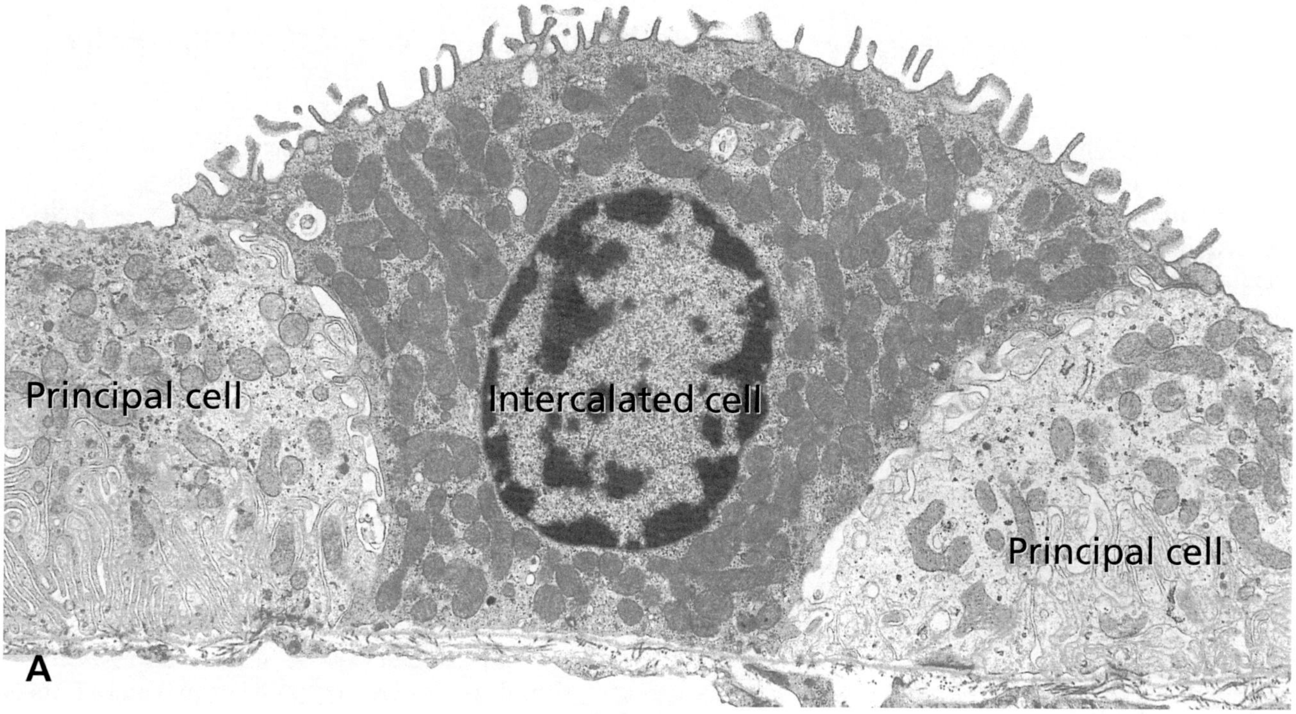
Principal cell
Intercalated cell
Principal cell
A

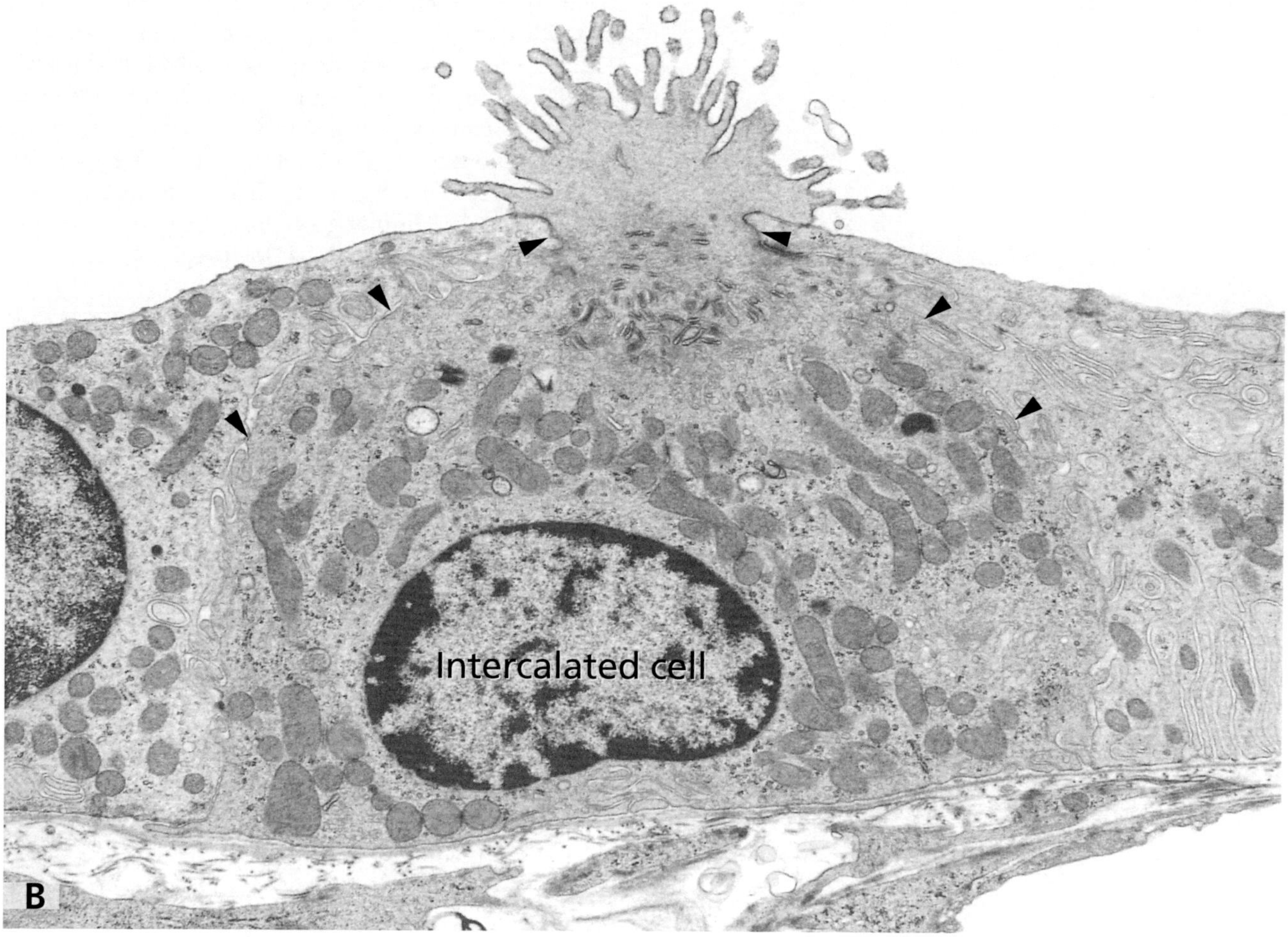
Intercalated cell
B

ENDOCRINE SECRETION: INSULIN-PRODUCING BETA CELLS OF ISLETS OF LANGERHANS

Endocrine cells can form distinct organs such as the thyroid and adrenal glands, a microorgan in a larger organ such as the islets of Langerhans in the exocrine pancreas, and exist as single cells dispersed in other organs such as the respiratory and digestive tract. They synthesise, store and secrete polypeptide or steroid hormones, which are important signalling molecules. Endocrine glands secrete their products in the blood circulation through which they are transported to distant target cells. Accordingly, endocrine glands are rich in capillaries. Upon binding of the secreted hormone to its receptor on the target cells, an intracellular signalling cascade is initiated that regulates essential functions such as proliferation, differentiation and metabolism.

The polypeptide hormone insulin plays a major role in the control of blood glucose homeostasis and is also involved in lipid and protein metabolism. Its target cells are hepatocytes, skeletal and cardiac muscle, adipocytes and fibroblasts. Insulin is synthesised by the beta cells of the islets of Langerhans. The cytoplasm of beta cells contains abundant rough endoplasmic reticulum, numerous mitochondria, a very well developed Golgi apparatus and plentiful secretory granules (SG), which is phenotypical of a professional secretory cell. The secretory granules consist of an electron dense core of insulin and a halo mainly of C-peptide. The anastomosing cords of the endocrine cells and the network of capillaries form a functional unit. Insulin becomes secreted by regulated exocytosis of secretory granules (cf. Fig. 44) and afterwards reaches the fenestrated capillaries.

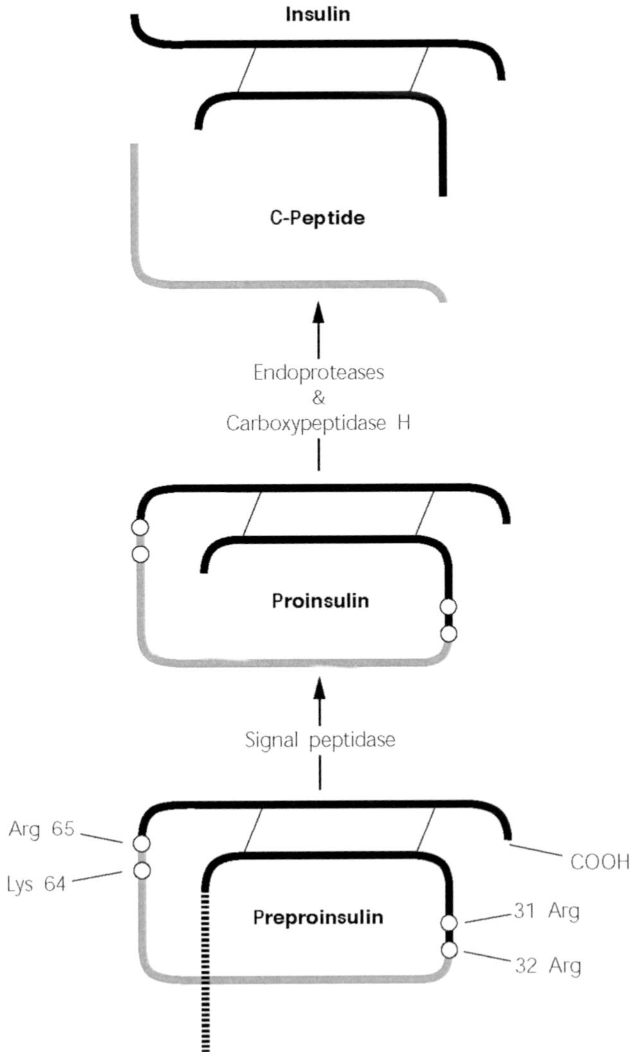

Insulin, like other polypeptide hormones, is synthesised as a large preproprotein, the preproinsulin. Removal of the signal peptide in the endoplasmic reticulum yields proinsulin, composed of an A and a B chain linked by two interchain disulphide bonds and connected by the C-peptide. Proinsulin becomes converted into insulin by the combined action of two endoproteases and carboxypeptidase H (see scheme). Endoprotease PC 2 cleaves the dibasic pair Arg_{31}-Arg_{32} at the B-chain/C-peptide junction of proinsulin and endoprotease PC 3 the dibasic pair Lys_{64}-Arg_{65} in the A-chain/C-peptide junction. The exposed dibasic amino acids are removed by carboxypeptidase H. The proinsulin-insulin conversion occurs in acidifying, clathrin-coated immature secretory granules in the *trans* Golgi region.

References

Bailyes E, Guest P, and Hutton J (1992) Insulin synthesis. In: Insulin. Molecular biology to pathology (FM A and SJH A, eds). Oxford New York Tokyo: IRL Press, pp 64

Loh Y, Beinfeld M, and Birch N (eds) (1993) Proteolytic processing of prohormones and proneuropeptides. Boca Raton Ann Arbor London Tokyo: CRC Press

Molinete M, Irminger JC, Tooze SA, and Halban PA (2000) Trafficking/sorting and granule biogenesis in the beta-cell. Semin Cell Dev Biol 11: 243

Orci L, Ravazzola M, Storch M-J, Anderson RGW, Vassalli J-D, and Perrelet A (1987) Proteolytic maturation of insulin is a post-Golgi event which occurs in acidifying clathrin-coated secretory vesicles. Cell 49: 865

Steiner DF, Smeekens SP, Ohagi S, and Chan SJ (1992) The new enzymology of of precursor processing endoproteases. J Biol Chem 267: 23435

Magnification: x 11,200

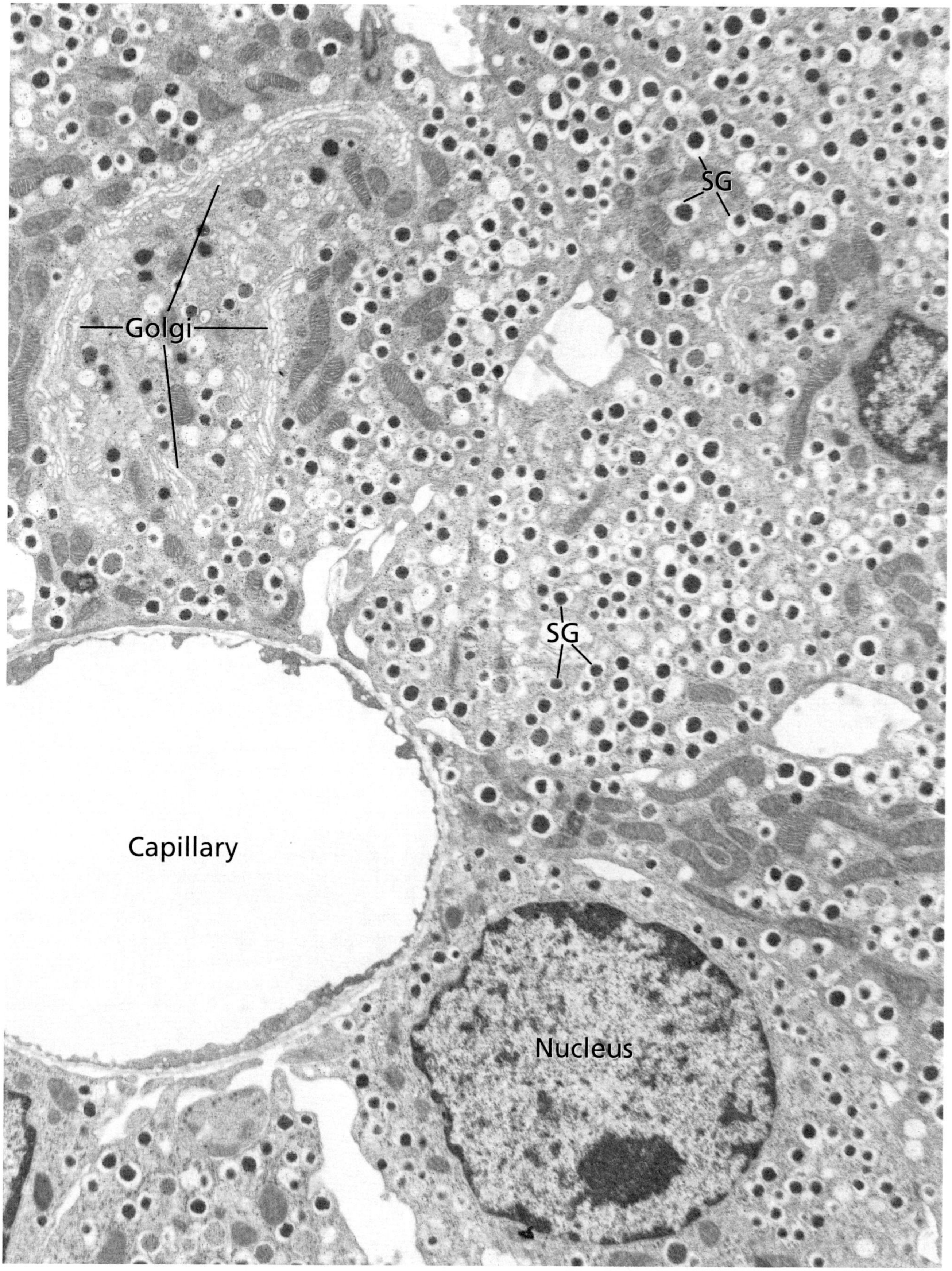
SG
Golgi
SG
Capillary
Nucleus

IMPAIRED INSULIN PROCESSING IN HUMAN INSULINOMA

Disorders of pancreatic beta cell function cause different clinically important diseases. Among the pancreatic endocrine tumours, functioning insulinomas give raise to persistent hyperinsulinemic hypoglycemia. This is due to inappropriate proinsulin-insulin conversion and faulty insulin secretion. However, like normal beta cells, functioning insulinomas form and store secretory granules which contain insulin and this is shown in panel A. The cytoplasm of this tumour cell contains numerous dense core secretory granules which are all positive for insulin as detected by immunogold electron microscopy. Despite this, the insulin-containing secretory granules differ ultrastructurally from those in normal human beta cells (cf. Fig. 44). Most of them have no halo and crystalloid cores are only occassionally detectable. Therefore, the secretory granules resemble immature secretory granules of normal beta cells. Patients with functioning insulinomas have not only high plasma insulin levels but also high plasma proinsulin levels. By immunoelectron microscopy, proinsulin in functioning human insulinomas as compared to normal beta cells was not only detectable in the Golgi apparatus and immature secretory granules, but additionally in secretory granules throughout the cytoplasm and beneath the plasma membrane. This is illustrated in panel B, which is an ultrathin section consecutive to the one in panel A showing the same region of an insulinoma cell. Obviously, many of the secretory granules exhibit immunogold labelling for proinsulin. This indicates that a fraction of the proinsulin was not converted to insulin but nonetheless was sorted and packed into secretory granules. Eventually, secretory granules positive for insulin and for proinsulin may undergo exocytosis. It has been shown for beta cells and other neuroendocrine cells that immature secretory granules may be released preferentially over mature granules. Enhanced proinsulin secretion may be also due to defective sorting in secretory granules. Carboxypeptidase E acts as a sorting receptor in the *trans* Golgi network of neuroendocrine cells. Mice deficient in carboxypeptidase E showed hyperproinsulinaemia caused by unregulated proinsulin secretion via the constitutive secretory pathway. The significance of this process in human insulinoma is uncertain.

In functioning insulinoma, the hormone conversion appears to be affected at different levels. It is not only incomplete but it already occurs in the Golgi apparatus and not first in immature secretory granules as in normal beta cells. The basis for these topographical abnormalities in hormone conversion in functioning insulinomas is unclear. There is evidence that the amounts of endoprotease PC3 are reduced in insulinoma cells, which resulted in a reduced rate of proinsulin-insulin conversion.

References

Cool DR, Normant E, Shen FS, Chen HC, Pannell L, Zhang Y, and Loh YP (1997) Carboxypeptidase E is a regulated secretory pathway sorting receptor: Genetic obliteration leads to endocrine disorders in Cpe(fat) mice. Cell 88: 73

Komminoth P, Heitz PU, and Roth J (1999) Human insulinoma: clinical, cellular, and molecular aspects. Endocr Pathol 10: 269

Neerman-Arbez M, Sizonenko SV, and Halban P (1993) Slow cleavage at the proinsulin B-chain/connecting peptide junction associated with low levels of endoprotease PC1/3 in transformed β-cells. J Biol Chem 268: 16098

Roth J, Komminoth P, and Heitz PU (1995) Topographic abnormalities of proinsulin – insulin conversion in functioning human insulinomas: Comparison of immunoelectron microscopic and clinical data. Am J Pathol 147: 489

Roth J, Komminoth P, Klöppel G, and Heitz PU (1995) Diabetes and endocrine pancreas. In: Anderson's pathology (Damjanov I, and Linder J, eds). St. Louis Chicago: Mosby Year Book, pp 2042

Smeekens S, Montag A, Gary T, Albiges-Rizo C, Carroll R, Benig M, Phillips L, Martin S, Ohagi S, Gardner P, et al. (1992) Proinsulin processing by the subtilisin-related proprotein convertases furin, PC2, and PC3. Proc Natl Acad Sci USA 89: 8822

Tooze S, Flatmark T, Tooze J, and Huttner WB (1991) Characterization of the immature secretory granule, an intermediate in granule biosynthesis. J Cell Biol 115: 1491

Magnification: x 38,500 (A, B)

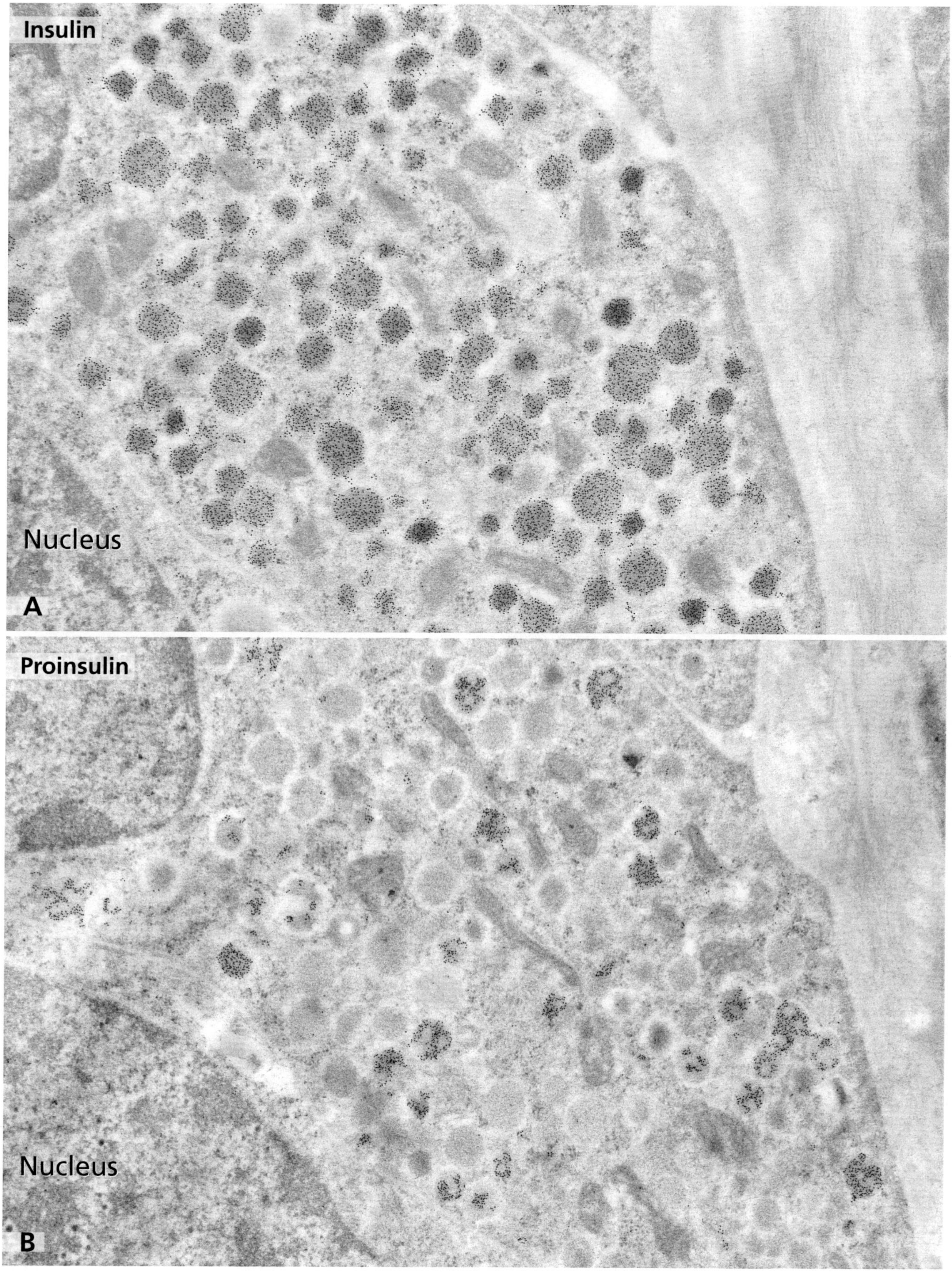
Insulin
Nucleus
A
Proinsulin
Nucleus
B

CELLS OF THE DISSEMINATED ENDOCRINE SYSTEM

Endocrine cells exist in the mucosa of the digestive, respiratory and uro-genital tract as well as in the skin either as single cells or in small clusters. Although the various cells of the disseminated (neuro-) endocrine system share many morphological and biochemical features, they differ not only greatly in their tissue distribution, but more importantly by the different peptides they synthesize and secrete. Despite the many-fold differences, they constitute a functionally related network of cells. In fact, the cells of the disseminated endocrine system of the digestive tract together represent the largest endocrine gland of the body. Like their physiology, the pathology of the disseminated endocrine system is of complex nature and is represented by hyperplastic changes and benign as well as malignant tumors, which may be associated with severe forms of hormonal dysregulation.

Panel A shows an example of two endocrine cells located in the duodenal mucosa and panel B exemplifies a single endocrine cell in a small exocrine pancreatic duct. The endocrine cells in panels A and B contain numerous dense core secretory granules, typically observed in (neuro-) endocrine cells (cf. Fig. 44). In panel A, a somatostatin-producing D-cell is seen together with a TG-cell, which produces a gastrin-cholecystokinin-like polypeptide. The D-cell represents the so-called closed-type of an intraepithelial endocrine cell without contact to the lumen (L). In contrast, the TG-cell represents an endocrine cell of the open-type since its apical, microvilli-rich plasma membrane (arrowheads) is in open contact with the lumen (L). Between the TG-cell and the neighboring non-endocrine, absorptive epithelial cells, junctional complexes exist (arrows). Of note, the same endocrine cell type can be either of the closed- or the open-type simply depending on its location. For instance, gastrin-producing G-cells located in the mucosa of the pyloric region of the stomach are of the open-type, whereas those present in the mucosa of the gastric body are of closed-type.

Panel B represents a solitary insulin-producing B-cell present in the mucosal lining of a small excretory pancreatic duct. The presence of insulin in the core of the secretory granules is demonstrated by immunogold labelling for insulin. In a single ultrathin section, the B-cell appears to be of the closed-type, which is covered by adjacent duct epithelia (arrowheads). However, as demonstrated in the inset, serial section analysis revealed that it is indeed of the open-type since its apical plasma membrane (arrowhead) is exposed to the duct lumen (L). Junctional complexes exist between the B-cell and duct epithelia (arrows in the inset).

Following a stimulus, the endocrine cells of open- or closed-type release their specific secretory products in the surrounding connective tissue from where they gain access to lymph or blood capillaries and consequently to the circulation. After their release, the polypeptides can act in a paracrine manner on resident cells such as other endocrine cell types, exocrine cells, muscle fibers and nerve cells. Furthermore, an autocrine feedback loop may be observed. Altogether, the diverse cells of the disseminated (neuro-)endocrine system produce more than 100 bioactive substances, which include typical polypeptide hormones such as insulin, somatostatin and gastrin, various neuropeptides and different growth factors as well as regulatory peptides such as orexin, leptin and ghrelin. Of note, different polypeptides may be synthesized by a single endocrine cell type.

References

Bordi C, D'Adda T, Azzoni C, and Ferraro G (2000) Classification of gastric endocrine cells at the light and electron microscopical levels. Microsc Res Tech 48: 258

Dockray G, Varro A, and Dimaline R (1996) Gastric endocrine cells: gene expression, processing, and targeting of active products. Physiol Rev 76: 767

Ito T (1999) Differentiation and proliferation of pulmonary neuroendocrine cells. Progr Histochem Cytochem 34: 245

Larsson L (2000) Developmental biology of gastrin and somatostatin in the antropyloric mucosa of the stomach. Micr Res Tech 48: 272

Montuenga L, Guembe L, Burell M, Bodegas M, Calvo A, Sola J, Sesma P, and Villaro A (2003) The diffuse endocrine system: from embryogenesis to carcinogenesis. Progr Histochem Cytochem 38: 153

Pearse A (1972) The diffuse neuroendocrine system: an extension of the APUD concept. In: Endocrinology (Taylor S, ed). London: Heinemann

Rehfeld J (1998) The new biology of gastrointestinal hormones. Physiol Rev 78: 1087

Rehfeld JF, Friis-Hansen L, Goetze JP, and Hansen TV (2007) The biology of cholecystokinin and gastrin peptides. Curr Top Med Chem 7: 1154

Rindi G and Solcia E (2007) Endocrine hyperplasia and dysplasia in the pathogenesis of gastrointestinal and pancreatic endocrine tumors. Gastroenterol Clin North Am 36: 851

Slominski A and Wortsman J (2000) Neuroendocrinology of the skin. Endocrine Rev 21: 457

Solcia E, Capella C, Fiocca R, Cornaggia M, and Bosi F (1989) The gastroenteropancreatic endocrine system and related tumours. Gastroenterol Clin North Am 18: 671

Magnification: x 5,800 (A,B) x 12,000 (inset in B)

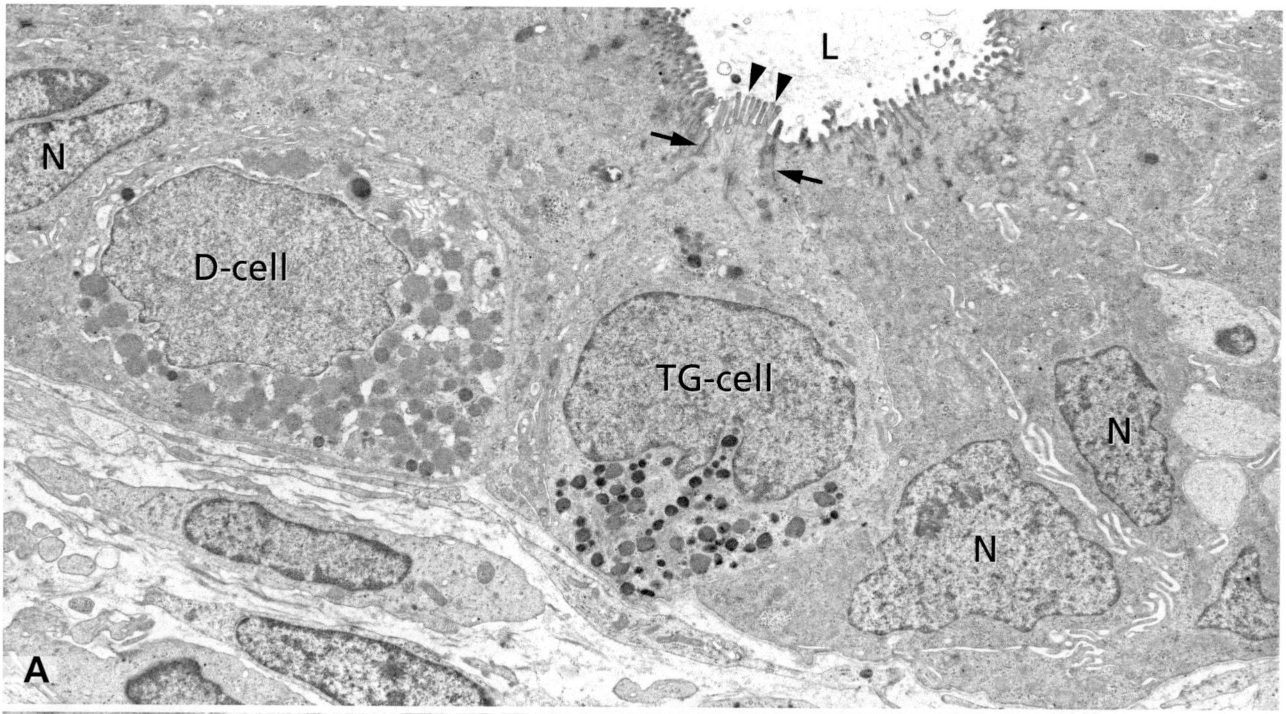
L
N
D-cell
TG-cell
N
N
A

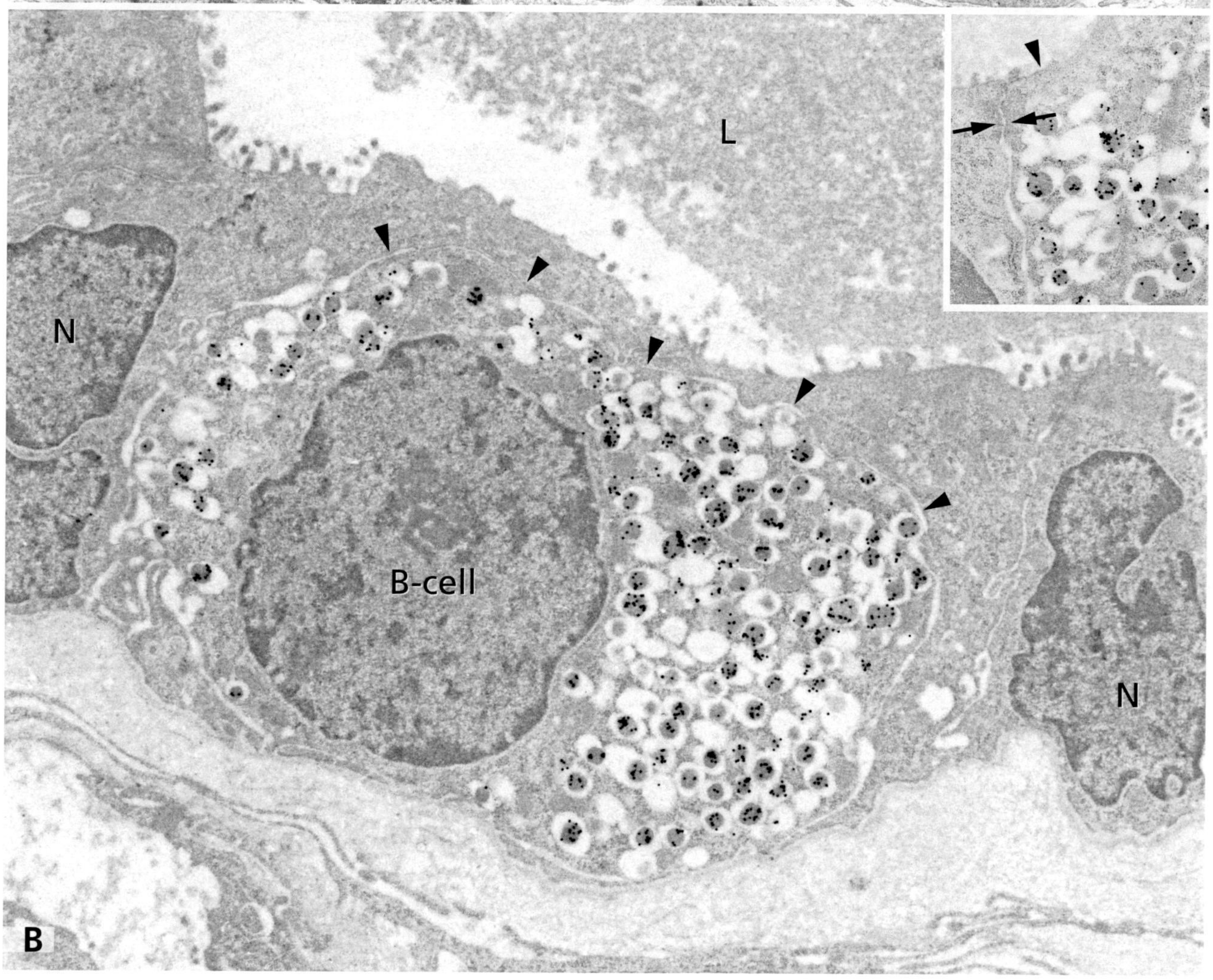
L
N
B-cell
N
B

LIVER

The epithelium of the liver exhibits a unique organisation that guarantees that this central organ of the body is equal to its multiple tasks. In the liver lobules, which are morphological and functional units of the liver tissue, the epithelial cells (hepatocytes) form branching plates, which are surrounded by spaces that contain the sinusoidal blood capillaries (sinus). The picture shows a flat section through the liver epithelium, in which the close relationships between hepatocytes and the capillaries are evident. The sinusoidal capillaries collect both venous blood from the portal vein enriched with substances absorbed in the intestine, and oxygen-rich blood from the hepatic artery. Within the liver lobules, the slow blood stream in the sinus facilitates uptake of substances from the blood into the liver cells for further metabolising. In a radial organisational structure, the sinus proceed to the centres of the lobules; the blood is collected in the central veins und further conducted via the large collecting veins to the vena cava inferior.

The picture shows several voluminous hepatocytes containing large nuclei that are surrounded by broad bands of cytoplasm, in which a large number of mitochondria (M) and accumulations of glycogen (Glyc) are apparent. The polarity of the cells is clearly visible; wide areas of the basal cell surfaces are directed towards the sinus, the lateral surfaces form the cell borders between the neighbouring cells (arrowheads), and the apical cell surfaces form the wall of the bile canaliculi; their lumina are labelled by asterisks. The liver epithelial cells have characteristics of both an endocrine gland, which exports products – e.g. plasma proteins, lipoprotein particles, and components of the blood clotting system – at the basolateral cell surfaces to be taken up into the blood, and characteristics of an exocrine gland secreting at the apical surface into an excretory duct. The exocrine secretory fluid of the liver is the bile, and the bile canaliculi form the first part of the excretory duct system of the liver. From the intercellular spaces, the bile canaliculi are sealed off by tight junctions (cf. Figs. 106, 107), thus preventing bile from entering the lumen of the sinus and mixing with the blood, which would cause jaundice. Tight junctions also constitute the borders between the apical and basolateral parts of the plasma membrane.

Very flat endothelial cells (End) line the capillaries. Only those cellular parts that bear the nucleus exhibit some height and protrude into the sinus lumen, as shown in the lower part of the picture, whereas the other parts of the endothelial cells are extremely flat and perforated by numerous large and small pores. Through these perforated parts of the endothelial cells blood plasma leaves the capillary lumen and enters the small space surrounding the endothelium, the space of Disse (cf. Figs. 106 and 107). This mechanism enables direct contact between blood and the basal cell surfaces of the hepatocytes facilitating traffic between blood and liver cells. The basal cell surfaces of the hepatocytes are enlarged by numerous microvilli and filamentous extensions protruding into the space of Disse; they reflect the intense endocytic and secretory traffic at these poles of the cells.

Magnification: x 4,500

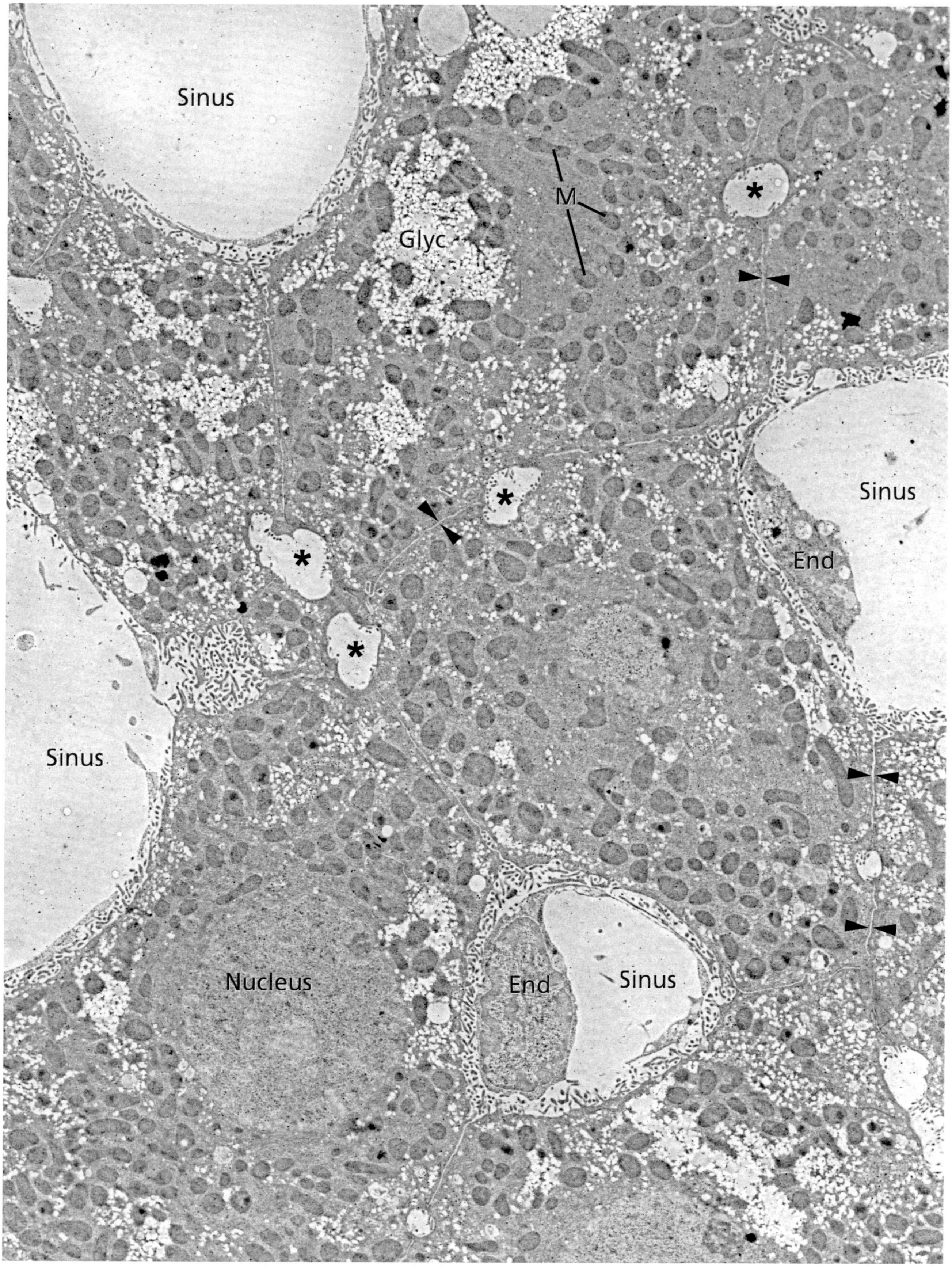
Sinus
M
Glyc
*
*
*
*
Sinus
End
Sinus
Nucleus
End
Sinus

LIVER: HEPATOCYTES, KUPFFER CELL, CELL OF ITO

The picture shows a part of the liver epithelium and its close relationship to the sinusoidal capillaries. In the right hand part of the micrograph, it is clearly visible that a liver cell plate is constructed of a one-cell layer. In the upper left corner of the micrograph, a flat section through the cell plate gives insight into the organisation of the apical liver cell domains that join to form the walls of the bile canaliculi (arrows).

Laterally, the plasma membranes contact the neighbouring cells and build up a system of tight, adhering and communicating junctions (cf. Fig. 107). The basal cell domains are directed toward the space of Disse, a narrow space in between liver epithelial cells (hepatocytes) and the endothelial cells lining the blood capillaries. Extended areas exist for transport and exchange of substances in both directions between hepatocytes and blood plasma. The sinusoidal lumen is lined by the sinus endothelium of specialised, extremely thin endothelial cells with a discontinuous basal lamina. Multiple small and large pores (arrowheads) perforate the endothelial cells and provide gates for the blood plasma to get in direct contact with the basal surfaces of the liver cells. The multitude of endocytosis and secretion events taking place at the basal cell surfaces is mirrored by the abundant microvilli extending into the space of Disse and being in contact with blood plasma.

Specialised mesenchymal cells, called cells of Ito, or fat-storing cells, or stellate cells reside in the space of Disse or are localised in between neighbouring hepatocytes, as shown in the upper right quarter of the micrograph. Cells of Ito are the major fibrinogenic cells of the liver. They produce extracellular matrix components (ECM), including procollagen, as well as ECM-degrading metalloproteinases, and they play a central role in ECM-remodelling. Multiple lipid droplets are accumulated in the Ito cells' cytoplasm (asterisks) mirroring their function as major storage cells of lipid-soluble vitamins. Containing over 80% of total vitamin A of the body, cells of Ito have an essential role in the maintenance of vitamin A homeostasis. Activated upon pathologic conditions, they transform into proliferative myofibroblastic cells that express type I collagen and alpha-smooth muscle actin. Cells of Ito play a pivotal role in development of liver fibrosis and cirrhosis.

By contrast to the cells of Ito, which reside outside of the sinusoids, Kupffer cells localise in the sinusoidal lumen. With their voluminous cell bodies and multiple processes, they are in close contact with the blood stream. Kupffer cells are phagocytic cells of the unspecific immune system that are able to catch substances and filter them out of the blood. Owing to their outstanding position, Kupffer cells can interact directly with circulating materials and cells, and may also be involved in immunomodulation. A Kupffer cell of typical star-like appearance with three processes bridging the sinusoidal lumen is shown in the left part of the micrograph. The processes are in contact with the sinus endothelium and, through the endothelial pores, enter the space of Disse.

References

Bedossa P and Paradis V (2003) Liver extracellular matrix in health and disease. J Pathol 200: 504

Sato M, Suzuki S, and Senoo H (2003). Hepatic stellate cells: Unique characteristics in cell biology and phenotype. Cell Struct Funct 28: 105

Sun ZL, Wada T, Maemura K, Uchokura K, Hoshino S, Diehl AM, and Klein AS (2003) Hepatic allocraft-derived Kupffer cells regulate T-cell response in rats. Liver Transpl 9: 489

Magnification: x 9,400

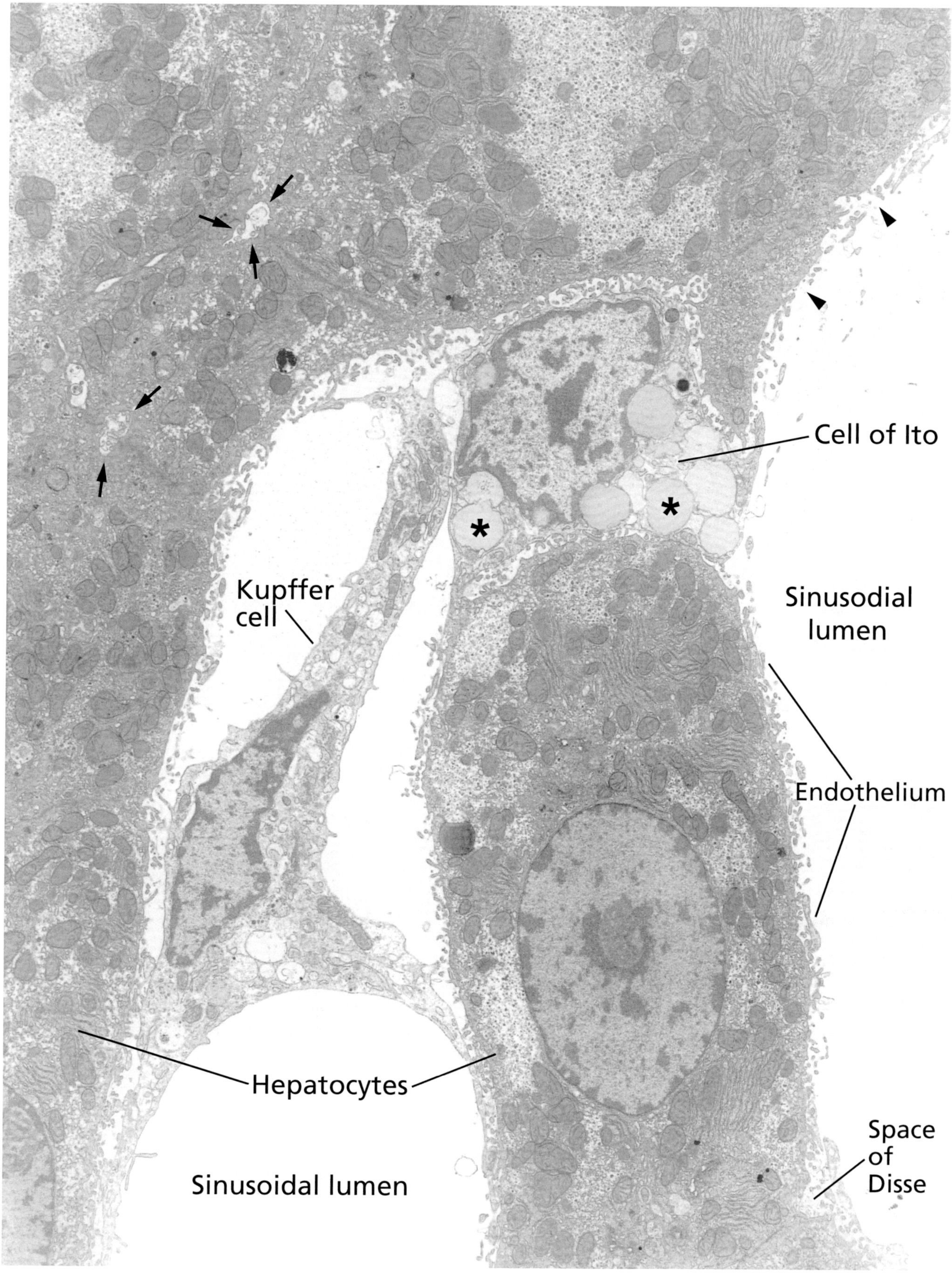
Cell of Ito
*
*
Kupffer
cell
Sinusodial
lumen
Endothelium
Hepatocytes
Sinusoidal lumen
Space
of
Disse

LIVER EPITHELIUM: BILE CANALICULI

The liver parenchyma functions as both an exocrine gland producing excretory products to be secreted into the biliary duct system, and an endocrine gland, synthesising products to be directly delivered to the blood. Delicate canaliculi (BC) bordered by the apical liver cell surfaces build up the initial parts of the bile transport system visible in the central part of both panels A and B.

The particular bile canaliculi organisation within a one-cell layer plate of hepatocytes is illustrated in the diagram below.

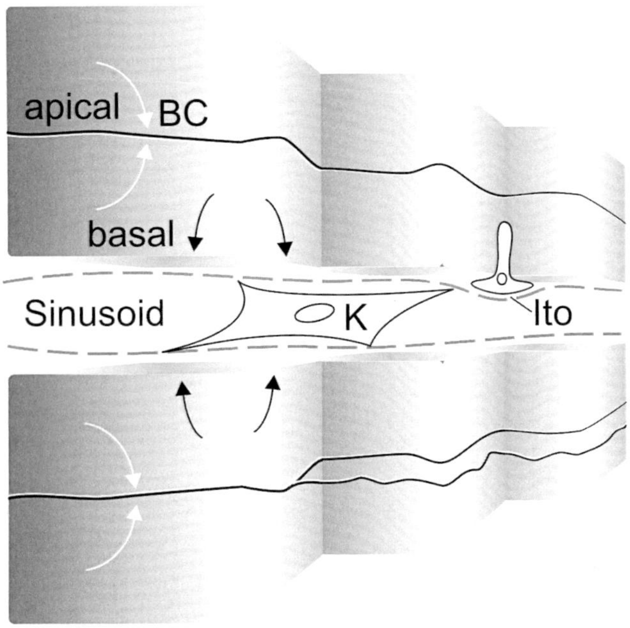

Branched bile canaliculi lined by the apical plasma membranes (BC) run around the cuboidal hepatocytes. The lateral and basal cell surfaces interact with neighbouring cells and are directed towards the sinusoidal capillaries (sinusoid), respectively. White curved arrows indicate the sites of exocrine secretion, black arrows label the sites of endocrine secretion. K – Kupffer cell, Ito – cell of Ito.

Bile is released into the narrow lumina of the bile canaliculi and, at the periphery of the liver lobules, drained by the ductuli of Hering, which further merge to form the larger bile ducts of the portal spaces.

Bile is the exocrine product of the hepatocytes and contains bile salts, cholesterol, phospholipids, conjugated bilirubin and electrolytes. Bile salts are necessary for emulging dietary lipids in the small intestine and are essential for regular fat absorption. Furthermore, through the bile metabolic products of drugs and heavy metals are excreted and IgA is transported to the intestinal mucosa.

The apical hepatocyte cell surfaces lining the canaliculi (BC) display multiple microvilli, possibly occupying major parts of the luminal space, as shown in panel A. The surrounding cytoplasm contains abundant membrane profiles of endoplasmic reticulum, Golgi apparatus stacks (Golgi), mitochondria (M), peroxisomes (PO), lysosomes and autophagic vacuoles (AV). Export of bile constituents takes place via ATP-dependent export pumps; hence, particular secretory granules are not visible. Molecular transporters for cholesterol, phospholipids, bilirubin and bile salts ("ABC transporters" – "ATP binding cassettes") make up essential components of the apical plasma membrane. The bile canalicular lumen is sealed by tight junctions forming an occluding belt (arrows in panel B), which prevents components of the bile from entering the intercellular space, the space of Disse, and the sinusoidal lumen. Defects of the occluding junctional system lead to leakages and allow bile to mix up with the blood, clinically resulting in jaundice.

In panel A, it is clearly visible that there exist different domains in the cytoplasm of the liver cells, with either rough or smooth endoplasmic reticulum dominating (cf. Fig. 21). Cisternae of the Golgi stacks are filled with multiple lipoprotein particles (cf. Fig. 108).

References

Wolkoff AW and Cohen DE (2003) Bile acid regulation of hepatic physiology. I. Hepatocyte transport of bile acids. Am J Physiol-Gastr L 284: G175

Magnification: x 24,000 (A), x 32,500 (B)

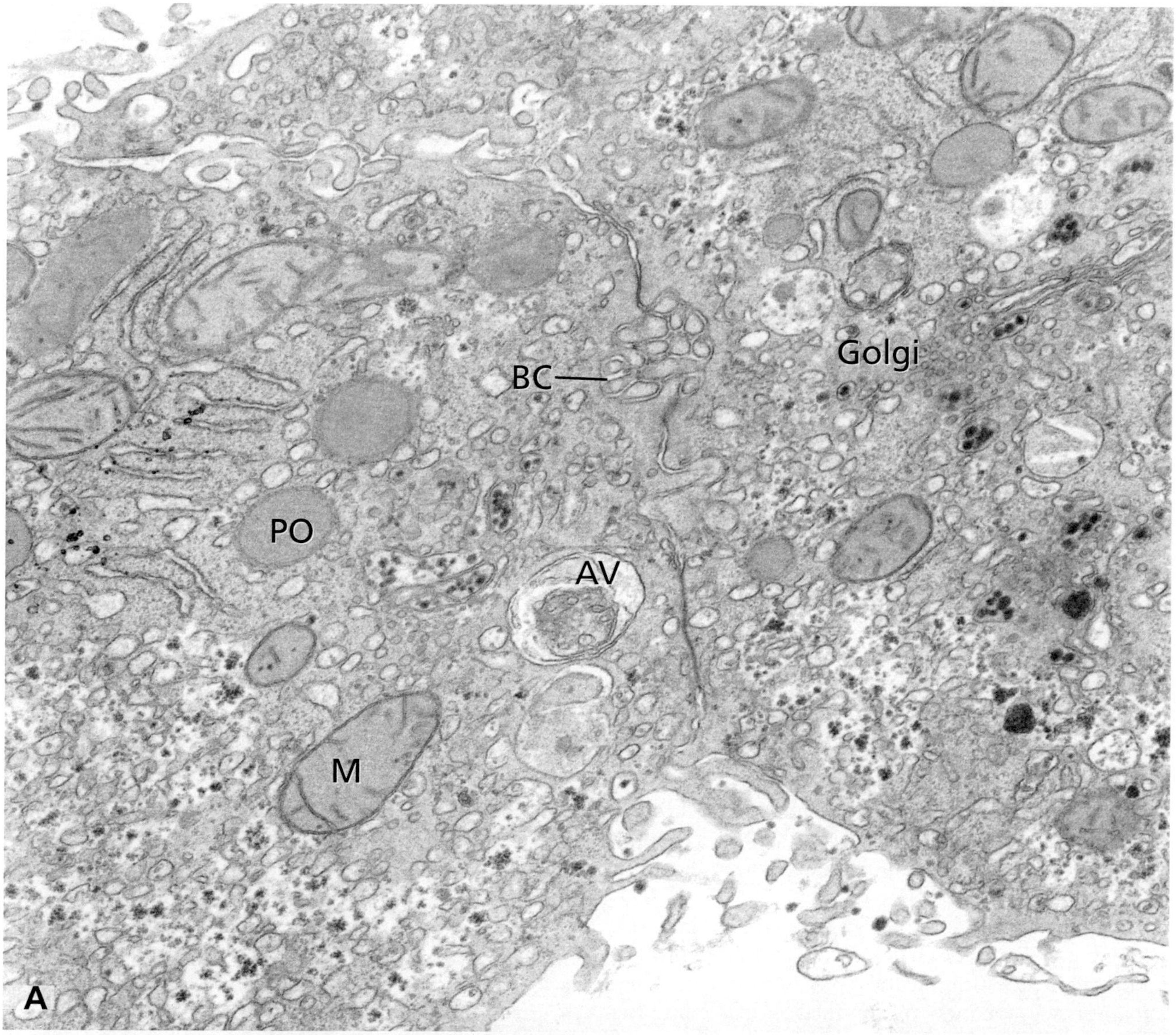
BC
Golgi
PO
AV
M
A

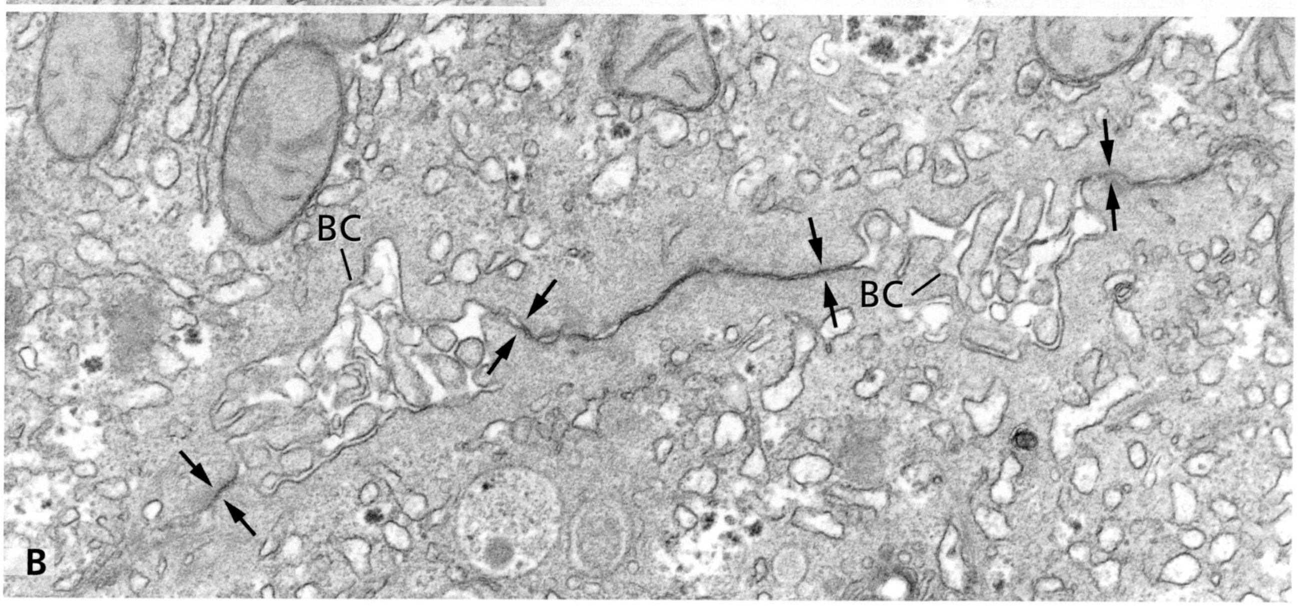
BC
BC
B

LIVER EPITHELIUM: PATHWAY OF SECRETORY LIPOPROTEIN PARTICLES

At their extended basal domains, the hepatocytes are in direct contact with blood plasma that enters the space of Disse via multiple pores perforating the endothelial cells (cf. Fig. 106). By constitutive secretion, multiple endocrine products of the hepatocytes including albumin and a series of plasma proteins required for blood clotting, such as prothrombin, fibrinogen, and coagulation factors V, VII, and IX, are delivered to the blood.

Lipoprotein particles (LPs) assembled within the endoplasmic reticulum (ER) lumen and crossing the Golgi apparatus stacks are transported to the basolateral cell domains by a microtubule-dependent mechanism. Lipoprotein particles produced by hepatocytes belong mainly to the class of very low density lipoproteins (VLDLs) measuring 50–80 nm in diameter. They are perfectly visible under the electron microscope because of their intense contrast after osmification, and their routes can be followed readily from the assembly sites in the ER (panel A) across the Golgi apparatus (panel B) to the exocytosis sites and their extracellular destination within the space of Disse (panel C).

Panel A shows a boundary region between rough (RER) and smooth endoplasmic reticulum domains of a liver cells. Located in close proximity to mitochondria (M), peroxisomes (PO), and glycogen particles multiple cross sections of LP-containing endoplasmic reticulum segments (arrows) are apparent.

LPs are taken up into the Golgi apparatus (panel B), where the major glycosylation of LP-associated proteins and further modifications take place. Within the Golgi apparatus, LPs (arrows) are confined to dilated cisternal subregions. Multiple LP-containing vesicles (arrows) presumably correspond to pre- and post-Golgi LP-carriers.

Close to the basolateral cell surface of a hepatocyte shown in the lower left corner of panel C, multiple LP-containing vesicles (arrows) are visible. They exclusively belong to a class of small vesicles enclosing only one LP. By exocytosis, the LPs are secreted into the space of Disse, where they enter the blood. Arrowheads indicate LPs visible in the space of Disse, where fine collagen fibrils (C) form a delicate network surrounding the sinusoidal endothelial cells. A continuous basal lamina is lacking.

References

Cenedella RJ, Crouthamel WG, and Mengolf HF (1974) Intestinal versus hepatic contribution to circulating triglyceride levels. Lipids 9: 35

Olofsson S-O, Boström P, Andersson L, Rutberg M, Perman J, and Borén J (2009) Lipid droplets as dynamic organelles connecting storage and efflux of lipids. Biochim Biophys Acta 1791: 448

Reaven EP and Reaven GM (1980) Evidence that microtubules play a permissive role in hepatocyte very low density lipoprotein secretion. J Cell Biol 84: 28

Siddiqi SA (2008) VLDL exits from the endoplasmic reticulum in a specialized vesicle, the VLDL transport vesicle, in rat primary hepatocytes. Biochem J 413: 333

Swift LL, Valyi-Nagy K, Rowland C, and Harris C (2001) Assembly of very low density lipoproteins in mouse liver: evidence of heterogeneity of particle density in the Golgi apparatus. J Lipid Res 42: 218

Magnification: x 33,800 (A), x 34,800 (B), x 26,500 (C)

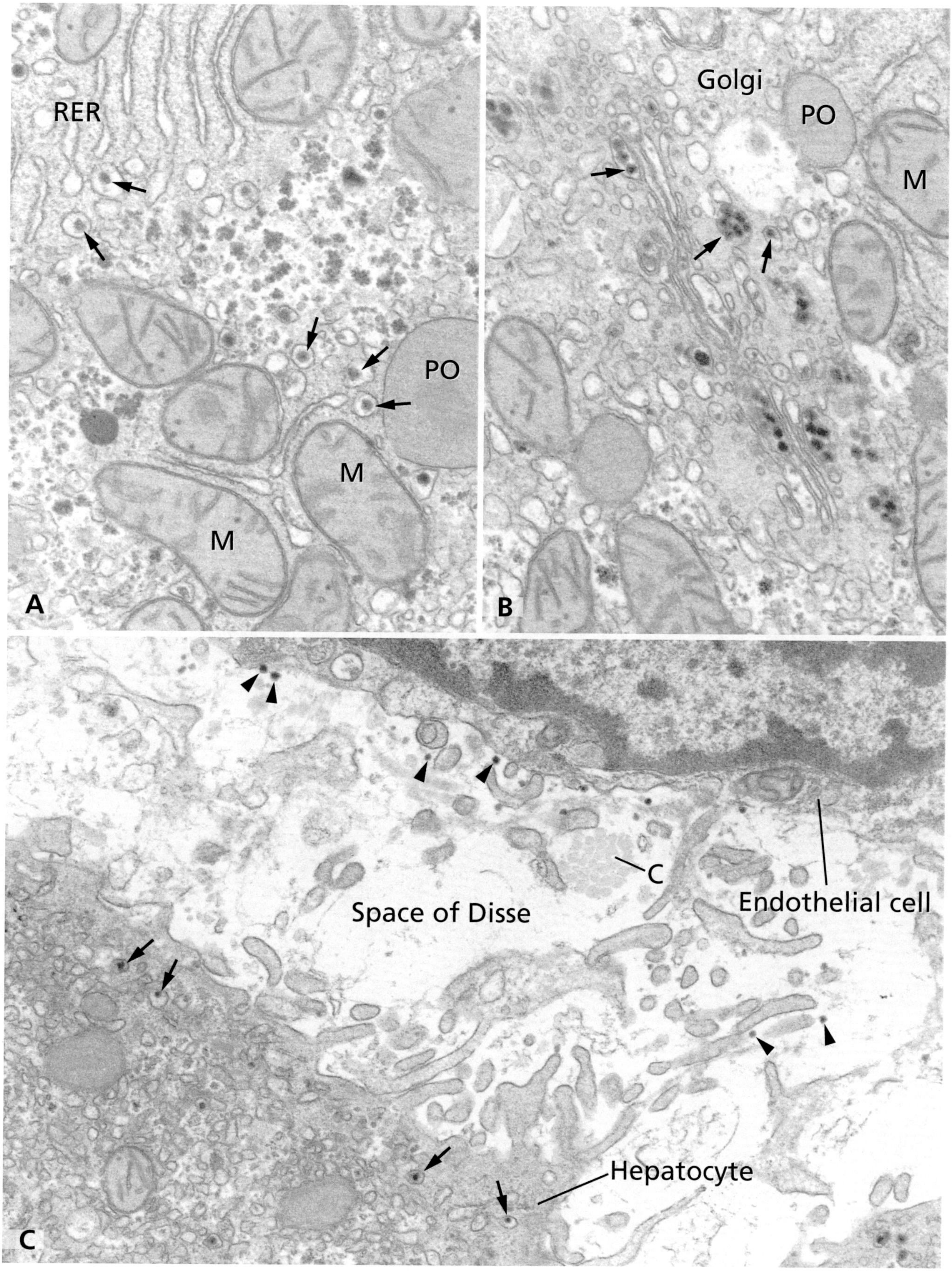
RER
PO
M
M
A
Golgi
PO
M
B
Space of Disse
C
Endothelial cell
Hepatocyte
C

CONGENITAL HEPATIC FIBROSIS

Autosomal recessive polycystic kidney disease/congenital hepatic fibrosis is an inherited disease characterized by a non-obstructive fusiform dilatation of collecting ducts in the kidney and ductal plate malformation in the liver. This results in enlarged spongiform kidneys and in congenital hepatic fibrosis. The abnormal ductal plate remodeling results in dilated bile ducts, which show abnormal branching and in fibrosis.

The electron micrograph from a liver biopsy of a patient shows the characteristic features such as a dilated bile duct with otherwise normal appearing epithelial lining. The increased number of bile ducts and their abnormal branching cannot be observed in this single ultrathin section. The excessive fibrosis is manifested by the presence of numerous large bundles of collagen fibers (asterisks). Numerous sheet-like, cytoplasmic processes of fibroblasts are obvious as well. These changes result in periportal thickening. Congenital hepatic fibrosis leads to portal hypertension. Furthermore, patients with congenital hepatic fibrosis are at increased risk of liver tumors, especially cholangiocarcinoma. Often, the liver of these patients has an abnormal large left lobe.

Autosomal recessive polycystic kidney disease/congenital hepatic fibrosis is caused by mutations in the *PKHD1* gene located on chromosome 6p12. The gene encodes for fibrocystin, which is a hepatocyte growth factor receptor-like protein with a function for primary cilia of collecting duct and biliary epithelia.

References

Bergmann C, Senderek J, Sedlacek B, Pegiazoglou I, Puglia P, Eggermann T, Rudnik-Schoneborn S, Furu L, Onuchic LF, De Baca M, et al. (2003) Spectrum of mutations in the gene for autosomal recessive polycystic kidney disease (ARPKD/PKHD1). J Am Soc Nephrol 14: 76

Desmet VJ (1992) Congenital diseases of intrahepatic bile ducts: variations on the theme "ductal plate malformation". Hepatology 16: 1069

Ward CJ, Hogan MC, Rossetti S, Walker D, Sneddon T, Wang X, Kubly V, Cunningham JM, Bacallao R, Ishibashi M, et al. (2002) The gene mutated in autosomal recessive polycystic kidney disease encodes a large, receptor-like protein. Nat Genet 30: 259

Magnification: x 5,800

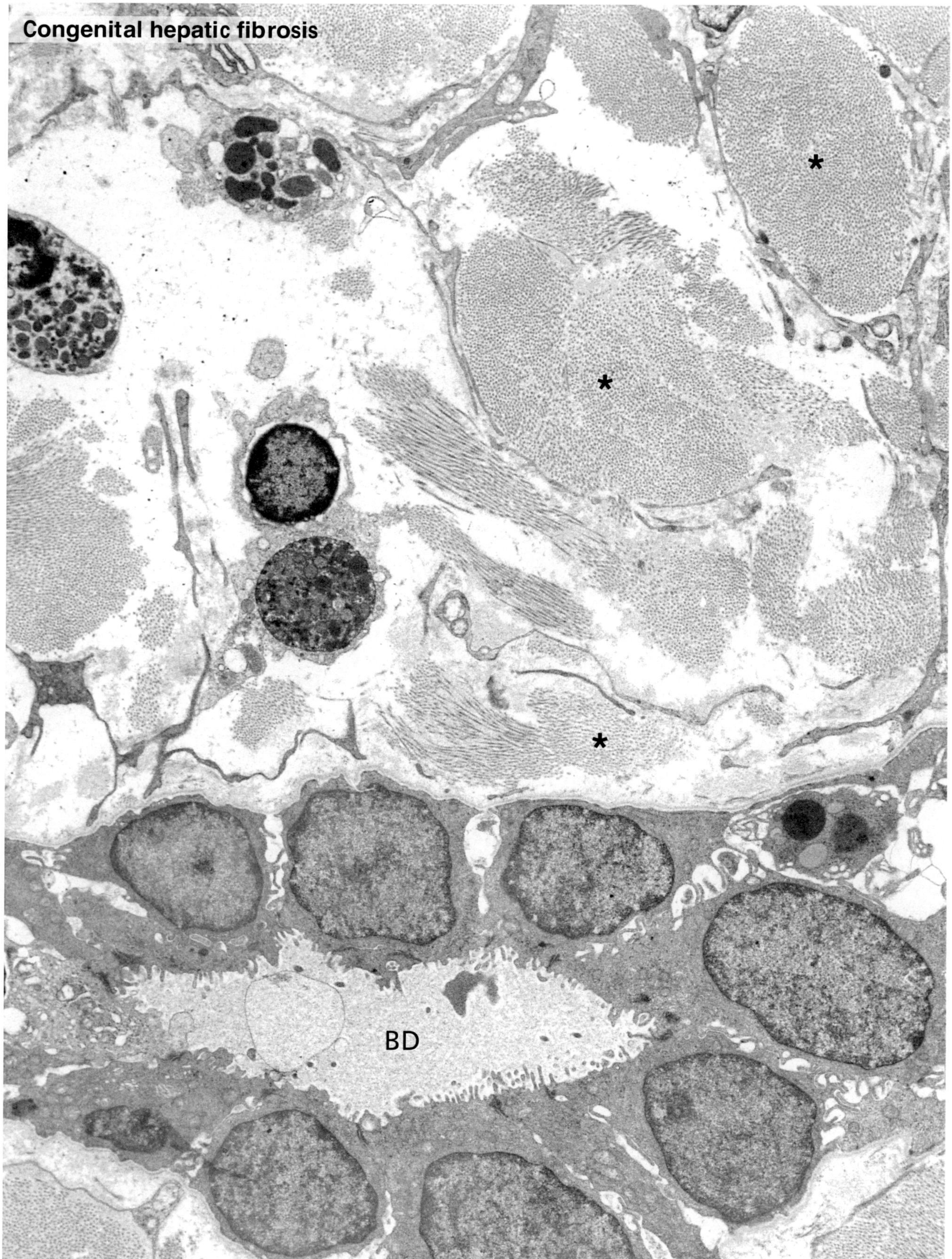
Congenital hepatic fibrosis
*
*
*
BD

CHOROID PLEXUS EPENDYMA

The ependymal cell layer that covers the surface of the ventricles of the brain and the central canal of the spinal cord consists of a simple cuboidal epithelium; the cells are linked by belt desmosomes. Specialised ependyma regions are present in the third ventricle, characterised by the presence of tanycytes, which contain basal processes that form end-feet on blood vessels and are attached to each other by zones of tight junctions building up a barrier.

The choroid plexus epithelium is another ependyma specialisation, which differentiates during development at contact sites with the highly vascularised meninges. Tela choroidea are formed in the roofs of the third and fourth and in the walls of the lateral ventricles. Here, the ependyma cells differentiate into cells with both secretory and resorptive functions and together with the loose connective tissue leading dense networks of blood vessels form the choroid plexus.

The choroid plexus is the site of production of the cerebrospinal fluid (CSF), which circulates through the brain ventricles and the subarachnoidal space around the central nervous system and forms a supporting coat that protects brain and spinal cord from external impact. The figure shows both components of the choroid plexus playing a main part in CSF production, the highly polarised choroid plexus ependyma and the blood capillary system located beneath, close to the basal lamina. The capillary endothelial cells (arrowheads) are fenestrated and lack tight junctions. In a first step of CSF-production, an ultrafiltrate of the blood plasma is built passing through the fenestrated capillary endothelium in the surrounding connective tissue. In a secondary step, this ultrafiltrate is transformed by the choroidal ependyma cells in an excretory product, the cerebrospinal fluid. Direct passage of the ultrafiltrate into the ventricle lumen is prevented by the choroid ependyma. Adjacent cells are extensively interdigitated and the intercellular space is sealed by occluding junctions (circle), forming part of the CSF barrier. The choroid ependyma cells contain abundant mitochondria, plenty of endoplasmic reticulum membranes, and a prominent Golgi apparatus (Golgi). They exhibit surface specialisations characteristic of epithelia involved in transcellular transport. Both the surfaces of the basolateral and the apical cell domains are enlarged by extended interdigitating folds and a brush border of densely packed microvilli (arrows), respectively. Components of the ultrafiltrate enter the cells at the basolateral surfaces. Na^+, K^+-ATPases, present in the membranes of the microvilli, pump Na^+ into the extracellular space, which produces an osmotic gradient facilitating the passage of water into the lumen of the ventricle. The brush border microvilli are also involved in endocytosis and form irregular convolutes on the ependyma cells' surfaces. The choroid plexus, like many other organs, expresses endocytic receptors, such as megalin.

Cells of the immune system known to be present in the CSF not only under pathological conditions, but also in healthy individuals, are assumed to enter CSF directly from the systemic circulation, and there is evidence for a passage through the meninges and choroid plexus.

References

Christensen E and Birn H (2002) Megalin and cubilin: multifunctional endocytic receptors. Nat Rev Mol Cell Biol 3: 258

Kivisäkk P, Mahad DJ, Callahan MK, Trebst C, Tucky B, Wei T, Wu L, Baekkevold ES, Lassmann H, Staugaitis SM, Campbell JJ, and Ransohoff RM (2003) Human cerebrospinal fluid central memory CD4+T cells: Evidence for trafficking through choroids plexus and meninges via P-selectin. Proc Natl Acad Sci USA 100: 8389

Magnification: x 11,000

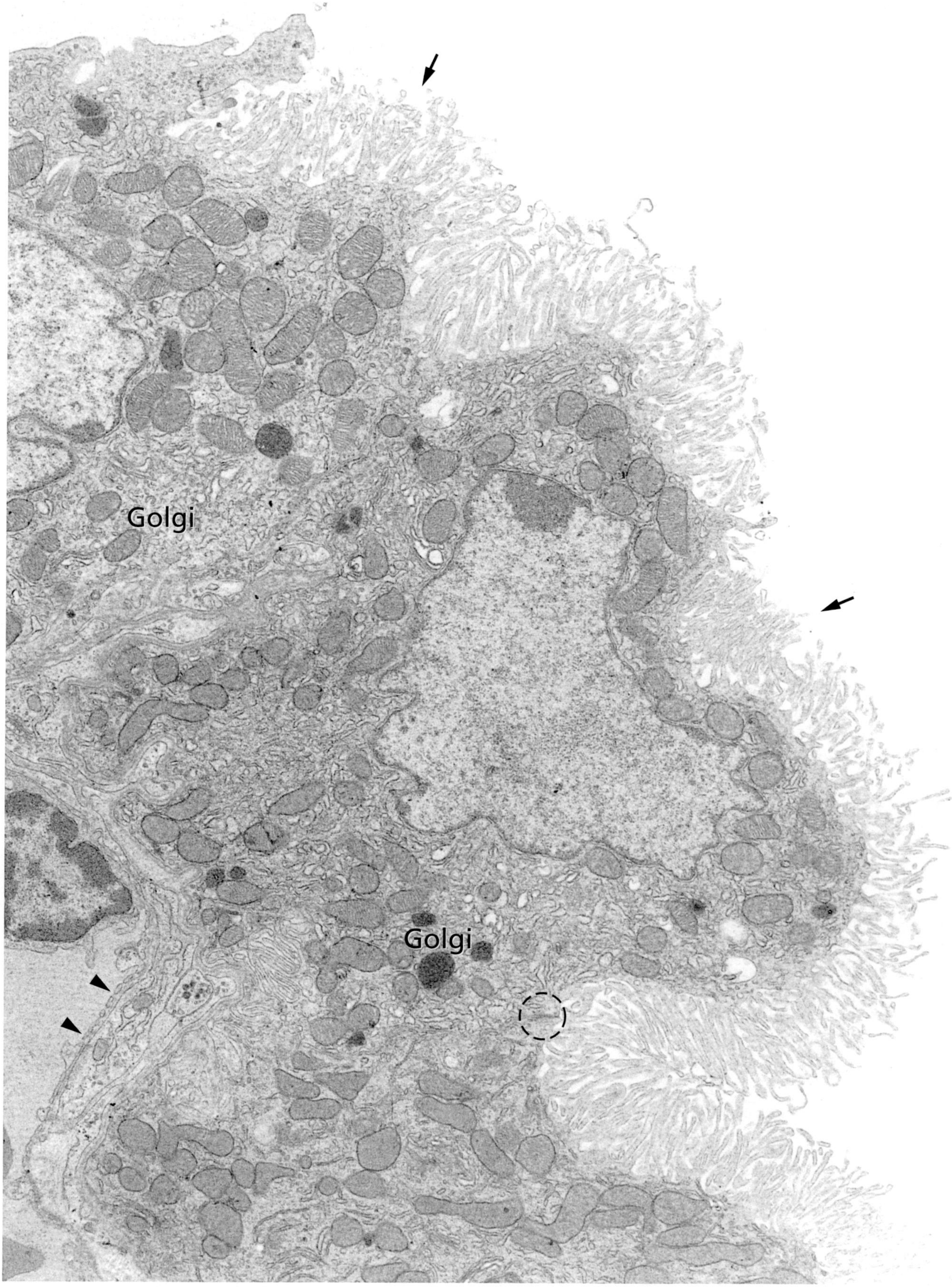
Golgi
Golgi

SMALL INTESTINE: ABSORPTIVE CELLS

The ultrastructures of small intestinal enterocytes, which line the villi and are continually renewed, mirror their main tasks in the absorption of digested food. Panels A and B show the apical cell domain of absorptive cells with the prominent brush border made up of numerous densely packed microvilli, which are the sites of final digestion and uptake of nutrients. Microvilli exist in an approximate number of 3000 per cell, which immensely increases the luminal cell surface. They measure 1–2 μm in length and contain a filamentous core, which is composed of actin filaments, crosslinked by fimbrin and villin, and attached to the plasma membrane by myosin I and calmodulin. The rootlets of the filament bundles project into the terminal web located beneath the brush border (tw). They are interconnected by intestine-specific spectrin, attach to cytokeratin intermediate filaments of the terminal web, and are part of the apical cytoskeletal apparatus that is responsible for maintaining the upright positions of the microvilli and the overall organisation of the brush border (cf. also Fig. 74). Components of the terminal web are associated with the junctional complex. In particular, actin filaments connected with the belt desmosome in the middle part of the junctional complex (panel B-2; cf. also Fig. 83) contribute essentially to the cells-cells expanding motion system. This is responsible for changes of the diameter of the apical cell domains, leading to a tilting of the brush border microvilli, which facilitates contact with the digested nutrients and supports absorption.

The luminal cell surface is lined by a 10 nm thick plasma membrane, clearly visible in panel B as a distinct three-lamellar structure covered by the fuzzy, filamentous network of the glycocalyx (cf. Fig. 80). Intramembraneous enzymes are responsible for the final breakdown of oligosaccharides and oligopeptides. The products, such as amino acids, di- and tripeptides, and sugars, are transported across the plasma membrane, which involves specific membrane channels and carrier systems.

Brush border enzymes and other membrane constituents have to be continually renewed and are inserted into the plasma membrane at areas located between the microvilli. These are the only membrane sites of the apical surface accessible for fusion and budding events and from here, membrane invaginations deeply extend into the terminal web region. They appear as pleiomorphic compartments and apical tubules close to the microvilli rootlets (cf. Fig. 74, panels A and D and Fig. 83, panel B). Their membranes exhibit lipid microdomains different to those of the microvilli membranes and are considered as functioning as a membrane reservoir for adaptative changes at the apical cell surface.

Regular absorption requires a clear cell polarity with distinct boundaries between apical and basolateral cell surfaces, which are constituted by tight junctions forming the most apical zone of the junctional complex that connects adjacent cells. Panel A shows the apical part of an absorptive cell with numerous mitochondria, lysosomes (Ly), and the Golgi apparatus in typical supranuclear position (Golgi). Bars indicate the profiles of the junctional complex, which is shown at higher magnification in panel B. Close to the apical cell surface, tight junctions form an occluding belt, which defines cell polarity, seals the intercellular spaces and controls the intercellular passage of substances (1; zonula occludens). The second and the third parts of the junctional complex are built up by adhering junctions forming a belt desmosome closely below the tight junctions (2; zonula adhaerens) and an additional circle of spot desmosomes (3; maculae adhaerentes). At more basal regions, adjacent cells are often interlocked by extended interdigitations (panel C).

References

Affleck JA, Helliwell PA, and Kellett GL (2003) Immunocytochemical detection of GLUT2 at the rat intestinal brush-border membrane. J Histochem Cytochem 51: 1567

Hansen GH, Pedersen J, Niels-Christiansen L, Immerdal L, and Danielsen EM (2003) Deep-apical tubules: dynamic lipid-raft microdomains in the brush-border region of enterocytes. Biochem J 373: 125

Louvard D, Kedinger M, and Hauri HP (1992) The differentiating intestinal epithelial cell: establishment and maintenance of functions through interactions between cellular structures. Annu Rev Cell Biol 8: 157

Sanco E, Batlle E, and Clevers H (2003) Live and let die in the intestinal epithelium. Curr Opin Cell Biol 15: 763

Magnification: x 12,500 (A), x 57,000 (B, C)

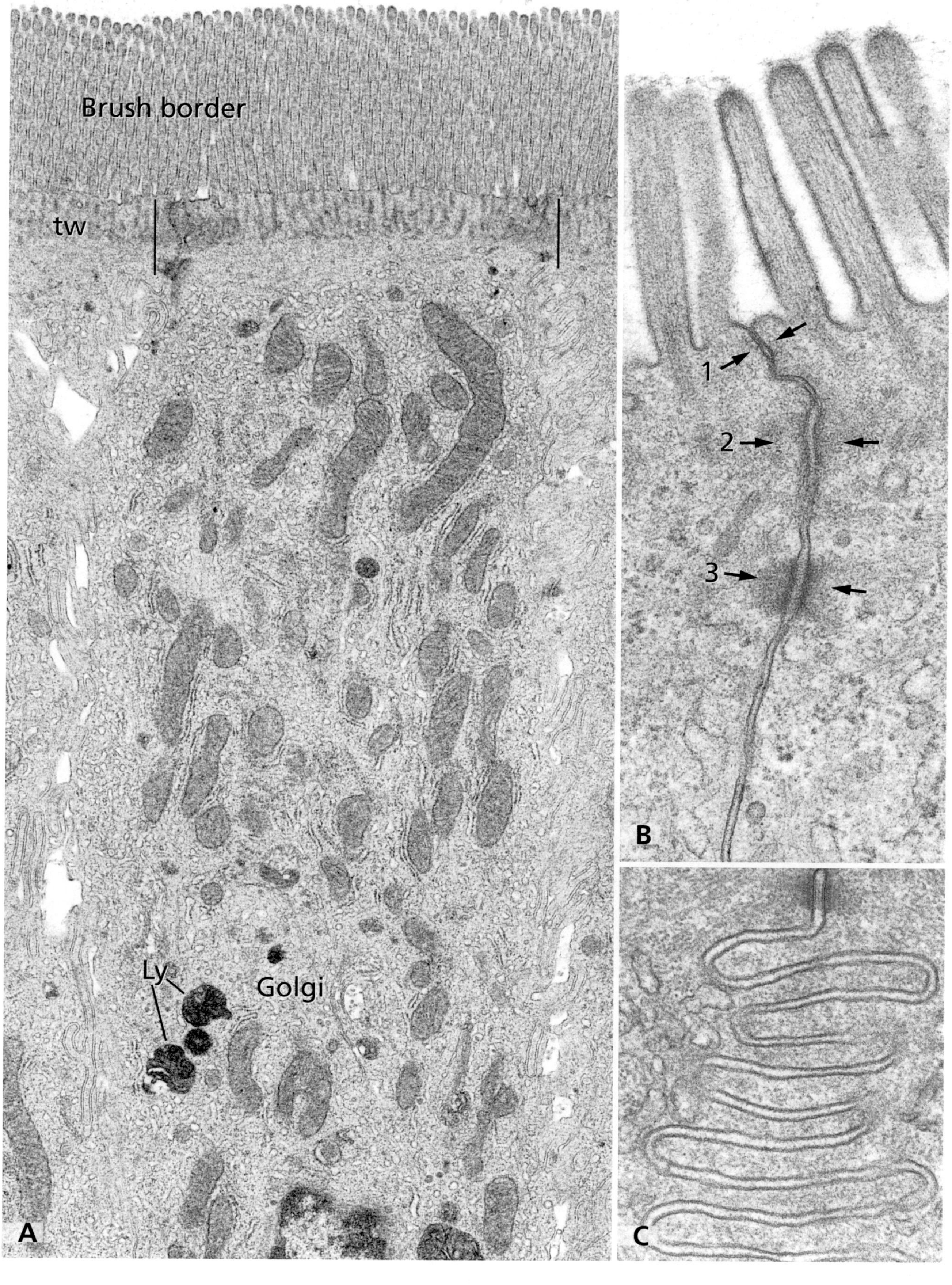
Brush border
tw
Ly
Golgi
A
1
2
3
B
C

SMALL INTESTINE: PATHWAY OF LIPIDS

Through the absorptive enterocytes, digested nutrients cross from the lumen of the gut to the connective tissue underlying the epithelium. After ingestion and specific processing by the absorptive cells (cf. Fig. 111), nutrient constituents are transported to the basolateral cell surfaces, where they again leave the cells and enter blood or lymph capillaries to be distributed in the body. A particular route across the absorptive cells is travelled by lipids, which in the majority are packed into lipoprotein particles, secreted by exocytosis, and moved into the lumen of lymphatics. Some stages of the transcellular and extracellular lipid pathways can readily be followed under the electron microscope.

In the gut lumen during fat breakdown, mixed micelles containing bile salts and the main products of fat digestion are assembled and have immediate access to the brush border through movements of villi and microvilli. Free fatty acids and monoglycerides liberated from the micelles diffuse into the microvilli and associate with fatty acid-binding proteins, to be transported into the apical cytoplasm. In the smooth endoplasmic reticulum, triglycerides and other lipids are resynthesised and packaged into lipoprotein particles (LPs). Subsequently exported out of the endoplasmic reticulum and transported to the Golgi apparatus, they are glycosylated and packaged into vesicles to be transported to the basolateral cell surfaces. Via exocytosis, they leave the cell into the intercellular spaces. Particularly large LPs, which are known as chylomicrons, are formed during postprandial lipid absorption. However, LPs are not only formed after intake of food. Lipids, mainly derived from the bile and shedded cells, are also absorbed during starvation. The LPs formed during starvation measure 50–80 nm in diameter and belong to the class of very low density lipoproteins (VLDLs; cf. Figs. 107 and 108).

Panels A to D show the intra- (A) and extracellular (B–D) pathways of the small intestinal VLDL particles in the mucosa of a starving rat. VLDL particles are visible in all compartments along the secretory pathway. They are particularly prominent in the dilated cisterns of the Golgi stacks (arrows in panel A), in which they are taken up after export out of the endoplasmic reticulum. Golgi cisternae are the sites of VLDL-glycosylation. In the Golgi apparatus, VLDL particles are packed into large carrier vesicles for transport to the lateral cell surface. One of the carriers is shown in panel A at the left side of the Golgi stack. Secretion of VLDLs is known to occur via exocytosis. Extracellularly, VLDL particles are apparent in the dilated region of the intercellular space (arrowheads in panel B) located between extended cell-cell interdigitations and are accumulated in the connective tissue of the mucosal lamina propria (arrowheads in panel C). They do not enter the blood but move into the lumen of the lymphatics. This is shown in panel D. Here, lipoprotein particles appear negatively stained, since all extracellular fluid of the lamina propria of this rat small intestinal mucosa contains peroxidase, applied intravenously in connection with transport experiments, and visualised by oxidation of diaminobenzidine. The lumina of blood capillaries are devoid of lipoprotein particles (panel C).

References

Ikeda I, Mitsui K, Matsuoka R, Hamada T, Imabayashi S, Uchino A, Yamada K, and Imaizumi K (2003) Cholesterol esterase bound to intestinal brush border membranes does not accelerate incorporation of micellar cholesterol into absorptive cells. Biosci Biotechnol Biochem 67: 2381

Iqbal J and Hussain MM (2009) Intestinal lipid absorption. Am J Physiol Endocrinol Metab 296: 1183

Jones AL and Ockner RK (1971) An electron microscopic study of endogenous very low density lipoprotein production in the intestine of rat and man. J Lipid Res 12: 580

Marcil V, Delvin E, Seidman E, Poitras L, Zoltowska M, Garofalo C, and Levy E (2002) Modulation of lipid synthesis, apolipoprotein biogenesis, and lipoprotein assembly by butyrate. Am J Physiol Gastrointest Liver Physiol 283: G340

Montoudis A, Delvin E, Menard D, Beaulieu J-F, Jean D, Tremblay E, Bendayan M, and Levy E (2006) Intestinal-fatty acid binding protein and lipid transport in human intestinal epithelial cells. Biochem Biophys Res Commun 339: 248

Pavelka M and Gangl A (1983) Effects of colchicine on the intestinal transport of endogenous lipid. Ultrastructural, biochemical, and radiochemical studies in fasting rats. Gastroenterol 84: 544

Pepper MS and Skobe M (2003) Lymphatic endothelium: morphological, molecular and functional properties. J Cell Biol 163: 209

Magnifications: x 33,500 (A), x 28,000 (B), x 18,300 (C), x 36,500 (D)

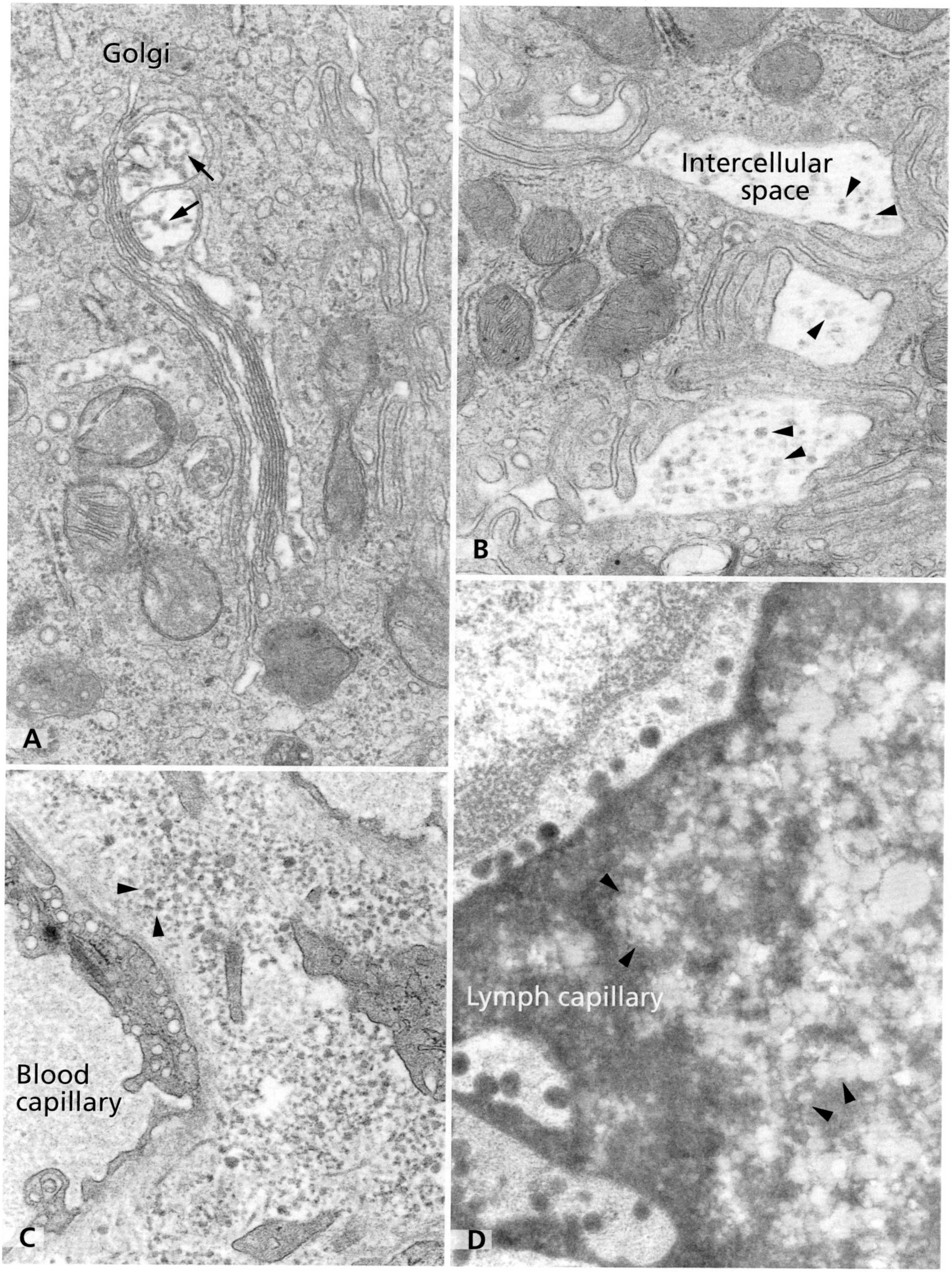
Golgi
A
Intercellular space
B
Blood capillary
C
Lymph capillary
D

RENAL PROXIMAL TUBULE: A REABSORPTION PLANT

The plasma ultrafiltrate generated in the glomeruli is extensively modified and concentrated along the renal tubule. In the proximal tubules, some 70% of the filtered water, glucose, ions, vitamins, low molecular mass proteins, drugs, and other substances are reabsorbed to the blood or degraded in lysosomes.

Proximal tubules have function-related structural specialisations: an apical brush border (cf. Fig. 74) for absorption from the lumen and a basal membrane labyrinth (cf. Fig. 89) for excretion in the extracellular space. These specialisations are important for transcellular transport since they provide vast membrane surfaces for receptors, carriers, and transporters. The apical and basolateral plasma membrane domains are functionally different and thereby establish the epithelial asymmetry. Besides the transcellular route, a paracellular route exists for water and ions, which pass the tight junctions by osmosis.

In proximal tubules, an apical endocytic apparatus occupies most of the apical cytoplasm (panel A), which consists of endocytic vesicles (arrow in B) and dense tubules (arrowheads in B). The sections shown in panels A and B have been stained to preferentially contrast the carbohydrate coat of the plasma membrane and of the apical endocytic apparatus. Various substances that undergo receptor – mediated endocytosis via coated pits situated between the bases of microvilli are routed to this endocytic apparatus and from there to lysosomes for degradation. The dense tubules represent structures for membrane recycling.

In proximal tubules two major endocytic receptors have been identified: megalin and cubilin. In panel C, megalin, as detected in an ultrathin frozen section by immunogold labelling, is present in brush border microvilli and the apical endocytic apparatus. The lateral plasma membrane (PM) is unlabelled. Megalin was initially described as the Heymann nephritis antigen gp 330. Many organs (e.g., inner ear, type II alveolar epithelial cells, thyroid, choroid plexus) in addition to kidney express megalin. Megalin belongs to the low density lipoprotein receptor family. It is a single-spanning type 1 membrane glycoprotein with a molecular mass of about 600 kDa and is structurally different from cubilin, which is a 460 kDa peripheral membrane glycoprotein. Both receptors contain a large extracellular domain, which functions in ligand binding, and they form a dual receptor complex. Since megalin binds a multitude of ligands, it is regarded as a scavenger receptor. Megalin is involved in vitamin homeostasis since it binds at least three vitamin-binding proteins (transcobalmin-vitamin B_{12}, retinol-binding protein, vitamin D-binding protein). Further ligands for megalin involve different carrier proteins (e.g., albumin, lactoferrin, transthyretin), various lipoproteins, hormones (e.g., insulin, parathyroid hormone, epidermal growth factor), enzymes and enzyme inhibitors. Polybasic drugs such as the antifibrinolytic aprotinin, polymyxin B, and aminoglycosides such as gentamycin are reabsorbed by megalin and accumulate in lysosomes of proximal tubules. This is most probably the basis for the known nephrotoxicity of polymyxin B and gentamycin.

References

Christensen E and Birn H (2002) Megalin and cubilin: multifunctional endocytic receptors. Nat Rev Molec Cell Biol 3: 258
Christensen EI and Willnow TE (1999) Essential role of megalin in renal proximal tubule for vitamin homeostasis. J Amer Soc Nephrol 10: 2224
Farquhar MG, Saito A, Kerjaschki D, and Orlando RA (1995) The Heymann nephritis antigenic complex: Megalin (gp330) and RAP. J Amer Soc Nephrol 6: 35
Kerjaschki D and Farquhar MG (1982) The pathogenic antigen of Heymann nephritis is a membrane glycoprotein of the renal proximal tubule brush border. Proc Natl Acad Sci USA 79: 5557
Moestrup SK, Cui SY, Vorum H, Bregengard C, Bjorn SE, Norris K, Gliemann J, and Christensen EI (1995) Evidence that epithelial glycoprotein 330/megalin mediates uptake of polybasic drugs. J Clin Invest 96: 1404
Orlando RA, Rader K, Authier F, Yamazaki H, Posner BI, Bergeron JJM, and Farquhar MG (1998) Megalin is an endocytic receptor for insulin. J Amer Soc Nephrol 9: 1759

Magnification: x 20,000 (A); x 50,500(B); x 45,000 (C)

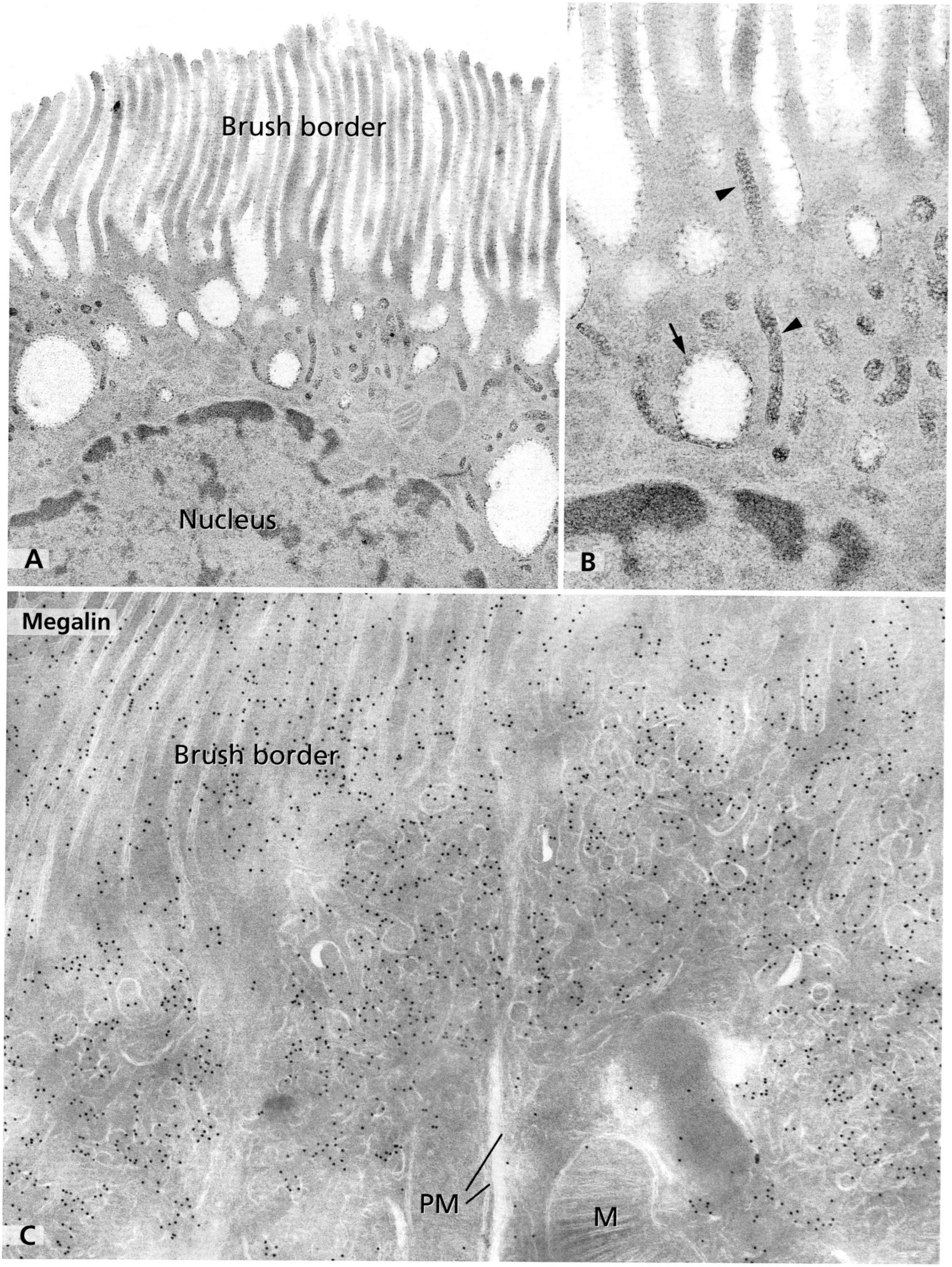
Brush border
Nucleus
A
B
Megalin
Brush border
PM
M
C

PARATHYROID HORMONE RESPONSE OF RENAL PROXIMAL TUBULES

The epithelia lining the proximal tubules must adapt to varying functional demands imposed by changes in the water-electrolyte homeostasis. Experimental studies in which animals have adapted to certain salt diets or have been subjected to hormone treatment have provided a mechanistic insight into these adaptive changes and will be illustrated by the example of phosphate ion reabsorption.

Phosphate ions (Pi) are filtered by the glomeruli, and approximately 80% of phosphate is reabsorbed in the proximal tubules. Reabsorption occurs through the brush border membrane and requires the presence of sodium ions. Among the three types of Na/Pi cotransporters, the Na/Pi-cotransporter NaPi-IIa has been shown to be the most important one for phosphate ion reabsorption in the proximal tubules. It is a polytope membrane glycoprotein with eight transmembrane domains, four extracellular loops and three intracellular loops. Intracellular loop 1 and extracellular loop 3 are involved in Na/Pi-cotransport. The cotransporter is found exclusively in the brush border of proximal tubules, and its amount in this location determines the capacity for the reabsorption of phosphate ions. NaPi-IIa cotransporter becomes rapidly downregulated in response to parathyroid hormone, which results in inhibition of phosphate ion reabsorption. Quantitative electron microscopy and immunoelectron microscopy revealed the steps resulting in the parathyroid hormone-induced NaPi-IIa downregulation. The brush border membrane became greatly depleted of cotransporter, which accumulated in the apical endocytic apparatus through an increased rate of endocytosis. This resulted in an enlargement of the apical endocytic apparatus. Panel A illustrates the steady state situation, whereas panel B shows the acute changes in the apical tubulovesicules in response to parathyroid hormone treatment. The difference in the extent of the apical endocytic apparatus is indicated by bars, and the greater abundance of endocytic structures, in particular the dense tubules, is clearly visible in panel B. This accumulation of the NaPi-IIa cotransporter was transient and followed by its degradation in lysosomes, which resulted in its rapid downregulation. It could also be shown that the lysosomal delivery of the cotransporter was dependent on a taxol-sensitive apical-to-basal rearrangement of microtubules.

In human genetic disorders, which result in renal phosphate wasting, the expression of the NaPi-IIa cotransporter seems to be decreased by serum factors such as elevated concentrations of fibroblast growth factor 23.

References

Beck L, Karaplis A, Amizuka N, Hewson A, Ozawa H, et al. (1998) Targeted inactivation of Npt2 in mice leads to severe phosphate wasting, hypercalciura and skeletal abnormalities. Proc Natl Acad Sci USA 95: 5372

Karim-Jimenez Z, Hernando N, Biber J, and Murer H (2001) Molecular determinants for apical expression of the renal type II Na/Pi-cotransporter. Pflügers Arch 442: 782

Lambert G, Traebert M, Hernando N, Biber J, and Murer H (1999) Studies on the topology of the renal type-II NaPi-cotransporter. Pflügers Arch 43: 972

Lötscher M, Scarpetta Y, Levi M, Halaihel N, Wang H, Zajicek HK, Biber J, Murer H, and Kaissling B (1999) Rapid downregulation of rat renal Na/P(i) cotransporter in response to parathyroid hormone involves microtubule rearrangement. J Clin Invest 104: 483

Murer H, Hernando N, Forster I, and Biber J (2003) Regulation of Na/Pi transporter in the proximal tubule. Annu Rev Physiol 65: 531

Shimada T, Mizutani S, Muto T, Yoneya T, Hino R, and et al. (2001) Cloning and characterisation of FGF23 as a causative factor of tumour-induced osteomalacia. Proc Natl Acad Sci USA 98: 6500

Tenenhouse H and Sabbagh Y (2002) Novel Pi regulating genes in the pathogenesis of renal Pi wasting disorders. Eur J Physiol 444: 317

Traebert M, Roth J, Biber J, Murer H, and Kaissling B (2000) Internalization of proximal tubular type-II Na/Pi-cotransporter by parathyroid hormone: an immunogold electron-microscopic study. Am J Physiol Renal Physiol 278: F148

Magnification: x 42,000 (A, B)

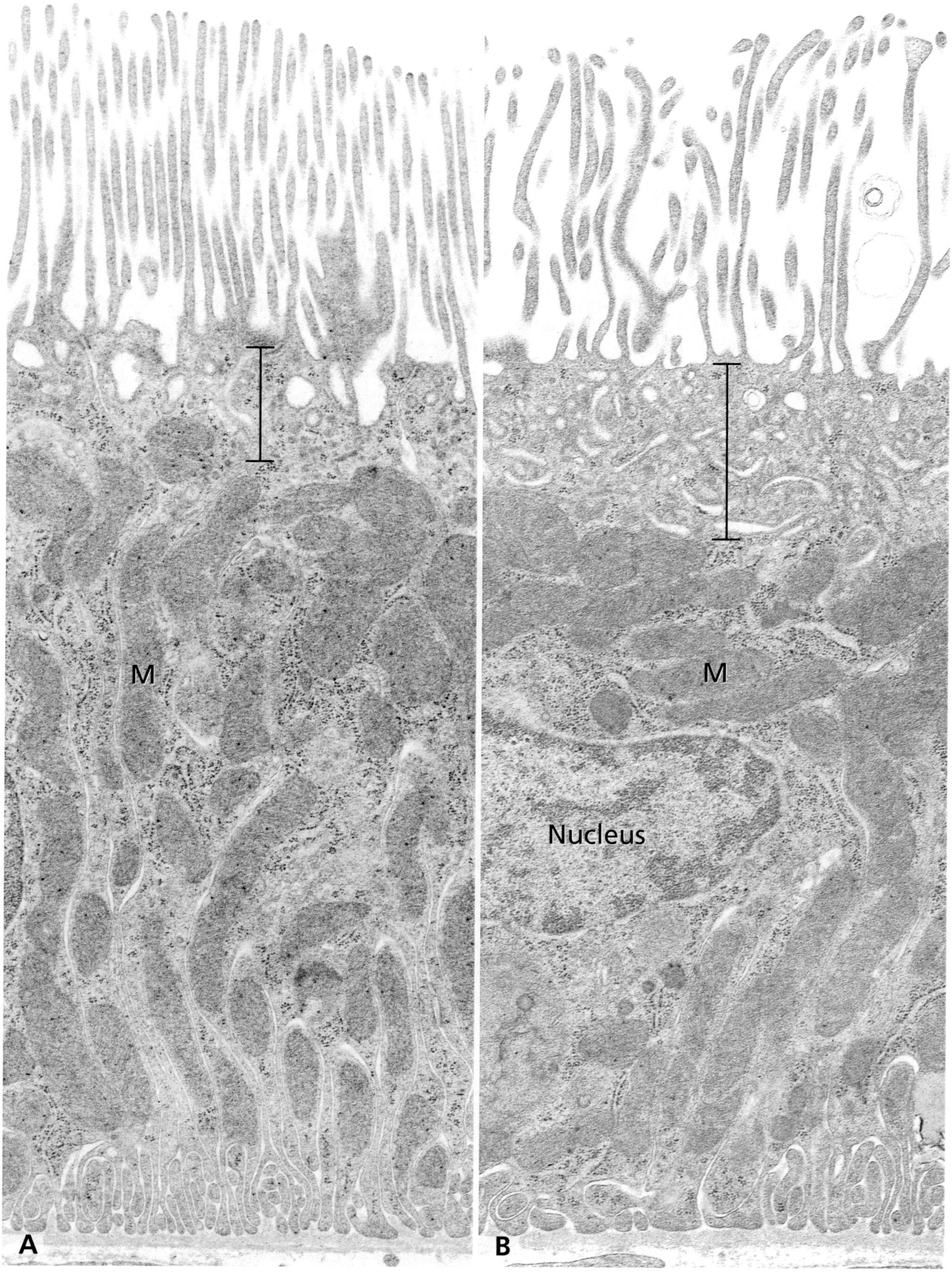
M
A
M
Nucleus
B

PHOTORECEPTOR CELLS OF THE RETINA: SIGNALLING OF LIGHT

Photoreceptor cells are part of the inner sensory retina and become activated by light. Two major types of photoreceptor cells exist, which reside in specific regions in the retina. The rod cells, which contain the photopigment rhodopsin and occupy the periphery of the retina, are exceptionally sensitive to low light levels and specialised for night vision. The cone cells occupy central parts of the retina and are specialised for high resolution at high light levels and for colour detection. Three types of cone cells exist based on the photopigment they contain, which permits discrimination of blue, green, and red.

Both rod and cone cells are elongated, highly polarised neurons. Their main structural and functional features are the inner and outer segments above the nuclear region. Infranuclear lies the synaptic region establishing contact with interneurons, which transmit light-induced electrical signals to the retinal ganglion cells. The border between the inner and outer segment in rod cells is indicated by open arrows in the micrograph and shown at higher magnification in the inset. Inner segments contain numerous elongated mitochondria in addition to endoplasmic reticulum and Golgi apparatus. They are connected to the outer segment by a modified cilium (9+0 symmetry) as shown at higher magnification in the inset (asterisk). In mammals, the outer segments of rods and cones are cylindrically shaped. Outer segments contain stacks of double membrane discs (details shown in Fig. 105), in which the photosensitive visual pigment molecules are embedded. The various components of the discs are synthesised in the inner segments and transported into the outer segments through the narrow cytoplasm surrounding the connecting cilium. The membrane discs originate from plasma membrane infolds of the connecting region from which they pinch off. The visual pigment present in the disc membranes of rods and cones is rhodopsin, a highly specialised G protein-coupled receptor. Rhodopsin is a multispanning membrane protein that detects photons. It bears two *N*-linked oligosaccharides that are required for its full signal transduction activity. Rhodopsin consists of opsin apoprotein to which the chromophore is covalently linked. The chromophore is vitamin A_1- derived 11 *cis*-retinal. Its extended structure accounts for its visible light absorption properties. Upon photon capture, the 11 *cis*-retinal is photoisomerised into 11 *trans*-retinal which changes the conformation of rhodopsin. This initiates a signalling cascade that results finally in the closing of gated Na^+ channels and hyperpolarisation of the plasma membrane.

The apical ends of the outer segments of cones and rods are embedded in the pigment epithelium that contains melanin granules, which absorb excess light and prevent its reflection. Between the two cell types lies the interphotoreceptor matrix, which functions as a glue. Separation of the two layers results in retinal detachment, with severe consequences for vision. The function of the pigment epithelium with regard to outer segment turnover will be discussed in Fig. 116. In addition to its nutritional function for the retina, the pigment epithelium has an essential role in metabolic rhodopsin regeneration. Bleached rhodopsin disassembles in opsin and all *trans*-retinal. All *trans*-retinal is reduced, transported to the pigment epithelium, enzymatically oxidised and re-isomerised to 11 *cis*-retinal. The chromophore transport to the pigment epithelium and back to the photoreceptor is mediated by the interstitial retinoid-binding protein.

References

Borhan B, Souto M, Imai H, Shichida Y, and Nakanishi K (2000) Movement of retinal along the visual transduction pathway. Science 288: 2209

Kaushal S, Ridge K, and Khorana H (1994) Structure and function in rhodopsin: the role of asparagine-linked glycosylation. Proc Natl Acad Sci USA 91: 4024

Masland R (2001) The fundamental plan of the retina. Nat Neurosci 4: 877

Menon S, Han M, and Sakmar T (2001) Rhodopsin: structural basis of molecular physiology. Physiol Rev 81: 1659

Rando R (1996) Polyenes and vision. Chem Biol 3: 255

Sakmar T (1998) Rhodopsin: a prototypical G protein-coupled receptor. Prog Nucleic Acid Res Mol Biol 59: 1

Wald G (1968) The molecular basis of visual excitation. Science 162: 230

Magnification: x 8,200; x 65,000 (inset)

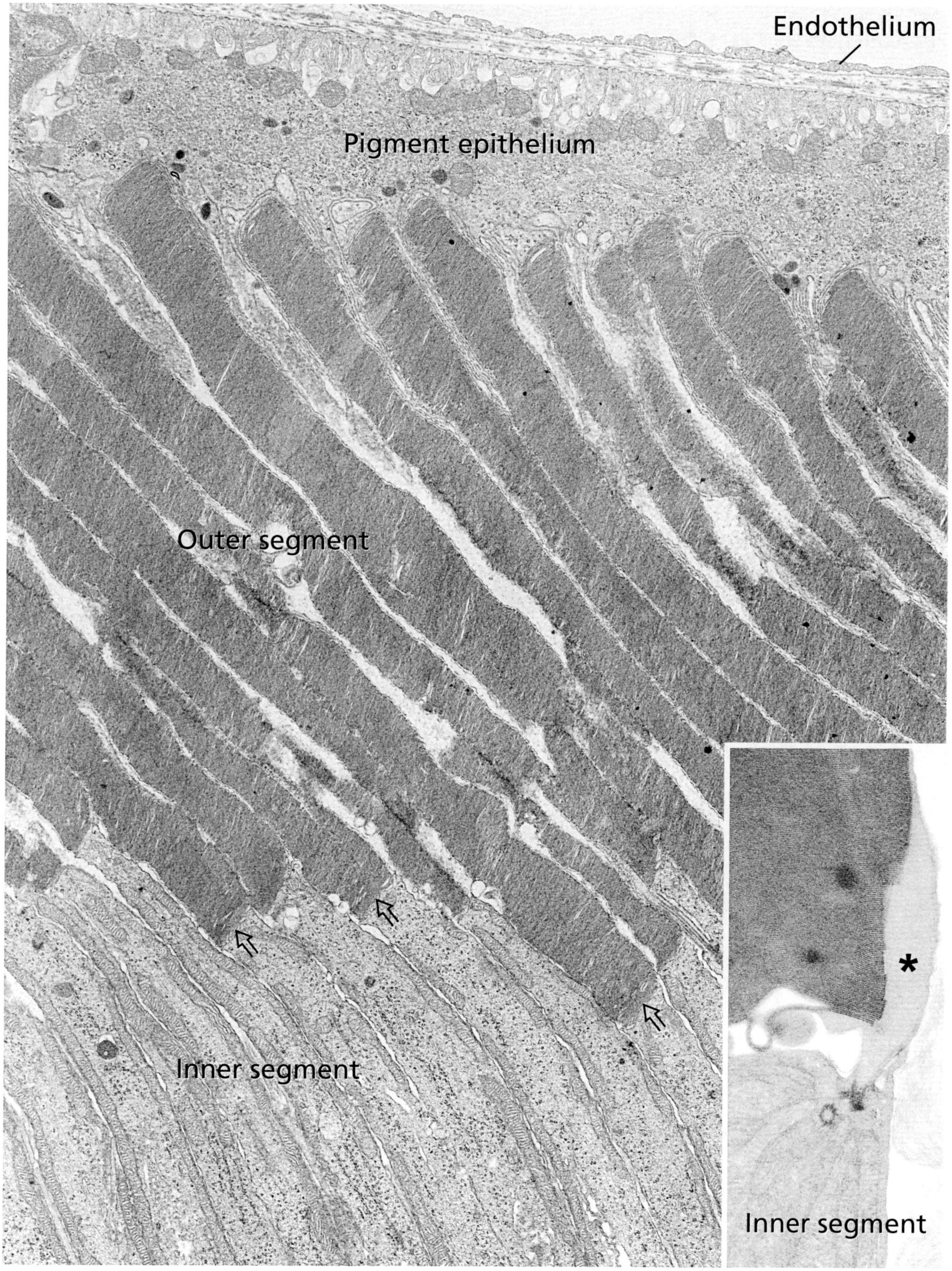
Endothelium
Pigment epithelium
Outer segment
Inner segment
*
Inner segment

PHOTORECEPTOR CELLS OF THE RETINA: LIGHT-INDUCED APOPTOSIS

Photoreceptor cells are highly polarised and specialised cells unable to undergo mitotic division. However, they show the remarkable phenomenon that part of it, the light sensitive rod and cone outer segments, are continuously renewed whereas the remainder of the cells is relatively stable. In panel A, the fine structural organisation of rod outer segments is shown. It consists of parallel arranged, densely packed membrane discs that in rods are separated from the plasma membrane (inset in panel A). The space between the apical parts of the outer segments is filled with processes of the pigment epithelium.

Metabolically pulse-labelled proteins detected by histoautoradiography could be followed in an orderly fashion from the Golgi apparatus of the inner segment into newly forming discs at the base of the outer segment and from there to the tip. Tips containing discs finally shed, are phagocytosed and degraded by the pigment epithelium. The pigment epithelium continuously degrades large amounts of photoreceptor discs and is involved in the recycling of their components. The renewal of the rod outer segments is regulated and follows a circadian rhythm. Related to their differential visual function, phagocytosis of rod discs commences shortly after the onset of light whereas phagocytosis of cone discs occurs at the onset of darkness.

The phagocytic activity of the pigment epithelium is also of importance when rapidly changing illumination conditions require adaptation and under conditions which result in photoreceptor damage and apoptosis. Not only in relation to the circadian rhythm but also in abruptly increasing light intensity, the light sensitivity of rods requires downregulation. This occurs at least in part through changes in the pattern of disc shedding from the outer segment tips and their phagocytosis by pigment epithelium. In addition, the rhodopsin levels are adapted to low and high light levels, respectively.

Panel B depicts the enormous structural alterations occurring in rod outer segments after acute bright light exposure (up to 1000 lx for 2 hours) of albino rat eye. At higher light intensities, apoptotic photoreceptor death occurs. White light and certain wavelengths of the visible spectrum, namely blue light, preferentially induce photoreceptor apoptosis in vertebrates. In mice, the availability of rhodopsin during light exposure is one determinant for the sensitivity to light-induced apoptosis. With some delay to photoreceptor cells, apoptosis of pigment epithelium takes place. Collectively these data indicate that retinal degeneration can result from prolonged exposure to intense light.

In experimental settings with mice, evidence for two different apoptotic pathways in light-induced retinal degeneration was obtained. The pathway triggered by bright light is independent of transducin-mediated signalling but requires activation of rhodopsin. The other one induced by low-intensity light is depending on transducin-mediated signalling.

References

Hafezi F, Marti A, Munz K, and Reme CE (1997) Light-induced apoptosis: differential timing in the retina and pigment epithelium. Exp Eye Res 64: 963

Hao W, Wenzel A, Obin MS, Chen CK, Brill E, Krasnoperova NV, Eversole-Cire P, Kleyner Y, Taylor A, Simon MI, et al. (2002) Evidence for two apoptotic pathways in light-induced retinal degeneration. Nat Genet 32: 254

La Vail M (1976) Rod outer segment disc shedding in the rat retina: relationship to cyclic lighting. Science 194: 1071

Marmor M and Wolferberger T (1998) The retinal pigment epithelium. Function and disease. New York Oxford: Oxford University Press

Remé CE, Grimm C, Hafezi F, Wenzel A, and Williams TP (2000) Apoptosis in the retina: The silent death of vision. News Physiol Sci 15: 120

Schremser J and Williams T (1995a) Rod outer segment (ROS) renewal as a mechanism for adaptation to a new intensity environment. II. Rhodopsin synthesis and packing density. Exp Eye Res 61: 25

Schremser J and Williams T (1995b) Rod outer segment renewal as a mechanism for adaptation to a new intensity environment, I: Rhodopsin levels and ROS length. Exp Eye Res 61: 17

Young R (1976) Visual cells and the concept of renewal. Invest Ophthalmol 15: 700

Magnification: x 35,000(A); x 73,000 (inset); x 36,000 (B)

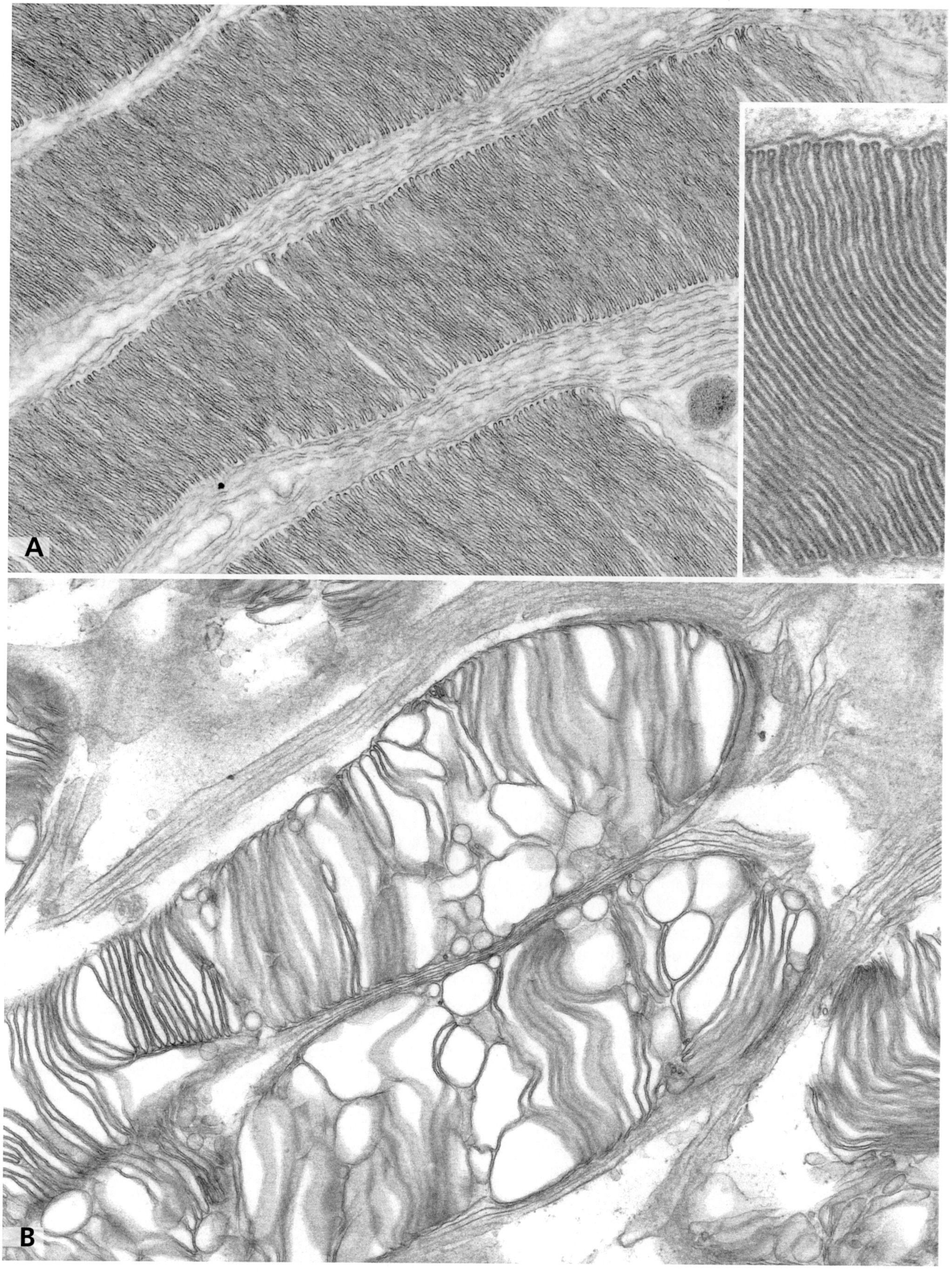
A
B

OLFACTORY EPITHELIUM

The olfactory epithelium localised in the olfactory area of the nasal cavities represents the sensory organ for the highly sensible odorant reception. The epithelium bears the bipolar sensory neurons, which are embedded in surrounding supporting cells. The neuronal dendrites form the characteristic knob-like sensory endings on the luminal side of the epithelium; the basal cell areas give rise to the axonal projections (neuritic processes), which leave the epithelium, become surrounded by extensions of olfactory myelin-producing cells, and form the fila olfactoria, which pierce the cribriform plate of the ethmoid bone and approach special synapse regions at the olfactory bulb. Basal cells, immature neurons and supporting cells are found in the basal and medial regions of the epithelium. The basal cells are mitotically active; their daughter cells differentiate into neurons during adult life. Two different types of microvilli-bearing epithelial cells exist, which in part are seen in connection with sensory properties and in part are comparable with brush cells (cf. Fig. 79), and are involved in transport processes responsible for an optimal ionic milieu at the epithelial surface.

The olfactory epithelium is similarly constructed in all vertebrates. Panels A and B show parts of the olfactory epithelium of the mouse as visualised by electron microscopy of thin sections, in panel C a freeze fracture preparation is on display. In the apical part of the olfactory epithelium shown in panel A, the knob-like endings of the cilia-bearing dendrites of three neurons are particularly prominent. By means of complexes of occluding and adhering junctions, the neurons are connected with the surrounding supporting cells (SC). In the most apical cytoplasmic area of the supporting cells, the light zone of the terminal network can be clearly distinguished from the cytoplasm lying beneath, in which abundant mitochondria and densely packed membranes of the endoplasmic reticulum are predominant. In panel C, arrowheads indicate the ridges of tight junctions that form the broad band of the occluding zone (cf. Fig. 84). The knob-like endings of the dendrites extend beyond the surface of the epithelium. They bear 10–20 modified cilia, in the plasma membrane of which the receptor molecules are localised. The dendritic knobs are surrounded by abundant microvilli of the supporting cells, which form a layer at the surface of the epithelium. Both the dendritic knobs of the neurons and the microvilli are floating in the fluids of the secretions of the Bowman's glands, in which odoriferous substances are dissolved. The olfactory glands of Bowman are situated in the subepithelial connective tissue and end in short ducts in the epithelium. In addition to protective molecules and immunoglobulin A, the secretions of the olfactory gland contain the odorant-binding protein with a strong binding affinity for numerous odorant molecules. With the aid of this protein, odorants are carried to the receptors in the membrane of the cilia and again removed after they have been sensed.

The basal bodies (kinetosomes) of the cilia are prominent structures in the cytoplasm of the knoblike dendrite endings (panels A and B); in the freeze-fracture preparation shown in panel C, the lace-like plasma membrane particles of the cilia, comparable to those of cilia in the ciliated epithelium of the respiratory tract, are clearly visible; the circular areas from where the long, whip-like cilia arise are labelled by arrows. In the cilia, the 9+2 arrangement of microtubules becomes irregular in the distal segments of the cilia; here, microtubule doublets are missing, as is visible in the cross-sectioned cilia in panel B.

References

Asan E and Drenckhahn D (2005) Immunocytochemical characterization of two types of microvillar cells in rodent olfactory epithelium. Histochem Cell Biol 123: 157

Barnea G, O'Donnell S, Mancia F, Sun X, Nemes A, Mendelsohn M, and Axel R (2004) Odorant receptors on axon termini in the brain. Science 304: 1468

Buck L and Axel R (1991) a novel multigene family may encode odorant receptors: a molecular basis for odor recognition. Cell 65: 175

Firestein S (2001) How the olfactory system makes sense of scents. Nature 413: 211

Kerjaschki D and Hörandner H (1976) The development of mouse olfactory vesicles and their contacts: A freeze-etching study. J Ultrastr Res 54: 420

Mombaerts P (2001) How smell develops. Nat Neurosci Suppl 4: 1192

Magnification: x 16,500 (A); x 39,000 (B); x 35,000 (C)

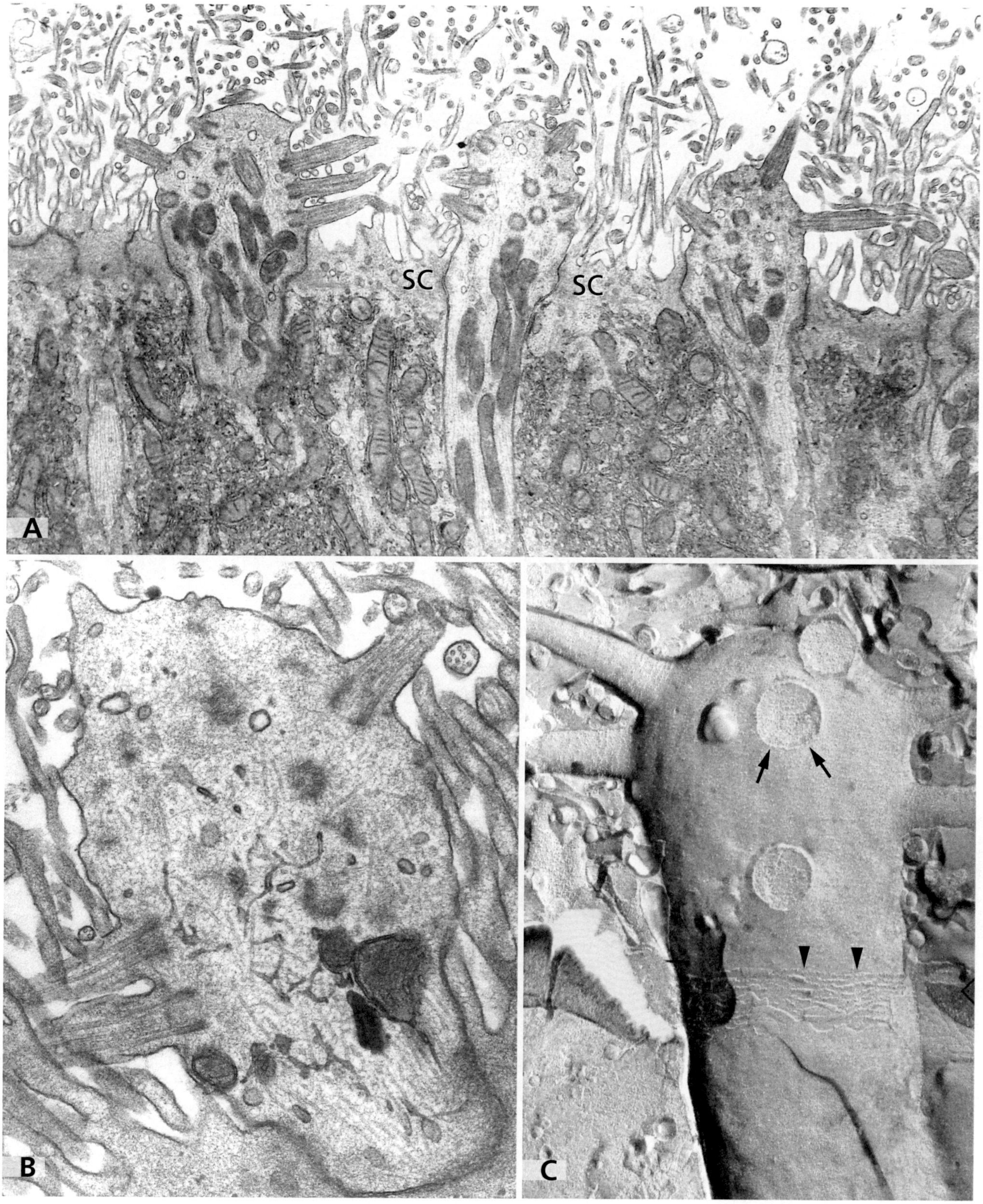
SC
SC
A
B
C

CORNEAL EPITHELIUM

The stratified epithelium of the cornea is part of the corneoscleral coat that forms the outer tunic of the eyeball, protecting its inner structures and, together with the pressure of the intraocular fluid, maintaining the eye's shape and consistency. The anterior surface of the cornea is built up by the corneal epithelium shown in this figure.

The corneal epithelium exhibits all characteristics of a simple stratified squamous epithelium of five to seven layers of cells. In the most basal layer, high polygonal cells are aligned along the basal lamina. They are anchored to the basal lamina and to the adjacent Bowman's layer by hemidesmosomes (cf. Fig. 141). The Bowman's layer represents the anterior part of the corneal stroma and is visible in the most basal section of the figure. Basal cells are mitotically active and replace the differentiated cells of the upper layers. The corneal epithelium has a remarkable wound healing capacity. Cytokeratin intermediate filaments are associated with multiple desmosomes that attach the cells to one another. In the basal layers, cells are connected by long bridges that span the wide intercellular spaces.

The cells change shapes during differentiation, migration, and transport to the upper layers, increasingly becoming flat. The squamous superficial cells in the outermost layer are particularly rich in cytokeratin filaments. They protect the cells lying beneath from the external environment. On their apical domains, short microvilli are present. They are in contact with a protective film of tear, by which the surface of the corneal epithelium is continually kept wet. Mucins in the tear film lubricate the epithelial surface during the blinking of the eyelid. They stabilise the tear film, preventing desiccation of the underlying epithelial cells, and form a barrier to penetration of pathogens.

A barrier leading to impermeability of the corneal epithelium to water soluble substances is formed by tight junctions between the superficial cells. Tight junctions are established during differentiation from the basal to the superficial cells and may regenerate within one hour after abrasion of the superficial cells.

For maintenance of corneal transparency, a precise regulation of water content is necessary (cf. Fig. 140). Recent data provide evidence that the water transporting proteins aquaporin 5 and 1, expressed in the corneal epithelium and in the corneal endothelium, respectively, provide main routes for water transport across the epithelial and endothelial barriers of the cornea.

References

Argüeso P and Gipson IK (2001) Epithelial mucins of the ocular surface: Structure, biosynthesis and function. Exp Eye Res 73: 281

Daniels JT, Dart JKG, Tuft SJ, and Khaw PT (2001) Corneal stem cells in review. Wound Rep Reg 9: 483

Kinoshita S, Adachi W, Sotozono C, Nishida K, Yokoi N, Quantock AJ and Okubo K (2001) Characteristics of the human ocular surface epithelium. Progr Ret Eye Res 20: 639

Verkman AS (2003) Role of aquaporin water channels in eye function. Exp Eye Res 76: 137

Magnification: x 4,300

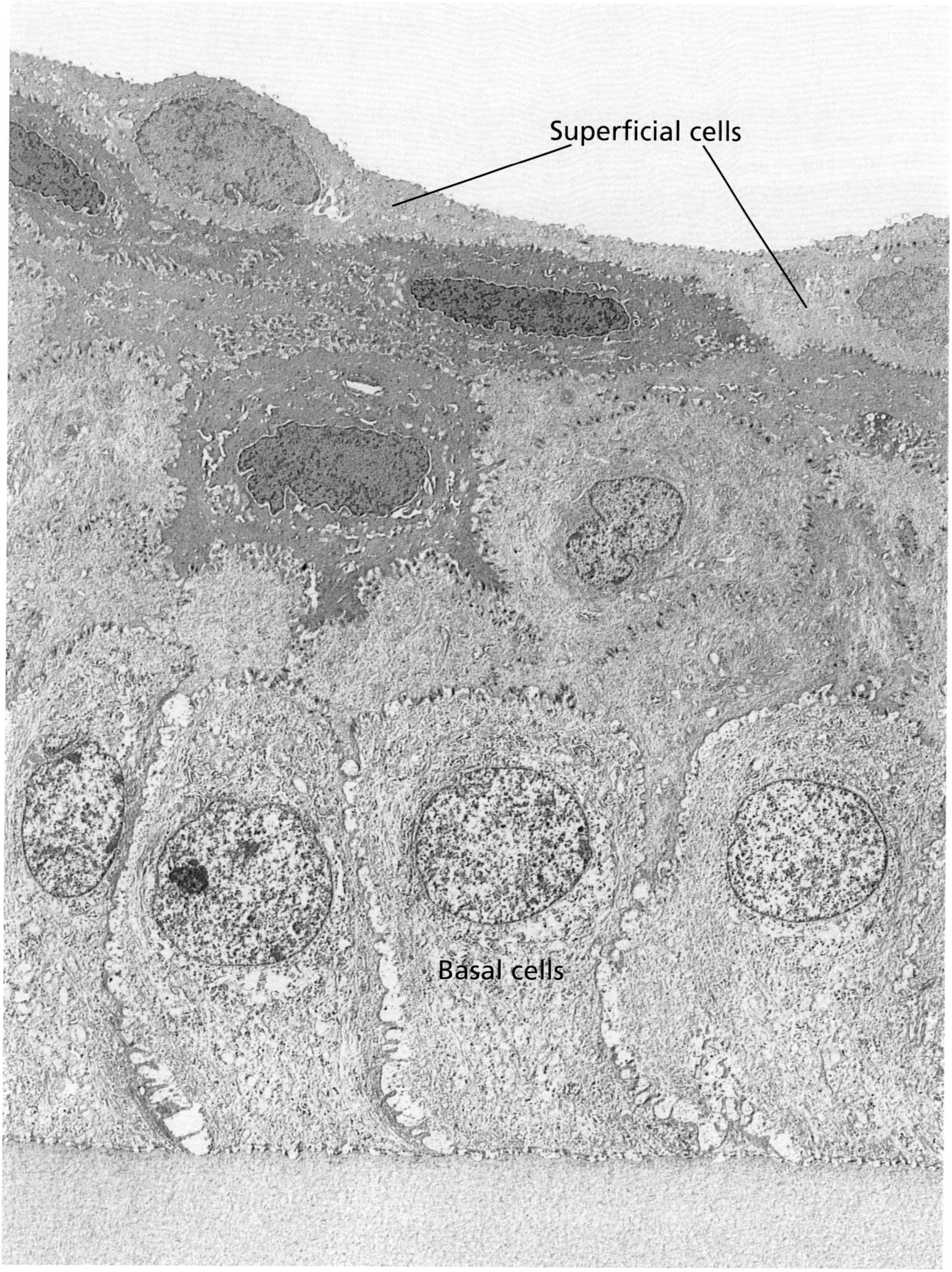
Superficial cells
Basal cells

EPIDERMIS

The stratified squamous keratinised epithelium of the epidermis, which forms the outermost layer of the skin, protects the body against various external influences, such as mechanical stress, radiation, microbial penetration, and exsiccation. The epidermis is self-renewing and differentiation of the cells from the innermost germinative layer, where cells undergo mitosis, to the surface is connected with keratinisation of the epithelium and the construction of a fluid barrier that is essential for terrestrial life form. The barrier consists of three main components, the terminally differentiated cells with their cell envelope, extracellular lipid and tight junctions.

Most epidermal cells belong to the class of keratinocytes, which are arranged in four layers (strata) reflecting the mechanical and barrier functions of the epithelium and the process of keratinisation. All four layers of the epidermis are shown in the figure. The epidermis covers the connective tissue of the dermis, which is apparent in the lowermost part of the figure.

The polygonal cells of the basal layer (stratum basale), resting on the basal lamina, contain plenty of cytokeratin intermediate filaments of the keratin types 5 and 14, in the low magnification micrographs appearing as dense fibrils (tonofibrils, cf. also Figs. 75 and 93). They are associated with desmosomes connecting adjacent cells and with hemidesmosomes anchoring the basal cell domains to the basement membrane, which forms the bordering layer between dermis and epidermis (cf. Fig. 93).

In the subsequent spinous layer (stratum spinosum), tonofibrils are particularly prominent. They protrude into fine cell processes bearing desmosomes. The processes connect the neighbouring cells, and like bridges span the dilated intercellular spaces. In the light microscope, these intercellular bridges appear as the "spines" leading to this layer's designation.

Both cells of the basal and of the spinous layers contain multiple melanin granules (cf. Fig. 93) appearing as particularly dense dots within the cytoplasm (arrows). Melanin is produced in melanocytes (M) by oxidation of tyrosine to 3,4-dihydroxyphenylalanin by the enzyme tyrosinase primarily stored in premelanosomes and subsequently transformed to melanin accumulating in melanosomes. Melanocytes originate in the neural crest and migrate into the epidermis. They settle in the basal layer and develop multiple cell processes extending between the keratinocytes. Melanosomes are transported into the melanocyte cytoplasmic extensions. They are transferred to the keratinocytes by uptake of the tips of the melanocyte extensions, a process termed cytocrine secretion. One melanocyte contacts approximately 36 keratinocytes, forming a functional unit: the epidermal melanin unit. In the micrograph, one melanocyte (M) is apparent in the central part of the basal layer. Multiple melanocyte processes are visible between the keratinocytes in Fig. 93.

The third layer is designated the granular layer (stratum granulosum) after the prominent keratohyalin granules occurring in the cytoplasm of the keratinocytes. At the transition from the spinous to the granular layer, the cells become flattened. They still are interconnected by multiple desmosomes. In addition, tight junctions are established in the granular layer. Together with the extracellular lipid layers formed by the secretions of the keratinocytes in the granular layer, tight junctions are important parts of the epidermal permeability barrier preventing exsiccation of the body.

The horny layer (stratum corneum) in its lower parts consists of transforming keratinocytes. It represents a particular stratum lucidum in the thick epidermis covering the palms of the hands and the soles of the feet. The outermost part of the horny layer is made up of terminally differentiated keratinocytes lacking cell organelles and nuclei. They form flattened squames with a highly resistant compound envelope (cf. Fig. 120).

References

Doering T, Holleran WM, Potratz A, Vielhaber A, Elias PM, Suzuki K, and Sandhoff K (1999) Sphingolipid activator proteins are required for epidermal permeability barrier formation. J Biol Chem 274: 11038

Furuse M, Hata M, Furuse K, Yoshida Y, Haratake A, Sugitani Y, Noda T, Kubo A, and Tsukita S (2002) Claudin-based tight junctions are crucial for the mammalian epidermal barrier: a lesson from claudin-1-deficient mice. J Cell Biol 156: 1099

Tsuruta D, Green KJ, Getsios S, and Jones JCR (2002) The barrier function of the skin: how to keep a tight lid on water loss. Trends Cell Biol 12: 355

Magnification: x 6,100

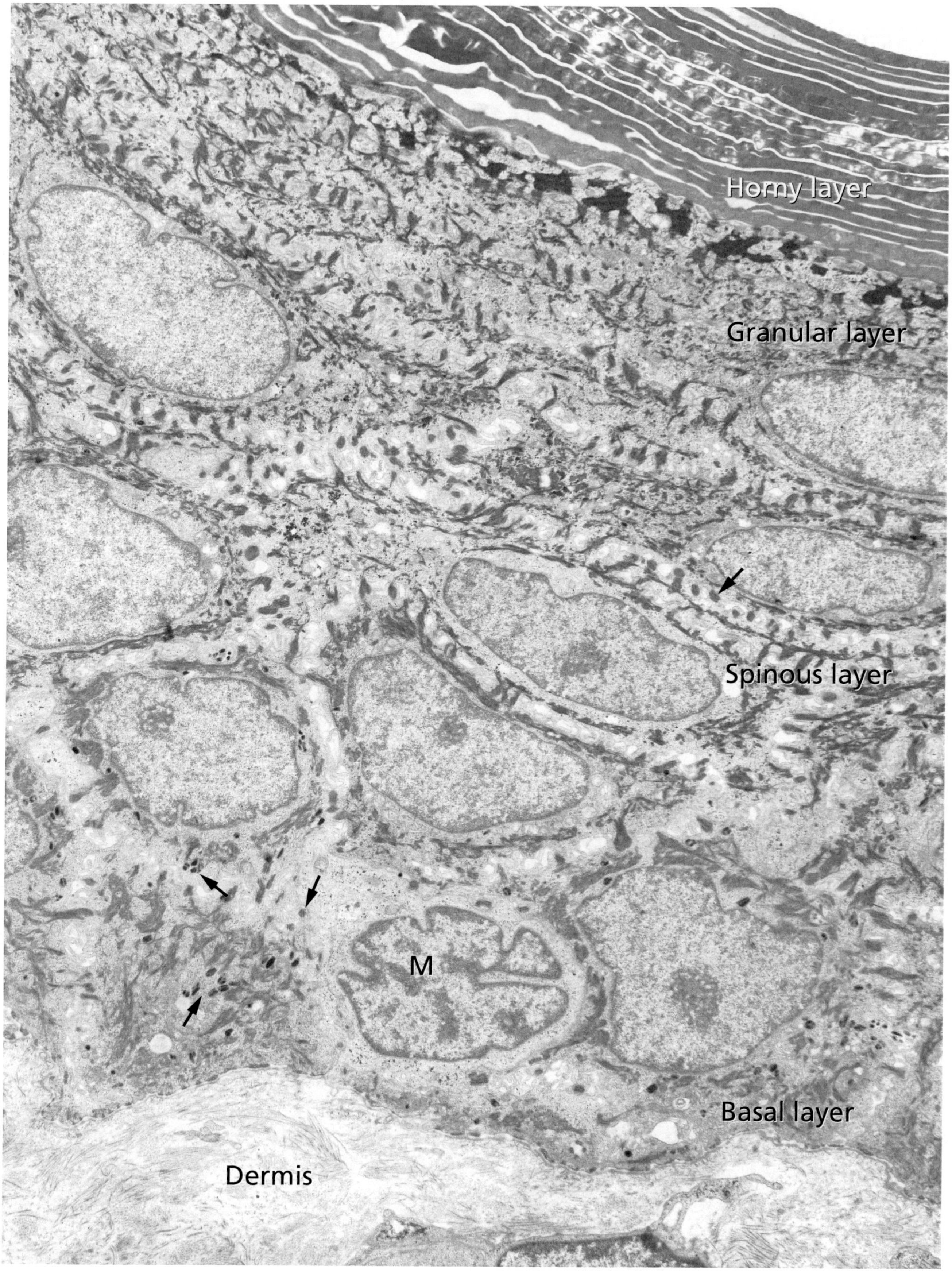
Horny layer
Granular layer
Spinous layer
M
Basal layer
Dermis

DIFFERENTIATION OF KERATINOCYTES AND FORMATION OF THE EPIDERMAL FLUID BARRIER

Keratinocytes undergo a specific programme of differentiation on their passage through the various epithelial layers from the basal regions of the epidermis to the outer surface. Regulated biochemical events result in characteristic cell transformations that finally lead to the keratinocytes' death. The dead cells embedded and sealed in their lipid secretions form the insoluble and fluid impermeable superficial horny layer of the epidermis.

Panel A shows keratinocytes of the spinous layer containing thick bundles of tonofilaments (tonofibrils) which extend into the spinous-like cell bridges and attach to the plaques of the spot desmosomes (maculae adhaerentes; cf. Fig. 86) connecting the neighbouring cells. The keratin types 5 and 14 present in the basal keratinocytes are replaced by keratins 1 and 10 and finally keratins 2e and 9 when the cells move from the basal over the spinous to the granular layer. Two desmosomes are shown at higher magnification in the inset. Both the dense cytoplasmic plaques containing desmoplakin and plakoglobin proteins and the intercellular dense materials composed of the glycosylated extracellular domains of the interlocking cadherins, such as desmoglein1 and desmocollin3 are well visible. Tonofilaments are anchored to the dense plaques. Some of the events leading to the characteristic cell changes during keratinisation start in the spinous layer.

Panel B shows part of the uppermost granular layer and part of the horny layer. In the granular layer, the keratohyalin granules (KG) make up the most prominent structures within the keratinocytes. They contain the protein filaggrin and are associated with keratin tonofilaments (arrowheads), which are anchored to desmosomes (arrows). In a stepwise process, the keratinocytes of the granular layer differentiate into the cornified flat squames of the horny layer, apparent in the upper half of panel B. Concomitantly, cell organelles and nuclei are degraded. A cell envelope consisting of a coat underneath the plasma membrane (arrowheads in panel C), composed of keratin-fillagrin complexes, involukrin and the plakin proteins envoplakin and periplakin forms. It becomes further reinforced by insertion of additional structural proteins such as loricrin and small prolin-rich proteins. Desmosomes seem to nucleate the initial processes leading to the formation of this scaffold underneath the plasma membrane. During this process, cells become permeable and by Ca^{2+}-influx, transglutaminases are activated which irreversibly cross-link the scaffold proteins resulting in the formation a continuous layer at the inner surface of the plasma membrane (arrowheads panel C). Numerous lamellar bodies (Odland bodies; arrows in panel C) are apparent in the cytoplasm of keratinocytes. They start to occur in the spinous layer. Their lamellar product, the glycoplipid acylglucosylceramide, is secreted into the intercellular space (asterisks in panel C) and covalently linked to involucrin. The cell envelope together with the multilamellar extracellular lipid constitutes the compound cell envelope. Under the electron microscope, the extracellular both lucid and multilamellar material and the dense compact cell surface correspond to the two parts of the compound cell envelope (arrowheads and asterisks in panel B).

Changes of the keratin-filaggrin complexes still occur after the cells have lost their transcriptional ability. This is mirrored by the ultrastructural changes during progression from the inner to the outer horny layers, as shown in the upper part of panel B. The outermost cells are continuously shedd. Cell desquamation is connected with a progressive loss of desmosomes of which remnants can be seen in the horny layer.

References

Doering T, Holleran WM, Potratz A, Vielhaber A, Elias PM, Suzuki K, and Sandhoff K (1999) Sphingolipid activator proteins are required for epidermal permeability barrier formation. J Biol Chem 274: 11038

Furuse M, Hata M, Furuse K, Yoshida Y, Haratake A, Sugitani Y, Noda T, Kubo A, and Tsukita S (2002) Claudin-based tight junctions are crucial for the mammalian epidermal barrier: a lesson from claudin-1-deficient mice. J Cell Biol 156: 1099

Segre J (2003) Complex redundancy to build a simple epidermal permeability barrier. Curr Opin Cell Biol 15: 778

Magnification: x 10,000 (A); x 21,000 (B); x 60,000 (C); x 74,000 (inset)

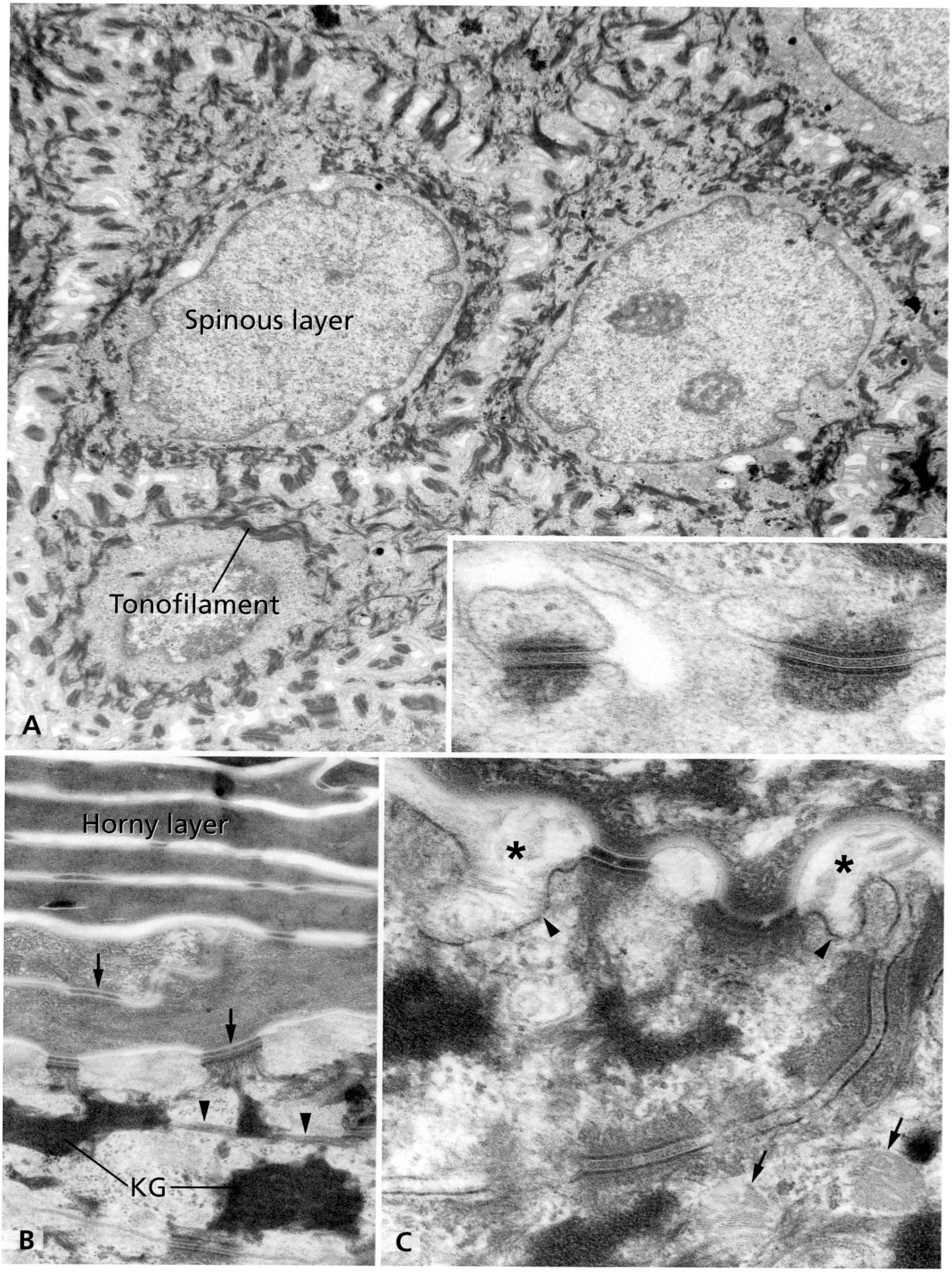
Spinous layer
Tonofilament
A
Horny layer
KG
B
*
*
C

THE TRACHEO-BRONCHIAL EPITHELIUM

The upper part of the respiratory tract functions as an air-conducting system. Here, warming and moistening of the inhaled air and the removal of inhaled particles occurs. It is lined by the pseudostratified respiratory epithelium, which consists of four cell types. (1) The columnar ciliated cells, which possess cilia that project from their apical plasma membrane and function in the autonomous, coordinated movement of luminal content. (2) The goblet cells (MC in A), unicellular glands synthesizing and secreting mucus, which forms part of the mucus blanket. The mucus blanket functions as a lubricant and a protective layer. (3) A small number of neuroendocrine cells (also referred to as Kultschinsky cells), dispersed in the epithelium. (4) The basal cells, which function as stem cells for the regeneration of the respiratory epithelium.

The cilia are motile cytoplasmic extensions of the ciliated cells and 4–6 μm long (panel A). Their core structure is the axoneme (AX in panel B), a microtubule-based structure, surrounded by the plasma membrane (panels B, C, D, and diagram). The axoneme is composed of one central microtubule pair (C in diagram) and nine peripheral arranged microtubule doublets, the so-called 9+2 arrangement. The peripheral microtubule doublets consist of one complete microtubule (A-microtubule composed of 13 protofilaments), which is attached to an incomplete microtubule (B-microtubule composed of 10 protofilaments). Toward the tips of the cilia, the axoneme consist of 9 single peripheral microtubules and a central pair (panel D).

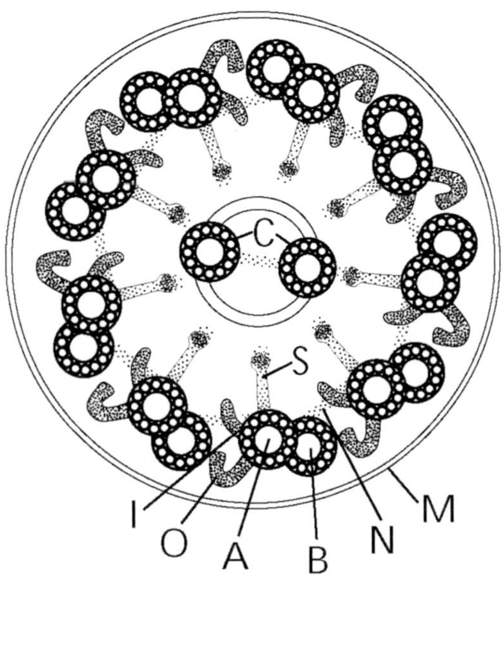

The 9+2 arrangement is joined and stabilised by accessory proteins. Radial spokes (S in diagram) extend from the A-microtubules and insert toward the inner sheet, which surrounds the central microtubule pair. They participate in the conversion of the sliding into a bending movement. The peripheral microtubule pairs are connected to each other by nexin filaments (N in diagram), which are important for the maintenance of the axoneme structure during sliding. From the A-microtubules, pairs of dynein arms (O and I in the diagram) extend. Dyneins are microtubule-associated ATPases and instrumental in microtubule sliding. Upon ATP hydrolysis by dyneins, the outer doublets slide relative to each other, and this results in bending of the cilia. Thus, sliding and bending is the basis for ciliary movement. Cilia move with a rapid effective forward stroke and a slow whip-like recovery stroke. All cilia beat in the same direction, slightly out of phase, and transport the mucus layer with entrapped particles toward the pharynx. In the absence of the outer dynein arms, slow ciliar beating is possible.

Axonemata are anchored to basal bodies (BB in panel B), which consist of nine microtubule triplets (panel E). Basal bodies are modified centrioles and act as growth templates for the outer microtubule pairs. At the distal ends of basal bodies, so-called roots can be recognised (R in panel B). Between the cilia, short, often branched, microvilli-like extensions of the plasma membrane can be observed (A, MV in panel B).

References

Gerdes JM, Davis EE, and Katsanis N (2009) The vertebrate primary cilium in development, homeostasis, and disease. Cell 137: 32

Goodenough UW and Heuser JE (1985) Substructure of inner dynein arms, radial spokes, and the central pair/projection complex of cilia and flagella. J Cell Biol 100: 2008

Pazour GJ, Agrin N, Leszyk J, and Witman GB (2005) Proteomic analysis of a eukaryotic cilium. J Cell Biol 170: 103

Salathe M (2007) Regulation of mammalian ciliary beating. Annu Rev Physiol 69: 401

Satir P and Christensen ST (2007) Overview of structure and function of mammalian cilia. Annu Rev Physiol 69: 377

Satir P and Christensen S (2008) Structure and function of mammalian cilia. Histochem Cell Biol 129: 687

Sleigh MA, Blake JR, and Liron N (1988) The propulsion of mucus by cilia. Am Rev Respir Dis 137: 726

Magnification: x 6,000 (A); x 25,000 (B); x 60,500 (C); x 68,000 (D); x 94,000 (E)

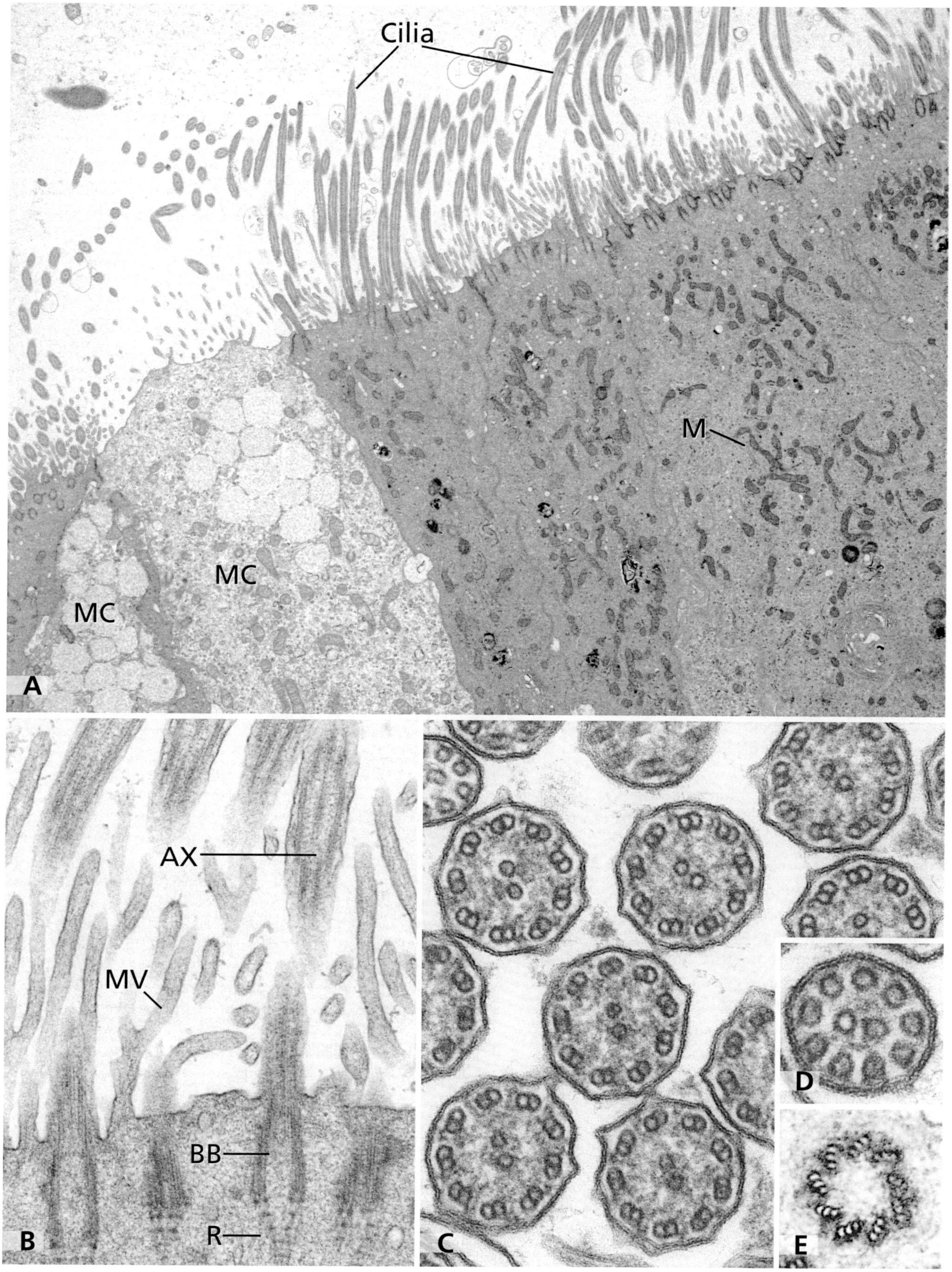
Cilia
M
MC
MC
A
AX
MV
BB
R
B
C
D
E

CILIARY PATHOLOGY: IMMOTILE CILIA SYNDROME AND KARTAGENER SYNDROME

Functional impairment or lack of the ciliary machinery of the respiratory epithelium results in serious chronic airway disease. Owing to lacking, deficient, or uncoordinated ciliar motility, mucociliary clearance is affected. This results in accumulation of mucus and particles, including bacteria and viruses. This condition is known as the immotile cilia syndrome or ciliary dyskinesia and includes the Kartagener syndrome.

The immotile cilia syndrome is a heterogeneous disease, and several different ultrastructural alterations of the cilia have been documented. However, the relation between functional impairment and altered ultrastructure of cilia is not always clear. In extreme cases, only isolated cilia (arrow in panel A) or no cilia at all may be found, as illustrated in panel B. Only microvilli-like structures exist (MV in panels A and B). Sometimes, remnants of basal body-like structures can be observed (arrowheads in panel A). More commonly observed are the following defects of cilia: (a) absence or reduction in number of outer and inner dynein arms, (b) outer or inner dynein arm deficiency, (c) shortening of outer dynein arms and absence of inner dynein arms (panel C), (d) spoke defects such as complete absence or missing spoke heads, (e) peripheral microtubule abnormalities with single microtubules (panel D), (f) missing pair of central microtubuli with transposition of a peripheral one (upper axoneme in panel E), (g) supernumerary pair of central microtubuli (upper axoneme in panel F), (h) or missing single central microtubulus (upper axoneme in panel G). It should be noted that cilia with normal ultrastructure may be immotile or dysmotile. Usually the correlation between ciliary and spermatozoal motility is good, and the same defects can be observed in ciliated cells and spermatozoa of the same patient.

The Kartagener syndrome is a subgroup of the immotile cilia syndrome and due to dynein deficiency. Clinically it is characterised by an situs inversus, bronchiectasis and chronic sinusitis. The relation between the immotile cilia syndrome and situs inversus is not entirely clear. However, it has been proposed that functional impairment of monocilia in early embryonic development is causative. Monocilia on embryonic cells, specifically the primitive node, rotate rapidly in a clockwise direction and are assumed to be involved in the establishment of the left-right asymmetry of the body. Due to immotile or dysmotile monocilia, there is a 50% chance for situs inversus.

The immotile cilia syndrome is an autosomal recessively inherited trait. Because many axonemal proteins are potentially involved, mutations in different genes must exist. Mutations in the gene coding for an intermediate-chain dynein located at chromosome 9p13-p21 have been shown to be associated with the absence of outer dynein arms.

References

Badano JL, Mitsuma N, Beales PL, and Katsanis N (2006) The ciliopathies: an emerging class of human genetic disorders. Annu Rev Genomics Hum Genet 7: 125

Carlen B and Stenram U (2005) Primary ciliary dyskinesia: a review. Ultrastruct Pathol 29: 217

Eggenschwiler JT and Anderson KV (2007) Cilia and developmental signaling. Annu Rev Cell Dev Biol 23: 345

Gerdes JM, Davis EE, and Katsanis N (2009) The vertebrate primary cilium in development, homeostasis, and disease. Cell 137: 32

Hirokawa N, Tanaka Y, Okada Y, and Takeda S (2006) Nodal flow and the generation of left-right asymmetry. Cell 125: 33

Jorissen M and Cassiman J (1991) Relevance of the ciliary ultrastructure in primary and secondary ciliary dyskinesia: a review. Am J Rhinol 5: 91

Pennarun G, Escudier E, Chapelin C, Bridoux AM, et al. (1999) Loss-of-function mutations in a human gene related to Chlamydomonas reinhardtii dynein IC78 result in primary ciliary dyskinesia. Am J Hum Genet 65: 1508

Schneeberger EE, McCormack J, Issenberg HJ, Schuster SR, and Gerald PS (1980) Heterogeneity of ciliary morphology in the immotile-cilia syndrome in man. J Ultrastruct Res 73: 34

Supp DM, Potter SS, and Brueckner M (2000) Molecular motors: the driving force behind mammalian left-right development. Trends Cell Biol 10: 41

Zariwala MA, Knowles MR, and Omran H (2007) Genetic defects in ciliary structure and function. Annu Rev Physiol 69: 423

Magnification: x 13,000 (A, B); x 121,000 (C); x 125,000 (D); x 110,000 (E); x 101,000 (F); x 67,000 (G)

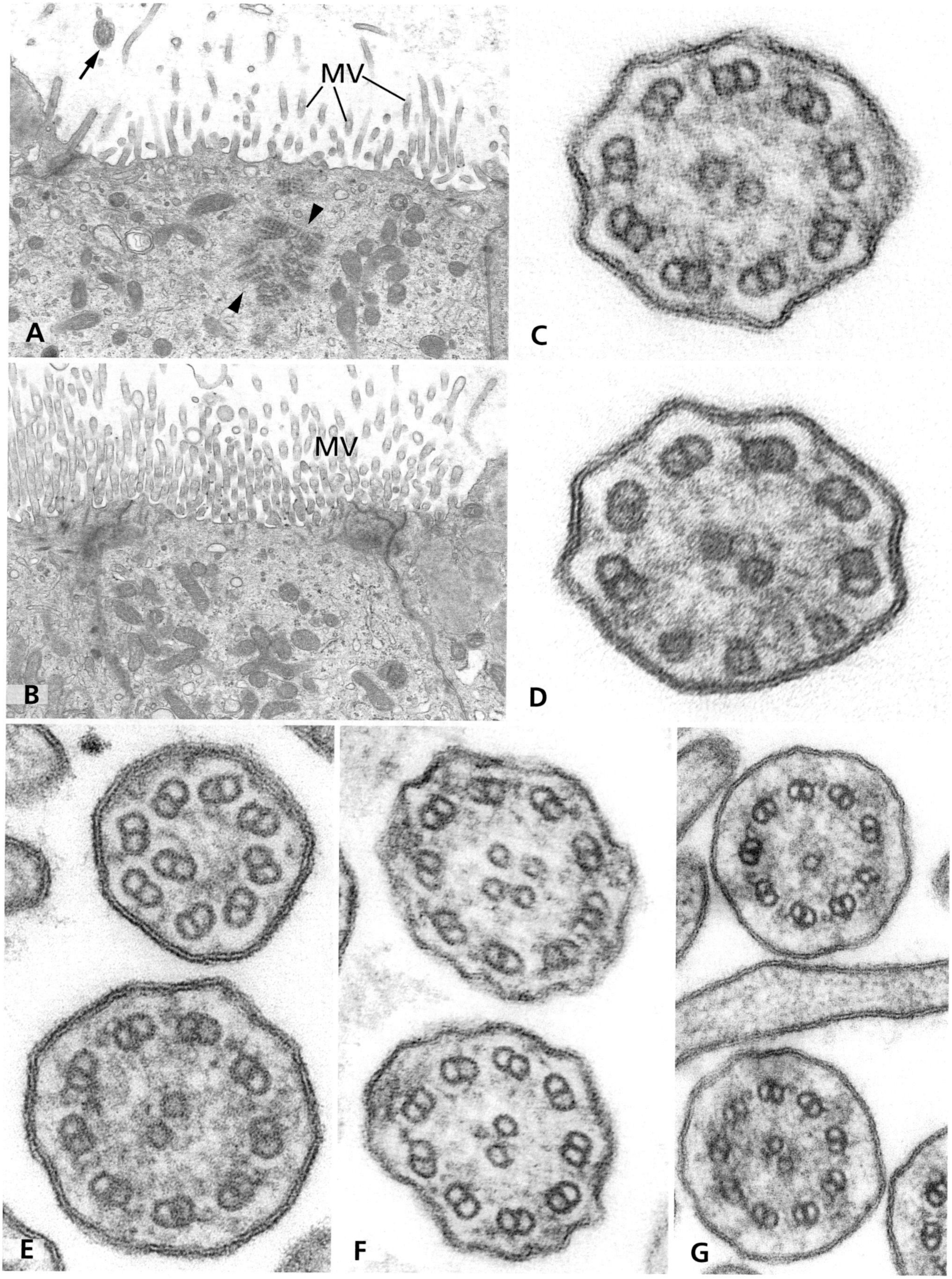
MV
A
MV
B
C
D
E
F
G

ALVEOLI: GAS EXCHANGE AND HOST DEFENSE

The alveoli as the most distal part of the respiratory tract function primarily in gas exchange. They provide a surface of approximately 75 m^2 in humans for oxygen and carbon dioxide exchange between air and blood, which occurs by passive diffusion. The gas exchanging part of the alveolar septum consists of (1) a thin walled continuous capillary, (2) a dual basal lamina synthesized by endothelia and type I alveolar epithelial cells, (3) thin cytoplasmic extensions of type I alveolar epithelial cells, and (4) red blood cells (RBC) (panels A and B). The flat type I alveolar epithelial cells represent about 40% of the alveolar epithelial cell population and cover some 90% of the alveolar surface. The cuboidal type II alveolar epithelial cells (panel C), the second alveolar epithelial cell type, cover only 10% of the alveolar surface and are found predominantly in the alveolar niches. They are indirectly involved in gas exchange by producing the pulmonary surfactant, which consists of phospholipids, cholesterol, and specific surfactant proteins. The active surfactant phospholipids, mainly dipalmitoylphosphatidylcholine, reduce the alveolar surface tension and prevent their collapse. Type II alveolar epithelial cells have secretory granules, the multilamellar bodies (MLB in C and D), which contain the pulmonary surfactant. Surfactant is secreted in the alveolar lumen (panel D) and spreads as a monolayer of phospholipids (arrows in D) over the aqueous hypophase. In the hypophase, secreted surfactant phospholipids are stored as tubular myelin (asterisk in D). The surfactant proteins B and C have most important roles in the organization of the multilamellar bodies and the tubular myelin, and in the formation of a stable, functional extracellular surfactant monolayer. Inherited deficiency of surfactant protein B results in lethal neonatal respiratory distress syndrome.

The surfactant proteins A and D are involved primarily in host defense, although surfactant protein A additionally functions in the organization and metabolism of the surfactant phospholipids. They belong to the collectin protein family, which possess a C-type lectin domain at their C-terminus that binds oligosaccharides found on the surface of bacteria (e.g., group B *Streptococcus, Pseudomonas aeruginosa*), viruses (e.g. respiratory syncytial virus, *Haemophilus influenzae*), fungi and mycobacteria. These pathogens become opsonized by the surfactant proteins A and D, which facilitates their phagocytosis by pulmonary macrophages and monocytes. However, surfactant proteins A and D also have direct antimicrobial effects and inhibit the growth of Gram negative bacteria.

Type II alveolar epithelial cells are the stem cells of the alveolar epithelium and thus progenitor of type I alveolar epithelial cells, and involved in repair after lung injury. Both type I and II alveolar epithelial cells function in ion and water transport and therefore are important for the homoestasis of the surfactant hypophase.

References

Beers MF and Mulugeta S (2005) Surfactant protein C biosynthesis and its emerging role in conformational lung disease. Annu Rev Physiol 67: 663

Crouch E and Wright JR (2001) Surfactant proteins A and D and pulmonary host defense. Annu Rev Physiol 63: 521

Dietl P and Haller T (2005) Exocytosis of lung surfactant: from the secretory vesicle to the air-liquid interface. Annu Rev Physiol 67: 595

Eaton DC, Helms MN, Koval M, Bao HF, and Jain L (2009) The contribution of epithelial sodium channels to alveolar function in health and disease. Annu Rev Physiol 71: 403

Fehrenbach H (2001) Alveolar epithelial type II cell: defender of the alveolus revisited. Respir Res 2: 33

Hawgood S and Poulain FR (2001) The pulmonary collectins and surfactant metabolism. Annu Rev Physiol 63: 495

Hermans C and Bernard A (1999) Lung epithelium-specific proteins: characteristics and potential applications as markers. Am J Respir Crit Care Med 159: 646

Mason RJ and Crystal RG (1998) Pulmonary cell biology. Am J Respir Crit Care Med 157: S72

Kuronuma K, Sano H, Kato K, Kudo K, Hyakushima N, Yokota S, Takahashi H, Fujii N, Suzuki H, Kodama T, et al. (2004) Pulmonary surfactant protein A augments the phagocytosis of Streptococcus pneumoniae by alveolar macrophages through a casein kinase 2-dependent increase of cell surface localization of scavenger receptor A. J Biol Chem 279: 21421

Nogee L, Garnier G, Dietz H, Singer L, and Murphy A (1994) A mutation in the surfactant protein B gene responsible for fatal neonatal respiratory in multiple kindreds. J Clin Invest 93: 1860

Weaver T and Conkright J (2001) Functions of surfactant proteins B and C. Annu Rev Physiol 65: 555

Williams M (2003) Alveolar type I cells: molecular phenotype and development. Annu Rev Physiol 65: 669

Wu H, Kuzmenko A, Wan S, Schaffer L, Weiss A, Fisher J, Kim K, and McCormack F (2003) Surfactant proteins A and D inhibit the growth of Gram-negative bacteria by increasing membrane permeability. J Clin Invest 111: 1589

Magnification: x 2,200 (A); x 48,000 (B); x 10,200 (C); x 36,000 (D)

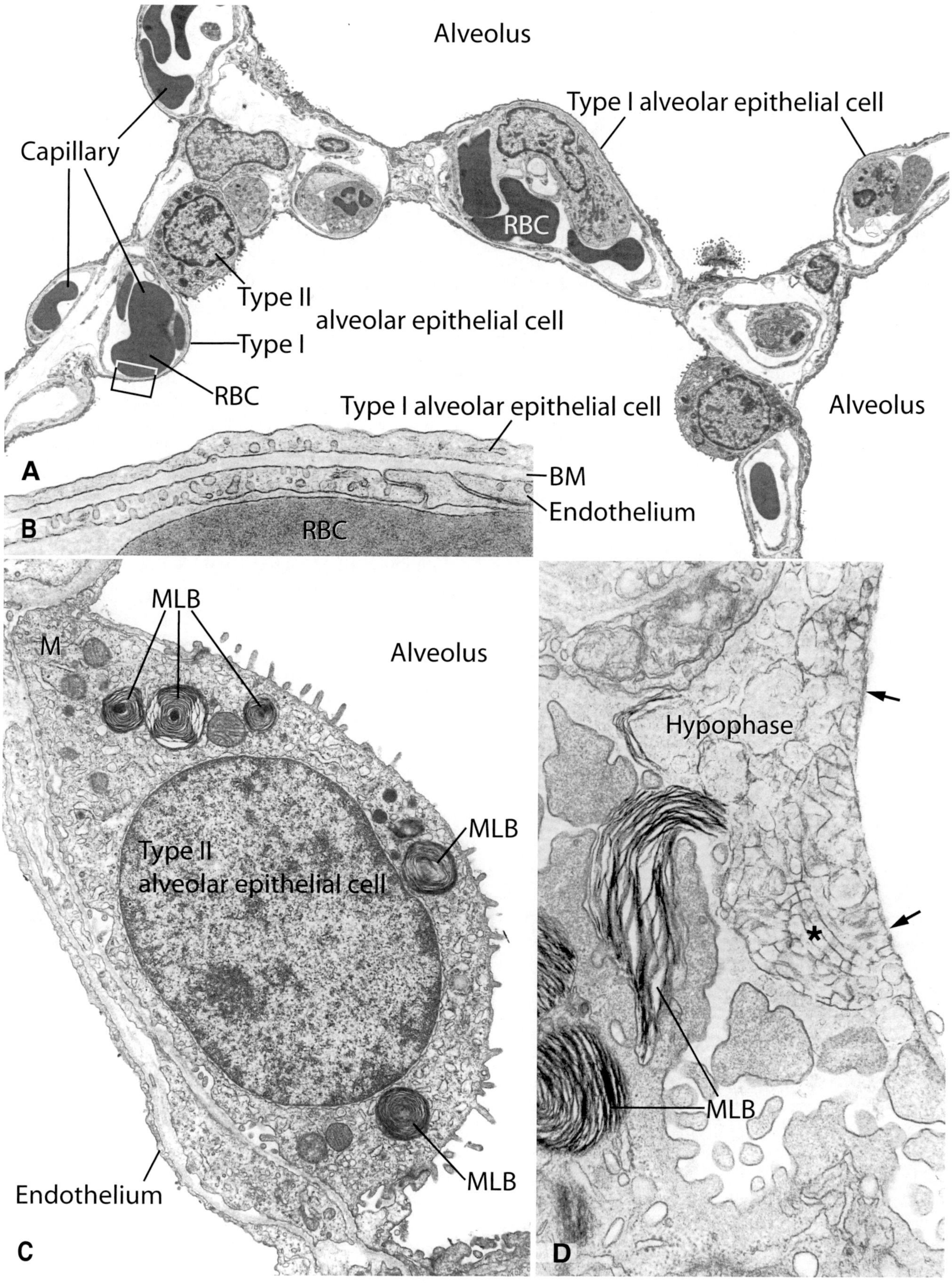
Alveolus
Type I alveolar epithelial cell
Capillary
RBC
Type II
alveolar epithelial cell
Type I
RBC
Type I alveolar epithelial cell
Alveolus
A
BM
Endothelium
B
RBC
MLB
M
Alveolus
MLB
Type II
alveolar epithelial cell
MLB
Endothelium
C
Hypophase
*
MLB
D

UMBRELLA CELL – SURFACE SPECIALISATIONS

The wall of the urinary passages including the surfaces of the renal pelvis, the ureters, the urinary bladder, and proximal parts of the urethra is covered by a unique specialised epithelium, the transitional epithelium or urothelium. The urothelium is stratified and composed of three types of cells: basal precursor cells, intermediate cells, and large superficial umbrella cells. The latter line the lumina of the organs and are responsible for the main specific urothelial functions. The urothelium enables the retention of urine and forms a barrier that makes it impermeable for water and movement of ions and metabolites. The various types of superficial cells in the distinctive parts of the urinary tract epithelium differ considerably in the expression of uroplakins, cytokeratins, and in the structure of the apical cell surface. In the bladder, the urothelium adapts to the cyclical changes of luminal contents and must maintain the permeability barrier under variations in pressure during filling and voiding. The permeability barrier ("blood-urine barrier") between tissue fluids and urine depends on the high-resistance tight junctions between the superficial cells, and on their unique apical plasma membrane visible under the electron microscope.

The survey electron micrograph of a segment of the mouse urothelium in panel A shows three cell layers containing basal cells in the lower part of the picture, neighboured by intermediate cells in the middle part of the micrograph, and an umbrella cell in superficial position. The wrinkly apical surface of the umbrella cell visible in the transmission electron micrograph in panel A is shown in scanning micrographs in panels B and C. The arrows in the rectangle in panel B label the borders between three neighbouring umbrella cells. The wrinkly character of the luminal surfaces of the umbrella cells, indicated by arrowheads in the circle, is achieved by the scalloped membrane formations shown at higher magnification in panel C. A particular area marked by a rectangle is further enlarged in the inset. Multiple ridges and microplicae are visible corresponding to the "hinge" regions of the scalloped formations, which are seen as small concave areas. The scalloped membrane formations are covered almost entirely with plaques consisting of two-dimensional crystals of hexagonally packed 16 nm particles composed of uroplakins (UP), a family of at least five proteins that include the tetraspan proteins UPIa and UPIb, as well as the type I single-span proteins UPII, UPIIIa, and UPIIIb. The formation of correct heterodimers (UPIa/UPII, and UPIb/UPIII) is required for uroplakins to exit from the endoplasmic reticulum on their way to the cell surface. In the *trans* Golgi network, heterodimers interact to form heterotetramers (UPIa/UPII-UPIb/UPIII) before they leave the Golgi apparatus to form 16 nm particles. Morphologically, plaque areas are characterised by an asymmetric unit membrane (AUM), shown in the inset in panel A. Because of the particular shapes and locations of the uroplakin particles, the outer membrane leaflet seems about twice as thick as the inner leaflet. The plaques constitute a main part of the barrier system of the urothelium. In the centre of the scanning micrograph in panel C, the borders between three adjacent cells are visible as thin lines. Here, the cells are connected by complexes of tight and adhering junctions.

Inactivation of the uroplakin III gene in the mouse results in vesico-ureteral reflux (VUR), and urothelial abnormalities by uroplakin defects can lead to renal failure. The bladder urothelium of patients suffering from primary VUR is heterogeneous. Panel D shows the apical surface of immature superficial cells of the urothelium of a VUR patient exhibiting abundant microvilli but only a small number of ridges and microplicae.

References

Apodaca G (2004) The uroepithelium: Not just a passive barrier. Traffic 5: 117

Hu P, Meyers S, Liang FX, Deng FM, Kachar B, Zeidel M, and Sun TT (2002) Role of membrane proteins in permeability barrier function: uroplakin ablation elevates urothelial permeability. Am J Physiol 283: F1200

Jezernik K and Pipan N (1993) Blood-urine barrier formation in mouse urinary bladder development. Anat Rec 235: 533

Min G, Zhou G, Schapira M, Sun T-T, and Kong X-P (2003) Structural basis of urothelial permeability barrier function as revealed by cryo-EM studies of the 16 nm uroplakin particle. J Cell Sci 116: 4987

Romih R, Korosec P, Jezernik K, Sedmak B, Trisinar B, Deng FM, Liang FX, and Sun TT (2003) Inverse expression of uroplakins and inducible nitric oxide synthase in the urothelium of patients with bladder outlet obstruction. BJU Int 91: 507

Truschel ST, Ruiz WG, Shulman T, Pilewski J, Sun TT, Zeidel ML, and Apodaca G (1999) Primary uroepithelial cultures: a model system to analyse umbrella cell barrier function. J Biol Chem 274: 15020

Veranic P, Romih R, and Jezernik K (2004) What determines differentiation of urothelial umbrella cells? Eur J Cell Biol 83: 27

Magnification: x 3,500 (A), x 193,000 (inset), x 3,700 (B), x 5,200 (C), x 10,000 (inset), x 1,870 (D)

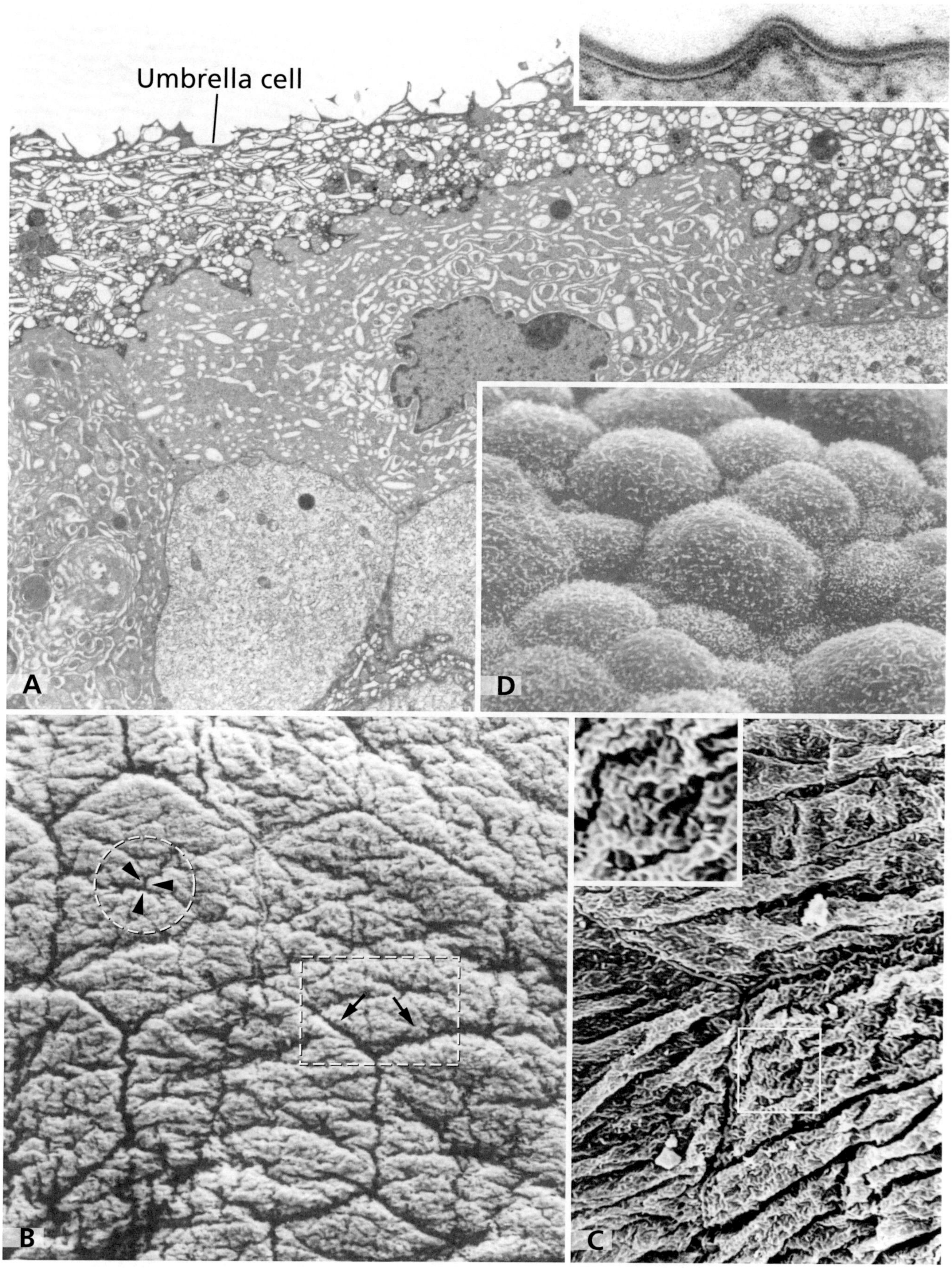
Umbrella cell
A
D
B
C

UMBRELLA CELL – FUSIFORM VESICLES

Plaques are not restricted to the apical plasma membrane of the umbrella cells (cf. Fig. 124) but also found in special membrane vesicles, termed "fusiform vesicles", which are abundant in the cytoplasm of these cells; they are assumed to have an important role in the formation of the umbrella cells' specialised asymmetric membrane areas. Fusiform vesicles are accumulated in the apical cytoplasm close to the apical surface. They may occur as single vesicles but are often arranged in groups or stacks closely associated with components of the cytoskeleton, which include a trajectorial network of cytokeratins, especially of cytokeratin 20, in the subapical cytoplasm of the cells. Panels A and B show fusiform vesicles in umbrella cells of the mouse urothelium in an ultrathin section and in a freeze fracture replica, respectively. The asymmetric membrane (AUM) of plaques is visible in the high magnification electron micrograph in the inset of panel B. As in the apical plasma membrane, the asymmetric membrane character is due to the presence of uroplakin particles. In the freeze fracture replica of panel B, exoplasmic fractured surfaces of fusiform vesicle plaques are shown containing arrays of densely packed uroplakin particles (arrowheads). Fusiform vesicles presumably are preformed in the Golgi apparatus of the umbrella cells, and transport uroplakins to the apical cell surface. In the highly differentiated umbrella cells, the uroplakin-positive membrane regions of the apical plasma membrane are excluded from internalisation, suggesting that uroplakin plaques hinder endocytosis.

AUM particles and plaques are not rigid but change their shapes, and it is assumed that interactions of the head domains may have a crucial role in determining shapes and sizes and allowing morphological alterations.

References

Kachar B, Liang F, Lins U, Ding M, Wu X-R, Stoffler D, Aebi U, and Sun T-T (1999) Three-dimensional analysis of the 16 nm urothelial plaque particle: Luminal surface exposure, preferential head-to head interaction, and hinge formation. J Mol Biol 285: 595

Kreft ME, Jezernik K, Kreft M, and Romih R (2009) Apical plasma membrane traffic in superficial cells of bladder urothelium. Ann NY Acad Sci 1152: 18

Min G, Zhou G, Schapira M, Sun T-T, and Kong X-P (2003) Structural basis of urothelial permeability barrier function as revealed by cryo-EM studies of the 16 nm uroplakin particle. J Cell Sci 116: 4087

Romih R, Korosec P, de Mello W Jr, and Jezernik K (2005) Differentiation of epithelial cells in the urinary tract. Cell Tiss Res 320: 259

Veranic P and Jezernik K (2002) Trajectorial organization of cytokeratins within the subapical region of umbrella cells. Cell Motil Cytoskeleton 53: 317

Magnification: x 40,500 (A), x 80,000 (B), x 194,000 (inset)

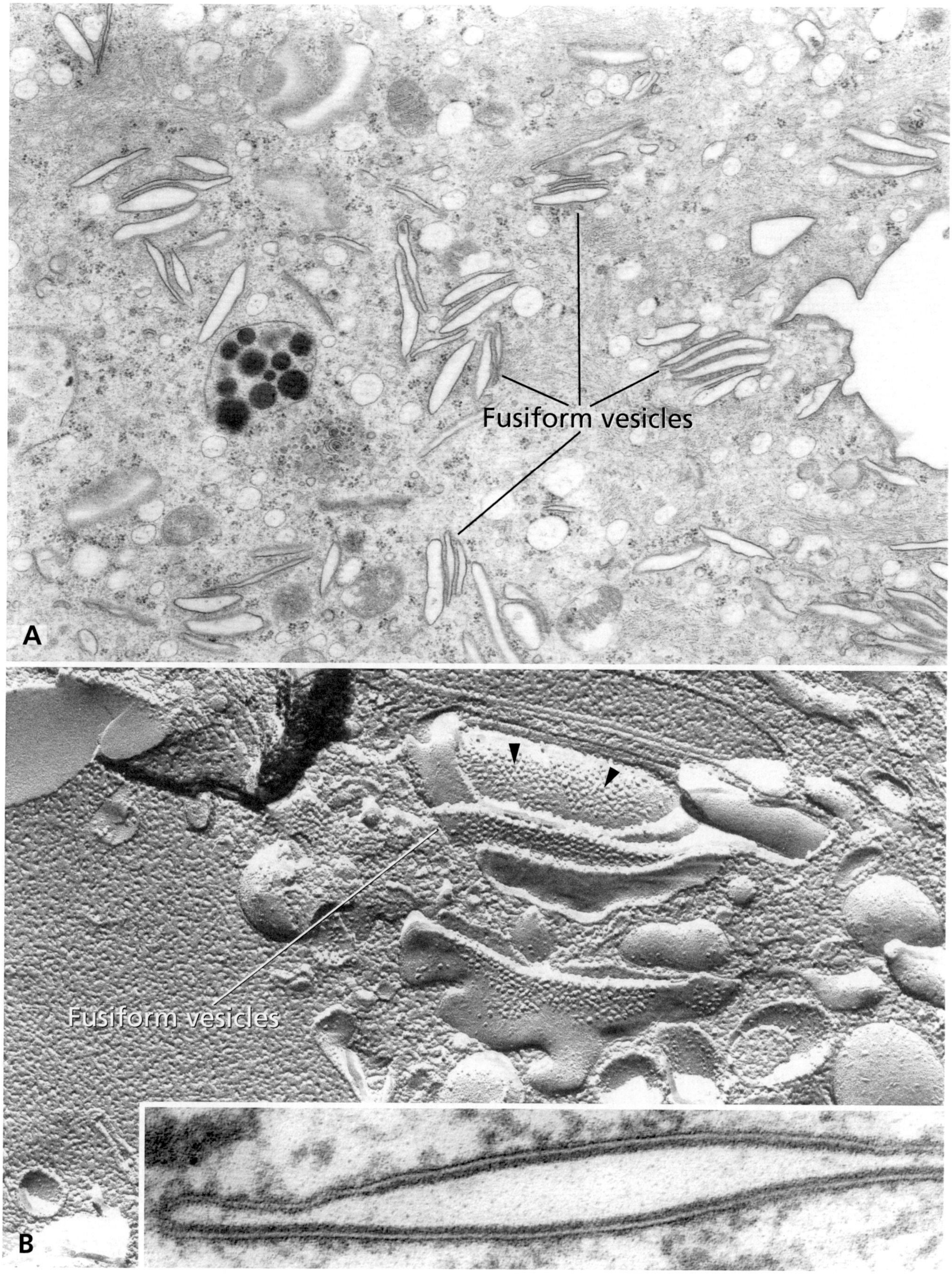
Fusiform vesicles
A
Fusiform vesicles
B

CONTINUOUS CAPILLARY, WEIBEL-PALADE BODIES

The inner lining of blood and lymphatic vessels is a simple, squamous epithelium, designated as "endothelium". The endothelium holds multiple tasks and is diversely developed according to the different functions in the various segments of the macro- and microvascular systems. It may be tight or more or less open for traffic between the lumen and the tissue outside in one or both directions, with specific transendothelial transport mechanisms existing in capillaries. Capillaries are exchange vessels and can be classified in three main types, continuous capillaries (panels A and B), discontinuous capillaries and fenestrated types (cf. Figs. 106 and 128, respectively).

Panel A presents a continuous capillary with the complete endothelium (E) and an embracing pericyte. Details of the endothelium are on display in panel B. The adjacent endothelial cells are overlapping. They are in contact and linked to each other (arrows in B) by both tight and adhering junctions. A barrier is formed, which impedes intercellular transport. In continuous capillaries, the multitude of transendothelial traffic including fluids, solutes, and macromolecules occurs by transcellular transport via small vesicles, which are transported in both directions from apical to basolateral cell surfaces and vice versa. In the endothelial cells shown in panels A and B, numerous caveolae are visible close to the cell surfaces, and small transport vesicles are abundant in the cytoplasm. There is evidence that the endothelial barrier function in part is regulated by the transcellular transport of albumin and other macromolecules via caveolae (caveolae – cf. Fig. 51). In microvascular endothelial cells, interactions between a 60-kDa endothelial cell surface albumin-binding protein and caveolin-1 have been identified being supposed to activate signalling pathways eventually leading to vesicle formation and transendothelial vesicle transport. Other proteins, such as transferrin, are transported via clathrin-dependent mechanisms (cf. Fig. 46). The endothelial cells form prominent basal processes, which in places accompany the capillary and build up a second endothelial layer between the basal lamina and the pericyte (panel A). The complete endothelial part of the capillary is surrounded by a continuous basal lamina (arrowheads in A and B), which also continuously covers the surface of the pericyte embracing the endothelium.

Within endothelial cells, a number of bioactive substances including hormones, adhesive molecules, and factors of the coagulation system, are stored and on disturbance are promptly delivered to the surface of the cells. By this mechanism, endothelial cells are able to change the microenvironment of perturbed regions and modulate and control inflammatory and haemostasis processes. The insets in panel B show endothelial cell-specific storage vesicles, the Weibel-Palade bodies, in cross section and longitudinal section (insets a and b, respectively). In both the typical tubular substructures are visible. Weibel-Palade bodies represent the storage compartments for von Willebrand factor, which has a pivotal role in controlling adhesion and aggregation of platelets at sites of vascular injury. Recently, new insights into the biogenesis of Weibel-Palade bodies have been provided by advanced cryotechnologies and three-dimensional analyses.

References

Baumgartner W, Hinterdorfer P, Schindler H, and Drenckhahn D (2000) The cadherin mediated cell-cell adhesion of vascular endothelium cells. Biophys J 78: 2630

Berriman JA, Li S, Hewlett LJ, Wasilewski S, Kiskin FN, Carter T, Hannah MJ, and Rosenthal PB (2009) Structural organization of Weibel-Palade bodies revealed by cryo-EM of vitrified endothelial cells. Proc Natl Acad Sci USA 106: 17407

Minshall RD, Tiruppathi C, Vogel SM, and Malik AB (2002) Vesicle formation and trafficking in endothelial cells and regulation of endothelial barrier function. Histochem Cell Biol 117: 105

Neumüller J, Neumüller-Guber S, Lipovac M, Mosgoeller W, Vetterlein M, Pavelka M, and Huber J (2006) Immunological and ultrastructural characterization of endothelial cell cultures differentiated from human cord blood derived endothelial progenitor cells. Histochem Cell Biol 126: 649

Reininger AJ (2008) Function of von Willebrand factor in haemostasis and thrombosis. Haemophilia 14: 11

Valentijn KM, Valentijn JA, Jansen KA, and Koster AJ (2008) A new look at Weibel-Palade body structure in endothelial cells using electron tomography. J Struct Biol 161: 447

Van Mourik JA, de Wit TR, and Voorberg J (2002) Biogenesis and exocytosis of Weibel-Palade bodies. Histochem Cell Biol 117: 113

Weibel ER and Palade GE (1964) New cytoplasmic components in arterial endothelia. J Cell Biol 23: 101

Zenner HL, Collinson LM, Micheaux G, and Cutler DF (2007) High-pressure freezing provides insights into Weibel-Palade body biogenesis. J Cell Sci 120: 2117

Magnification: x 16,000 (A), x 29,500 (B), x 65,000 (inset a), x 90,000 (inset b)

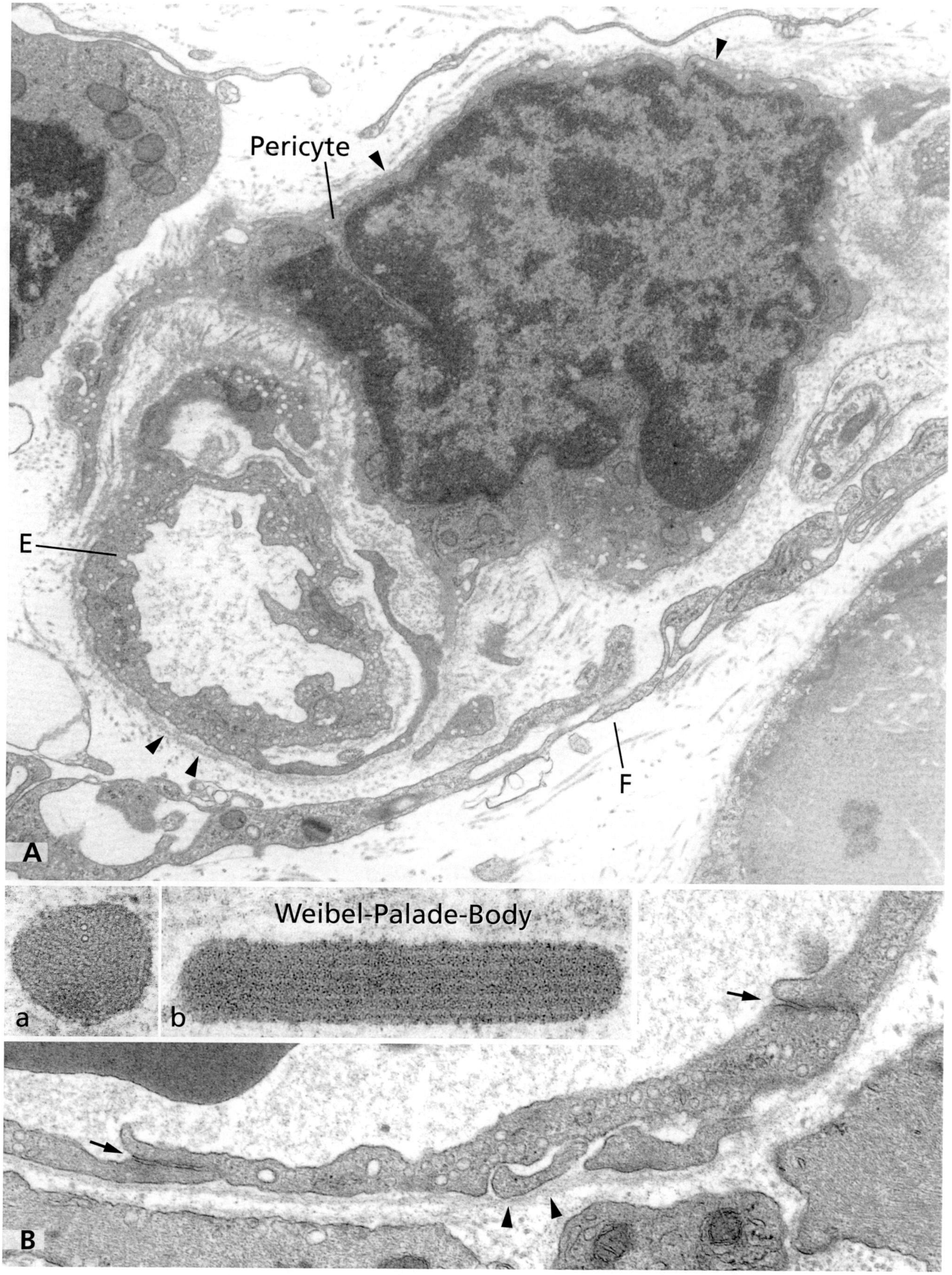
Pericyte
E
F
A
Weibel-Palade-Body
a
b
B

HYALINE ARTERIOLOSCLEROSIS

Hyaline arteriolosclerosis is a common pathological lesion of arterioles and results in a thickening of the vessel wall. The deposition of hyaline material begins as a focal process that eventually involves the entire circumference of the vessel and may extend into the media. The hyaline material is composed of precipitated plasma proteins, a major component being the inactive form of complement C3b. It therefore is a consequence of leakage of plasma proteins across the endothelium, which become trapped, and of synthesis of increased amounts of basement membrane components by the smooth muscle cells.

In the electron micrograph, which shows a cross-sectioned arteriole, the deposition of hyaline, an amorphous, slightly electron dense material (asterisks) is obvious. Hyaline deposits appear in several layers around the entire circumference of the vessel. The endothelium of the arteriole is indicated by arrows. Often, a narrowing of the vessel lumen can be observed, which results in ischemic atrophic changes such as loss of nephrons in the affected kidney. The vessel lumen in the electron micrograph shown, however, appears normal.

Hyaline arteriolosclerosis is observed in different settings such as aging, hypertension, diabetes mellitus and focal segmental glomerulosclerosis. In cerebral arterioles it is usually related to age and in renal arterioles to hypertension. It is more prevalent and more severe in association with diabetes mellitus, which is a general predisposing factor.

References

Gamble C (1986) The pathology of hyaline arteriosclerosis. Am J Pathol 122: 140
Hill G, Heudes D, and Bariety J (2003) Morphometric study of arterioles and glomeruli in aging kidney suggests focal loss of autoregulation. Kidney Int 63: 1027
Hotta O, Yoshizawa N, Oshima S, Takeuchi A, Kawamura O, Kondo S, Kubota T, and Niwa H (1987) Significance of renal hyaline arteriolosclerosis and tubulo-interstitial change in IgA glomerulonephropathy and focal glomerular sclerosis. Nephron 47: 262
Kumar V, Cotran, RS, and Robbins SL (2007) Basic pathology. 8th edn. Philadelphia, PA: Saunders
Moritz A and Oldt M (1937) Arteriolar sclerosis in hypertensive and nonhypertensive individuals. Am J Pathol 13: 679
Olson J (2003) Hyaline arteriolosclerosis: New meaning for an old lesion. Kidney Int 63: 1162

Magnification: x 4,800

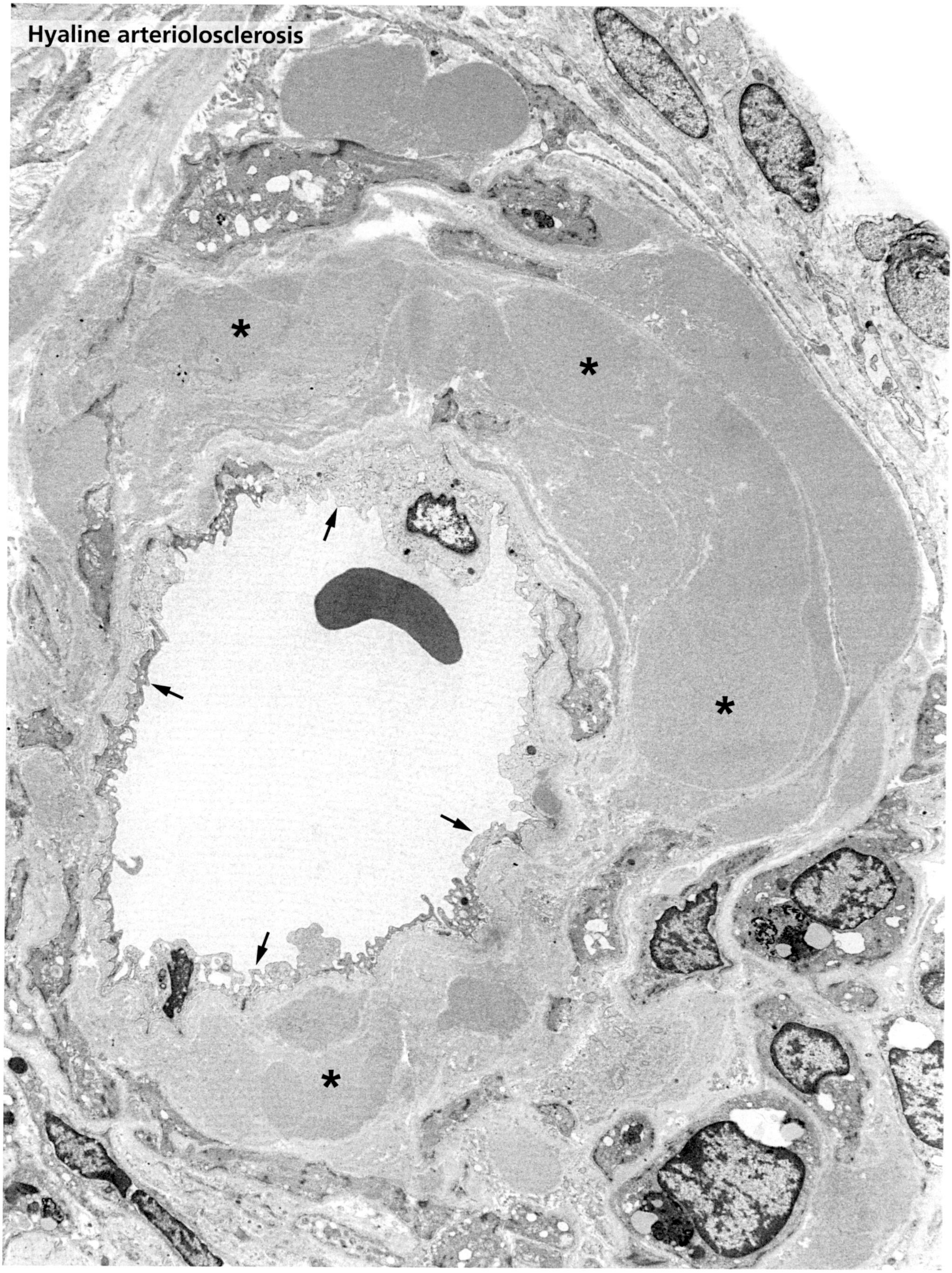
Hyaline arteriolosclerosis

FENESTRATED CAPILLARY

Fenestrated capillaries are characterised by the existence of pores within the endothelial cells and form specialised regions of the capillary bed in the mucosa of the intestinal tract, in the pancreas, in endocrine organs, in the choroid plexus and in the ciliary processes of the eye. The pores, however, are not free for passage of blood plasma, as is true for the endothelial pores and gaps in the liver sinus endothelium (cf. Fig. 106). The pores in the fenestrated capillaries like windows *(fenestrae)* are spanned by a diaphragm formed by radially oriented fibrils. They are homogenous in diameter (70–80 nm) and constitute specific transport sites within the endothelial cells, also referred to as filtration pores. The basal lamina is complete, which again contrasts to the discontinuous sinusoidal capillaries in the liver, spleen and bone marrow possessing only rudiments of basal laminae.

The micrograph of a fenestrated capillary of the rat pancreas in panel A shows that the endothelial cells are overlapping and linked to each other. Different cell domains are apparent. At the lefthand side, the endothelial cell shows abundant caveolae and transcytosis vesicles resembling the situation in continuous type capillaries (cf. Fig. 126). At the lower righthand side, the endothelium is extremely flat and typical *fenestrae* are inserted (arrowheads). The basal lamina is not interrupted and continuously covers fenestrated and non-fenestrated endothelial cell domains. Panels B and C show that the extremely flat fenestrated endothelial cell domains (arrowheads) reside side by side with higher parts of the endothelium characterised by the presence of multiple transcytosis vesicles. Caveolae show characteristic narrow neck regions, as they occur during their budding to form transport vesicles by fission from the plasma membrane. In panel C, an unmyelinated nerve fibre of the vegetative nerve system is visible in between the pancreatic secretory epithelium on the lefthand and the fenestrated endothelium on the righthand side. Abundant synaptic vesicles are accumulated in the axon (A), which is embedded in the cytoplasm of a Schwann cell.

References

Minshall RD, Tiruppathi C, Vogel SM, and Malik AB (2002) Vesicle formation and trafficking in endothelial cells and regulation of endothelial barrier function. Histochem Cell Biol 117: 105

Oh P, McIntosh DP, and Schnitzer JE (1998) Dynamin at the neck of caveolae mediates their budding to form transport vesicles by GTP-driven fission from the plasma membrane of endothelium. J Cell Biol 141: 101

Simionescu M and Simionescu N (1991) Endothelial transport of macromolecules: Transcytosis and Endocytosis. Cell Biol Rev 25: 1

Vogel SM, Easington CR, Minshall RD, Niles WD, Tiruppathi C, Hollenberg SM, Parrillo JE, and Malik AB (2001) Evidence of transcellular permeability pathway in microvessels. Microvasc Res 61: 87

Magnification: x 28,400 (A), x 55,600 (B), x 35,000 (C)

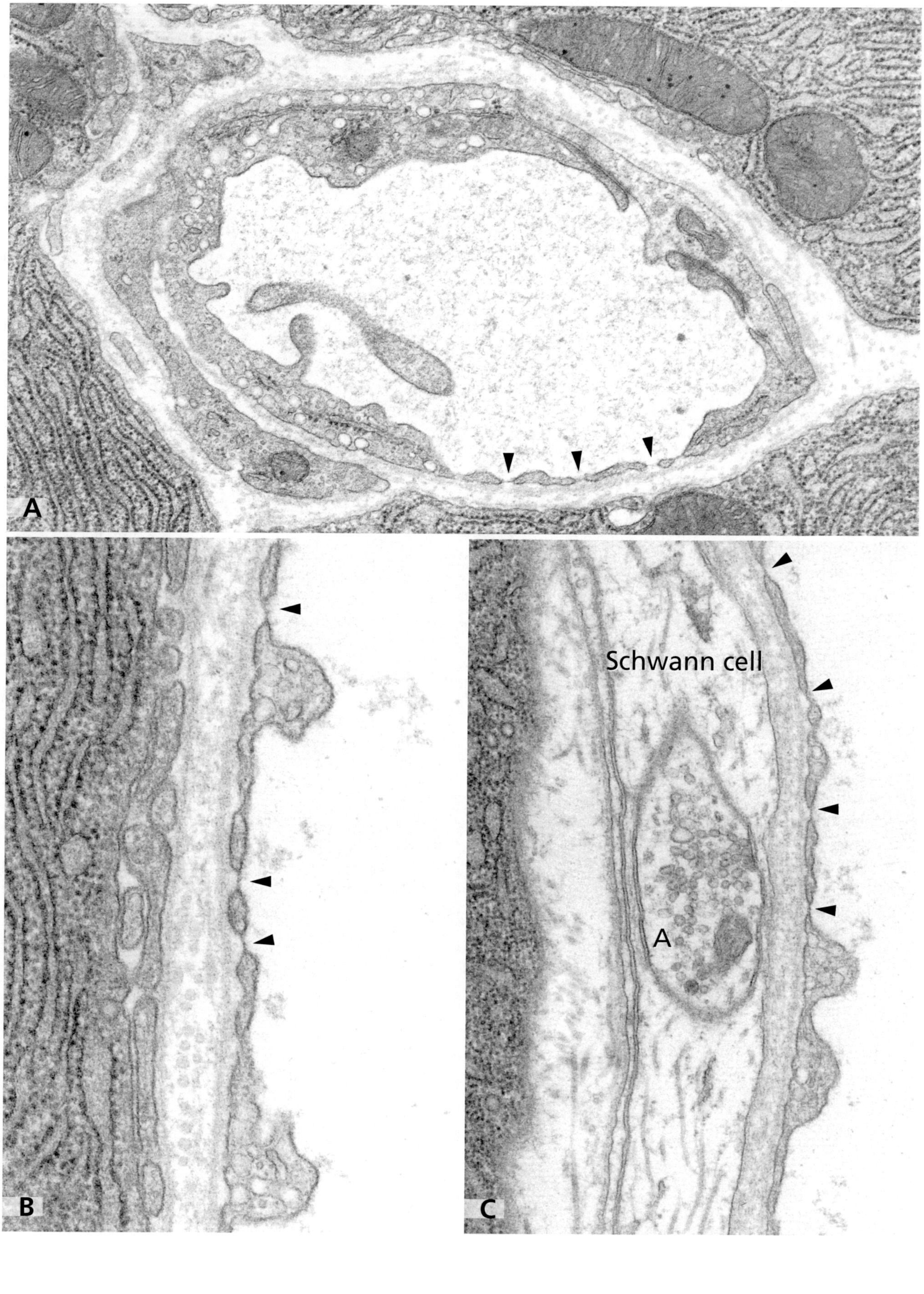
A
Schwann cell
A
B
C

ENDOTHELIO-PERICYTE AND ENDOTHELIO-SMOOTH MUSCLE CELL INTERACTIONS

The ultrastructures of endothelial cells of a capillary and a large artery shown in panels A and B, respectively, provide visual evidence that the vascular endothelium is not a simple inner lining of vascular tubes but is involved in multiple functions. All compartments of the biosynthetic apparatus and secretory system are well developed. Endothelial cells possess abundant ribosomes, extended rough and smooth endoplasmic reticulum, Golgi apparatus, storage granules and multiple mitochondria, indicating a high functional cell activity. Endothelial cells not only trigger blood coagulation (Weibel-Palade bodies, cf. Fig. 126) but also are involved in preventing intravascular clot formation. By secretion of smooth muscle cell relaxing and contraction factors, such as nitric acid and endothelin 1, respectively, endothelial cells modulate the activities of smooth muscle cells. They also are involved in the regulation of transendothelial migration of inflammatory cells and control vascular cell growth. Endothelial cells reorient in response to shear stress, a process connected with extensive remodelling of the actin cytoskeleton.

The uneven apical microarchitecture of the cell surfaces mirror the high dynamics connected with the multiple exocytosis and endocytosis events. Endothelial cells also show bizarre basal surfaces, reflecting their interactions with other components of the vascular wall. At the basal domains, endothelial cells form processes, which penetrate the basal lamina and interact with matrix components and neighbouring cells. In capillaries, feet-like protrusions of endothelial cells contact the surrounding pericytes (panel A). Particularly extended protrusions possibly form additional endothelial sheets (cf. Fig. 126). In the arterial wall shown in panel B, several prominent long, thin cell processes (arrows) penetrate the broad internal elastic lamina. They stick in basal lamina sheets, which cover the innermost smooth muscle cells and are continuous with the endothelial cells' basal lamina. Each smooth muscle cell is embedded in a basal lamina coat. The panel B micrograph also shows that the subendothelial layer is thin and the space between the endothelium and smooth muscle cells is mainly occupied by the internal elastic lamina. In the wall of vessels, precursors of elastic fibres (cf. Fig. 138) are smooth muscle cell products. Arrowheads label a spur-like dense area in the innermost muscle cell (cf. also Fig. 155).

Endothelio-matrix interactions are crucial for the vessel wall organisation and new formation of vascular tubes during both vasculogenesis in the embryo and angiogenesis from pre-existing vessels, as occurs during wound healing, chronic inflammation, development of collateral circulation and vascularisation of tumours.

References

Brooke BS, Karnik SK, and Li DY (2003) Extracellular matrix in vascular morphogenesis and disease: structure versus signal. Trends Cell Biol 13: 51

Drenckhahn D and Ness W (1997) The endothelial contractile cytoskeleton. In: Vascular endothelium. Physiology, pathology and therapeutic opportunities (Born GVR and Schwarz CJ, eds). Stuttgart: Schattauer, pp 1–27

Nelson WJ (2003) Tube morphogenesis: closure, but many openings remain. Trends Cell Biol 13: 615

Rong JX, Shapiro M, Trogan E, and Fisher EA (2003) Transdifferentiation of mouse aortic smooth muscle cells to a macrophage-like state after cholesterol loading. Proc Natl Acad Sci USA 100: 13531

Tsilibary EC (2003) Microvascular basement membranes in diabetes mellitus. J Pathol 200: 537

Woijciak-Stothard B and Ridley AJ (2003) Shear stress-induced endothelial cell polarization is mediated by Rho and Rac but not Cdc42 or PI 3-kinases. J Cell Biol 161: 429

Magnification: x 31,200 (A), x 18,700 (B)

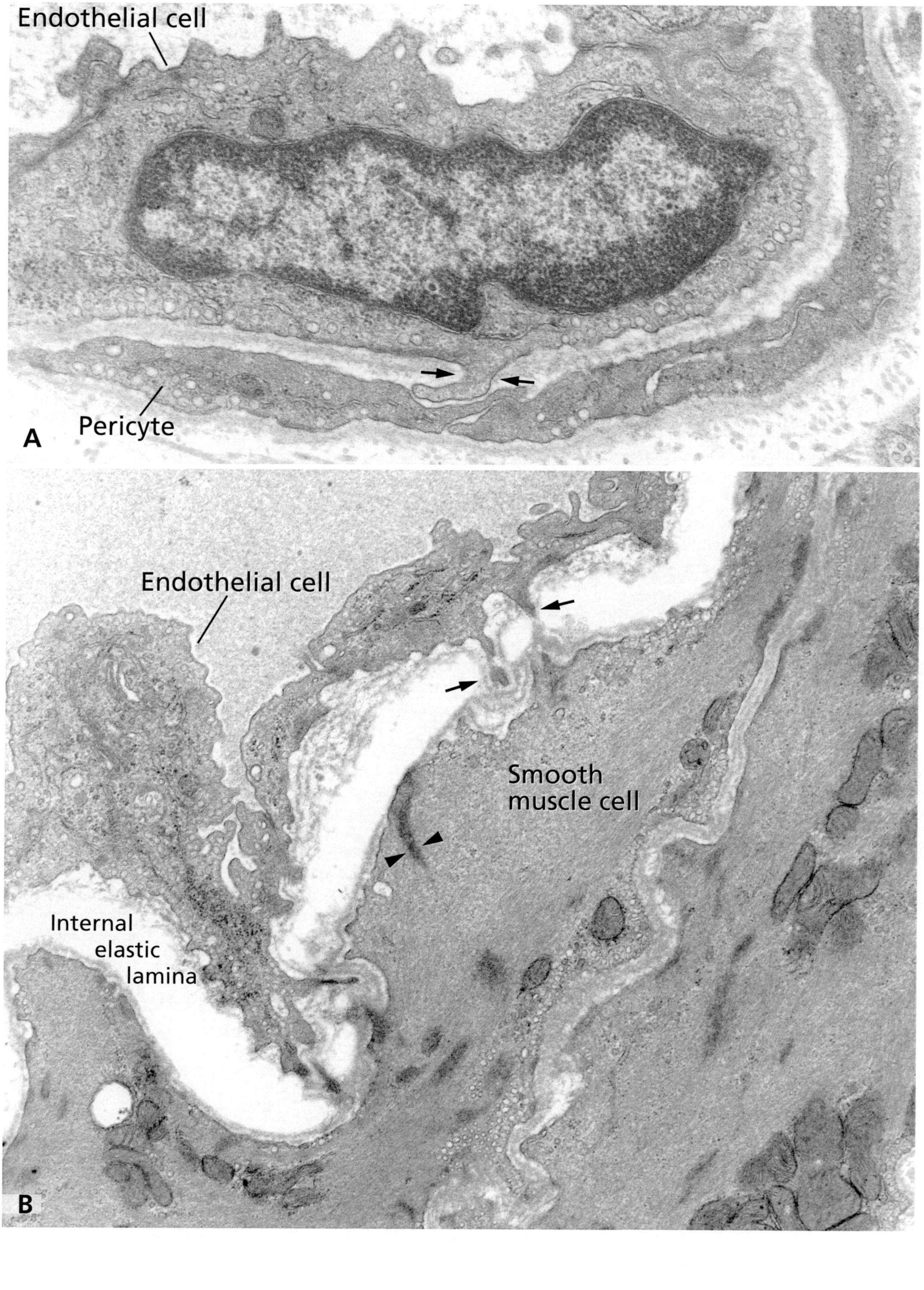
Endothelial cell
Pericyte
A
Endothelial cell
Smooth muscle cell
Internal elastic lamina
B

GLOMERULUS: A SPECIALISED DEVICE FOR FILTERING

The glomeruli of kidney are selective filtering corpuscles composed of a capillary network, podocytes, and mesangial cells (panel A). The glomerular basement membrane (GBM) is situated between capillary endothelia and podocytes (cf. Fig. 91). All three cell types and the GBM are important for the ultrafiltration of plasma. The fenestrated endothelia (arrowheads in panel B) face the blood and differ from other fenestrated endothelia (cf. Fig. 128) for the larger size of pores and lack of diaphragms. The podocytes protrude in the urinary space (US), and their peculiar shape is best appreciated by scanning electron microscopy (panel C). From their cell body, long ramifications originate, the primary processes, that embrace the capillaries. From them interdigitating secondary processes originate, the foot processes, which form the narrow filtration slits (30–40 nm wide), which are bridged by slit diaphragms (arrows in panel B). Slit diaphragms resemble adherens intercellular junctions and mark the border between the apical and basal podocyte plasma membrane. The foot process base inserts into the GBM. The transmembrane proteins podocalyxin and podoplanin are important for the shape of podocytes.

Podocytes possess a glycocalyx rich in negatively charged sialic acids, the glomerular epithelial polyanion, which is formed mainly by podocalyxin. The filtration slits stay open by repulsion through negatively charged sialic acids of podocalyxin, which acts as an antiadhesin. Nephrin, which belongs to the immunoglobulin superfamily of adhesion molecules, is found exclusively in the slit diaphragm and is an important component of an isoporous zipper-like filter structure. Nephrin is linked to the cytoskeleton through ezrin.

Accordingly, the filtration machinery is constituted by endothelia, the highly hydrated GBM, and the diaphragms bridging the filtration slits. Podocalyxin, nephrin, podocin, and CD2-AP represent its main molecular components in addition to negatively charged GBM components. The filter is highly permeable for water, and only the slit diaphragm poses a barrier. The negative charges of endothelia and the GBM are important to filter macromolecules according to their charge, whereas the selection according to molecular mass and configuration occurs in the GBM and at the level of the slit diaphragms. Through their endocytic activities, podocytes and mesangial cells contribute to the functional maintenance of the glomerular filter by removing escaped proteins from the slit diaphragms and from the subendothelial space.

References

Kerjaschki D, Sharky DJ, and Farquhar MG (1984) Identification and characterisation of podocalyxin – the major sialoprotein of the renal glomerular epithelial cell. J Cell Biol 98: 1591

Kriz A and Kaissling B (2000) Structural organisation of the mammalian kidney. In: The kidney: Physiology and pathophysiology (Seldin D, and Giebisch G, eds). New York Lippincott, Williams and Wilkins, pp 587

Matsui K, Breitender-Geleff S, Soleiman A, Kowalski H, and Kerjaschki D (1999) Podoplanin, a novel 43-kDa membrane protein, controls the shape of podocytes. Nephrol Dial Transplant 14 Suppl 1: 9

Mundel P and Shankland SJ (2002) Podocyte biology and response to injury. J Am Soc Nephrol 13: 3005

Orlando RA, Takeda T, Zak B, Schmieder S, Benoit VM, McQuistan T, Furthmayr H, and Farquhar MG (2001) The glomerular epithelial cell anti-adhesin podocalyxin associates with the actin cytoskeleton through interactions with ezrin. J Amer Soc Nephrol 12: 1589

Pavenstädt H, Kriz W, and Kretzler M (2003) Cell biology of the glomerular podocyte. Physiol Rev 83: 253

Ruotsalainen V, Ljungberg P, Wartiovaara J, Lenkkeri U, Kestila M, Jalanko H, Holmberg C, and Tryggvason K (1999) Nephrin is specifically located at the slit diaphragm of glomerular podocytes. Proc Natl Acad Sci USA 96: 7962

Takeda T, Go WY, Orlando RA, and Farquhar MG (2000) Expression of podocalyxin inhibits cell-cell adhesion and modifies junctional properties in Madin-Darby canine kidney cells. Mol Biol Cell 11: 3219

Tryggvason K (1999) Unraveling the mechanisms of glomerular ultrafiltration: nephrin, a key component of the slit diaphragm. J Am Soc Nephrol 10: 2440

Magnification: x 7,000 (A); x 45,000 (B); x 15,000 (C)

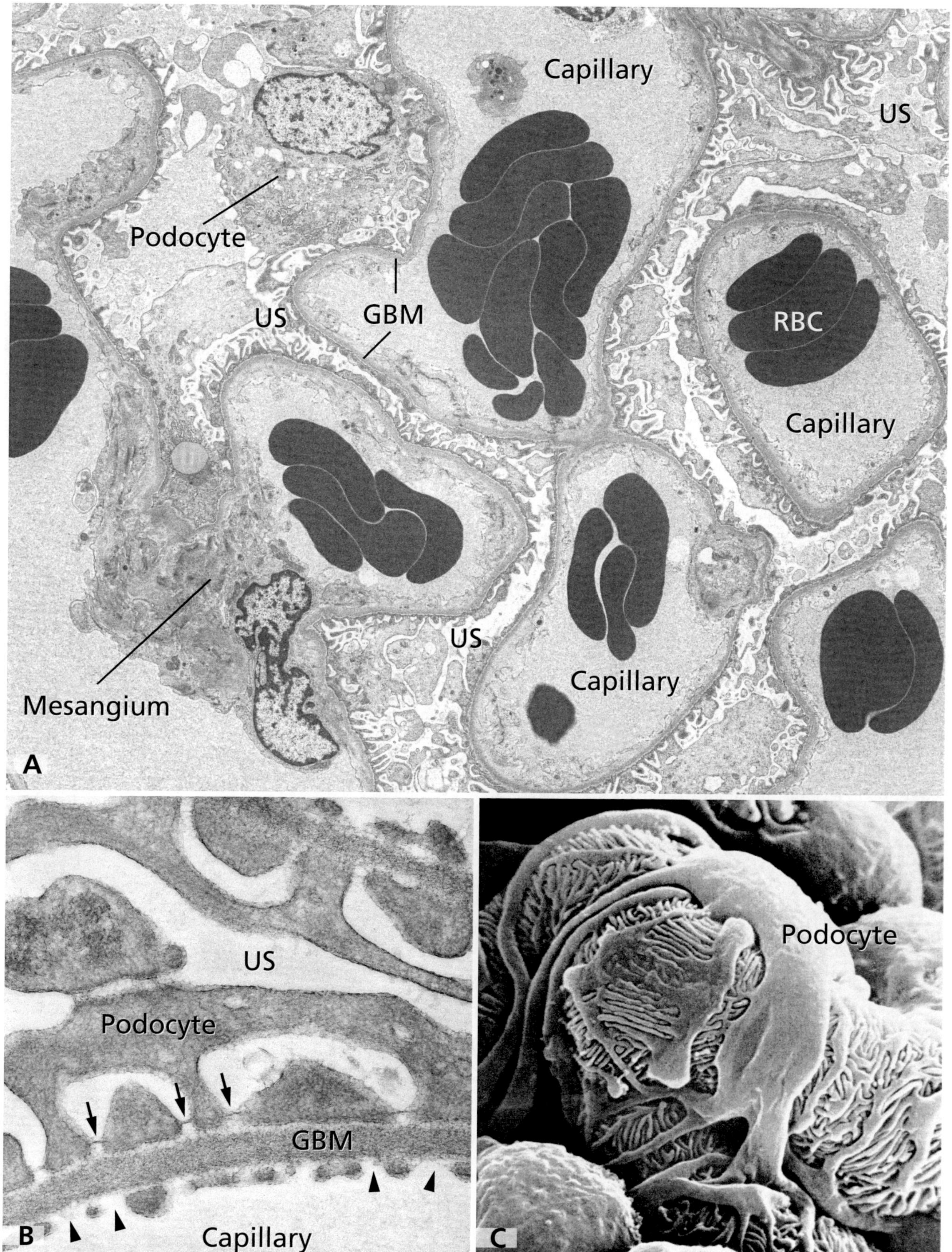
Capillary
US
Podocyte
GBM
US
RBC
Capillary
US
Capillary
Mesangium
A
US
Podocyte
GBM
B
Capillary
Podocyte
C

PATHOLOGY OF THE GLOMERULAR FILTER: MINIMAL CHANGE GLOMERULOPATHY AND CONGENITAL NEPHROTIC SYNDROMES

The impairment of the glomerular filter occurring in various acquired and inherited diseases results in a nephrotic syndrome characterised by proteinuria, oedema, hypoalbuminaemia, and hyperlipidaemia. In minimal change glomerulopathy, diffuse effacement of podocyte foot processes (panels A, B), absence of slit diaphragms (arrows in B), lipid droplets in podocytes (asterisks in B), and microvilli hypertrophy (arrowheads in panels A and B) are observed. Podocyte detachment from the glomerular basement membrane (GBM) may occur. Protein reabsorption by proximal tubular epithelia results in cytoplasmic protein-filled vacuoles (open asterisks in panel C). In puromycin-induced nephrotic syndrome, a rat model of human minimal change glomerulopathy, reduced sialylation of podocalyxin and downregulation of podoplanin was observed.

Various congenital nephrotic syndromes result in similar fine structural changes, and few of the causative genes have been identified. The congenital nephrotic syndrome of the Finnish type is due to mutations of the *NPHS1* gene coding for nephrin. In a less severe, steroid resistant, familial nephrotic syndrome, the causative gene, *NPHS2*, was mapped and identified to encode for podocin. These pathological conditions show the importance of nephrin and podocin in normal glomerular filtration.

An immunoglobulin superfamily protein, the CD2-associated protein (CD2-AP), known to be involved in contact stabilisation between T-cells and antigen presenting cells, has been shown to cause a congenital nephrotic syndrome in mice that are deficient in CD2-AP. In normal mice, CD2-AP is expressed in podocytes and associated with nephrin present in the slit diaphragm.

References

Boute N, Gribouval O, Roselli S, Benessy F, Lee H, Fuchshuber A, Dahan K, Gubler MC, Niaudet P, and Antignac C (2000) NPHS2, encoding the glomerular protein podocin, is mutated in autosomal recessive steroid-resistant nephrotic syndrome. Nat Genet 24: 349

Breiteneder-Geleff S, Matsui K, Soleiman A, Meraner P, Poczewski H, Kalt R, Schaffner G, and Kerjaschki D (1997) Podoplanin, novel 43-kd membrane protein of glomerular epithelial cells, is down-regulated in puromycin nephrosis. Am J Pathol 151: 1141

Dickersin G (2000) Diagnostic electron microscopy. A text/atlas. New York: Springer

Kerjaschki D, Vernillo A, and Farquhar M (1985) Reduced sialylation of podocalyxin – the major sialoprotein of the rat kidney glomerulus – in aminonucleoside nephrosis. Am J Pathol 118: 343

Kestila M, Lenkkeri U, Mannikko M, Lamerdin J, McCready P, Putaala, H, Ruotsalainen V, Morita T, Nissinen M, Herva R, et al. (1998) Positionally cloned gene for a novel glomerular protein - nephrin- is mutated in congenital nephrotic syndrome. Mol Cell 1: 575

Lenkkeri U, Mannikko M, McCready P, Lamerdin J, Gribouval O, Niaudet PM, Antignac CK, Kashtan CE, Homberg C, Olsen A, et al. (1999) Structure of the gene for congenital nephrotic syndrome of the finnish type (NPHS1) and characterisation of mutations. Am J Hum Genet 64: 51

Liu G, Kaw B, Kurfis J, Rahmanuddin S, Kanwar YS, and Chugh SS (2003) Neph1 and nephrin interaction in the slit diaphragm is an important determinant of glomerular permeability. J Clin Invest 112: 209

Mundel P and Shankland SJ (2002) Podocyte biology and response to injury. J Am Soc Nephrol 13: 3005

Olson J and Schwartz M (1998) The nephrotic syndrome: minimal change disease, focal segmental glomerulosclerosis, and miscellaneous causes. In: Heptinstall's pathology of the kidney (Jennette J, Olson, and Silva F, eds). Philadelphia: Lippincott-Raven, pp 187

Shih NY, Li J, Karpitskii V, Nguyen A, Dustin ML, Kanagawa O, Miner JH, and Shaw AS (1999) Congenital nephrotic syndrome in mice lacking CD2-associated protein. Science 286: 312

Tryggvason K, Ruotsalainen V, and Wartiovaara J (1999) Discovery of the congenital nephrotic syndrome gene discloses the structure of the mysterious molecular sieve of the kidney. Int J Dev Biol 43: 445

Yoshikawa N, Cameron AH, and White RH (1982) Ultrastructure of the non-sclerotic glomeruli in childhood nephrotic syndrome. J Pathol 136: 133

Magnification: x 8,000 (A); x 14,000 (B); x 17,500 (C)

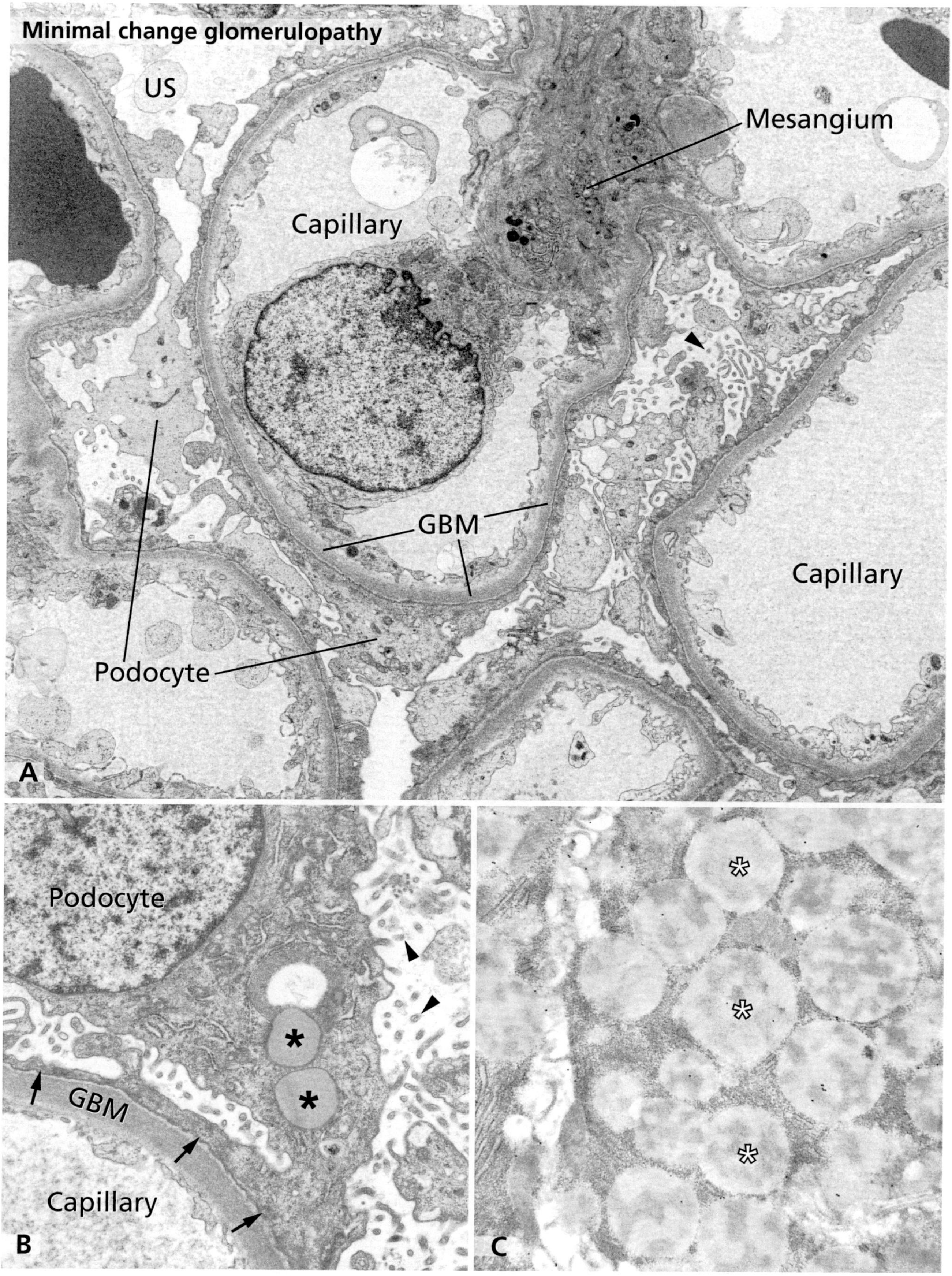
Minimal change glomerulopathy
US
Mesangium
Capillary
GBM
Capillary
Podocyte
A
Podocyte
*
*
GBM
Capillary
B
*
*
*
C

PATHOLOGY OF THE GLOMERULUS: MEMBRANOUS GLOMERULONEPHRITIS

Membranous glomerulonephritis represents an immune complex-mediated disease and is associated with the nephrotic syndrome or isolated severe proteinuria. Primary idiopathic forms are caused by unknown antigens and secondary forms are associated with infections, such as hepatitis B and syphilis, autoimmune diseases, such as systemic lupus erythaematosus and rheumatoid arthritis, or various drugs, such as penicillamine and gold. Depending on the structural alterations, four stages of membranous glomerulonephritis can be distinguished.

The fine structural hallmark of membranous glomerulonephritis are dense amorphous deposits that are located between the outer aspect of the glomerular basement membrane and the podocytes (asterisks). In the postinfectious forms, the deposits are dome-shaped humps (see also inset), and polymorphonuclear neutrophilic granulocytes (PMN) can be detected in the capillary lumen. Because of their location, they are called subepithelial deposits and found along virtually all glomerular capillaries as can be appreciated in the electron micrograph. The subepithelial deposits are not the only fine structural change, since extensive effacement of podocyte foot processes occurs as well.

The deposits are composed of immune complexes, which can be detected by immunofluorescence and are positive for IgG and other components such as C3 and light chains (primary forms) or C1q (secondary forms with systemic lupus erythaematosus). The immune complexes are probably formed *in situ* by circulating antibodies binding to an unidentified glomerular antigen of the capillary walls or binding to ill-defined extrinsic antigens deposited here.

An example for the *in situ* immune complex formation is the experimental Heyman nephritis. The nephritogenic antigen megalin (formely gp330) is located in coated pits along the free surface and the bases of the podocyte foot processes facing the glomerular basement membrane, and is also found in the brush border of proximal tubular epithelia. Circulating anti-megalin antibodies bind *in situ* to the antigen, and this results in progressive formation of subepithelial immune complexes.

References

Cavallo T (1994) Membranous nephropathy. Insights from Heymann nephritis. Am J Pathol 144: 651

Cohen A and Nast C (1996) Kidney. In: Anderson's pathology (Damjanov I, and Linder J, eds). St. Louis: Mosby, pp 2073

Collins AB, Bhan AK, Dienstag JL, Colvin RB, Haupert GT Jr, Mushahwar IK, and McCluskey RT (1983) Hepatitis B immune complex glomerulonephritis: simultaneous glomerular deposition of hepatitis B surface and E antigens. Clin Immunol Immunopathol 26: 137

Dickersin G (2000) Diagnostic electron microscopy. A text/atlas. New York: Springer

Kerjaschki D and Farquhar M (1983) Immunocytochemical localisation of the Heymann nephritis antigen (gp 330) in glomerular epithelial cells of normal Lewis rats. J Exp Med 157: 667

Kerjaschki D and Farquhar MG (1982) The pathogenic antigen of Heymann nephritis is a membrane glycoprotein of the renal proximal tubule brush border. Proc Natl Acad Sci USA 79: 5557

Kerjaschki D, Miettinen A, and Farquhar MG (1987) Initial events in the formation of immune deposits in passive Heymann nephritis. gp330-anti-gp330 immune complexes form in epithelial coated pits and rapidly become attached to the glomerular basement membrane. J Exp Med 166: 109

Zollinger H and Mihatsch M (1978) Renal pathology in biopsy. Berlin Heidelberg New York: Springer

Magnification: x 8,800; x 10,500 (inset)

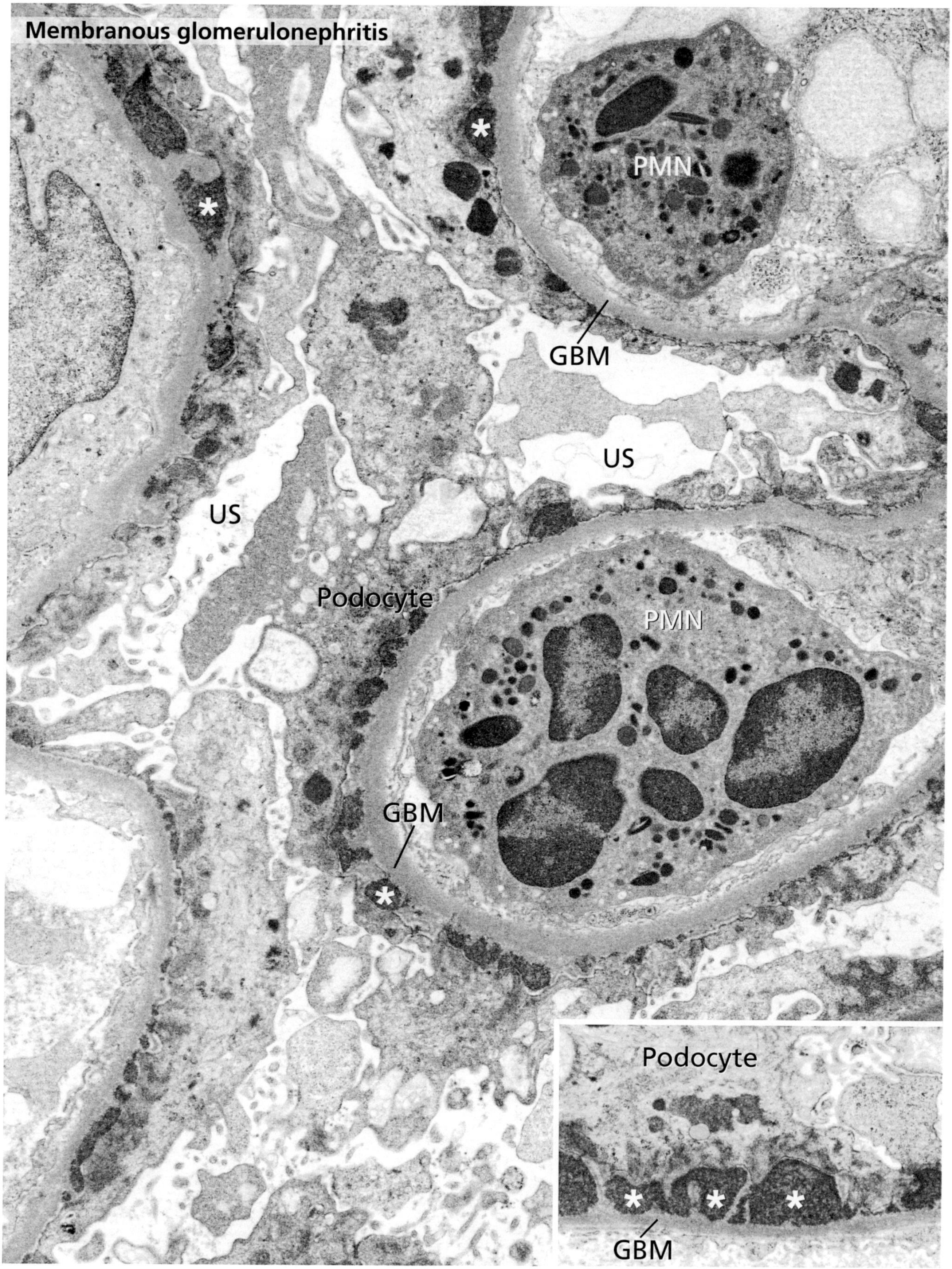
Membranous glomerulonephritis
*
*
PMN
GBM
US
US
Podocyte
PMN
GBM
*
Podocyte
* * *
GBM

PATHOLOGY OF THE GLOMERULUS: MEMBRANOPROLIFERATIVE GLOMERULONEPHRITIS

Membranoproliferative glomerulonephritis is representative of different disease entities with different aetiologies and clinical signs. The type-1 membranoproliferative glomerulonephritis is an immune complex-mediated disease that can be classified in primary (idiopathic) forms caused by unknown antigens and secondary forms. The latter are associated with different diseases such as infections, tumours, and systemic immune abnormalities. Among the infectious diseases, hepatitis C is the predominant cause. The causative factors of the membranoproliferative glomerulonephritis are circulating immune complexes, and this results in the secondary activation of the classical complement pathway.

The fine structural hallmark of the membranoproliferative glomerulonephritis type-1 consists of dense amorphous deposits located between the inner aspect of the glomerular basement membrane (GBM) and the endothelia (asterisks and inset). Because of their location, they are called subendothelial deposits and found along virtually all capillaries in a given glomerular lobulus. These deposits can be massive, as illustrated in the micrograph. The deposits are composed of C3, IgG and IgM and can be visualised by immunofluorescence. In addition to the subendothelial deposits, deposits can be also present in the mesangial matrix, and scattered subepithelial deposits (between the outer aspect of the glomerular basement membrane and podocytes) can be occasionally observed as shown in the micrograph. The occurrence of glomerular basement duplications is another characteristic feature, which, however, is not evident in the micrograph. Furthermore, podocyte effacement can be observed. In the micrograph, the narrowed lumen of the capillary caused by swelling of the endothelia and expansion of mesangial cells is evident.

References

Burstein DM and Rodby RA (1993) Membranoproliferative glomerulonephritis associated with hepatitis C virus infection. J Am Soc Nephrol 4: 1288

Cohen A and Nast C (1996) Kidney. In: Anderson's pathology. (Damjanov I and Linder J, eds). St. Louis: Mosby, pp 2073

Dickersin G (2000) Diagnostic electron microscopy. A text/atlas. New York: Springer

West CD (1992) Idiopathic membranoproliferative glomerulonephritis in childhood. Pediatr Nephrol 6: 96

Zollinger H and Mihatsch M (1978) Renal pathology in biopsy. Berlin Heidelberg New York: Springer

Magnification: x 7,000; x 9,500 (inset)

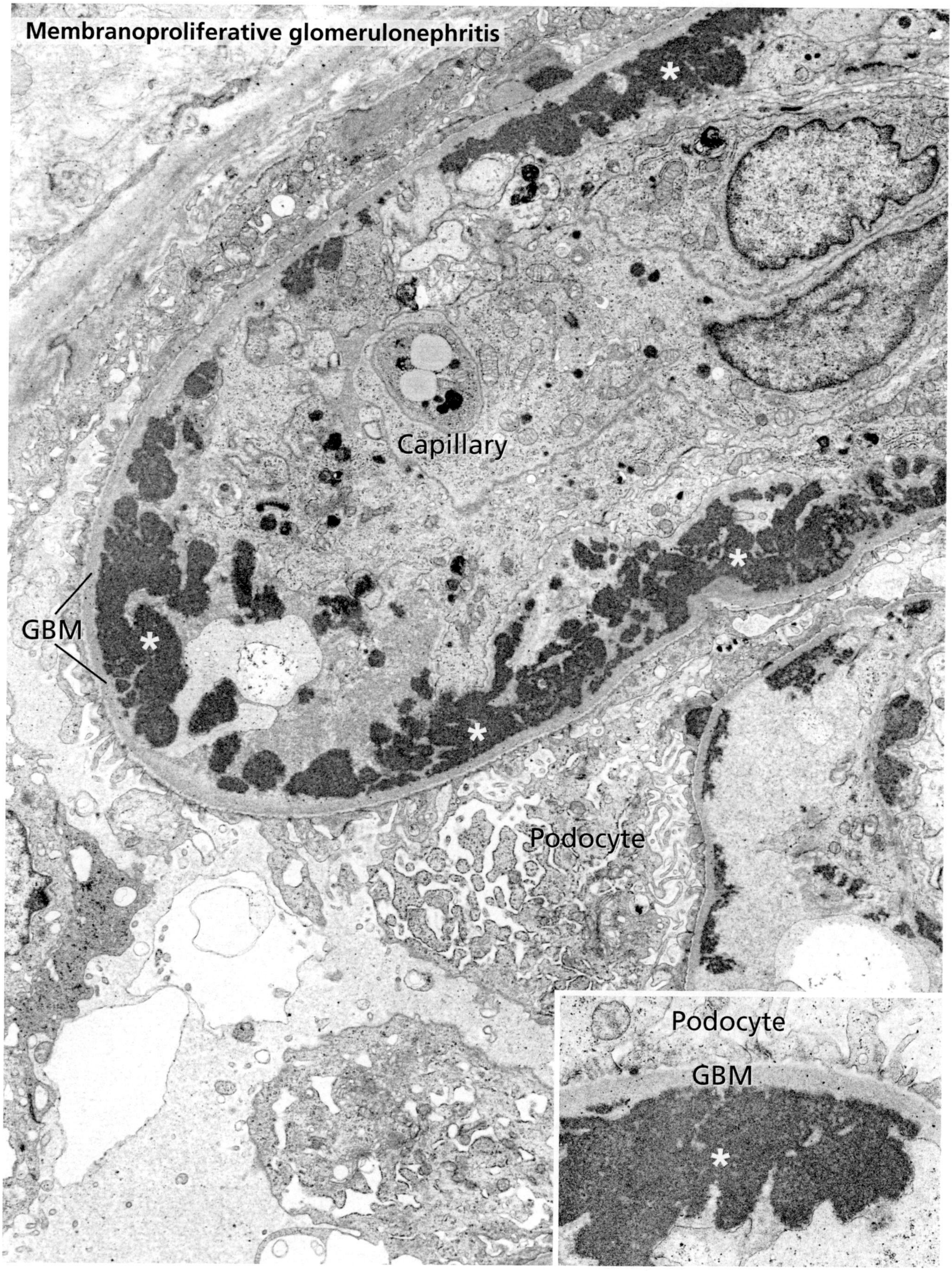
Membranoproliferative glomerulonephritis
*
Capillary
GBM
*
*
*
Podocyte
Podocyte
GBM
*

PATHOLOGY OF THE GLOMERULUS: IgA NEPHROPATHY (BERGER)

IgA nephropathy (Berger's disease) is an immune complex-mediated glomerulonephritis, which is the most common primary glomerulopathy world-wide. Despite, it is an idiopathic disorder and many factors are believed to contribute to its development. Most prominent in the deposited immune complexes is IgA either as the sole or the dominant immunoglobulin. In about 50% of the patients, the synthesis of IgA is increased. Furthermore, its glycosylation may be abnormal, which would hinder its clearance. Specifically, an abnormality in the glycosylation of the hinge region of the IgA1 isotype of IgA appears to be fundamental.

Characteristically, the IgA-containing immune complexes localise preferentially in the mesangium as shown in the electron micrograph. Activation of the alternative complement pathway subsequently results in proliferation of mesangial cells and increased mesangial matrix. By electron microscopy, electron-dense, large deposits of round shape are prominent in the mesangium and some of them are marked by asterisks in the micrograph. Notably, there are no capillary wall deposits. Furthermore, in the biopsy from the early stage case shown, glomerular podocytes are not affected and exhibit numerous foot processes in the urinary space (US). During the progression of the disease, secondary changes of the glomeruli and tubuli can be observed.

IgA nephropathy most often affects children and young adults with a history of upper respiratory tract infection and begins with gross hematuria, which may recur every few months and which may be associated with proteinuria.

References

Beerman I, Novak J, Wyatt RJ, Julian BA, and Gharavi AG (2007) The genetics of IgA nephropathy. Nat Clin Pract Nephrol 3: 325

Cohen A and Nast C (1996) Kidney. In: Anderson's pathology (Damjanov I and Linder J, eds). St. Louis: Mosby, pp 2073

Cook HT (2007) Interpretation of renal biopsies in IgA nephropathy. Contrib Nephrol 157: 44

Eijgenraam JW and van Kooten C (2008) IgA1 glycosylation in IgA nephropathy: as sweet as it can be. Kidney Int 73 :1106

Gharavi AG, Moldoveanu Z, Wyatt RJ, Barker CV, Woodford SY, Lifton RP, Mestecky J, Novak J, and Julian BA (2008) Aberrant IgA1 glycosylation is inherited in familial and sporadic IgA nephropathy. J Am Soc Nephrol 19: 1008

Moura IC, Benhamou M, Launay P, Vrtovsnik F, Blank U, and Monteiro RC (2008) The glomerular response to IgA deposition in IgA nephropathy. Semin Nephrol 28: 88

Oortwijn BD, Eijgenraam JW, Rastaldi MP, Roos A, Daha MR, and van Kooten C (2008) The role of secretory IgA and complement in IgA nephropathy. Semin Nephrol 28: 58

Sawanobori E, Terada N, Fujii Y, Higashida K, Nakazawa S, and Ohno S (2008) Ultrastructural study of human glomerular capillary loops with IgA nephropathy using quick-freezing and deep-etching method. Histol Histopathol 23: 297

Suzuki H, Fan R, Zhang Z, Brown R, Hall S, Julian BA, Chatham WW, Suzuki Y, Wyatt RJ, Moldoveanu Z, et al. (2009) Aberrantly glycosylated IgA1 in IgA nephropathy patients is recognized by IgG antibodies with restricted heterogeneity. J Clin Invest 119: 1668

Suzuki Y, and Tomino Y (2007) The mucosa-bone-marrow axis in IgA nephropathy. Contrib Nephrol 157: 70

Tomino Y (2007) Pathogenesis of IgA nephropathy. Contrib Nephrol 157: 1

Tomino Y (2008) IgA nephropathy: lessons from an animal model, the ddY mouse. J Nephrol 21: 463

Zollinger H and Mihatsch M (1978) Renal pathology in biopsy Berlin Heidelberg New York: Springer

Magnification: x 8,500

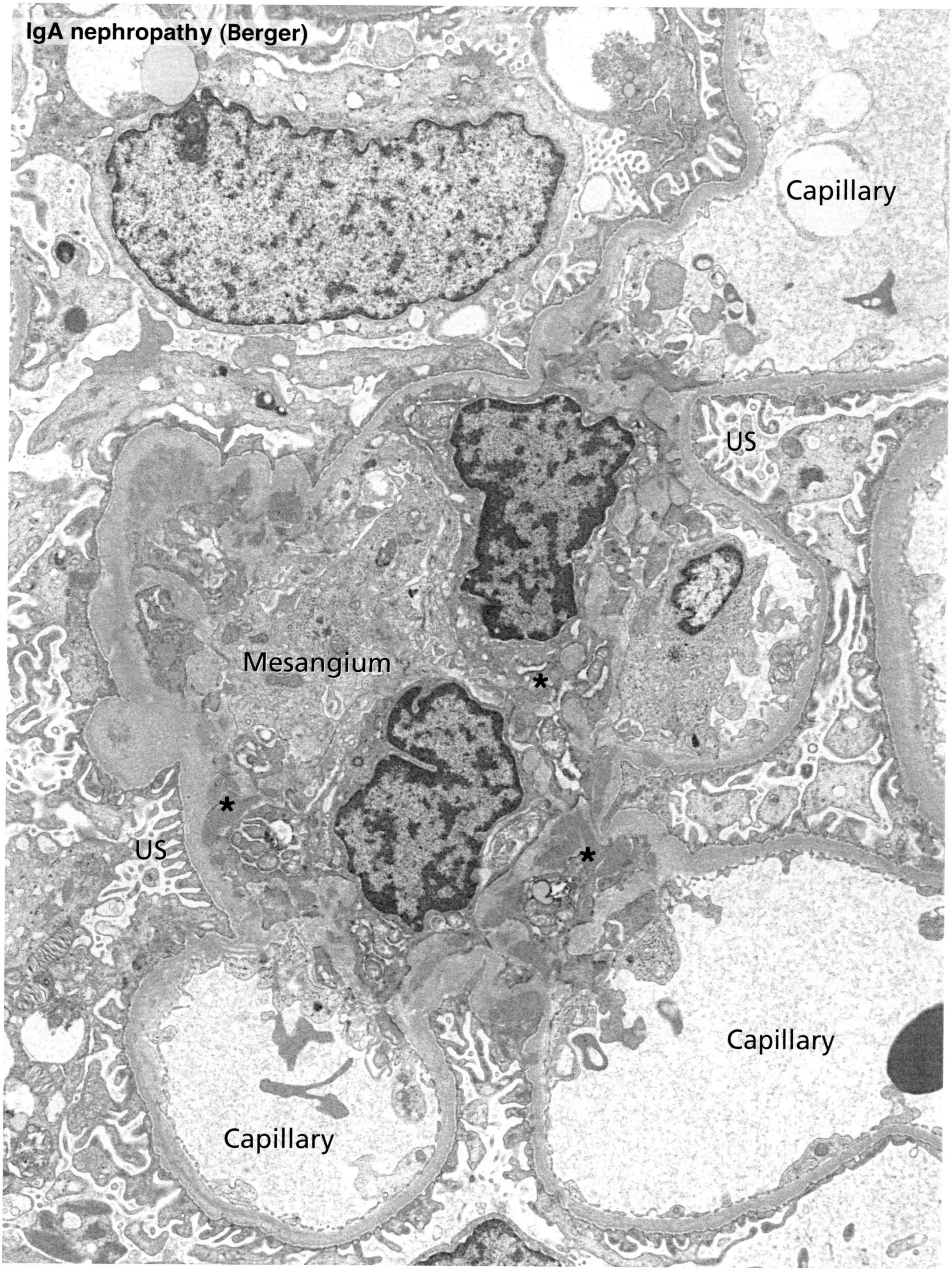
IgA nephropathy (Berger)
Capillary
US
Mesangium
*
*
*
US
Capillary
Capillary

PATHOLOGY OF THE GLOMERULUS: CHRONIC ALLOGRAFT GLOMERULOPATHY

The morphology of the chronic rejection reaction is highly complex and occurs secondary to immunological events affecting the renal vasculature and interstitium. Functionally, renal performance declines continuously.

The chronic allograft glomerulopathy represents only one aspect of the chronic transplant rejection and can occur in the sequel of acute allograft glomerulopathy. The fine structural alterations affect all parts of the capillary wall as can be well appreciated from the electron micrograph. The glomerular basement membrane is prominently thickened, exhibits duplications, and is laminated (arrowhead). In the subendothelial regions of the glomerular basement membrane, depositions of amorphous material can be present over long distances (asterisks). The lamina rara interna is expanded and can contain fragments of interposed cells, vesicular structures, and membrane profiles (open arrow). Segmental effacement of the podocyte foot processes occurs in advanced stages. The endothelia are affected as well and show morphological signs of dedifferentiation with loss of fenestrations. The capillary lumen is usually narrow. By immunofluorescence, IgM, C1q, and C3 can be detected in the capillary wall and in the expanded mesangium.

References

Briner J (1987) Transplant glomerulopathy. Appl Pathol 5: 82

Cohen A and Nast C (1996) Kidney. In: Anderson's pathology (Damjanov I and Linder J, eds). St. Louis: Mosby, pp 2073

Dickersin G (2000) Diagnostic electron microscopy. A text/atlas. New York: Springer

Olsen S, Bohman SO, and Petersen VP (1974) Ultrastructure of the glomerular basement membrane in long term renal allografts with transplant glomerular disease. Lab Invest 30: 176

Zollinger H and Mihatsch M (1978) Renal pathology in biopsy Berlin Heidelberg New York: Springer

Magnification: x 8,000

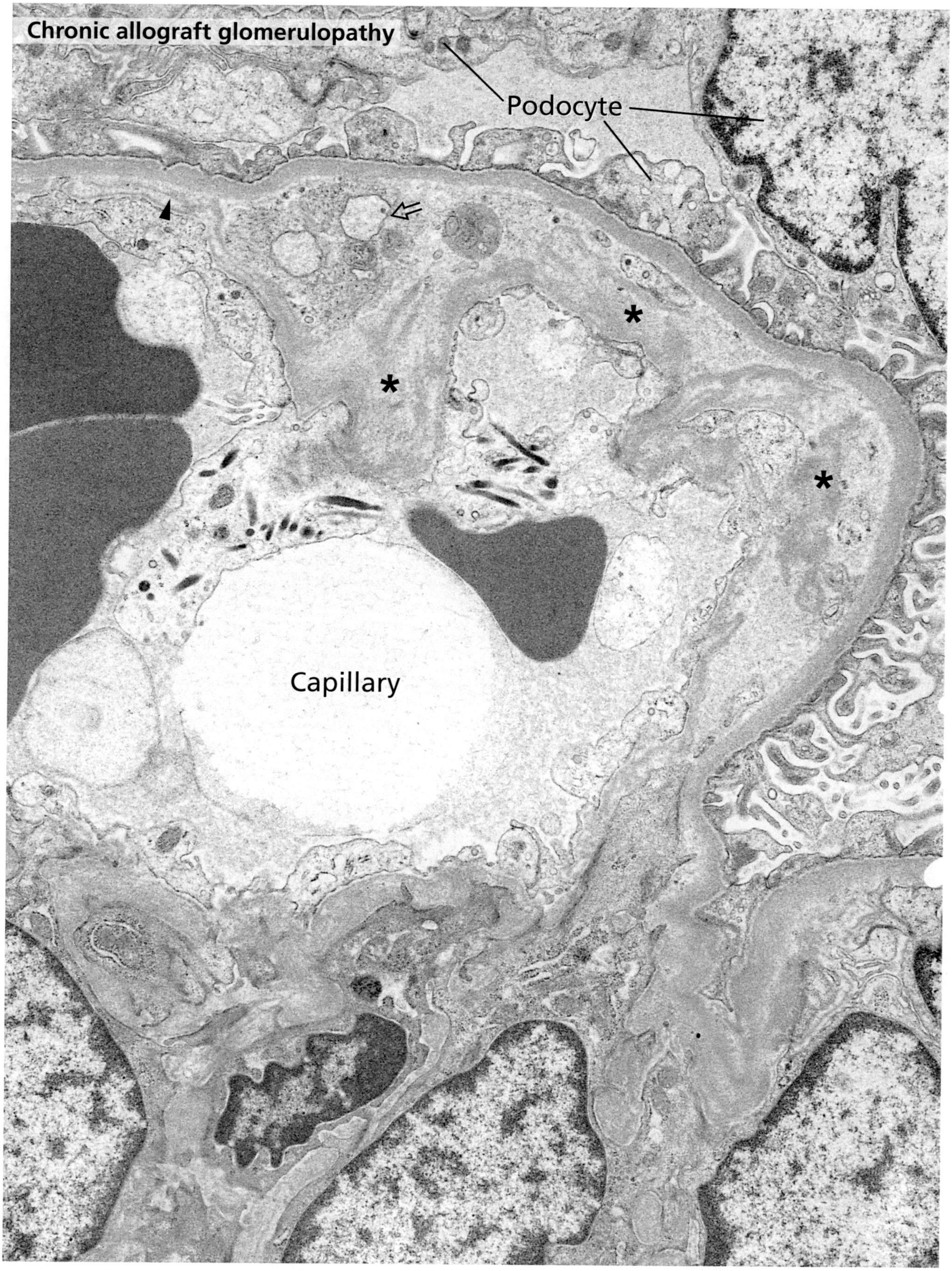
Chronic allograft glomerulopathy
Podocyte
*
*
*
Capillary

LOOSE CONNECTIVE TISSUE

The connective tissue, composed of cells and extracellular matrix, forms a framework designated as stroma, which connects and supports all other tissues of the body. However, it is not an inert scaffolding to stabilise other tissues and organs only, but is dynamic and the site of multiple regulatory processes involved in tissue organisation, development, wound healing, immune response and organ repair. The extracellular matrix is constantly renewed and molecules of the extracellular matrix modulate the functional activities of cells. They also play key roles in disease processes, including inflammatory and degenerative diseases and cancer. The matrix is involved in tumour invasion, metastasis and tumour angiogenesis.

The extracellular matrix is produced by fibroblasts, and mainly consists of fibrils and fibres of collagen and elastin embedded in a ground substance of non-collagenous glycoproteins and proteoglycans. Fibrils, fibres and components of the ground substance differ between the different types of connective tissue. Contents and architectures of the diverse collagen and elastic fibres determine the tissue's properties, such as consistency and elasticity. This in turn forms the basis for the classification in loose connective tissue shown here in this figure, in dense irregular and dense regular connective tissues (cf. Fig. 140), and in reticular and elastic tissues. Collagen is constituent of thick collagen fibres (type I collagen, cf. Fig. 138) and of fine reticular fibres (mainly type III collagen) present in the fibrillar layers of basement membranes (cf. Fig. 90), and in the reticular connective tissue of the bone marrow and lymphatic organs.

Loose connective tissue is shown in this survey micrograph. It is found in multiple organs surrounding blood vessels, lymphatics, nerves and muscles. The figure shows a section of the wall of the rat small intestine. A group of smooth muscle cells of the muscle layer is on display in the left lower corner. The main part of the micrograph presents the submucosal loose connective tissue with sections of cell bodies of a fibroblast and a macrophage. The fibroblast cytoplasm is stuffed with rough endoplasmic reticulum. The prominent Golgi region (G) is in perinuclear position. It is clearly visible that the main architecture of the tissue is made up by two types of networks, the one consisting of fine fibroblast processes (F), the other built by collagen fibrils and fibres (C). Longitudinal and cross sections of collagen fibres are apparent side by side indicating the reticular arrangement. Collagen fibrils also accompany multiple unmyelinated nerve fibres of the vegetative nerve system (asterisks).

References

Bosman FT and Stamenkovic I (2003) Preface to extracellular matrix and disease. J Pathol 200: 421

De Wever O and Mareel M (2003) Role of tissue stroma in cancer cell invasion. J Pathol 200: 429

Stamenkovic I (2003) Extracellular matrix remodelling: the role of matrix metalloproteinases. J Pathol 200: 448

Magnification: x 7,100

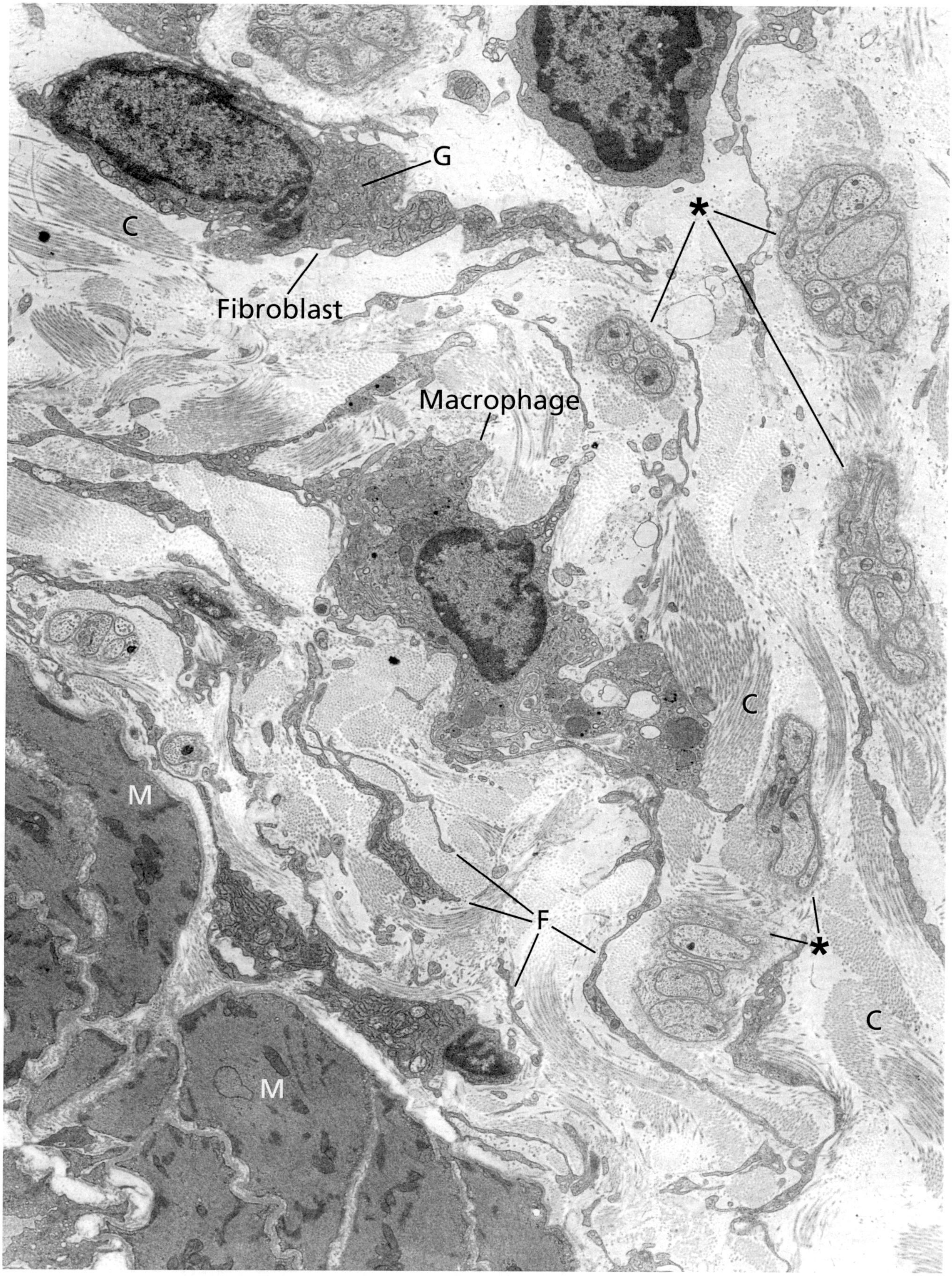
G
C
Fibroblast
Macrophage
*
C
M
F
*
C
M

FIBROBLAST, FIBROCYTE, MACROPHAGE

In panels A and B, an active fibroblast and its inactive counterpart, referred to as fibrocyte, are on display, respectively. Fibroblasts constitute the main resident cells of the connective tissue. They synthesise and secrete both the components of the connective tissue ground substance and the precursor molecules of various types of collagen and elastic fibres. The secretory program of the fibroblasts determines the composition of the extracellular matrix and, as a consequence, provides the basis for the construction of a certain type of connective tissue. Fibroblasts are spindle-shaped and show all the characteristics of cells active in protein synthesis. In the karyoplasm, nucleoli are prominent. The long cytoplasmic extensions with multiple surface folds contain densely packed rough endoplasmic reticulum (ergastoplasm). In the Golgi apparatus, the secretory proteins are modified and, after completion, packaged into small vesicles or other post-Golgi carriers to be subsequently transported to the cell surfaces and exported via exocytosis. All components of the ground substance, as well as procollagens and components of elastic fibres are secreted constitutively. Unlike the events during regulated secretion, cell products are continuously packaged at the Golgi apparatus and subsequently exported; they are not accumulated in the cells. Both transport of nascent secretory molecules across the Golgi stacks and their pathways from the Golgi apparatus to the cell surface were subjects of extensive studies. Procollagen molecules traverse the Golgi apparatus and move synchronously through the Golgi stacks together with other cargo proteins without leaving the cisternal lumen. The extremely fine end pieces of the fibroblasts' processes (arrows) are in contact with each other and build up a network, which, together with bundles of collagen fibrils (C), make up the basic architecture of the connective tissue.

In the inactive fibrocytes (panel B), the cytoplasm is less voluminous compared with the active fibroblasts. Fibrocytes lack ergastoplasm. Their cell processes like wings surround collagen fibres (C) leading to a compartmentalised organisation of the tissue. In tendons, the respective fibroblasts are called "wing cells".

In panel C, a higher magnification of the macrophage shown in Fig. 136 is on display. The surface is jagged and shows deep invaginations of the plasma membrane and multiple endocytosis profiles. Macrophages are mobile cells deriving from monocytes in the blood (cf. Fig. 173). They are equipped with extensive phagocytic properties and contain abundant lysosomes and phagosomes. As in other tissues and organs, macrophages fulfil important tasks during unspecific and specific immune reactions. In the connective tissue, macrophages also are involved in the turnover of extracellular matrix materials and senescent fibres. Elastic fibres (E) and collagen fibrils (C) are apparent close to the surface of macrophages and are present in the spaces that are formed and surrounded by deep plasma membrane invaginations.

References

Mironov AA, Beznoussenko GV, Nicoziani P, Martella O, Trucco A, Kweon H-S, Di Giandomenico D, Polishchuk RS, Fusella A, Lupetti P, Berger EG, Geerts WJC, Koster AJ, Burger KNJ, and Luini A (2001) Small cargo proteins and large aggregates can traverse the Golgi by a common mechanism without leaving the lumen of cisternae. J Cell Biol 155: 1225

Magnification: x 13,400 (A), x 7,600 (B), x 18,200 (C)

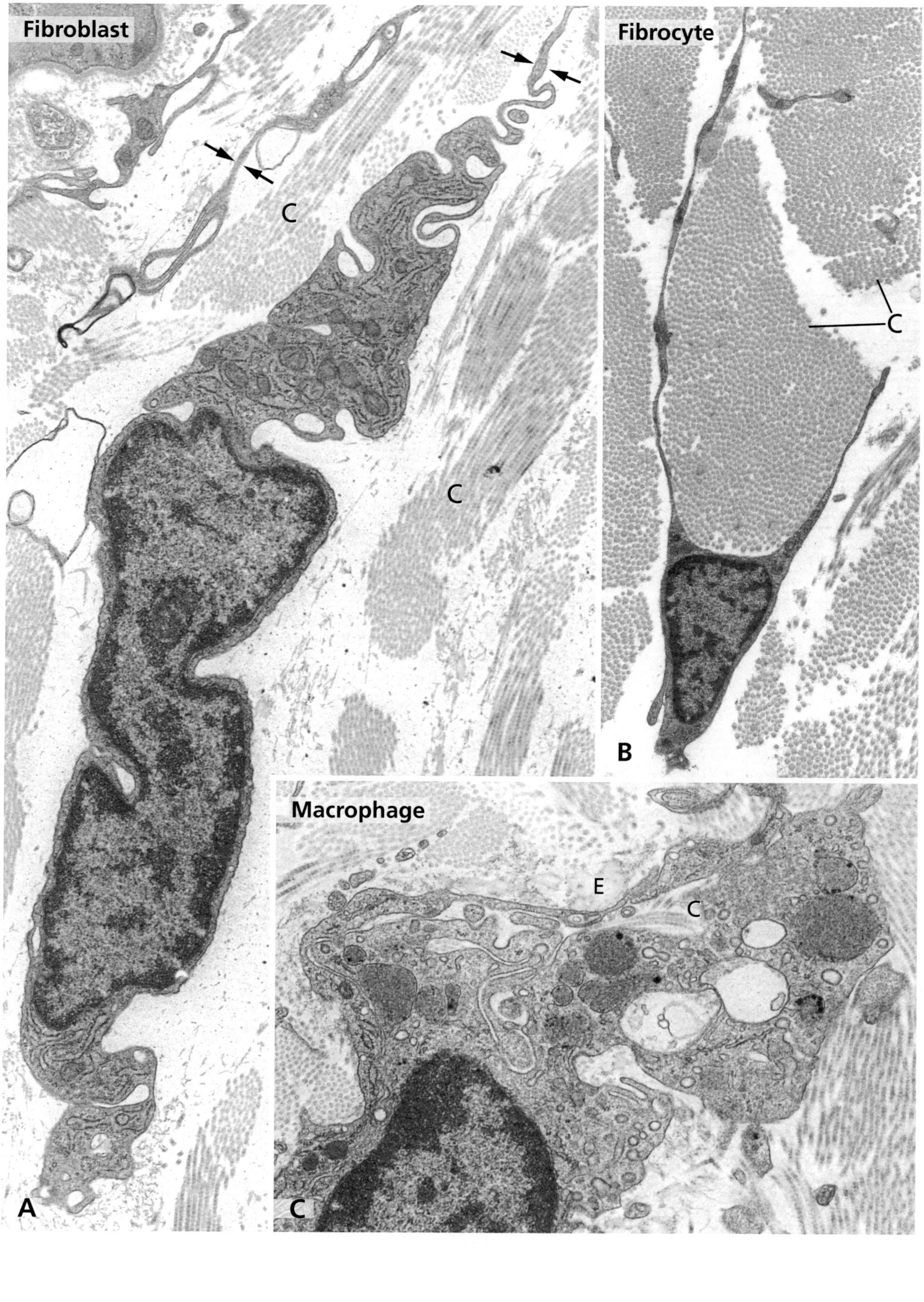
Fibroblast
C
C
A
Fibrocyte
C
B
Macrophage
E
C
C

COLLAGEN AND ELASTIC FIBRES

Both the precursors of collagen and elastic fibres are synthesised and secreted by fibroblasts and assemble extracellularly to form fibrils and fibres. Panel A shows collagen (C) and elastic fibres (E) in the loose connective tissue of the small intestinal wall. It is the type I collagen that forms particularly thick fibres measuring up to 20 micrometers in diameter. Collagen fibres are composed of collagen fibrils (C in panel B), which show a characteristic striated pattern effected by differentiated staining due to the staggered array of the individual tropocollagen molecules (diagram). Arrows in panel A indicate unmyelinated nerve fibres.

Formation of collagen fibres is summarised in the diagram:

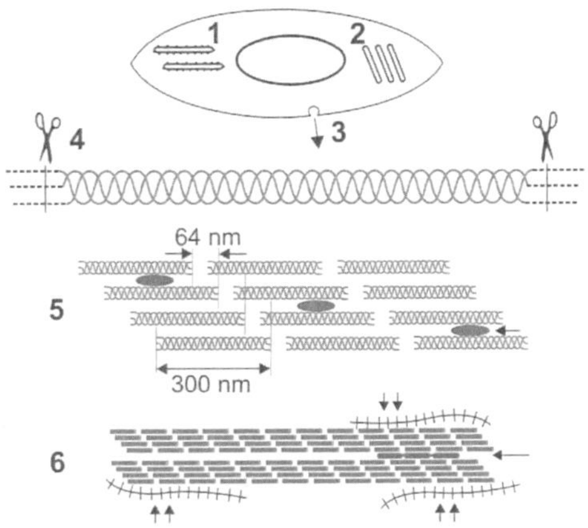

1. Procollagen synthesis on ribosomes bound to the RER, hydroxylation of prolin and lysine, initial glycosylation and formation of triple helices.
2. Terminal glycosylation in the Golgi apparatus.
3. Packaging into secretory vesicles and procollagen exocytosis.
4. Extracellular removal of the non-helical domains of procollagen by action of procollagen peptidase, formation of tropocollagen.
5. Formation of collagen fibrils by self-aggregation of tropocollagen molecules in a staggered array. Cross-links between tropocollagen molecules are catalysed by lysyl oxidase.
6. Side-by-side cross-linking of collagen fibrils to form a collagen fibre mediated by proteoglycans (double arrows) and FACIT collagens (arrow, FACIT stays for fibril associated collagens with interrupted triple helices, types IX, XII, and XIV collagens). For correct assembly and/or turnover of collagen fibrils, other molecules, such as tenascin-X, also are required.

   The diagram is drawn with modifications according to Kierszenbaum (2002).

Elastic fibres (E in panels A, C and D) have unique properties. Like a rubber, they recoil passively after tissues have been stretched. They are made up of elastin and microfibrils (arrows in panel D), composed of fibrillins 1 and 2 and associated glycoproteins. Microfibrils initiate formation of elastic fibres and form a scaffolding for elastin. Panel D shows that elastin and microfibrils are closely associated. Microfibrils are present in the fibre's interior and form an enclosing layer outside (arrows). The elastic fibre in panel D is surrounded by fine branching processes of fibroblasts (F).

Precursors of elastic fibres are produced by fibroblasts, as well as by chondroblasts in the elastic cartilage, and by smooth muscle cells in the walls of blood vessels. All three components of elastic fibres – tropoelastin, fibrillins 1 and 2, and microfibril-associated glycoproteins – are synthesised on the rough endoplasmic reticulum and, after glycosylation in the Golgi apparatus, packaged into secretory vesicles and exported via exocytosis. Extracellularly, tropoelastin assembly is initiated by bundles of microfibrils. Cross-linking of tropoelastin is catalysed by lysyl oxidase. Oxydised lysins condensate into a desmosin ring that covalently cross-links tropoelastin molecules to each other. Fibrillins 1 and 2 provide a force-bearing support and regulate further tropoelastin assembly. Immature elastic fibres formed initially aggregate to build up mature elastic fibres. The two amino acids desmosin and isodesmosin are characteristic for elastin; they are necessary for cross-linking mature elastic fibres and are responsible for their "elastic" properties, allowing stretching and recoil.

References

Chiquet-Ehrismann R and Chiquet M (2003) Tenascins: regulation and putative functions during pathological stress. J Pathol 200: 488

Kierszenbaum AL (2002) Histology and cell biology. An introduction to pathology. St. Louis: Mosby

Kreis T and Vale R (1999) Guidebook to the extracellular matrix and adhesion proteins, 2nd edn. Oxford New York: Oxford University Press

Rosenblom J, Abrams WR, and Mecham R (1993) Extracellular matrix 4: The elastic fiber. FASEB J 7: 1208

Timpl R, Sasaki T, Kostka G, and Chu M-L (2003) Fibulins: A versatile family of extracellular matrix proteins. Nat Rev Mol Cell Biol 4: 479

Magnification: x 16,000 (A), x 85,000 (B), x 77,000 (C), x 30,000 (D)

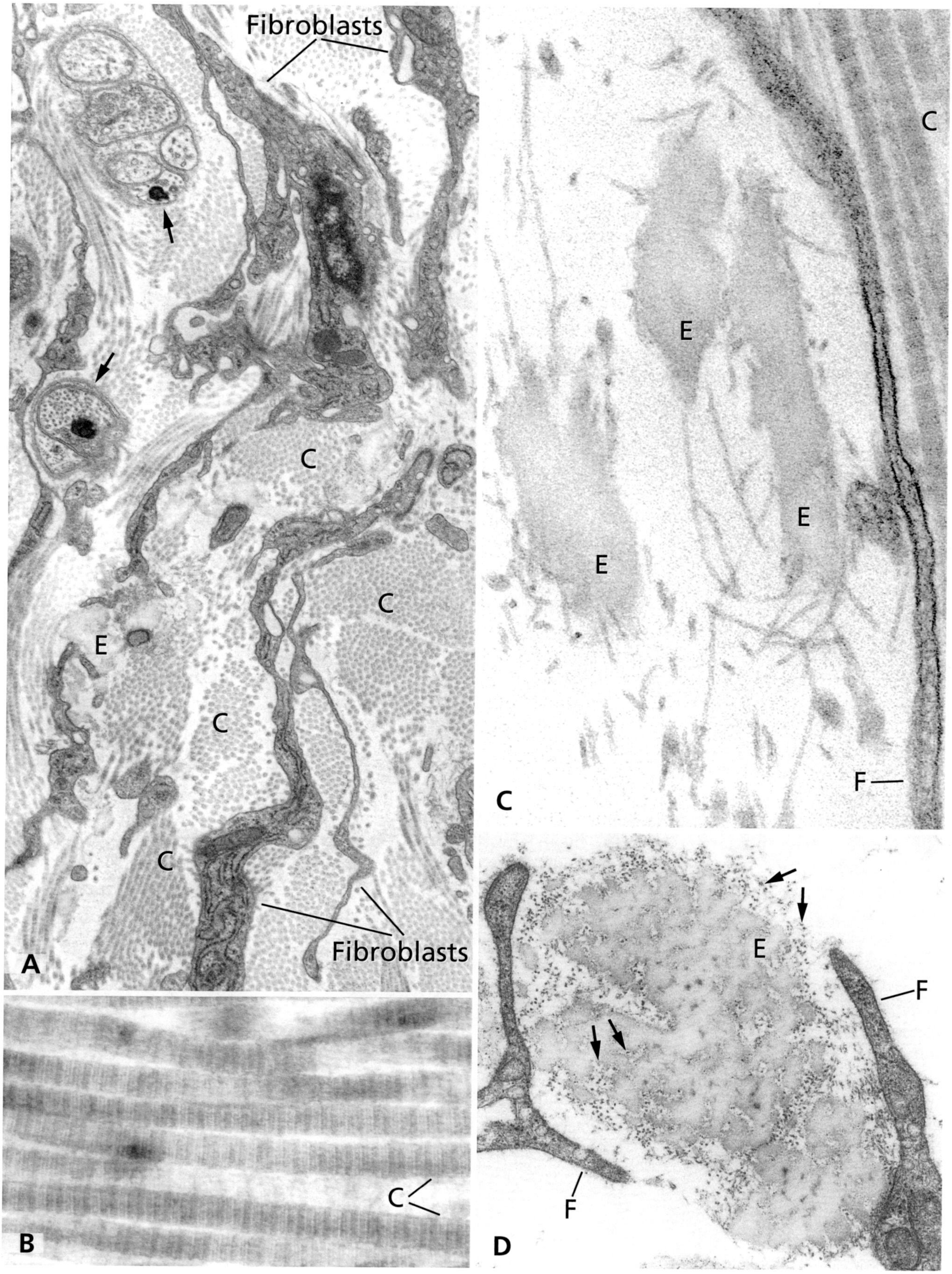
Fibroblasts
C
E
C
C
C
Fibroblasts
A
C
B
C
E
E
E
F
C
E
F
F
D

EOSINOPHILIC GRANULOCYTE, PLASMA CELL, MACROPHAGE, MAST CELL

In contrast to fibroblasts and fibrocytes, which are resident cells, a range of connective tissues cells are blood- and bone marrow-derived and secondarily migrate into the connective tissues. They are commonly designated as "mobile connective tissue cells" and altogether fulfil key roles in the specific and unspecific immune system. Mobile cells are particularly abundant at all sites where contacts with harmful substances and microbial invasion are facilitated. Particularly rich in mobile cells are the mucosal connective tissues of the respiratory and the alimentary tracts.

Panel A shows an eosinophilic granulocyte (E) and a plasma cell (P) embedded in collagen fibrils and in close contact to processes of fibroblasts (F). Eosinophilic granulocytes make up approximately 2–4% of the total leukocytes in the blood (cf. Fig. 171) and in connective tissues, are particularly prominent in the lamina propria of the small intestine. The specific eosinophilic granules, because of their crystalline centre (asterisk), show a unique appearance under the electron microscope. The main component of the crystalline center is the "major basic protein", which disrupts the membranes of parasites. After binding to the surface of parasites, the granule contents are released directly onto the parasites' membranes. Other cell products modulate mast cell activities and protect against potentially harmful effects of mast cell degranulation. The eosinophilic granules, such as those of mast cells and other cells derived from the haematopoietic lineage, are also designated as secretory lysosomes (cf. Figs. 53, 54 and 171).

Plasma cells are the immunoglobulin-producing cells of the specific humoral immunity. They derive from the differentiation of B lymphocytes after antigen stimulation. Plasma cells synthesise and secrete antibodies that bind specifically to that antigen, which originally has activated the precursor B lymphocyte. Plasma cells remain in the tissue, while their products, the antibodies, are distributed throughout the body. The plasma cell nucleus exhibits a typical cartwheel configuration, of which one "spoke" is visible in the cell shown in panel A. The cytoplasm is stuffed with rough endoplasmic reticulum (ergastoplasm) reflecting the cells' engagement in the synthesis of high amounts of glycoproteins. Another segment of a plasma cell (P) is on display in panel C.

Panel B shows a macrophage (MA) with huge lysosomes (Ly) accumulated in the cytoplasm. Its ultrastructural appearance mirrors the crucial functions of macrophages in the immune defence connected with phagocytosis, breakdown of engulfed substances in lysosomes and presentation of antigens and degradation and turnover of extracellular matrix materials and senescent fibres.

A mast cell (M) together with a part of a plasma cell (P) are on display in panel C. Mast cells have a crucial role in the normal defence system and also are involved in pathologic immune reactions, such as allergies. Products of mast cells facilitate the movement of cells and molecules into sites of foreign invasion. Histamine released from the storage granules upon mast cell activation increases vascular permeability and mast cell-derived chemotactic mediators attract mobile cells, such as monocytes, neutrophilic and eosinophilic granulocytes to the site of mast cell activation. In the cytoplasm of non-activated cells, storage granules (asterisks) are accumulated. They mainly contain histamine, heparin, proteoglycans, and a set of proteases, such as tryptase and chymase. The granule contents are released on activation of the cells, together with inflammatory activators produced at the time of activation. Mast cells are activated by binding of a specific antigen, which bridges two immunoglobulin E molecules anchored to receptors in the plasma membrane. A particularly rapid release of mast cell granule contents is possible by compound exocytosis. Upon stimulation, granules fuse with each and form channels to the surface for extrusion of the contents. Recent findings indicate that microtubule-dependent movement of the secretory granules plays an important role in the mast cell exocytic response.

References

Smith AJ, Pfeiffer JR, Zhang J, Martinez AM, Griffiths GM, and Wilson BS (2003) Microtubule-dependent transport of secretory vesicles in RBL-2H3 cells. Traffic 4: 302

Swanson MS and Fernandez-Moreira E (2003) A microbial strategy to multiply in macrophages: A pregnant pause. Traffic 3: 170

Magnification: x 10,400 (A); x 7,500 (B); x 11,600 (C)

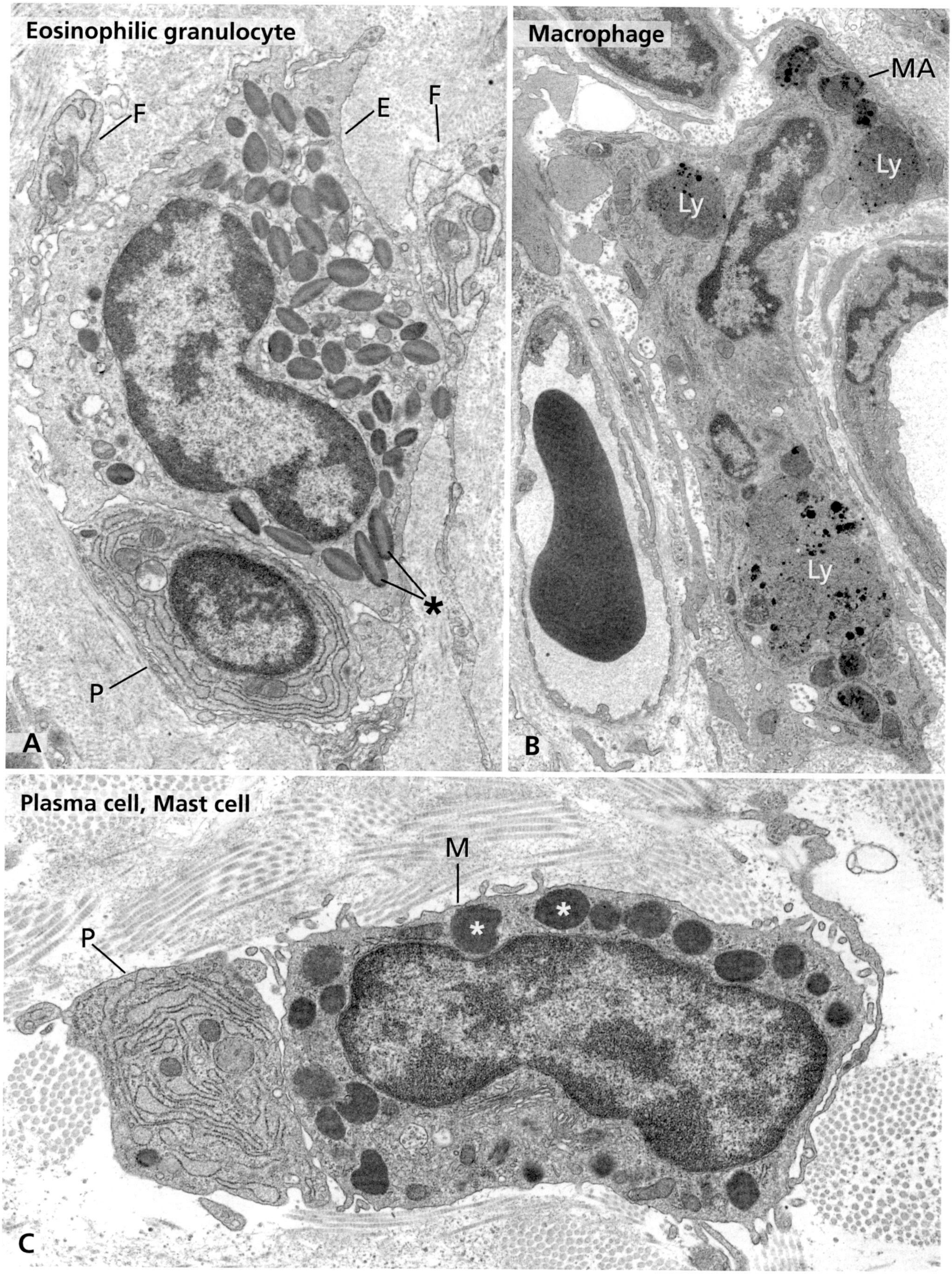
Eosinophilic granulocyte
F
E
F
*
P
A
Macrophage
MA
Ly
Ly
Ly
B
Plasma cell, Mast cell
M
*
*
P
C

DENSE CONNECTIVE TISSUE: COLLAGEN BUNDLES IN THE CORNEA

The cornea forms the front part of the outer tunic of the eyeball, which protects the inner structures and helps to maintain the eye's shape. The corneal centre coincides with the anterior pole of the eyeball. Being part of the optic apparatus, the cornea is transparent, leads numerous nerve fibres, but lacks blood vessels and lymphatics. Immune elements are thus prevented from invading the cornea, which makes the cornea an ideal organ for transplantation, with minimal risk of being rejected by the immune system of the host.

The anterior surface of the cornea is covered by a stratified epithelium (cf. Fig. 118). The basal cells are attached to a thin basal lamina being composed of an inner rare and an outer dense part (lamina rara and lamina densa). The latter is connected with the neighbouring Bowman's layer or Bowman's membrane (cf. Fig. 141), which lacks characteristics of a membrane but is the anteriormost part of the corneal stroma.

The corneal stroma (substantia propria) represents 90% of the thickness of the cornea. It is made up of a particularly dense connective tissue mainly consisting of types I and V collagens. The collagen fibrils are arranged regularly in thin and flat layers. In the successive planes, they cross at various angles, thus forming a complex fibrillar lattice, which makes the cornea highly resistant to any kind of deformation or trauma. The micrograph shows a segment of the middle part of the corneal stroma, displaying one of the stromal fibroblasts and a very thin process of another one. The cells are extremely flat and spanned between the collagen lattices that occupy the entire extracellular space.

Both cross sectioned and longitudinally sectioned segments of the layers of collagen fibrils show that there are regular distances between the individual collagen fibrils. The regularity and spacing of collagen fibrils are closely related to the water content of the cornea and this in turn is connected with the corneal transparency and dependent on the contents and qualities of proteoglycans. For transparency of the cornea, a precise regulation of the water content is required. Collagen fibrils are embedded in a ground substance enriched in proteoglycans containing keratan sulfate and dermatan sulfate. It is the high abundance of keratan sulfate in the corneal stroma, which seems to be pivotal in the precise regulation of the corneal water content and in the maintenance of a level of tissue hydration critical for transperancy. The keratan sulfate linked proteins mainly are lumican, keratokan, and mimecan. Keratan sulfate and dermatan sulfate linked proteoglycans possess distinct water binding properties. Whereas dermatan sulfate is fully hydrated at hydration levels typical for normal corneal tissue, keratan sulfate is only partially hydrated and seems to function as a buffer for hydration.

References

Funderburgh JL (2000) Keratan sulfate: structure, biosynthesis, and function. Glycobiology 10: 951

Verkman AS (2003) Role of aquaporin water channels in eye function. Exp Eye Res 76: 137

Magnification: x 30,000

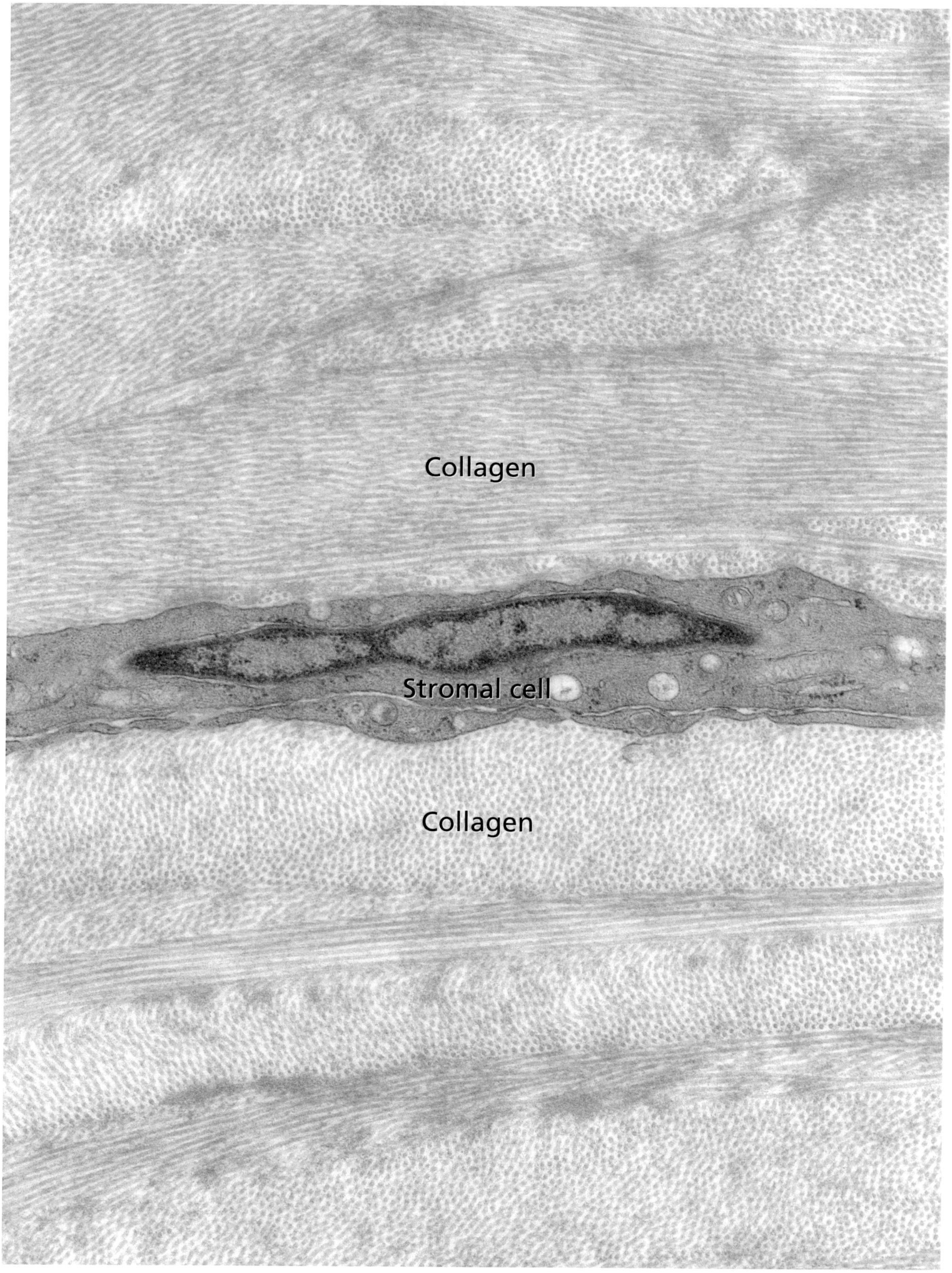
Collagen
Stromal cell
Collagen

BOWMAN'S LAYER

Originally, the layer of Bowman was termed Bowman's membrane analogously to Descemet's membrane at the inner side of the cornea. However, the layer of Bowman does not correspond to a basement membrane as is true for the membrane of Descemet (cf. Fig. 92). The layer of Bowman is the anteriormost part of the corneal stroma, although differently organised. It is 6 to 9 µm thick and formed by densely packed mainly type I collagen fibrils. The micrographs of a human cornea in panels A and B show segments of the basal cells of the corneal epithelium (cf. Fig. 118) and the associated basal lamina with the lamina rara close to the plasma membrane of the epithelial cells and the adjacent lamina densa labelled by asterisks in panel B. The lamina densa continues into the Bowman's layer, which is on display in the middle part of the survey micrograph in panel A. The different organisation of the collagen fibrils in the Bowman's layer in comparison with their organisation in the stroma shown in the lower part of panel A is evident. Whereas in the corneal stroma collagen fibrils are piled up forming a lattice of thin and flat layers, they are organised in a dense network in the layer of Bowman presented at higher magnification in panel B. Bowman's layer is transparent as are all parts of the cornea and builds up a protective barrier to trauma and microbial invasion.

References

Abrams GA, Bentley E, Nealey PF, and Murphy CJ (2002) Electron microscopy of the canine corneal basement membranes. Cells Tiss Org 170: 251

Magnification: x 15,000 (A), x 91,000 (B)

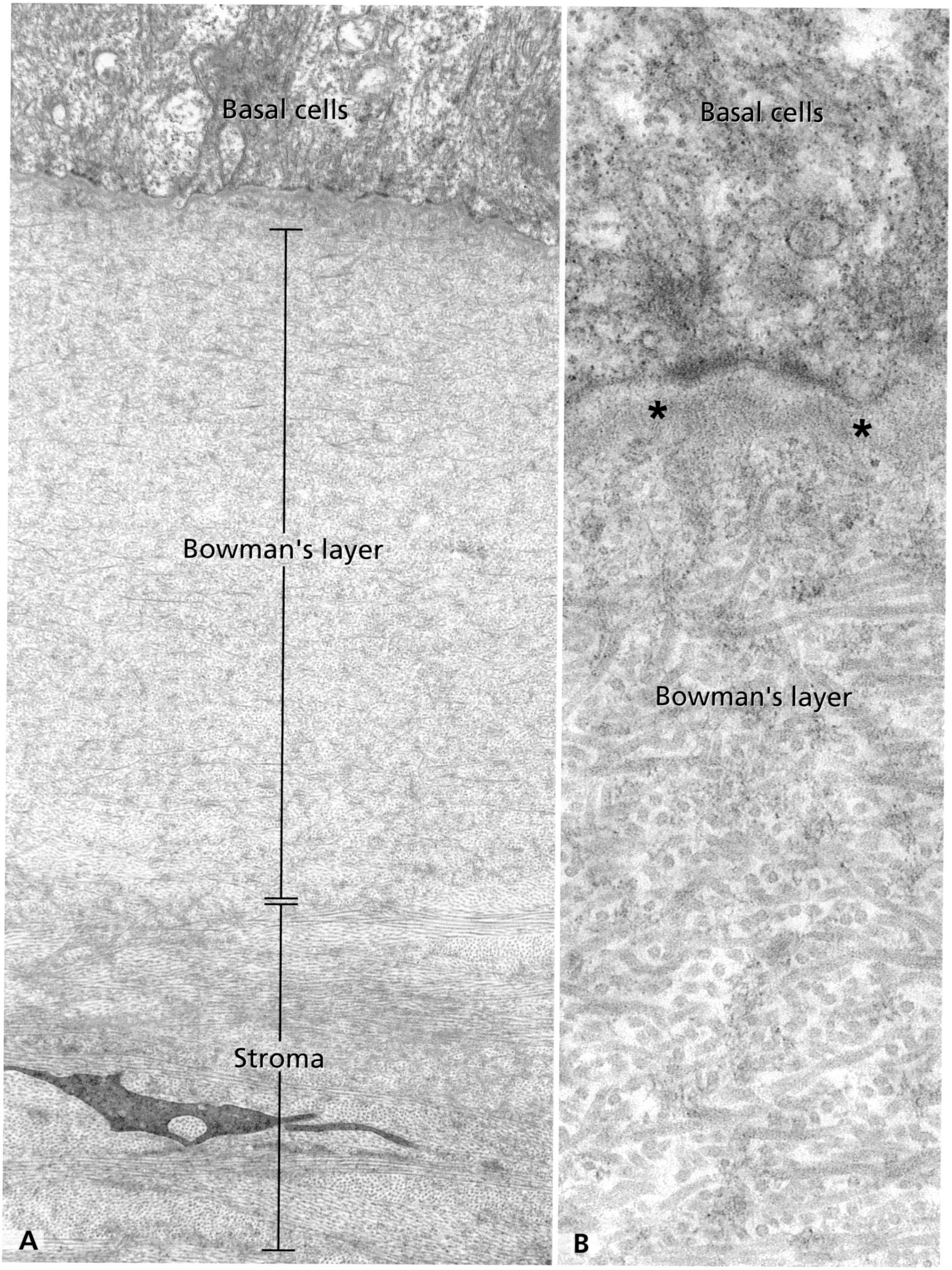
Basal cells
Bowman's layer
Stroma
A
Basal cells
*
*
Bowman's layer
B

AMYLOIDOSIS OF KIDNEY

Amyloidosis is a term to describe the deposition of proteins in a large variety of diseases. The deposited proteins form β-pleated sheet fibrils of 75 to 100 Å in diameter that accumulate in the extracellular spaces. The ordered deposits of amyloid fibres are birefringent and have a selective affinity for the histochemical dye Congo red, which is of diagnostic importance for the differentiation from other fibrillar protein deposits.

Amyloidosis can be an inherited or an acquired disease affecting either single organs or whole organ systems. Several types of amyloidosis can be distinguished depending on the composition of the fibril subunits. The most common forms are the amyloidosis of Ig light chains (AL amyloidosis), in which the subunit protein is the variable portion of monoclonal immunoglobulin light chains, and the reactive amlyoidosis (AA amyloidosis), in which the subunit protein is the amyloid A. As the consequence of progressive amyloid deposits, the normal tissue structure and function becomes disrupted.

In renal amyloidosis, the AA type is prevailing (about half of the cases) but the AL type is also often observed. Amyloid depositions can be found in glomeruli, blood vessels, the interstitium, and peritubular basement membranes. In the glomeruli, amyloid deposits are present in the mesangial matrix (asterisks) and along, as well as inside, the glomerular basement membrane. Podocytes usually show effaced foot processes and are focally detached from the glomerular basement membrane, which results in the extension of amyloid fibrils in the urinary space. The amyloid deposits may result in lamination of the glomerular basement membrane or its complete replacement. The inset illustrates the presence of large masses of amyloid fibrils (asterisk) between an endothelial cell and a podocyte, which in this instance have resulted in the replacement of the glomerular basement membrane. These apparently unbranched fibrils are haphazardly arranged for the most part. At the podocyte aspect of the glomerular wall, spicular arrangements of amyloid fibres can be often observed. Such amyloid spicules represent morphological signs of active amyloid deposition and are observed more often in so-called primary amyloidosis. They disappear with clinical signs of regression of renal amyloidosis or are primarily not observed in cases with glomerular basement lamination. The distortion of the normal glomerular architecture by the amyloid deposits resulting in the expansion of the mesangium, widening of glomerular basement membranes, and narrowing capillary lumen is obvious in the electron micrograph. Evidently, such structural alterations are incompatible with normal glomerular function.

References

Benson M (2001) Amyloidosis. In: The metabolic and molecular bases of inherited diseases (Scriver C, Beaudet A, Valle D, and Sly WS, eds). New York: McGraw-Hill, pp 5345

Dickersin G (2000) Diagnostic electron microscopy. A text/atlas. New York: Springer

Moss J, Shore I, and Woodrow D (1994) AA glomerular amyloid. An ultrastructural immunogold study of the colocalisation of heparan sulphate proteoglycan and P component with amyloid fibrils together with changes in distribution of type-IV collagen and fibronectin. Histopathology 24: 427

Dikmann S, Churg J, and Kahn T (1981) Morphologic and clinical correlates in renal amyloidosis. Hum Pathd 12: 160

Zollinger H and Mihatsch M (1978) Renal pathology in biopsy. Berlin Heidelberg New York: Springer

Magnification: x 6,300 (A); x 29,000 (inset)

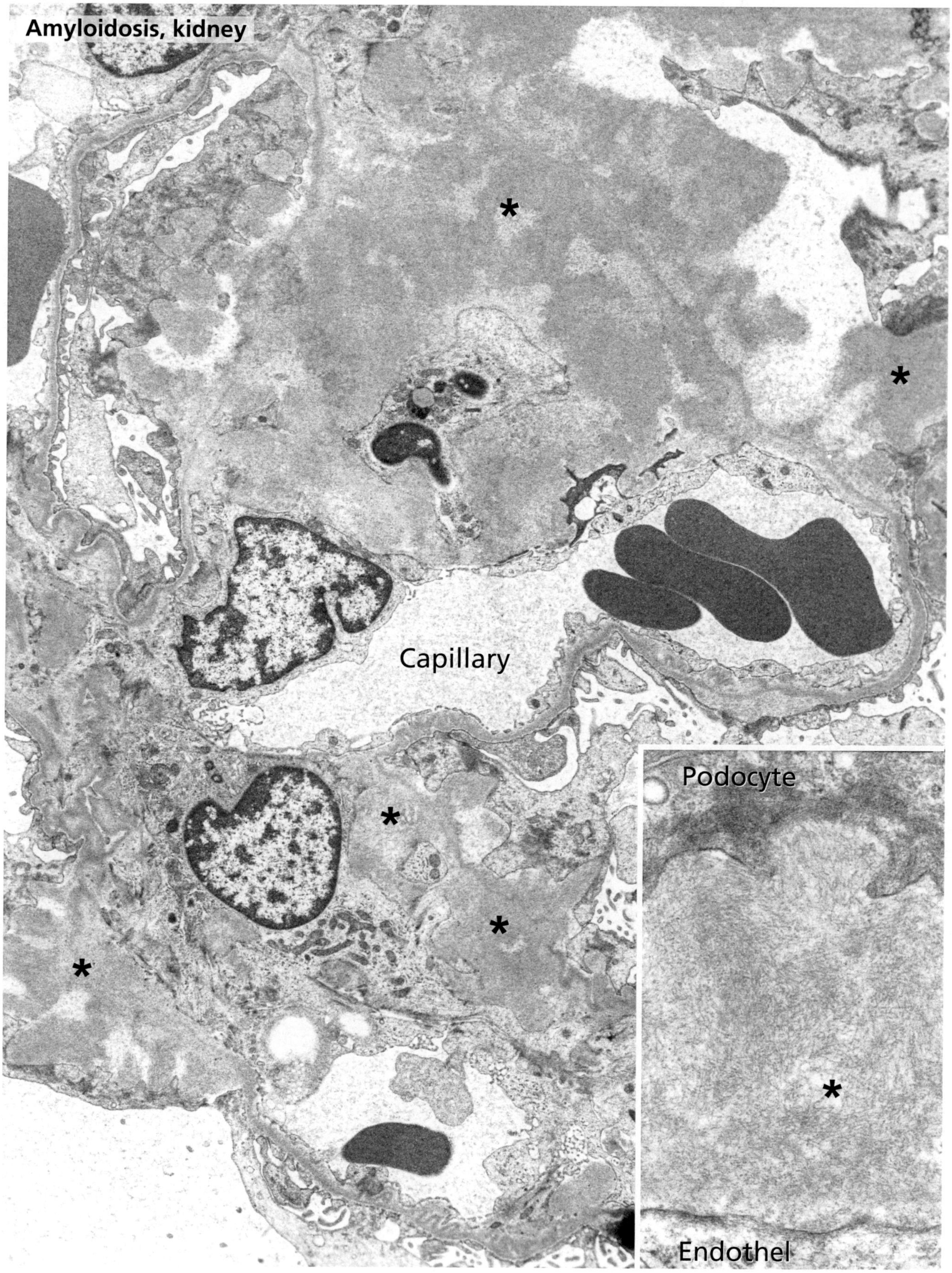
Amyloidosis, kidney
*
*
Capillary
Podocyte
*
*
*
*
Endothel

AMYLOID FIBRILS: GROWTH AS SEEN BY TIME LAPSE, ATOMIC FORCE MICROSCOPY

Amyloid deposits are composed of fibrils when investigated by transmission electron microscopy. The fibrils are insoluble and highly stable structures. Amyloid fibrils can be formed *in vitro* by using synthetic proteins such as amylin, which is the subunit protein of amyloid deposits observed in human type-2 diabetes mellitus.

The atomic force microscopy is an imaging technique that can be applied to study single macromolecules at high resolution in an aqueous environment. As such, this technique is a useful complementary tool to electron microscopy since it permits studies by time lapse imaging under quasi-physiological conditions. The example shown here concerns time lapse imaging by atomic force microscopy of amyloid fibril growth.

Panel A shows protofibrils formed on a mica surface by amylin in a buffer solution. Such protofilaments are assembled by the stacking of twisted β-sheets perpendicular to the fibril axis and represent the building blocks from which the formation of higher-order fibrils takes place. Panels B, C and D show the growth of individual amylin protofilaments on a mica surface. The same field was recorded over periods of times of several hours. Such imaging as shown here has allowed not only to study individual fibrils but also to monitor their growth. From the time lapse observation, growth rates of the fibrils could be calculated and were found to be 1.1 (SD = 0.5) nm/min. In addition to single fibrils and their growth profiles, the formation of protofilaments from nucleation centres seen as small dots on the mica surface has been possible. Individual fibrils and protofilaments could be distinguished unequivocally by their different diameters. Such experiments, visualising the formation and growth of protofilaments and the assembly in fibrils, may be applicable to study the effect of drugs developed to interfere with the process of fibril formation.

References

Binnig G (1992) Force microscopy. Ultramicroscopy 42: 7

Binnig G, Quate CF, and Gerber C (1986) Atomic force microscope. Phys Rev Lett 56: 930

Blackley HK, Sanders GH, Davies MC, Roberts CJ, Tendler SJ, and Wilkinson MJ (2000) In-situ atomic force microscopy study of beta-amyloid fibrillization. J Mol Biol 298: 833

Goldsbury C, Aebi U, and Frey P (2001) Visualizing the growth of Alzheimer's A beta amyloid-like fibrils. Trends Mol Med 7: 582

Goldsbury CS, Cooper GJ, Goldie KN, Muller SA, Saafi EL, Gruijters WT, Misur MP, Engel A, Aebi U, and Kistler J (1997) Polymorphic fibrillar assembly of human amylin. J Struct Biol 119: 17

Jena B and Hörber H (2002) Atomic force microscopy in cell biology. In: Meth Cell Biol, vol. 68. Amsterdam: Academic Press

Ionescu-Zanetti C, Khurana R, Gillespie JR, Petrick JS, Trabachino LC, Minert LJ, Carter SA, and Fink AL (1999) Monitoring the assembly of Ig light-chain amyloid fibrils by atomic force microscopy. Proc Natl Acad Sci U S A 96: 13175

Muller DJ, Schabert FA, Buldt G, and Engel A (1995) Imaging purple membranes in aqueous solutions at sub-nanometer resolution by atomic force microscopy. Biophys J 68: 1681

Shirahama T and Cohen A (1967) High-resolution electron microscopic analysis of the amyloid fibril. J Cell Biol 33: 679

Stolz M, Stoffler D, Aebi U, and Goldsbury C (2000) Monitoring biomolecular interactions by time-lapse atomic force microscopy. J Struct Biol 131: 171

Walsh DM, Hartley DM, Kusumoto Y, Fezoui Y, Condron MM, Lomakin A, Benedek GB, Selkoe DJ, and Teplow DB (1999) Amyloid beta-protein fibrillogenesis. Structure and biological activity of protofibrillar intermediates. J Biol Chem 274: 25945

Magnification: x 50,000 (A-D)

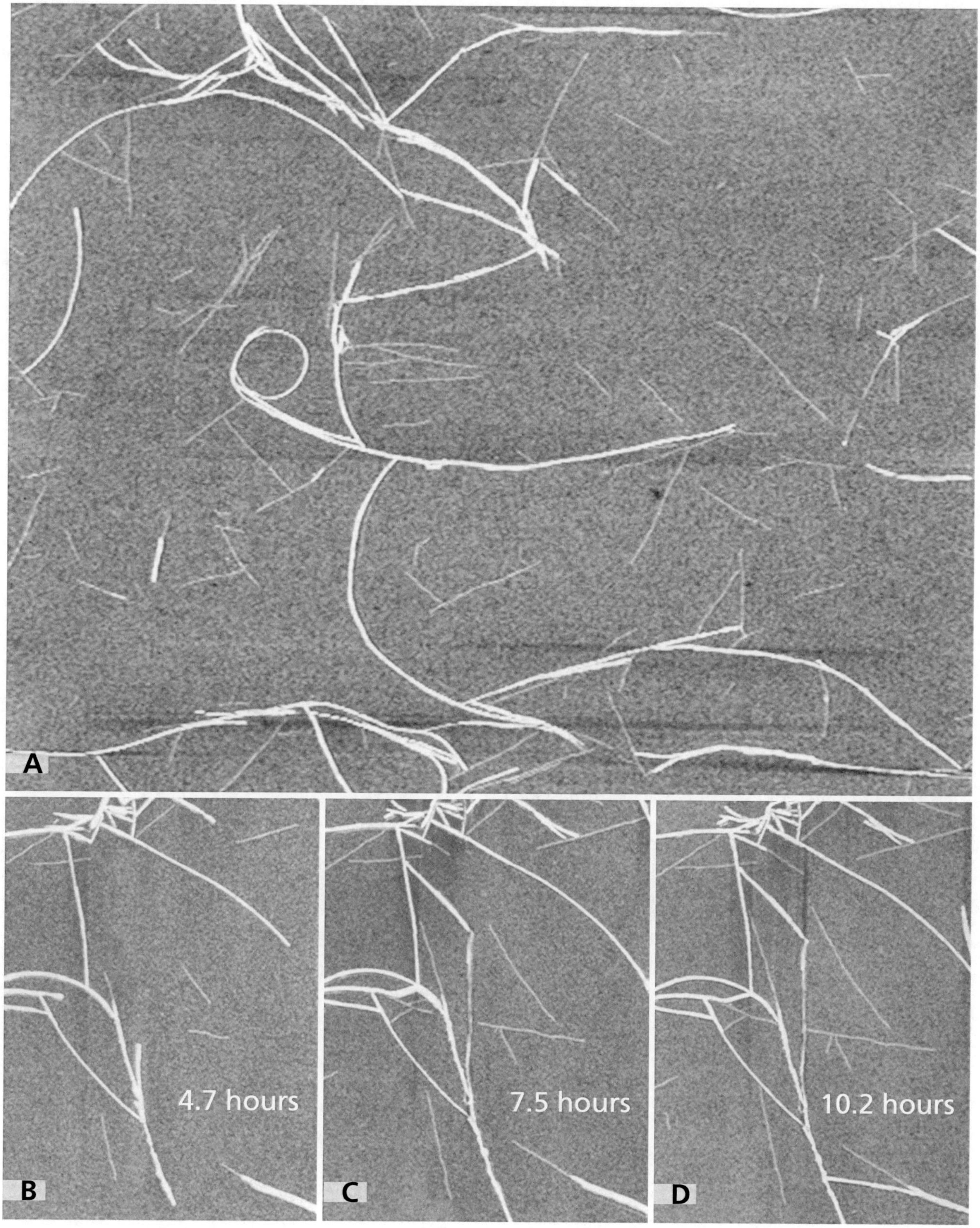
A
B
4.7 hours
C
7.5 hours
D
10.2 hours

WHITE ADIPOSE TISSUE

Adipose tissue exist in two different forms which can be distinguished with the free eye by their color: white adipose tissue and brown adipose tissue. Both have a highly differing fine structure. Furthermore, their location differs significantly. White adipose tissue is found in the subcutaneous tissue where it exists mainly as single adipocytes or in the peritoneal cavity where it forms a compact tissue. The adipocytes of the white adipose tissue contain a single lipid droplet and therefore are referred to as univacuolar (or unilocular) adipocytes. Since the size of the lipid droplet can vary considerably, white adipocytes may measure between 40 and 120 μm.

In panel A, an entire white adipocyte and parts of adjacent white adipocytes can be seen as they exist in the subcutaneous tissue. The predominant feature is the single, slightly electron dense lipid droplet (L), which determines the spherical shape of the adipocyte and gives them the appearance of signet ring. The adipocytes are surrounded by a basement membrane and its dense lamina is well visible in panel B (Lb). The cytoplasm exists only as a small peripheral rim, which contains a flattened nucleus. The area marked in panel A is shown at higher magnification in panel B. Besides part of the lipid droplet, a representative region of the cytoplasm with part of the nucleus is seen. In the juxtanuclear region, numerous mitochondria (M) of the crista type are located (cf. Fig. 64). In addition, the cisternal stack of the Golgi apparatus with a well-developed *trans* Golgi network and clathrin-coated vesicles can be appreciated. The cytoplasm contains several cisternae of the rough endoplasmic reticulum and numerous free ribosomes. Along the plasma membrane, caveolae (arrows) exist, which give raise to numerous smooth-surfaced vesicles. The lipid droplet, which consists to about 95% of triglycerides, is not separated by a membrane from the cytoplasm. However, along the border between the cytoplasm and the lipid droplet, a flat network of intermediate filaments exists, which can be only fully appreciated in grazing sections.

The white adipose tissue represents an important source for energy. Depending on the energy need of the organism, triglycerides of the lipid droplet will be metabolized or stored. Therefore, the white adipose tissue represents a highly dynamic and active tissue.

In addition to its main function in the energy metabolism, white adipose tissue also acts as an insulator and represents a storage site for water. From a mechanical point of view, it is important for the correct positioning of various organs such as the kidneys and eyes and functions as an elastic cushion at sites of high mechanical stress such as the soles and palmes as well joints.

The white adipose tissue is also active as an endocrine organ by secreting various adipocytokines such as the leptin, adiponectin, resistin, visfatin, tumor necrosis factor α and plasminogen activator inhibitor-1. The adipocytokines play important roles in metabolic homeostasis and their dysregulation results in metabolic disease and arteriosclerosis.

References

Brasaemle DL (2007) Thematic review series: adipocyte biology. The perilipin family of structural lipid droplet proteins: stabilisation of lipid droplets and control of lipolysis. J Lipid Res 48: 2547

Cinti S (1999) The adipose organ: Milan: Kurtis SRL

Cinti S (2001) The adipose organ: morphological perspectives of adipose tissues. Proc Nutr Soc 60: 319

Cushman S (1970) Structure-function relationship in the adipose cell: I. Ultrastructure of the isolated adipose cell. J Cell Biol 46: 326

Fukuhara A, Matsuda M, Nishizawa M, Segawa K, Tanaka M, Kishimoto K, Matsuki Y, and Murakami M (2005) Visfatin: a protein secreted by visceral fat that mimics the effects of insulin. Science 307: 426

Klaus S (2004) Adipose tissue as a regulator of energy balance. Curr Drug Targets 5: 241

Novakofski J (2004) Adipogenesis: usefulness of in vitro and in vivo experimental models. J Anim Sci 82: 905

Rayner DV (2001) The sympathetic nervous system in white adipose tissue regulation. Proc Nutr Soc 60: 357

Trayhurn P and Wood IS (2004) Adipokines: inflammation and the pleiotropic role of white adipose tissue. Br J Nutr 92: 347

Waki H and Tontonoz P (2007) Endocrine functions of adipose tissue. Annu Rev Pathol 2: 31

Magnification: x 3,700 (A); x 27,500 (B)

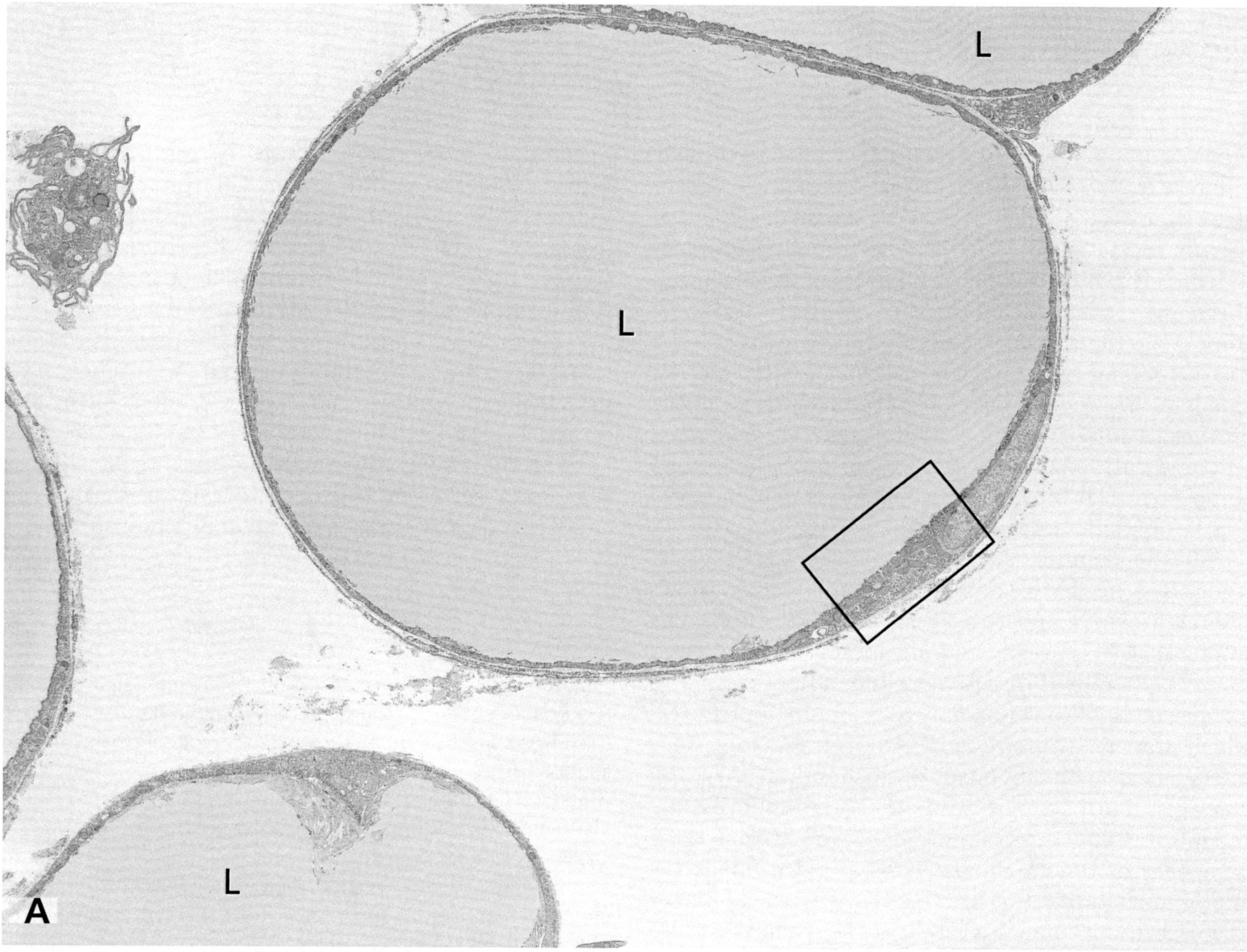
L
L
L
A

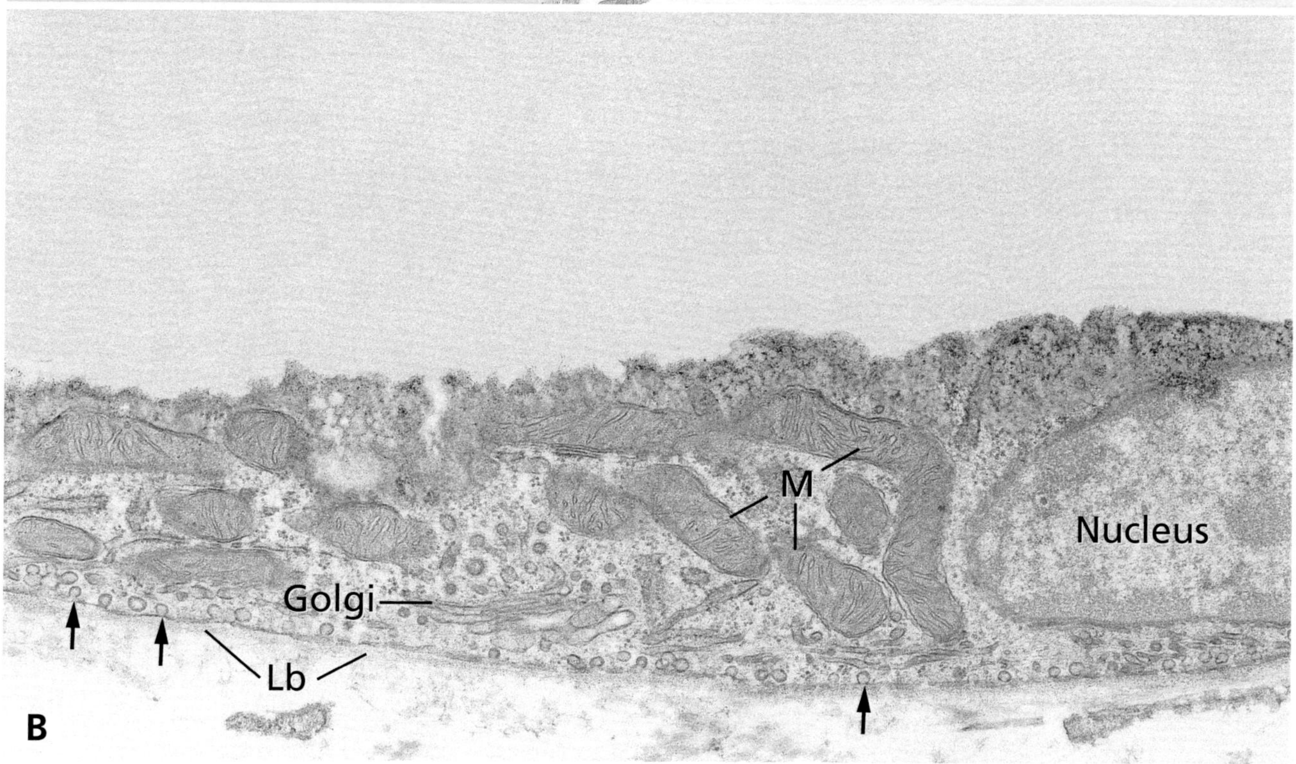
M
Nucleus
Golgi
Lb
B

BROWN ADIPOSE TISSUE

Brown adipose tissue is structurally and functionally different from white adipose tissue. Although abundant at birth, in the adult it exists only as small depots at specific locations such as the interscapular region of rodents. The brown colour of the tissue is due to an abundance of mitochondria. In contrast to white adipocytes, the adipocytes forming the brown adipose tissue contain numerous lipid droplets which differ greatly in size and are therefore called plurivacuolar (or plurilocal) adipocytes. Brown adipocytes are polygonal and as a rule smaller in size (20–40 μm) than white adipocytes. The brown adipose tissue is compact and rich in capillaries and sympathetic nerve fibres.

All the features mentioned above can be easily recognised in the electron micrograph which shows several brown adipocytes at low magnification. The round nucleus is usually centrally located, in contrast to the peripherally located of white adipocytes. The cytoplasm is filled with numerous lipid droplets (L), which vary greatly in size. Furthermore, the cytoplasm is interspersed with abundant mitochondria (M) and contains both rough and smooth endoplasmic reticulum. The inset shows the two characteristic organelles of brown adipose tissue, lipid droplets (L) and mitochondria (M) of the crista type, at higher magnification. Numerous capillaries (Cap) in very close contact with the adipocytes are another hallmark of brown adipose tissue. Some exhibit a hypetrophic endothel (Cap*). The cytoplasm contains also peroxisomes, usually arranged in small clusters (Po).

The fatty acids derived from triglycerides of the brown adipose tissue, in contrast to the white adipose tissue, are used for the generation of heat. The thermogenic process becomes activated when the body temperature falls below thermoneutrality. The thermogenesis is mainly controlled by noradrenalin which binds to β-adrenergic receptors of the brown adipocytes. The close topographic relation between the lipid droplets and the mitochondria, as seen in the inset, is important for the thermogenesis. The free fatty acids in the cytosol generated by the lipolysis of triglycerides of the lipid droplets are uitilised by mitochondria for β-oxidation. The proton gradient in the mitochondria generated by this process is used for thermogenesis instead of ATP synthesis. The characteristic brown adipose tissue uncoupling protein-1 (UPC1, thermogenin) of the inner mitochondrial membrane is critical for this process. Thus, brown adipose tissue is of vital importance under conditions of need of extra heat such as acute and chronic cold exposure, after birth, and during arousal from hibernation.

Like the white adipose tissue, the brown adipose tissue is a secretory organ. Autocrine, paracrine and endocrine factors are secreted upon different stimuli.

References

Cannon B and Nedergaard J (2004) Brown adipose tissue: function and physiological significance. Physiol Rev 84: 277

Cinti S (2001) The adipose organ: morphological perspectives of adipose tissues. Proc Nutr Soc 60: 319

Cinti S (2002) Adipocyte differentiation and transdifferentiation: plasticity of the adipose organ. J Endocrinol Invest 25: 823

Cypess A, Lehman S, Williams G, Tal I, Rodman D, Goldfine A, Kuo F, Palmer E, Tseng Y-H, Doria A, et al. (2009) Identification and importance of brown adipose tissue in adult humans. New Engl J Med 360: 1509

Havel P (2004) Update on adipocyte hormones. Regulation of energy balance and carbohydrate/lipid metabolism. Diabetes 53: S143

Klaus S (2004) Adipose tissue as a regulator of energy balance. Curr Drug Targets 5: 241

Klingenberg M and Huang S-G (1999) Structure and function of the uncoupling protein from brown adipose tissue. Biochim Biophys Acta 1415: 271

Lafontan M and Berlan M (1993) Fat cell adrenergic receptors and the control of white and brown fat cell function. J Lipid Res 34: 1057

Nedergaard J, Bengtsson T, and Cannon B (2007) Unexpected evidence for active brown adipose tissue in adult humans. Am J Physiol Endocrinol Metab 293: E444

Trayhurn P and Nicholls D (1986) Brown adipose tissue: London: Edward Arnold

Magnification: x 10,300 (A); x 33,000 (inset)

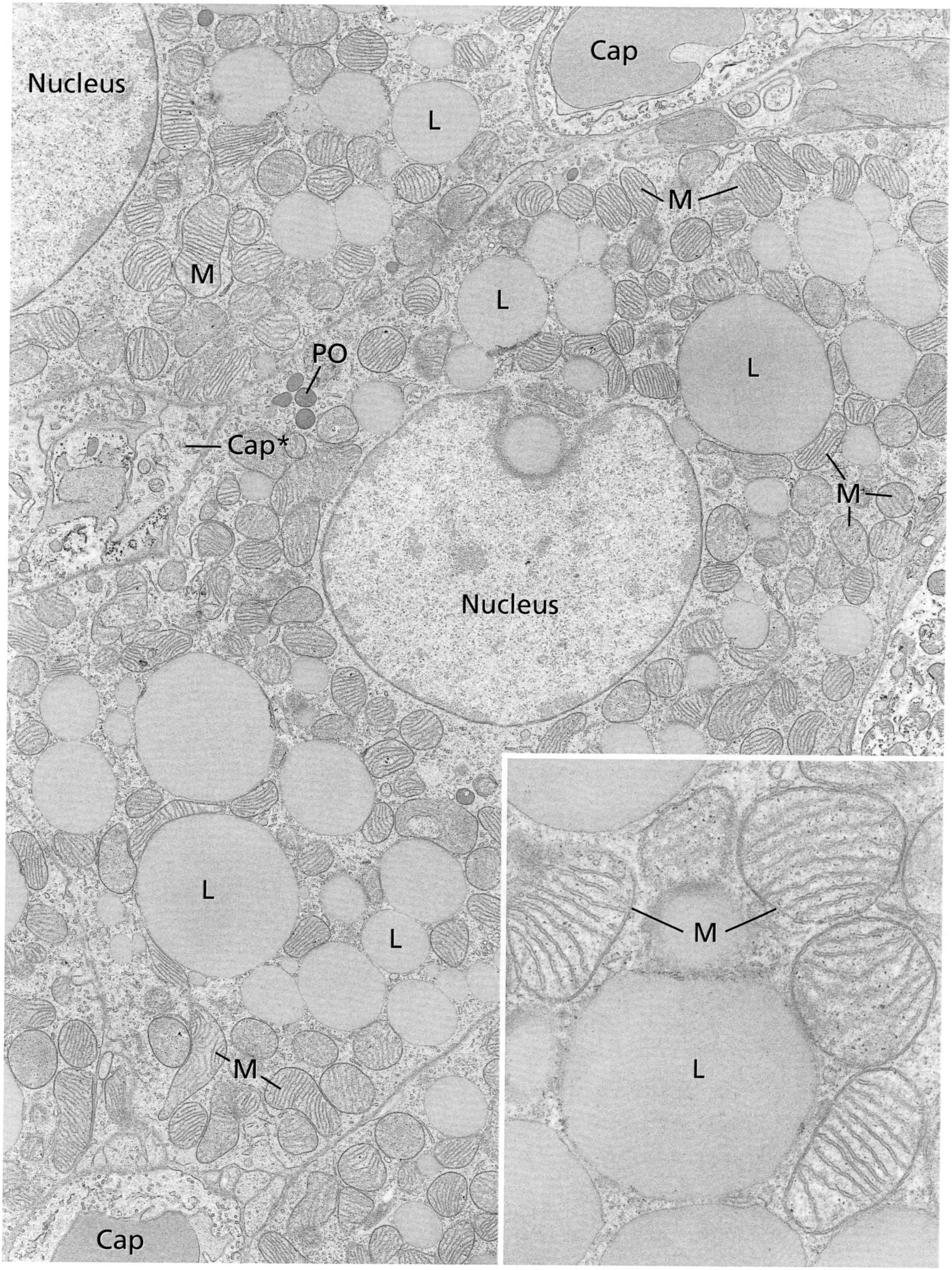
Nucleus
Cap
L
M
M
L
PO
L
Cap*
M
Nucleus
L
L
M
Cap
M
L

ARTICULAR CARTILAGE

Cartilage represents highly specialised connective tissues. Three types are distinguished, which are found in different locations and differ with regard to the amount and arrangement of chondrocytes and fibres: hyaline (articular), elastic, and fibrocartilage. All three have in common a low metabolic rate, the absence of blood vessels, and the ability to grow. Cartilage consists of chondrocytes and a matrix, the latter is the bulk of cartilage mass. The matrix consists of collagens and proteoglycans and has a 60–80% water content. Cartilage is elastic and flexible and has a high tensile strength. This characteristic is mainly due to the matrix composition and architecture. Hyaline cartilage is the most abundant type of cartilage and, for example, covers the bone surfaces of joints, hence the name articular cartilage.

A most detailed analysis of articular cartilage, especially its matrix, became possible through high pressure freezing, which preserved the cartilage in a near native vitrified state. Panels A and B show electron micrographs of bovine articular cartilage fixed by high pressure freezing followed by freeze substitution and Epon embedding. Pairs of chondrocytes (the so-called chondrons) can be seen. They are actively synthesising and secreting the matrix components. Young, as compared to old chondrocytes contain abundant endoplasmic reticulum and a well developed Golgi apparatus. Mature chondrocytes, as seen in panel A, are round and lie in nests.

The matrix synthesised by the chondrocytes consists of three compartments that differ in the architecture of their collagen fibres. The chondrocytes are surrounded by the pericellular matrix (PM in panel A), which is quite variable in extent. It seems homogenous because of the absence of collagen fibres and the presence of randomly oriented cross-banded filaments of 10–15 nm diameter, also found in the two other matrix compartments. The pericellular matrix is followed by the territorial matrix (TM in panel A) and the interterritorium (ITM in panel A). The territorial matrix contains basket-like arrangements of collagen fibres. The interterritorial matrix represents the bulk mass of the matrix, and collagen fibrils are found in two different orientations. One is in the form of parallel fibre arrangements, which give rise to a larger arcade-like construction, and the other in a more random organisation of fibres. The spaces between the collagen fibres are rich in proteoglycans. Panel B shows details of the interterritorial matrix. Longitudinally sectioned (arrows) and cross-sectioned (open arrow) collagen fibres of varying diameters exist. In between, the proteoglycans form a fine granular mass (G in panel B). It is now recognised that the previously described fine reticular network represents a segregation artefact.

Panel C shows foci of mineralisation in the interterritorium of human articular cartilage. This specimen was conventionally immersion-fixed in glutaraldehyde and osmium tetroxide and not by high pressure freezing.

References

Ghadially F (1983) Fine structure of synovial joints. A text and atlas of normal and pathological articular tissues. London: Butterworth

Hunziker E (1992) Articular cartilage structure in humans and experimental animals. In: Articular cartilage and osteoarthritis (Kuettner K, Schleyerbach R, Peyron J, and Hascall V, eds). New York: Raven, pp 183

Hunziker EB, Herrmann W, Schenk RK, Mueller M, and Moor H (1984) Cartilage ultrastructure after high pressure freezing, freeze substitution, and low temperature embedding. I. Chondrocyte ultrastructure-implications for the theories of mineralisation and vascular invasion. J Cell Biol 98: 267

Hunziker EB, Michel M, and Studer D (1997) Ultrastructure of adult human articular cartilage matrix after cryotechnical processing. Microsc Res Tech 37: 271

Hunziker EB and Schenk RK (1984) Cartilage ultrastructure after high pressure freezing, freeze substitution, and low temperature embedding. II. Intercellular matrix ultrastructure – preservation of proteoglycans in their native state. J Cell Biol 98: 277

Poole C (1993) The structure and function of articular cartilage matrices. In: Joint cartilage degradation. Basic and clinical aspects (Woessner J and Howell ED, eds). New York: Marcel Dekker, pp 1

Magnification: x 4,000 (A); x 34,000 (B); x 26,000 (C)

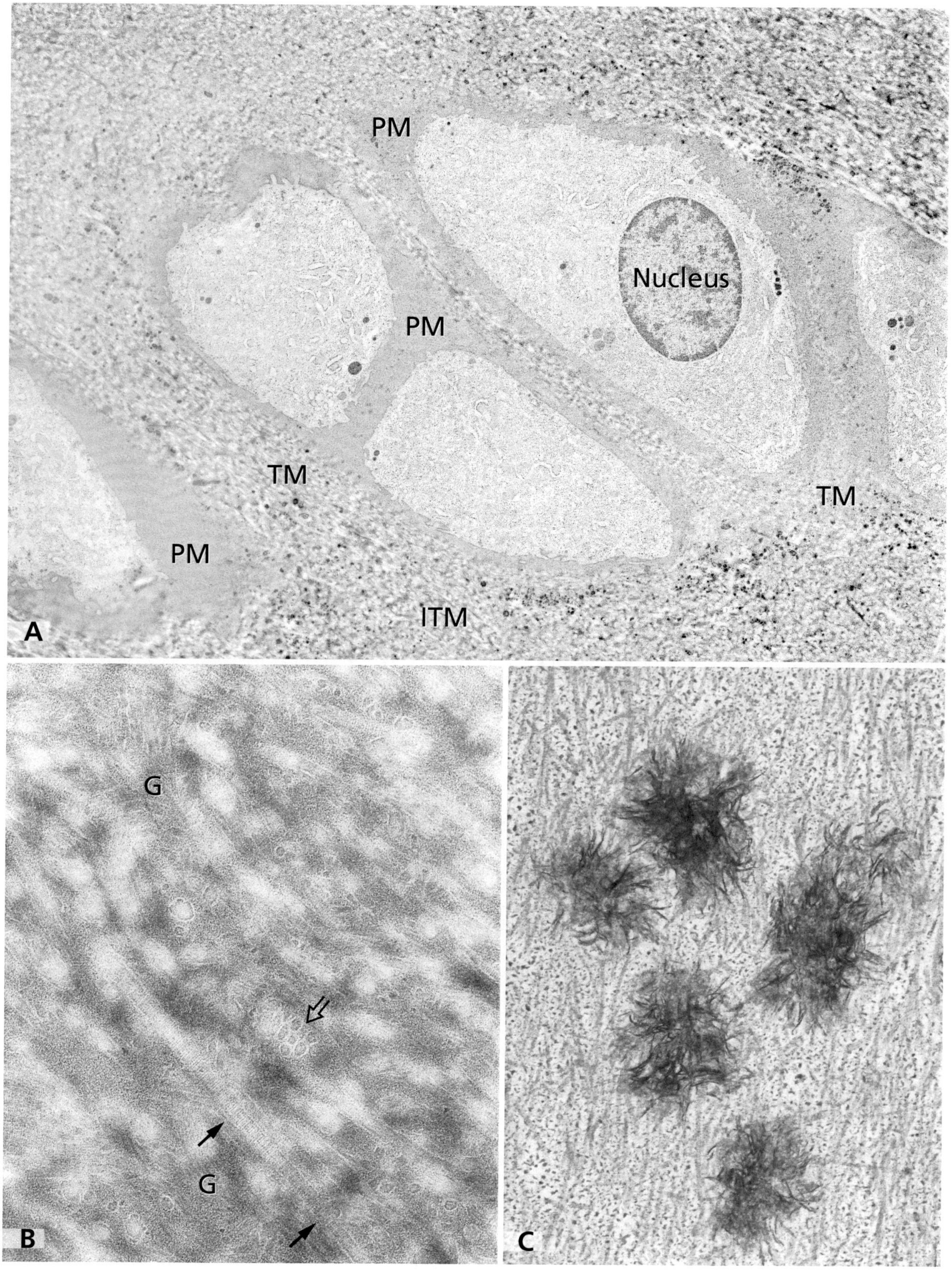
PM
PM
Nucleus
TM
TM
PM
ITM
A
G
G
B
C

OSTEOBLASTS AND OSTEOCYTES

Bone is a specialised form of the connective tissue characterised by impregnation of the extracellular matrix with salts of calcium and phosphate. Mineralisation is connected mainly with the apposition of hydroxyapatite crystals onto both components of the ground substance and collagen fibrils and leads to a special strength and stability of the tissue making it particularly qualified for providing support and protection for the body and its organs. Another important function of bone is to establish a reservoir for calcium and phosphate ions. Bone is highly vascularised and holds a crucial role in the regulation of blood calcium levels. Although it appears rigid and inflexible, it is metabolically very active and sensitive to functional alterations and changes of load, which result in a reorganisation of the tissue and reconstruction of the bone skeleton. Bone undergoes continuous remodelling associated with resorption, new production of mineralised matrix, and neovascularisation not only during development but also in the adult bone skeleton according to functional conditions.

Panels A, B, and C show segments of bone tissue derived from distal femur and proximal epiphyses of the mouse. In panel A, a vascular channel is shown, leading a capillary (Cap) embedded in loose connective tissue. Fibroblasts and osteoprogenitor cells are assembled. Active osteoblasts (Obl) are lined up forming an epithelial-like monolayer of polarised cells, which produce and deposit osteoid, the non-mineralised organic bone matrix consisting mainly of type I collagen fibrils, to a lesser extent of type V collagen, and a range of ground substance components including glycosaminoglycans, and glycoproteins such as osteocalcin, osteopontin, and osteonectin. Osteoid is shown in panel A, forming a light zone around the osteoblast layer. The neighbouring, extremely electron dense masses represent already mineralised matrix.

The osteoblasts in panel C show characteristics of highly active secretory cells with large nucleoli and the cytoplasm stuffed with rough endoplasmic reticulum (ergastoplasm). One of the osteoblasts has already left the epithelial-like osteoblast layer, and has lost its polar organisation. It is in part surrounded by its product, the osteoid (O), and forms thin projections, one of which (arrow) gets in contact with one of the processes of the "young" osteocyte (arrow) visible in the lowermost part of the figure. This young osteocyte is already surrounded by mineralised matrix (M) but it is still highly active, as can be seen by the abundance of rough endoplasmic reticulum. In this feature, it contrasts to the osteocyte shown in panel B. This "older" osteocyte is located in a small cave and is already completely embedded in calcified bone matrix. The small cytoplasm around the nucleus contains a reduced number of organelles. A thin cell process projects into a fine channel (arrow). The mineralised matrix is interspersed with numerous of such channels (canaliculi) providing a communicating nutritive system within the hard bone tissue. Nutrients diffuse from the blood vessels in the vascular channels through the canaliculi into the osteocyte caves. The channels also lead osteocyte processes, by which the cells communicate via gap junctions.

In the osteoid layer (O) in panel C, fine collagen fibrils are discernible that are further organised into bundles, concomitantly with the processes leading to transformation of osteoblasts into osteocytes. Electron dense spots and tufts indicate the onset of mineralisation that is initiated and controlled by the osteoblasts.

References

Boskey AL (1992) Mineral-matrix interactions in bone and cartilage. Clin Orthop Rel Res 281: 244

Ducy P, Schinke T, and Karsenty G (2000) The osteoblast: A sophisticated fibroblast under central surveillance. Science 289: 1501

Lorenz M and Plenk H Jr (1977) Die Ultrastruktur der Zellen des Knochengewebes. Acta Med Austr 4: 148

Olsen BR, Reginato AM, and Wang W (2000) Bone development. Annu Rev Cell Dev Biol 16: 191

Ortega N, Behonick DJ, and Werb Z (2004) Matrix remodelling during enchondral ossification. Trends Cell Biol 14: 86

Magnification: x 3,000 (A), x 1,800 (B), x 9,400 (C)

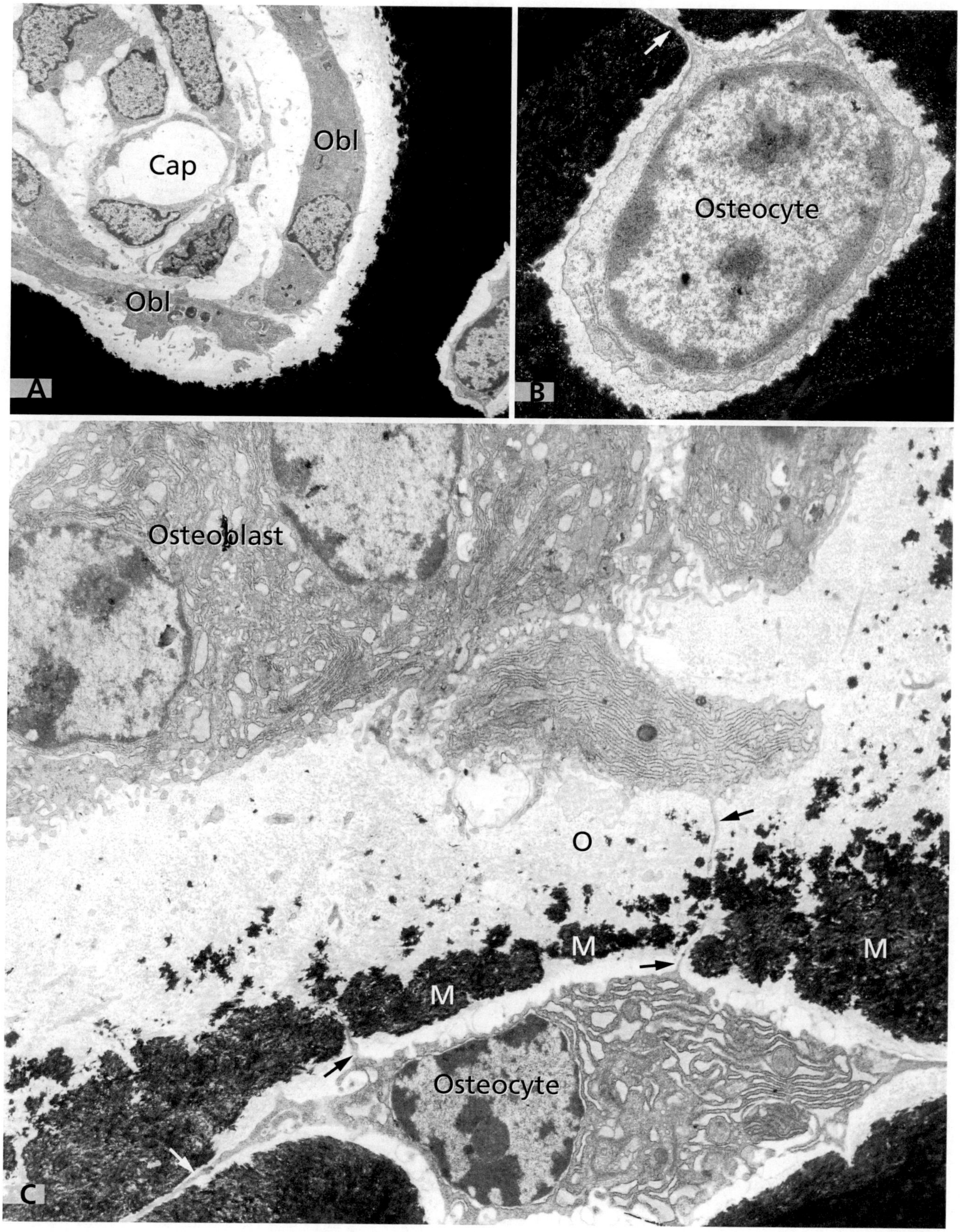
Cap
Obl
Obl
A
Osteocyte
B
Osteoblast
O
M
M
M
M
Osteocyte
C

OSTEOCLAST

Osteoclasts are terminally differentiated large, multinucleated cells associated with the removal and absorption of mineralised bone. They are formed by maturation and fusion of precursor cells, which derive from bone marrow cells of the monocyte/macrophage lineage. Osteoclastic bone resorption is necessary for normal growth and formation of the bone skeleton during development, as well as bone remodelling in the adult skeleton in response to altered functional conditions and changes of load. Osteoclast activity also is essential for the regulation of blood calcium concentrations. Osteoclasts rest directly on the bone tissue at sites where bone is being removed. Resulting from osteoclast activity, a resorption bay, designed Howship's lacuna, can be seen in the bone beneath the osteoclast. In panel A, a large osteoclast in the bone of the proximal tibia metaphysis of the mouse is shown residing in a resorption bay formed in the mineralised bone matrix. Three of the nuclei are sectioned. At the cell surface, where the osteoclast is apposed to the bone, a "ruffled border" (RB) has developed consisting of numerous plasma membrane infoldings. Panel B presents a higher magnification of the ruffled border of an osteoclast in apposition to partially resorbed mineralised bone matrix. Abundant fine needles of mineral crystals released from the bone are visible at the resorption front between the multiple plasma membrane infoldings. Bone resorption involves demineralisation, the dissolution of the inorganic constituents of the bone mediated by H^+-ATPase within an acidic environment, and enzymatic degradation of the organic components by lysosomal proteases. The bone area to be resorbed is exposed to the lysosomal enzymes, e.g. cathepsin K, that are packed into vesicles at the *trans* Golgi side and delivered by exocytosis into the clefts between the cytoplasmic processes of the ruffled border. The ruffled border membrane containing vacuolar H^+-ATPases shows several features comparable with the membrane of late endosomal compartments. The production of hydrochloric acid leads to the creation of a local acid environment favouring the action of the lysosomal enzymes. The area of bone resorption characterised by the ruffled border is demarcated by a ring-like perimeter of cytoplasm, the clear zone, containing abundant actin microfilaments but essentially lacking other organelles. Segments of the clear zone (CZ) are visible in panel B, demarcating the ruffled border at both sides.

Degraded bone matrix is internalised by the osteoclasts for further processing, enters transcytotic vesicles, and is finally exocytosed to the intercellular space via the functional secretory domain at the basolateral membrane opposite the bone surface. As recent studies have shown, in the ruffled border, secretion and endocytosis take place in a highly ordered manner. The ruffled border possesses two subdomains, a peripheral vesicle fusion zone and a central matrix uptake zone. It has been shown that endocytosis and transcytosis from the ruffled border are fast processes that involve microtubules, and are dependent on the acidification of the extracellular space. The half-life of endocytosed material inside the cells has been estimated to be 22 minutes.

References

Holtrop ME and King GJ (1977) The ultrastructure of the osteoclast and its functional implications. Clin Orthop 123: 177

Mulari MTK, Zhao H, Lakkakorpi PT, and Väänänen HK (2003) Osteoclast ruffled border has distinct subdomains for secretion and degraded matrix uptake. Traffic 4: 113

Ortega N, Behonick DJ, and Werb Z (2004) Matrix remodelling during enchondral ossification. Trends Cell Biol 14: 86

Salo J, Lehenkari P, Mulari M, Metsikko K, and Vaananen HK (1997) Removal of osteoclast bone resorption products by transcytosis. Science 276: 270

Stenbeck G and Horton MA (2003) Endocytic trafficking in actively resorbing osteoclasts. J Cell Sci 117: 827

Teitelbaum SL (2000) Bone resorption by osteoclasts. Science 289: 1504

Toyomura T, Murata Y, Yamamoto A, Oka T, Sun-Wada G-H, Wada Y, and Futai M (2003) From lysosomes to the plasma membrane. Localization of vacuolar H^+-ATPase with the *a*3 isoform during osteoclast differentiation. J Biol Chem 278: 22023

Magnification: x 8,600 (A), x 21,800 (B)

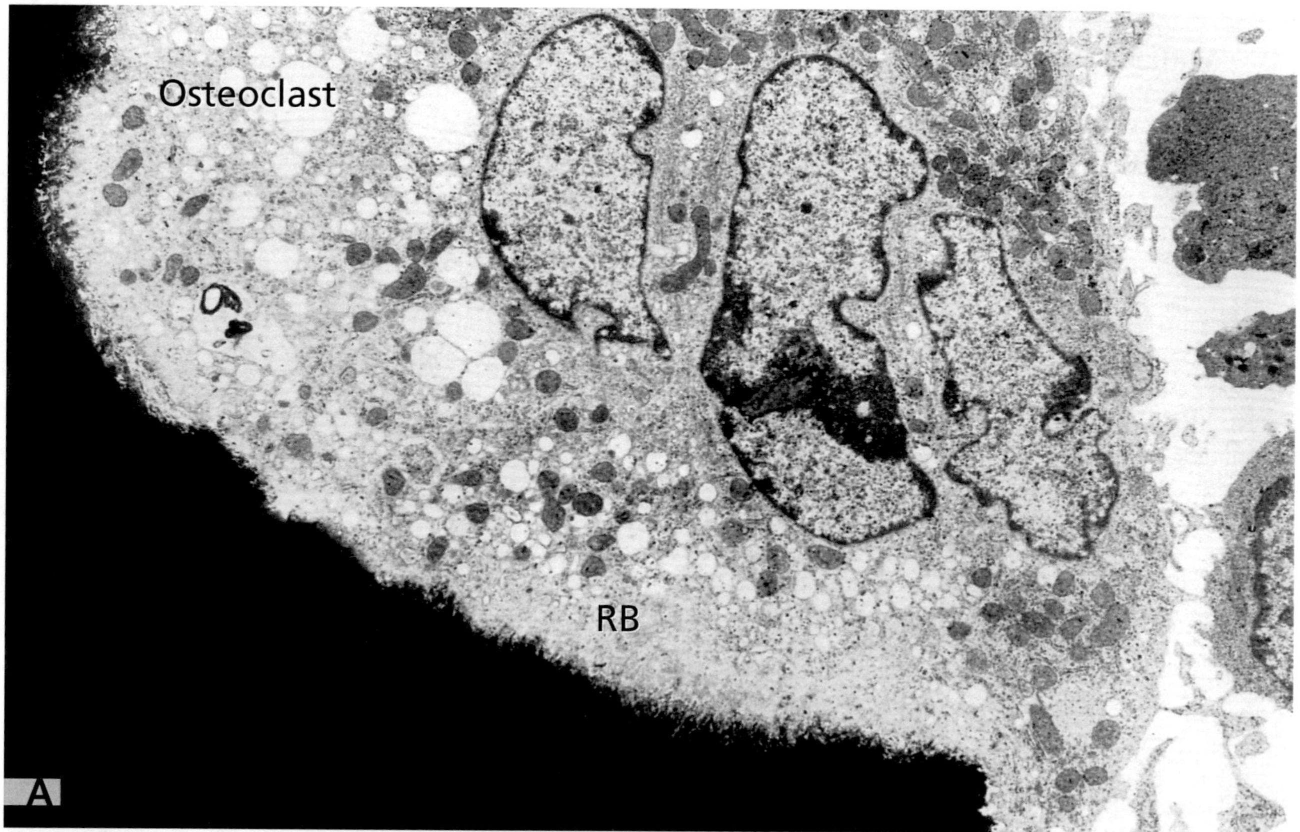
Osteoclast
RB
A

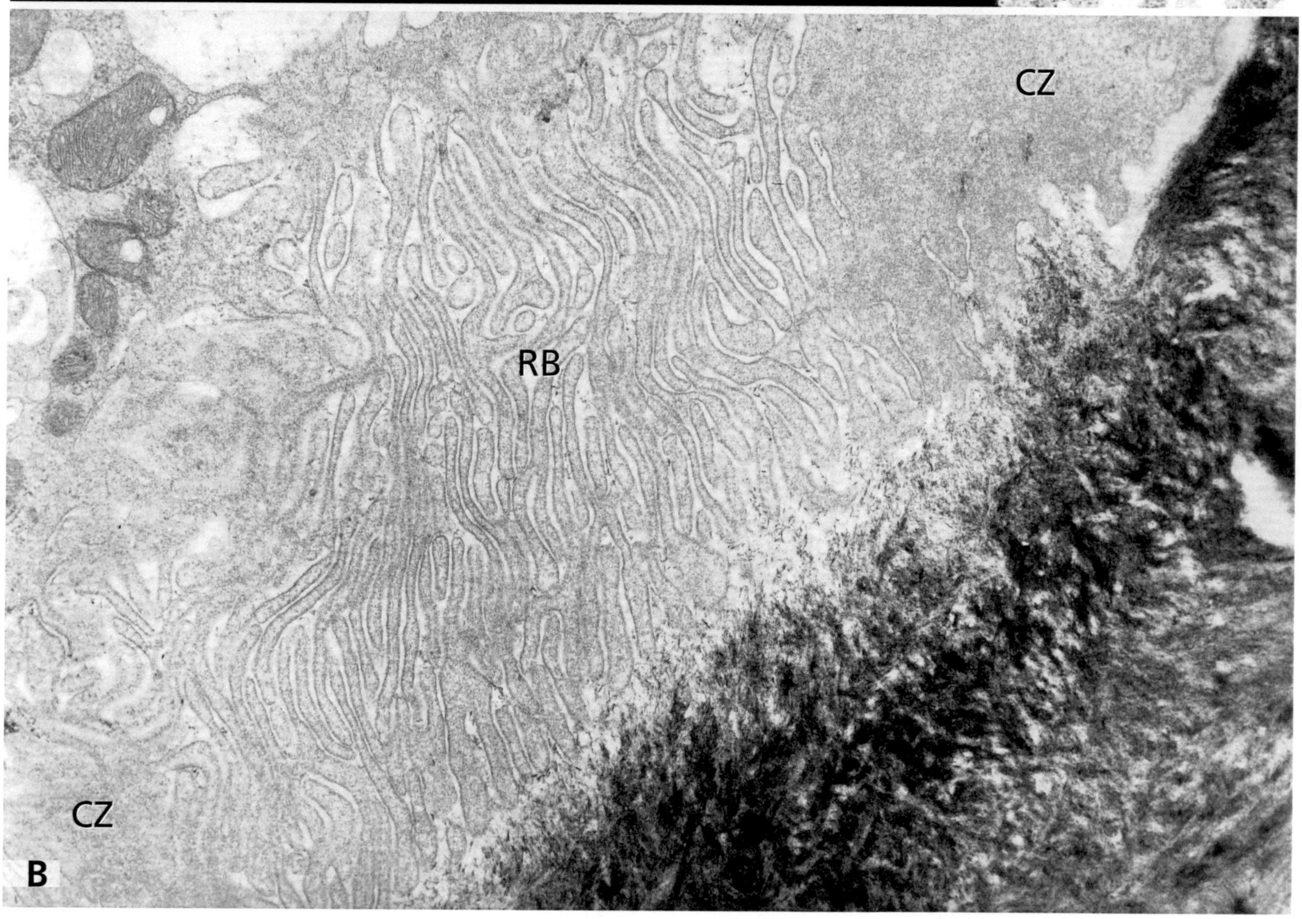
CZ
RB
CZ
B

MYOFIBRILS AND SARCOMERE

Skeletal muscle cells, more commonly called muscle fibres, are multinucleated syncytia formed during development by fusion of mononucleated precursor cells, the myoblasts. They measure 10-100 μm in diameter and can have lengths from a few millimeters up to almost a meter. The hundreds of nuclei are located in the cells' periphery close to the plasma membrane, whereas most of the cytoplasm is occupied by the longitudinally arrayed myofibrils composed of the myofilaments: thick filaments assembled from myosin II molecules containing rod-shaped segments and globular heads projecting out of the filaments, and thin filaments composed of actin. The electron micrograph shows part of the cytoplasm of a striated muscle fibre with parallel arrays of myofibrils. Mitochondria are lined in the small cytoplasmic cords between the myofibrils, where glycogen is accumulated. Profiles of the sarcoplasmic reticulum forming triads (arrows; cf. also Fig. 150) are visible.

The cross-banded pattern apparent in the cytoplasm of the muscle fibre reflects the arrangement, in register, of the myofibrils, and the banded pattern of the myofibrils reflects the arrangement of the myofilaments. Intensely stained A-bands (A – anisotropic), for which the myosin filaments account, and less intensely stained I-bands (I – isotropic), composed mainly of actin filaments, can be discriminated. The dense Z-lines in the centre of the I-bands represent sections of disk-shaped platforms, where the actin filaments are anchored by alpha-actinin. Nebulin wraps around the actin filaments and assists alpha-actinin in anchoring the actin filaments. Further accessory proteins are tropomodulin, an actin-capping protein, which regulates the length of the actin filaments, and titin, a giant elastic molecule that connects the thick myosin filaments with the Z-disk. In the centre of each A-band, a less dense band, the H-band, corresponds to the central part of the myosin filaments being bare of myosin head projections. The H-band is bisected by a narrow dense line, the M-line built by myosin-binding proteins that hold the thick filaments in register. An intermediate filament lattice of desmin forms stabilising cross-links between neighbouring myofibrils, which are connected with the plasma membrane and extracellular matrix via costameres, juxtaposed to the Z- and M-lines. Dystrophin, a rod-shaped, dimeric protein localised beneath the plasma membrane, links the cytoskeleton of the muscle fibre to the extracellular matrix by binding actin and a complex of transmembrane proteins, the dystroglycans and sarcoglycans, which are connected with laminin and agrin in the basal lamina.

A myofibril segment delineated by two adjacent Z-lines is defined as a sarcomere, the basic contractile unit of the striated muscle shown in the inset. Excluding the H-band, actin and myosin filaments interdigitate. Rapid contraction cycles of well defined subsequent stages involving attachment, release, bending, force generation and reattachment of myosin heads to actin molecules move the thin actin filaments along the thick myosin filaments leading to a shortening of the sarcomeres and contraction of the muscle. Tropomyosin and three troponin molecules have pivotal roles in the initiation of contraction triggered by binding of Ca^{2+} to one of the troponins. This leads to an uncovering of the myosin binding sites located at the actin molecules. Contraction of adjacent sarcomeres occurs with a short time delay.

References

Clark KA, McElhinny AS, Meckerle MC, and Gregorio CC (2002) Striated muscle cytoarchitecture: an intricate web of form and function. Annu Rev Cell Dev Biol 18: 637

Gordon AM, Homsher E, and Regnier M (2000) Regulation of contraction in striated muscle. Physiol Rev 80: 853

Towler MC, Kaufman SJ, and Brodsky FM (2004) Membrane traffic in skeletal muscle. Traffic 5: 129

Tskhovrebova L and Trinick J (2003) Titin: Properties and family relationships. Nat Rev Mol Cell Biol 4: 679

Magnification: x 33,000, x 48,500 (inset)

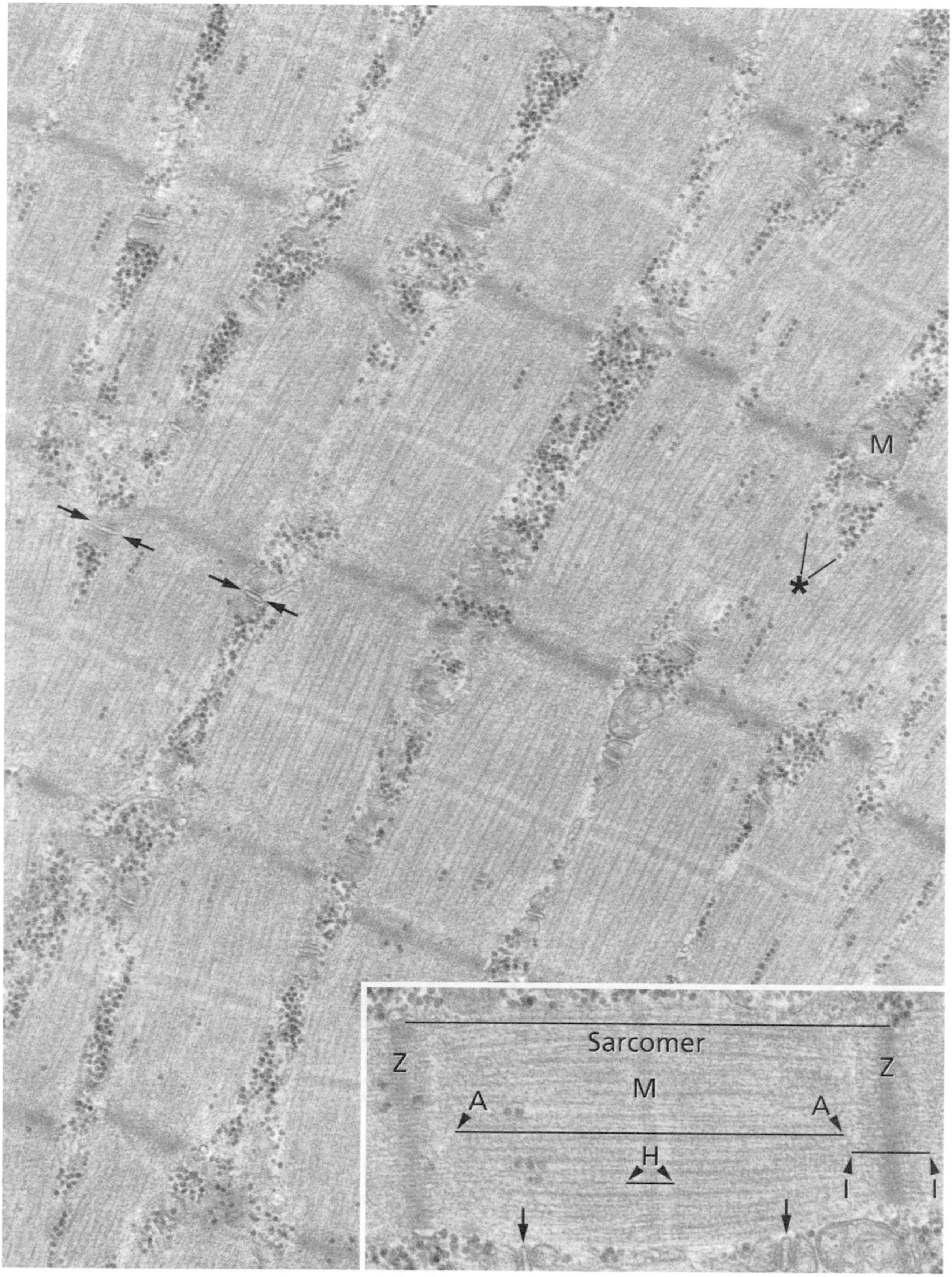
M
*
Sarcomer
Z
Z
M
A
A
H
I
I

SARCOPLASMIC RETICULUM, TRIAD, SATELLITE CELL

For muscle contraction, Ca^{2+} must be available as it is required for the binding between myosin and actin, and after contraction, Ca^{2+} must be removed. The rapid delivery and removal of Ca^{2+} is effected by a complex membrane system that surrounds each myofibril and consists of two parts: a longitudinally oriented part, the longitudinal or L-system, a muscle cell specific endoplasmic reticulum, called sarcoplasmic reticulum, that forms networks of cisternae around the myofibrils and is the major intracellular calcium store, and a transversely oriented part, the transverse tubule or T-system representing tubular projections of the plasma membrane that encircle each myofibril. At particular sites, the two parts of the membrane system come together forming specialised signal transduction organelles, the triads. Triads consist of two terminal cisterns of the L-system associated with a central T-tubule segment. The main function of the triads is to translate the action potential from the plasma membrane to the sarcoplasmic reticulum, effecting calcium flow into the cytoplasm and the initiation of muscle contraction.

In both panels A and B, triads are visible. In panel A indicated by arrows, triads are apparent in cytoplasmic cords between superficially localised myofibrils. At each side, the central T-tubule is accompanied by a globular profile of a terminal cistern of the L-system. Flatly sectioned triads are visible in panel B. It is evident that terminal cisterns of the L-system are associated with segments of T-tubules at particular sites. The inter-membraneous narrow cytoplasmic slits between the T-tubule and L-cistern membranes contain dense materials visible as a fine line, which may correspond to parts of the protein complexes involved in signal transmission. Voltage sensor proteins in the T-tubule membrane of the triads are activated when the plasma membrane depolarises leading to conformational changes of the proteins that, in turn, opens gated Ca^{2+}-release channels in the membrane of the adjacent terminal cisterns of the L-system effecting rapid release of Ca^{2+} into the cytoplasm. Simultaneously, Ca^{2+} is transported back into the terminal cisterns by Ca^{2+}-activated ATPases in the L-system membrane. In the skeletal muscle fibres, triads are regularly localised at the A-I-band junctions.

Close spatial relations between a T-tubule and caveolae-like vesicles are visible in panel A. The T-tubule membranes are enriched in cholesterol, similarly to the membranes of caveolae (cf. Fig. 51). Caveolae are assumed to be involved in the formation of the T-tubule system during development. It has been shown that caveolin 3, a muscle-specific isoform of caveolin associates with developing T-tubules. Arrowheads label the basal lamina that covers the entire surface of the muscle fibre.

Panel C shows a satellite cell located closely apposed to the plasma membrane of the muscle fibre visible in the upper part of the picture. Muscle fibre and satellite cell are in a common sheath of basal lamina (arrowheads). The cytoplasm of the satellite cell seems to be intensely involved in vesicle transport but lacks myofibrils. This is consistent with its function as precursor cell. Adult skeletal muscle fibres possess capacity for regeneration in response to damages due to injuries or progressive myopathies. Satellite cells are normally quiescent but are activated by damage and stimulated to proliferate and differentiate. By fusion, newly formed multinucleated myofibres are created, either *de novo* or from preexisting muscle fibres.

References

Franzini-Armstrong C, Protasi F, and Ramesh V (1998) Comparative ultrastructure of Ca^{2+} release units in skeletal and cardiac muscle. Ann NY Acad Sci 853: 20

Mazzarello P, Calligaro A, Vannini V, and Muscatello U (2003) The sarcoplasmic reticulum: Its discovery and rediscovery. Nat Rev Mol Cell Biol 4: 69

Nicole S, Desforges B, Millet G, Lesbordes J, Cifuentes-Diaz C, Vertes D, Cao ML, de Backer F, Languille L, Roblot N, Joshi V, Gillis JM, and Melki J (2003) Intact satellite cells lead to remarkable protection against Smn gene defect in differentiated skeletal muscle. J Cell Biol 161: 571

Towler MC, Kaufman SJ, and Brodsky FM (2004) Membrane traffic in skeletal muscle. Traffic 5: 129

Wagenknecht T and Rademacher M (1997) Ryanodine receptors: Structure and macromolecular interactions. Curr Opin Struct Biol 7: 258

Magnification: x 43,000 (A), x 69,000 (B), x 12,500 (C)

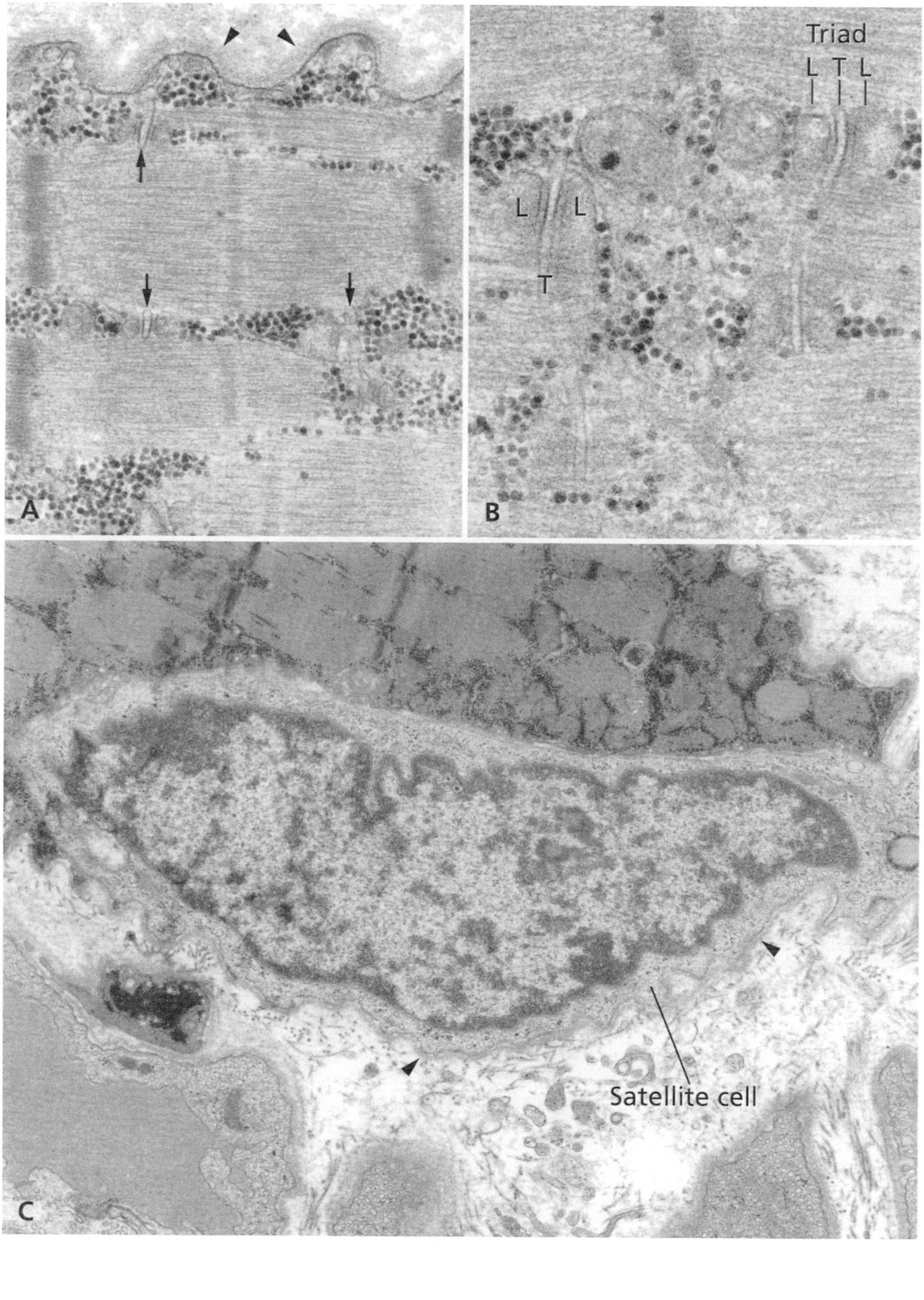
Triad
L T L
L L
T
A
B
Satellite cell
C

NEUROMUSCULAR JUNCTION

Neuromuscular junctions, also called motor end plates, are specialised chemical synapses formed at the sites where the terminal branches of the axon of a motor neuron contact a target muscle cell. Innervation is achieved by interaction of acetylcholine released from the axon endings with its receptor in the plasma membrane of the muscle cell. Neuromuscular junctions show unique architectures. Ramifications of the axon terminate in knob-like end pieces, each of which is embedded in a shallow depression on the surface of the muscle fibre. The myelin sheath covering the axon (cf. Fig. 162) ends, and the basal lamina of the cell of Schwann continues into that of the muscle fibre, and also extends into the synaptic cleft between the plasma membrane of the depressed end knob of the axon and the surface of the muscle fibre. The axon end knob represents the presynaptic part of the neuromuscular junction. It contains multiple mitochondria and synaptic vesicles where the transmitter acetylcholine is stored. The muscle cell plasma membrane underlying the synaptic cleft forms the postsynaptic membrane. It is extensely folded, leading to an increase of surface and the formation of narrow troughs that alternate with crest-like formations. In the postsynaptic membrane, receptors and ion channels are localised in segregated arrangements. The acetylcholine receptors are concentrated at the crests, and Na^+ channels located in the troughs. This allows differentiated actions in response to a nerve impulse that prompts release of acetylcholine into the synaptic cleft. Binding of acetylcholine to the receptors leads to an opening of the cation channels. Influx of Na^+ effects localised membrane depolarisation that is transmitted along the plasma membrane to the T-tubules to reach the triades, where Ca^{2+}-release from the sarcoplasmic reticulum is triggered initiating muscle contraction. Continued stimulation is prevented by the enzyme acetycholine esterase, by which acetylcholine is rapidly broken down.

The electron micrograph presents a survey view of a neuromuscular junction area. Nuclei are accumulated three of them being sectioned. The cytoplasm is enriched in mitochondria, endoplasmic reticulum, free ribosomes, Golgi apparatus, and glycogen. However, myofibrils (asterisks) are sparse in the inner junction area but are densely packed in the cytoplasm outside, as shown in the lower right part of the picture. A neuromuscular junction with all its components is visible in the left lower part of the micrograph. The end knob of an axon depressed in the cytoplasm of the muscle fibre contains mitochondria and abundant small synaptic vesicles. It is surrounded by the synaptic cleft and multiple junctional folds (arrows). Being branched they form a system of labyrinth-like membrane convolutes, also visible in the left upper part of the micrograph (arrows). It is evident that the multiple foldings lead to an extended augmentation of the surface area of the postsynaptic membrane. Basal lamina materials line the entire surface within the cleft including the deep trough regions. The entire muscle fibre is covered by a basal lamina and embedded in loose connective tissue, the endomysium.

The organisation of neuromuscular junctions and the extended postsynaptic apparatus is a complex process, influenced by extracellular signals from the motor neuron. It involves synthesis and secretion of the extracellular matrix protein agrin from the neurons. By agrin, a muscle-specific receptor tyrosine kinase is activated, which is necessary for the concentration of the acetylcholine receptor proteins at the developing synapse by the cytoplasmic linker protein rapsin. Subsequently, Na^+ channels are clustered. During maturation of the neuromuscular junctions, the membranes undergo essential reorganisation.

References

Burden SJ (2002) Building the vertebrate neuromuscular synapse. J Neurobiol 53: 501

Miana-Mena FJ, Roux S, Benichou JC, Osta R, and Brulet P (2002) Neuronal activity-dependent membrane traffic at the neuromuscular junction. Proc Natl Acad Sci USA 99: 3234

Sanes JR and Lichtman JW (1999) Development of the vertebrate neuromuscular junction. Annu Rev Neurosci 22: 389

Sanes JR and Lichtman JW (2001) Induction, assembly, maturation and maintenance of a postsynaptic apparatus. Nat Rev Neurosci 2: 791

Towler MC, Kaufman SJ, and Brodsky FM (2004) Membrane traffic in skeletal muscle. Traffic 5: 129

Magnification: x 9,800

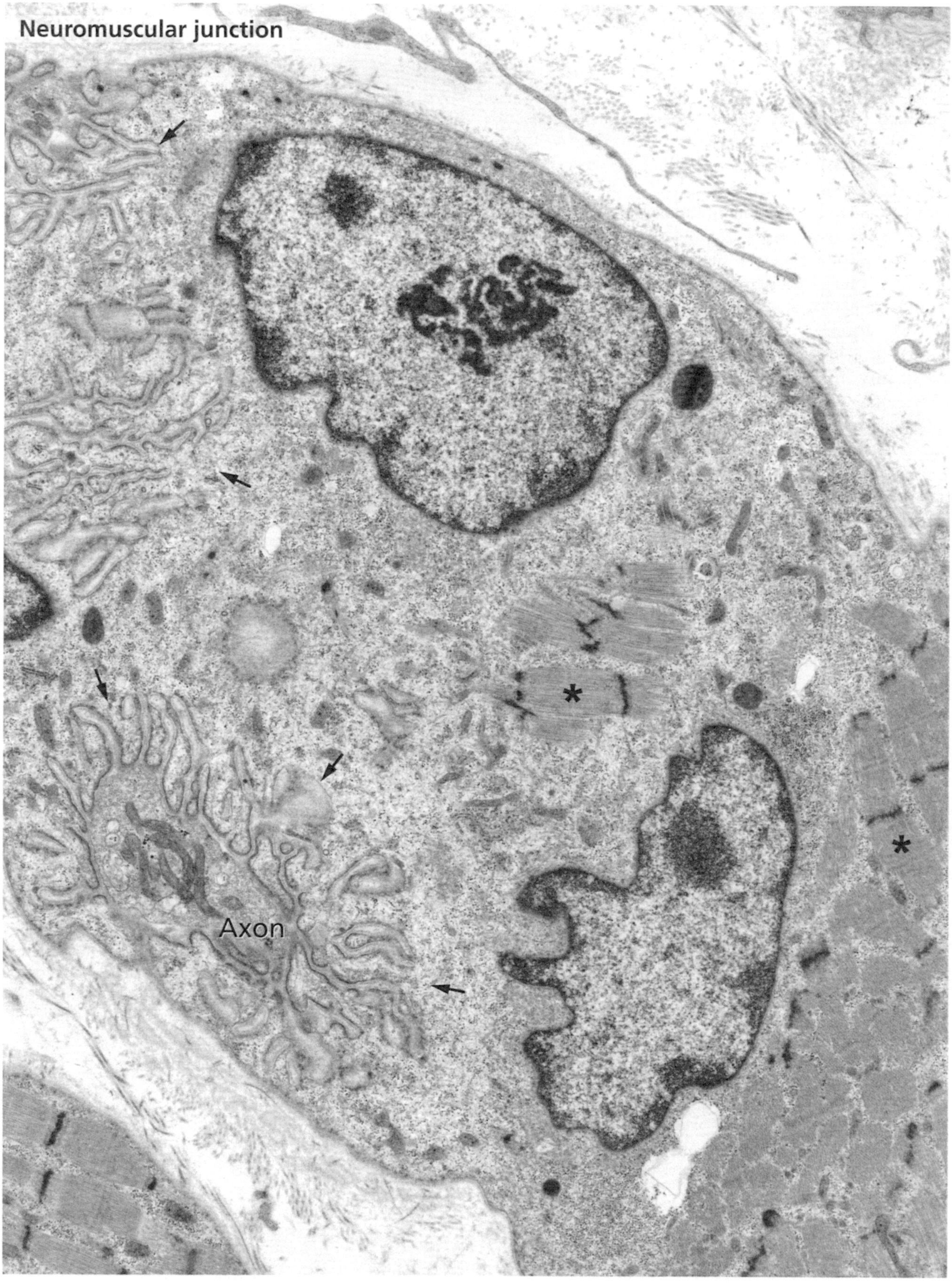
Neuromuscular junction
*
*
Axon

MUSCULAR DYSTROPHIES

Muscular dystrophies are a group of congenital muscular diseases characterized by muscle weakness due to progressive muscle wasting and elevated serum concentrations of creatine kinase due to muscle fibre damage and necrosis. The different types of muscular dystrophies are caused by mutations in genes encoding dystrophin, components of the dystrophin-glycoprotein complex, the sarcoglycan complex, and laminin 2. Impairment of any of these components results in the disruption of the mechanical link between the actin cytoskeleton and the basal lamina through the sarcolemmal membrane.

Duchenne muscular dystrophy (DMD) and Becker muscular dystrophy are X-linked, recessively inherited and either due to complete absence or presence of low levels of truncated dystrophin, which is an actin binding cytoskeletal protein. Fukuyama congenital muscular dystrophy and muscle-eye-brain disease are caused by hypoglycosylation of dystroglycan, which is a central component of the dystrophin-glycoprotein complex. Mutations in sarcoglycans, which are transmembrane glycoproteins, are causative in some forms of limb-girdle muscular dystrophies and mutations in laminin 2 (merosin) cause merosin negative congenital muscular dystrophy.

In Duchenne muscular dystrophy, degenerating or necrotic muscle fibres are found arranged in groups surrounded by normal muscle fibres. Panel A shows part of a nearly normal fibre adjacent to a dystrophic one. The changes in dystrophic muscle fibres are complex and primarily involve structural and functional alterations of the sarcolemma. Very early, focal breaches of the sarcolemma occur, followed by membrane dissolution. Wedge shaped areas of abnormal sarcoplasm are a common finding. In panels A and B, the highly altered sarcomer structure is obvious. Only remnants of Z-bands (arrows) are visible. Myofibrils can be recognized but are irregularly arranged. Asterisks denote glycogen particles. Thus, the sarcolemma is fragile and permeable for extracellular proteins. Furthermore, the calcium homeostasis is disturbed and increased intracellular calcium results in hyper-contraction of fibres and enhanced proteolysis through calcium activation of proteases.

At early stages of Duchenne muscular dystrophy, fibre regeneration secondary to fibre necrosis is possible. Regenerating muscle fibres have a smaller diameter and a centrally, not peripherally located nucleus (panel C).

References

Blake DJ, Weir A, Newey S E, and Davies KE (2002). Function and genetics of dystrophin and dystrophin-related proteins in muscle. Physiol Rev 82: 291

Durbeej M and Campbell KP (2002) Muscular dystrophies involving the dystrophin-glycoprotein complex: an overview of current mouse models. Curr Opin Genet Dev 12: 349

Grewal PK, Holzfeind PJ, Bittner RE, and Hewitt JE (2001) Mutant glycosyltransferase and altered glycosylation of alpha-dystroglycan in the myodystrophy mouse. Nat Genet 28: 151

Hohenester E, Tisi D, Talts JF, and Timpl R (1999) The crystal structure of a laminin G-like module reveals the molecular basis of alpha-dystroglycan binding to laminins, perlecan, and agrin. Mol Cell 4: 783

Martin PT (2003) Dystroglycan glycosylation and its role in matrix binding in skeletal muscle. Glycobiology 13: 55R

McNally E and Pytel P (2007) Muscle diseases: the muscular dystrophies. Annu Rev Pathol 2: 87

Michele DE, Barresi R, Kanagawa M, Saito F, Cohn RD, Satz JS, Dollar J, Nishino I, Kelley RI, Somer H, et al. (2002) Post-translational disruption of dystroglycan-ligand interactions in congenital muscular dystrophies. Nature 418: 417

Wallace G and McNally E (2009) Mechanisms of muscle degeneration, regeneration, and repair in the muscular Ddystrophies. Annu Rev Physiol 71: 37

Worton R, Molnar, M, Brais B, and Karpati G (2001) The muscular dystrophies. In: The metabolic and molecular bases of inherited disease (Scriver C, Beaudet A, Valle D, Sly WS, Childs B, Kinzler K, and Vogelstein B, eds). New York: McGraw-Hill, pp 5493

Magnification: x 12,000 (A); x 44,000 (B); x 6,500 (C)

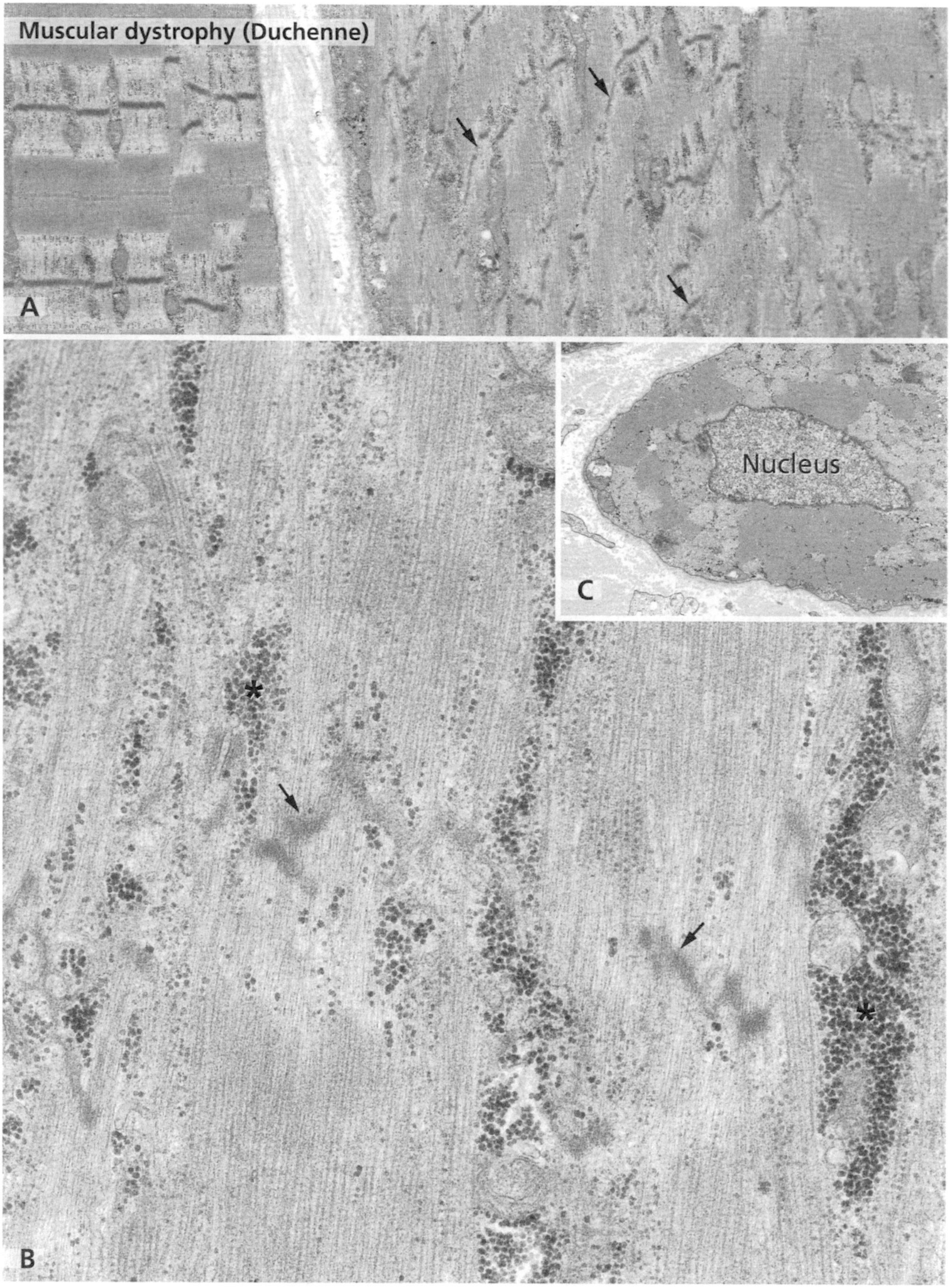
Muscular dystrophy (Duchenne)
A
B
Nucleus
C

GLYCOGENOSIS TYPE II (POMPE)

This glycogen storage disease is autosomal recessively inherited and also termed acid maltase deficiency or more commonly Pompe disease. The underlying pathogenesis is a deficiency of lysosomal acid α-glucosidase. The ultrastructural changes typically observed in hepatocytes of the liver of patients are illustrated by Fig. 61A.

Pompe disease is classically associated with myopathy, which is due to massive glycogen accumulation in muscle fibres. The glycogen depositions are shown in longitudinal (panel A) and cross (panel B) sections of skeletal muscle fibres. The glycogen deposits are mostly free in the cytoplasm (asterisks in panel A), although some are in membrane-limited vacuoles of unclear nature, which can be easily appreciated in the micrographs. Due to the massive glycogen accumulation, the sarcomers of the muscle fibres are diminished in size and numbers, which is most impressive in the cross-sectioned muscle fibres shown in panel B. The sarcomers may also contain lipid droplets, which is not the situation in the case shown here. In panel B, the muscle fibre on the left contains a high number of lysosomes. The degree to which these ultrastructural changes are observed can vary widely. Altogether, these structural changes manifest themselves clinically in muscle weakness with hypotonia and flaccidity. Despite the varying degree of sarcomer atrophy which can be observed by microscopical analysis, the muscles themselves are firm and may even appear hypertrophic, which is due to the presence of the glycogen deposits.

References

Engel AG, Gomez MR, Seybold ME, and Lambert EH (1973) The spectrum and diagnosis of acid maltase deficiency. Neurology 23: 95

Hirschhorn R and Reuser A (2001) Glycogen storage disease type II: acid α-glucosidase (acid maltase) deficiency. In: The metabolic and molecular bases of inherited diseases, vol III (Scriver C, Beaudet A, Valle D, Sly W, eds). New York: McGraw Hill, pp 3389

Isch F, Juif JG, Sacrez R, and Thiebaut F (1966) Muscular glycogenosis of myopathic form caused by acid maltase deficiency. Pediatrie 21: 71

Matsuishi T, Yoshino M, Terasawa K, and Nonaka I (1984) Childhood acid maltase deficiency. A clinical, biochemical, and morphologic study of three patients. Arch Neurol 41: 47

Temple JK, Dunn DW, Blitzer MG, and Shapira E (1985) The "muscular variant" of Pompe disease: clinical, biochemical and histologic characteristics. Am J Med Genet 21: 597

Magnification: x 8,000 (A); x 5,800 (B)

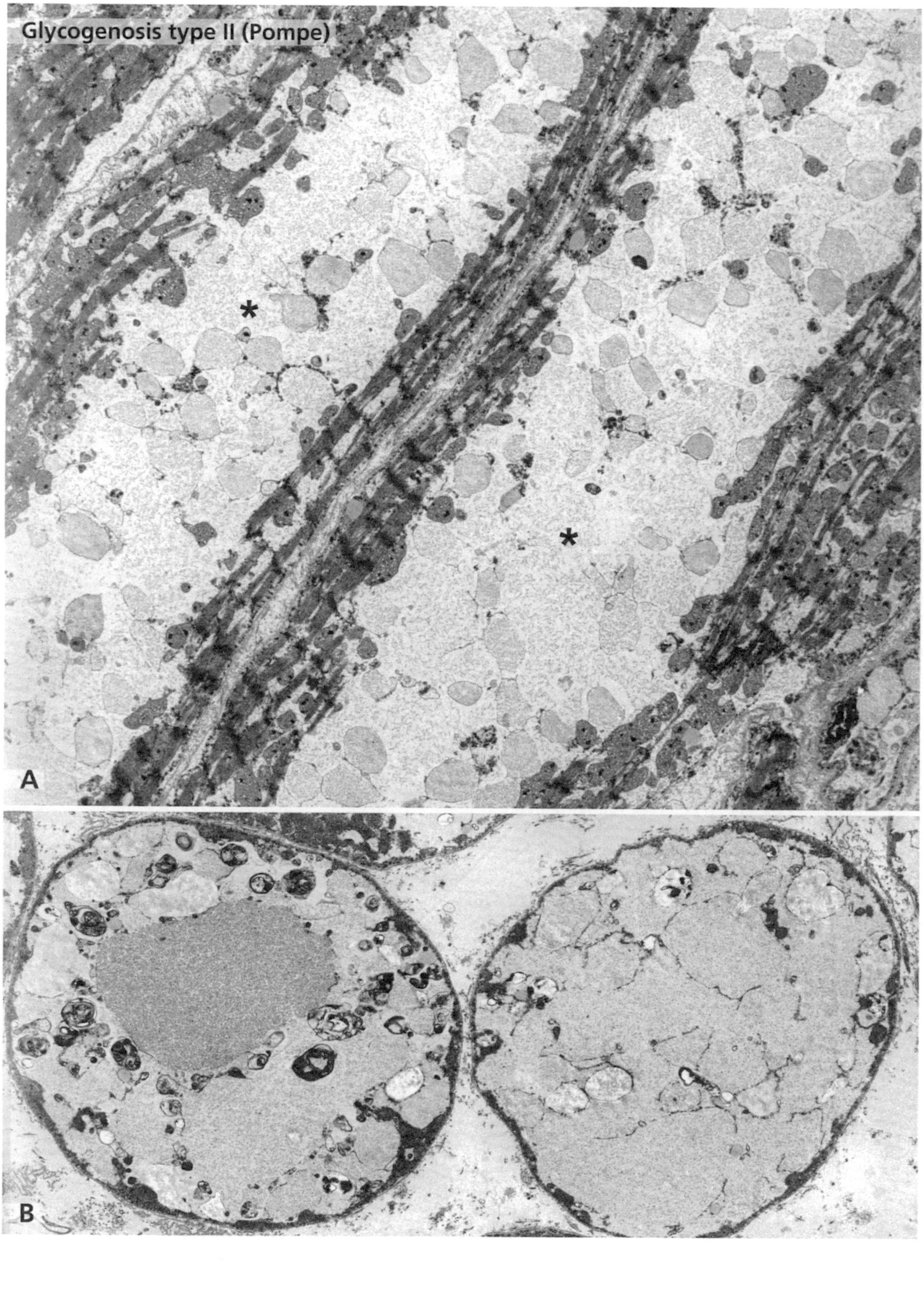
Glycogenosis type II (Pompe)
*
*
A
B

MYOFIBRILS, INTERCALATED DISK

Cardiac muscle belongs to the striated types of muscle containing the same arrangements of myofilaments as skeletal muscle. However, unlike in skeletal muscle no syncytia are built but cylindrical muscle cells containing one large cube-shaped nucleus in central position are arranged end to end and join with adjacent cells at special attachment sites, the intercalated disks. Intercalated disks represent the junctions at the boundaries between neighbouring cells. Since cardiac muscle cells end in a step-like fashion, the intercalated disks show also step-like arrangements with transverse components oriented at a right angle to the myofibrils, and lateral components oriented longitudinally in parallel position to the myofibrils. Cell-to-cell junctions located at the intercalated disks include three types: Fasciae adhaerentes (adhering junctions) are always located in the transverse parts of the intercalated disks. These are the sites where the actin filaments of the terminal sarcomeres are anchored and connected with the plasma membrane. Constituents and organisation of the fasciae adhaerentes can be compared with those of the adhering zone (belt desmosome) in epithelia (cf. Fig. 83). Maculae adhaerentes (spot desmosomes, cf. Fig. 86) found in both the transverse and lateral parts of the intercalated disks reinforce the fasciae adhaerentes and fix the adjacent cells to one another. Gap junctions (nexus, communicating junctions, cf. Fig. 84) are confined to the lateral parts of the intercalated disks. Gap junctions permit ionic traffic between the neighbouring cells and provide the base for functional cell coupling. Cardiac muscle behaves as a functional syncytium although it is composed of individual cells. At the lateral regions of the intercalated disks, gap junctions are protected from forces during contraction.

Panel A shows an intercalated disk in cardiac muscle of the rabbit including a characteristic "step"-region where the disc bends and the transversal part continues in the lateral part. Within the lateral component of the disk, a gap junction (arrows) is visible close to a neighbouring spot desmosome. Since the cell boundary is sectioned obliquely, only one of the two plaques of the desmosome can be seen. At the transverse part of the intercalated disk, the section leads through the central, very dense part of one of the plaques of a fascia adhaerens; since they are not included in the section, the plasma membranes of the adjacent cells and the intercellular space cannot be seen here but, at both sides of the intercalated disk, the neighbouring plasma membranes are visible. Both cells are covered by a basal lamina and show multiple caveolae-like vesicles close to the cell surfaces. Triads are not as regular and prominent as in skeletal muscle, and often, diads are formed instead of triads. However, anastomosing networks of the sarcoplasmic reticulum surround the myofibrils, to be seen in both panels A and B (arrowheads). Mitochondria (M) are numerous; they show densely packed cristae and occupy the cytoplasm between the myofibrils where also glycogen is stored. The energy storing, energy releasing and energy recapturing structures and organelles, glycogen particles and mitochondria, are clearly visible adjacent to the myofibrils where the energy is used for contraction. Recent three-dimensional studies of the membrane systems for Ca^{2+} signalling in cells of the mouse myocardium revealed detailed insights into the structures of diads, and showed frequent physical links between the outer mitochondrial membrane and the sarcoplasmic reticulum and T-tubules.

The area indicated by a rectangle in panel B is shown at higher magnification in the inset. The section leads through an A-band, where thick myosin and thin actin filaments overlap. Each thick myosin filament is located within the centre of a hexagonal array of thin actin filaments. Such a unit is encircled.

For adaptation according to the physiological demands, sensing of biomechanical signals is required involving the interfaces between myofibrils and the plasma membrane at the cell-to-cell junctions within the intercalated disks and the cell-matrix junctions at the costameres (cf. Fig. 149). It is assumed that the giant protein titin has a main role as a biomechanical sensor.

References

Franzini-Armstrong C, Protasi F, and Ramesh V (1998) Comparative ultrastructure of Ca^{2+} release units in skeletal and cardiac muscle. Ann NY Acad Sci 853: 20

Hayashi T, Martone ME, Yu Z, Thor A, Doi M, Holst MJ, Ellisman MH, and Hoshijima M (2008) Three-dimensional electron microscopy reveals new details of membrane systems for Ca^{2+} signaling in the heart. J Cell Sci 122: 1005

Miller MK, Granzier H, Ehler E, and Gregorio CC (2004) The sensitive giant: the role of titin-based stretch sensing complexes in the heart. Trends Cell Biol 14: 119

Severs NJ (2000) The cardiac muscle cell. Bioessays 22: 188

Magnification: x 35,500 (A), x 114,000 (B), x 273,600 (inset)

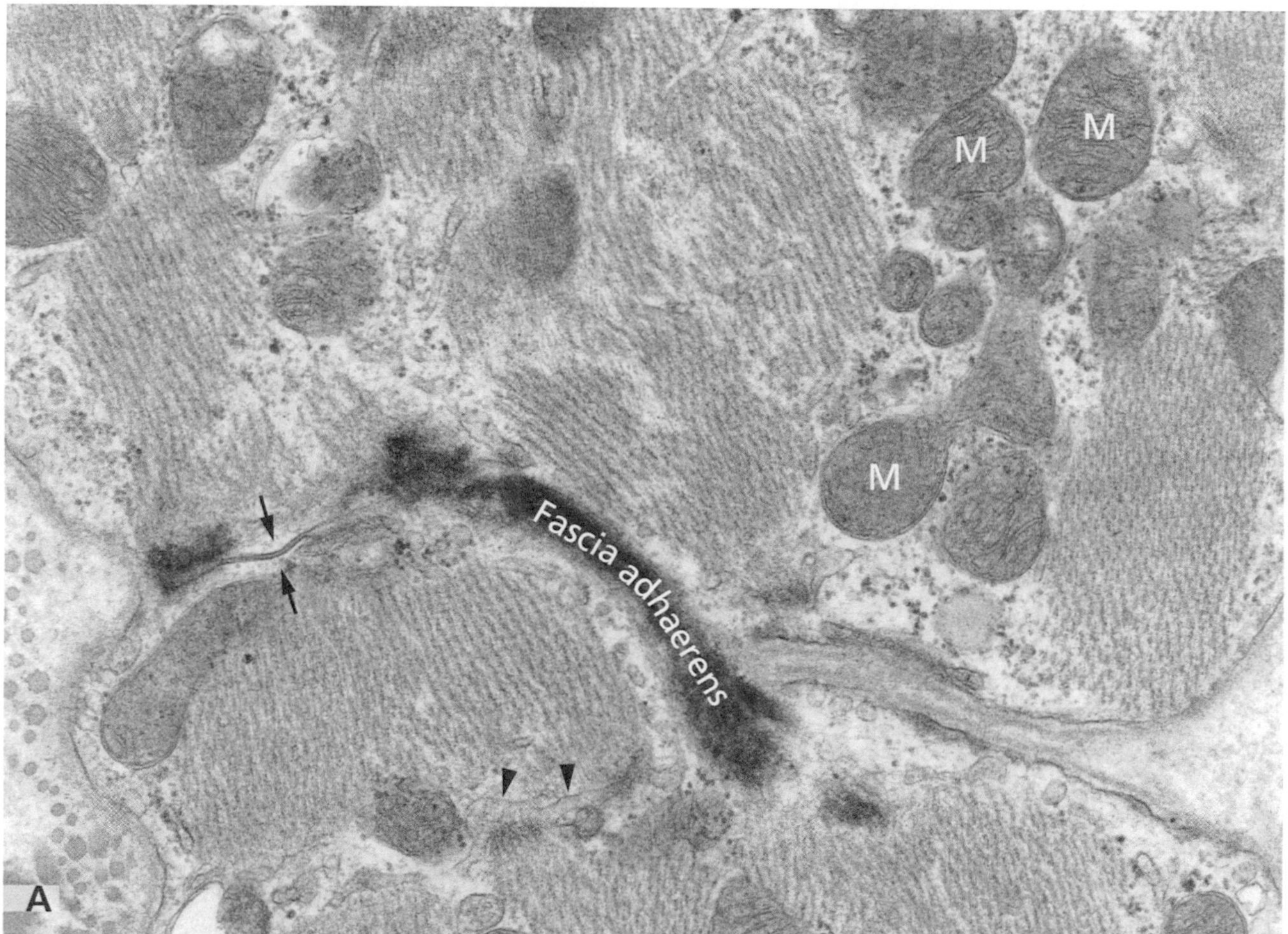
M
M
M
Fascia adhaerens
A

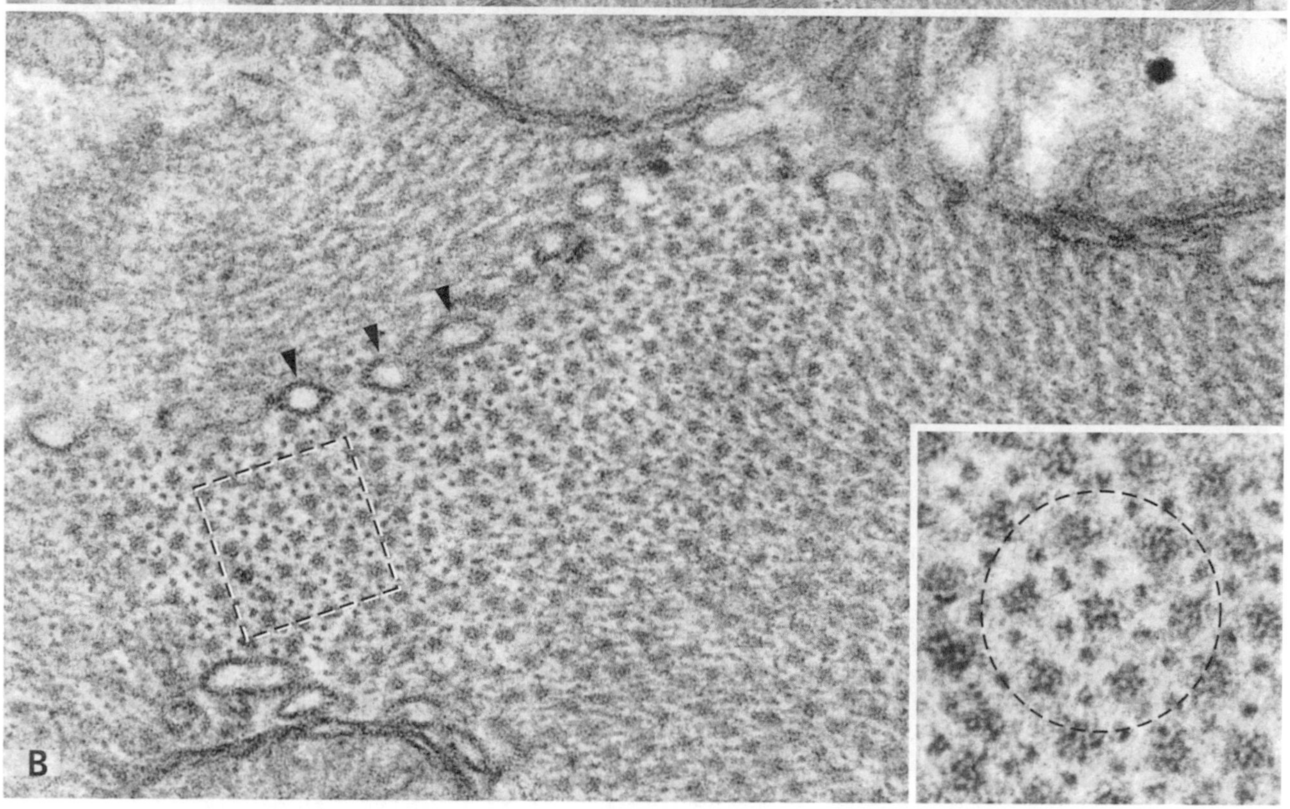
B

SMOOTH MUSCLE CELLS, SYNAPSE Á DISTANCE

Smooth muscle forming the contractile wall of inner hollow organs and vessels is composed of bundles or sheets of fusiform cells that measure in length between 20 μm in blood vessel walls and approximately 200 μm in the wall of the intestine; smooth muscle cells can reach lengths of 500 μm in the wall of the uterus during pregnancy. The cells possess usually one nucleus of elongated shape and tapering ends in central position. Excluding the nuclear pole regions where most organelles and compartments of the biosynthetic system are concentrated, the contractile apparatus occupies the vast majority of the cytoplasm. It consists of myosin II, actin, and associated proteins such as tropomyosin and caldesmon, which blocks the myosin binding sites. Calmodulin, a Ca^{2+}-binding protein, regulates the intracellular concentration of Ca^{2+}; binding to caldesmon, it effects its phosphorylation and release from F-actin. Unlike the onset of contraction in striated muscle where the troponin-tropomyosin complex has a pivotal role, contraction in smooth muscle is regulated by the Ca^{2+}-calmodulin/myosin light chain kinase system and initiated by phosphorylation of one of the myosin light chains in the head domain. Ultrastructurally, the thin actin filaments dominate. Via alpha-actinin, they are anchored in dense bodies and plaques attached to the plasma membrane, which are assumed to be part of an anastomosing network extending from the cell surface into the interior of the cells. The dense bodies and plaques are analogs of the Z disks in the striated muscle. Intermediate filaments consist of desmin or vimentin, in the case of vascular smooth muscle cells. Gap junctions permit communication between adjacent cells.

Panels A and B show smooth muscle cells in the small intestinal wall of the rat in a survey electron micrograph and at higher magnification, respectively. In panel A, smooth muscle cells are visible in longitudinal sections in the lower part and cross-sectioned in the upper part of the picture. The bulk of the cytoplasm is filled by the contractile filaments. Mitochondria (M) are interspersed and may be lined in the cell periphery but, together with abundant ribosomes and endoplasmic reticulum, are particularly crowded in cytoplasmic areas extending from the regions close to the nuclear poles, as shown in the micrograph in panel A in the uppermost smooth muscle cell sectioned longitudinally. Dense areas and dense bodies providing the platforms for the anchoring of actin filaments are conspicuous in the interior of the cells (asterisks) and are attached to the plasma membrane. Each smooth muscle cell is sheathed by a basal lamina (arrowheads in panel B) the constituents of which, such as type IV collagen, laminin, and proteoglycans, are products of the muscle cell. Both micrographs show groups of caveolae that occupy large areas of the smooth muscle cell surfaces. In smooth muscle cells, caveolae may be closely apposed to membranes of the endoplasmic reticulum (cf. panel A in Fig. 51). It is assumed that there may be analogies to the T- and L-systems in striated muscle fibres. Studies with vascular smooth muscle of the ferret aorta have revealed that the muscle cell contractility is regulated by caveolin-1.

Smooth muscle cells are surrounded by fine connective tissue, the endomysium, where an axon is visible embedded in a network of collagen fibrils (C) and neighboured by a process of a fibroblasts (F). The axon is bare of a glial cell sheath (cf. Fig. 160) and contains abundant small synaptic vesicles (arrows) accumulated close to the plasma membrane. Transmitters released from the axon have to diffuse across some distance to reach the muscle cell surface ("synapse á distance").

References

Baron CB and Coburn RF (2004) Smooth muscle raft-like membranes. J Lipid Res 45: 41

Bolton TB, Prestwich SA, Zholos AV, and Gordienko DV (1999) Excitation-contraction coupling in gastrointestinal and other smooth muscles. Annu Rev Physiol 61: 85

Isshiki M and Anderson RGW (2003) Function of caveolae in Ca^{2+} entry and Ca^{2+}-dependent signal transduction. Traffic 4: 717

Je H-D, Gallant C, Leavis PC, and Morgan KG (2004) Caveolin-1 regulates contractility in differentiated vascular smooth muscle. Am J Physiol Heart Circ Physiol 286: H91

Magnification: x 10,300 (A), x 52,000 (B)

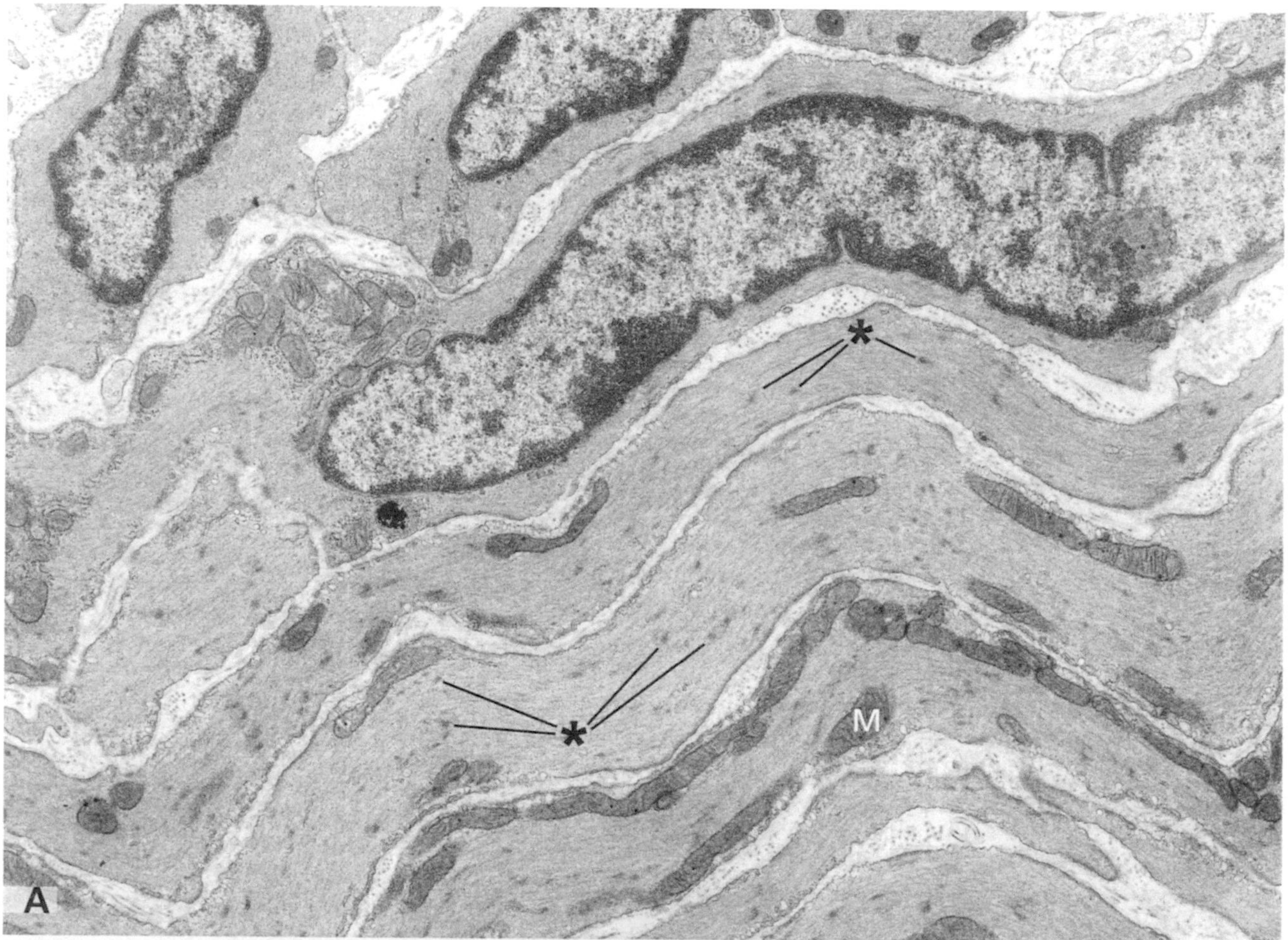
M
A

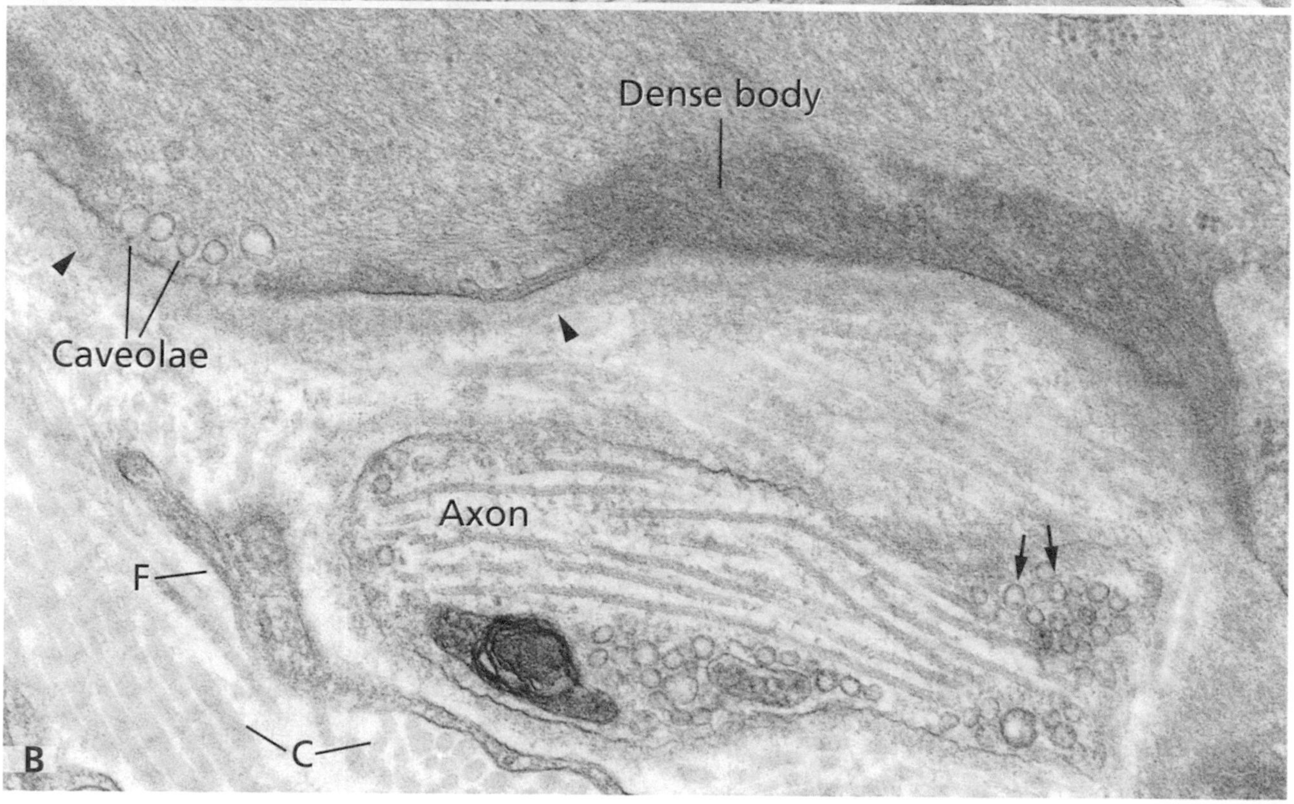
Dense body
Caveolae
Axon
F
C
B

CADASIL

CADASIL stands for Cerebral Autosomal Dominant Arteriopathy with Subcortical Infarcts and Leukoencephalopathy and is the most common form of hereditary stroke. This disease has an average age onset of 45 years and presents with repeated strokes, white matter changes, migraine, and progressive dementia. CADASIL is caused by mutations in the human *Notch3* gene found on chromosome 19q12. They entail either loss or gain of cysteine residues. Notch proteins regulate cell fates or are involved in controlling signals for cell proliferation and maturation.

A characteristic feature of CADASIL is a systemic arterial vasculopathy affecting the smooth muscle fibres of small arteries. It has been proposed that vascular smooth muscle fibres have a continued requirement for the Notch3 signalling pathway and that CADASIL results from a dominant inhibition of the Notch pathway. Of diagnostic importance is that the vasculopathy is not limited to small arteries of the brain white matter but also found in that of skin. Morphological manifestations of the vasculopathy comprise a narrowed lumen of small blood vessels because of intima thickening, progressive deterioration of smooth muscle fibres, and increased extracellular matrix. These changes are detectable both by light microscopy and by electron microscopy (panel A). However, there is a distinct and pathognomic change that can be detected only by electron microscopy and that permits the diagnosis of CADISIL. It consists of extracellular focal deposits of a granular electron dense material (arrows in panels A and B). These deposits are located between the plasma membrane of smooth muscle cells and the basement membrane. This is in contrast to the intracellular subplasma membrane localization of dense bodies representing attachment sites of myofilaments. The nature of the extracellular deposits in CADASIL is unknown at present. A possible relation to the extracellular domain of Notch3, which accumulates in CADASIL in the small arteries, remains to be established.

References

Baudrimont M, Chabriat H, Vahedi K, and Bousser M (1998) Diagnostic value of skin biopsies in cerebral autosomal dominant arteriopathy with subcortical infarcts and leukoencephalopathy (CADASIL). Neuropathol Appl Neurobiol 24: 148

Chabriat H, Joutel A, Dichgans M, Tournier-Lasserve E, and Bousser MG (2009) Cadasil. Lancet Neurol 8: 643

Ihalainen S, Soliymani R, Iivanainen E, Mykkanen K, Sainio A, Poyhonen M, Elenius K, Jarvelainen H, Viitanen M, Kalimo H, and Baumann M (2007) Proteome analysis of cultivated vascular smooth muscle cells from a CADASIL patient. Mol Med 13: 305

Joutel A, Andreux F, Gaulis S, Domenga V, Cecillon M, Battail N, Piga N, Chapon F, Godfrain C, and Tournier-Lasserve E (2000) The ectodomain of the Notch3 receptor accumulates within the cerebrovasculature of CADASIL patients. J Clin Invest 105: 597

Joutel A, Corpechot C, Ducros A, Vahedi K, Chabriat H, Mouton P, Alamowitch S, Domenga V, Cecillion M, Marechal E, et al. (1996) Notch3 mutations in CADASIL, a hereditary adult-onset condition causing stroke and dementia. Nature 383: 707

Kalimo H, Viitanen M, Amberla K, Juvonen V, Marttila R, Poyhonen M, Rinne JO, Savontaus M, Tuisku S, and Winblad B (1999) CADASIL: hereditary disease of arteries causing brain infarcts and dementia. Neuropathol Appl Neurobiol 25: 257

Lewandowska E, Leszczynska A, Wierzba-Bobrowicz T, Skowronska M, Mierzewska H, Pasennik E, and Czlonkowska A (2006) Ultrastructural picture of blood vessels in muscle and skin biopsy in CADASIL. Folia Neuropathol 44: 265

Ruchoux MM, Guerouaou D, Vandenhaute B, Pruvo JP, Vermersch P, and Leys D (1995) Systemic vascular smooth muscle cell impairment in cerebral autosomal dominant arteriopathy with subcortical infarcts and leukoencephalopathy. Acta Neuropathol (Berl) 89: 500

Sourander P and Walinder J (1977) Hereditary multi-infarct dementia. Morphological and clinical studies of a new disease. Acta Neuropathol (Berl) 39: 247

Tournier-Lasserve E, Joutel A, Melki J, Weissenbach J, Lathrop GM, Chabriat H, Mas JL, Cabanis EA, Baudrimont M, Maciazek J, et al. (1993) Cerebral autosomal dominant arteriopathy with subcortical infarcts and leukoencephalopathy maps to chromosome 19q12. Nat Genet 3: 256

Magnification: x 6,500 (A); x 36,000 (B)

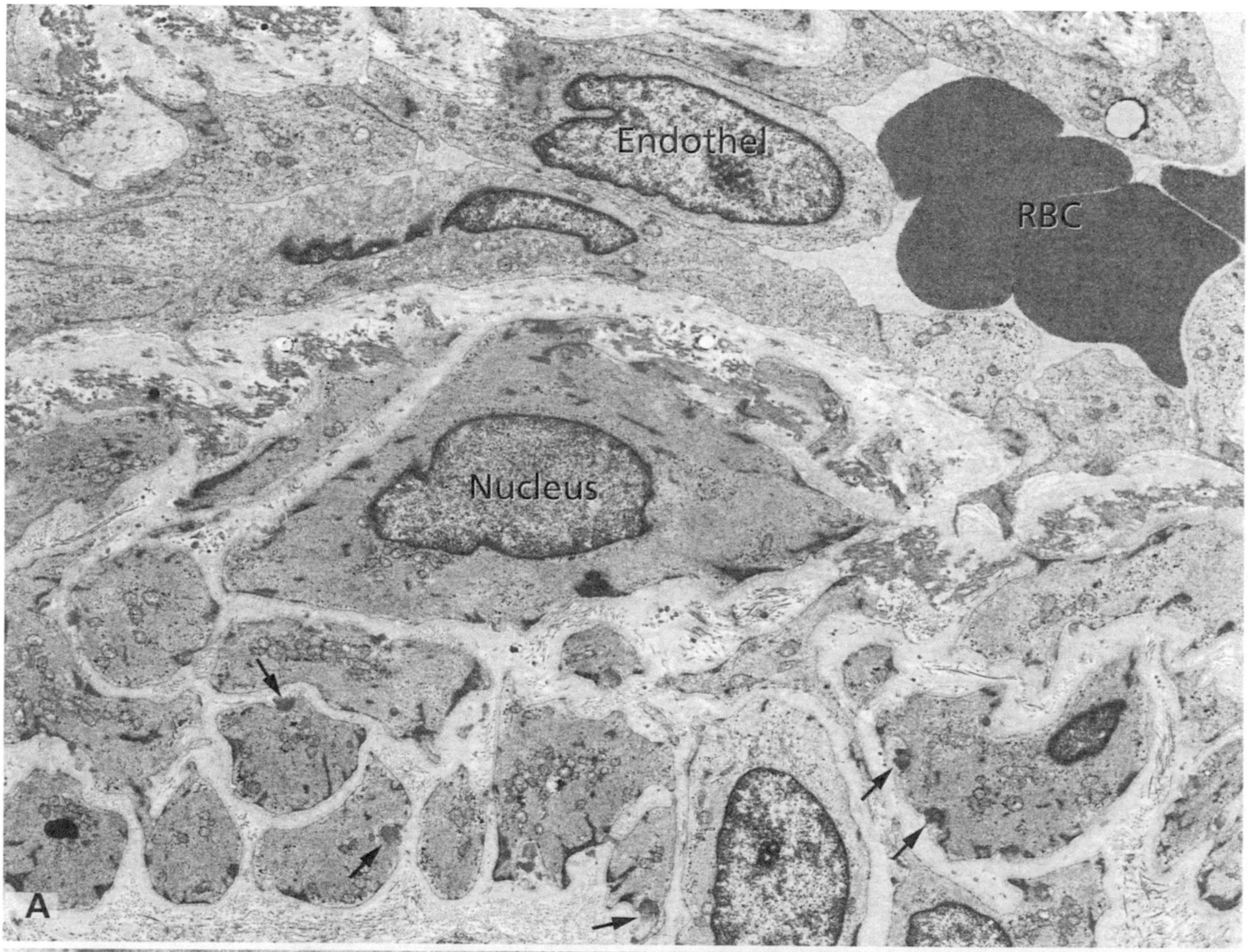
Endothel
RBC
Nucleus
A

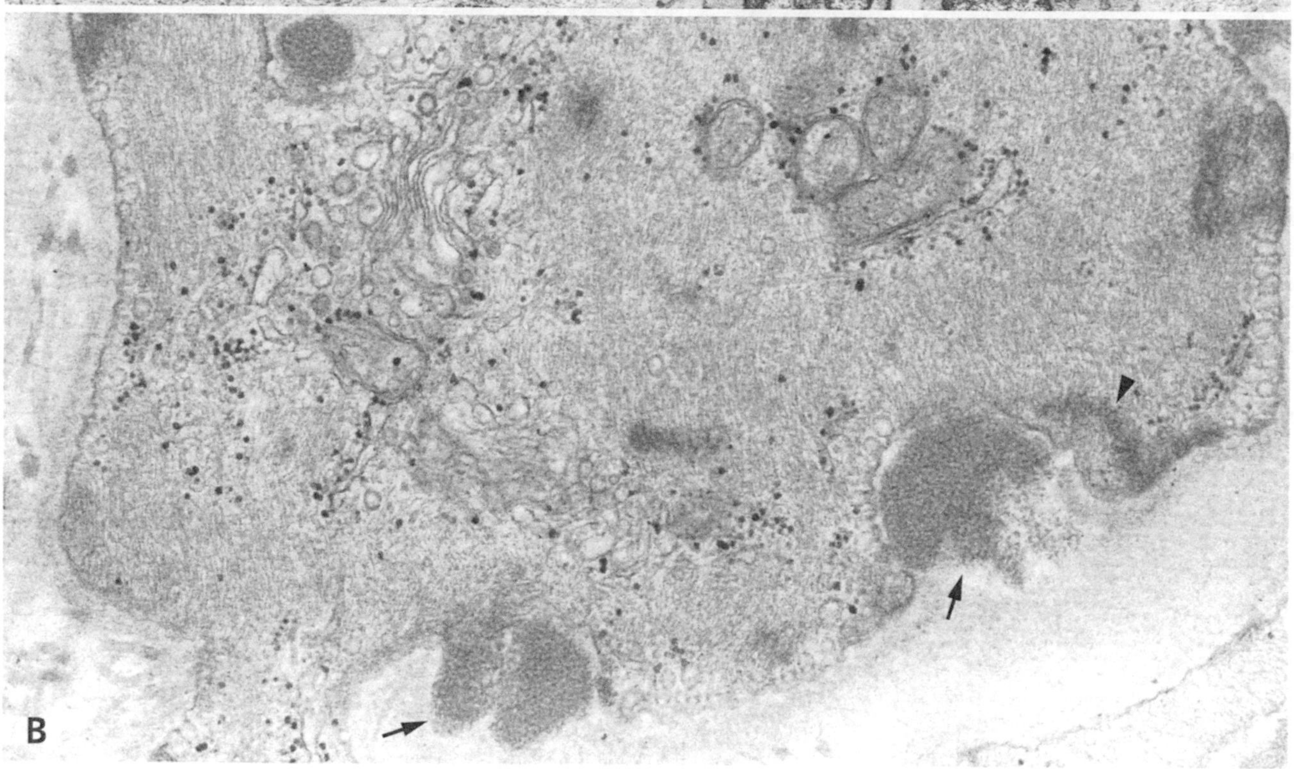
B

CENTRAL NERVOUS SYSTEM: NEURON, GLIAL CELLS

Providing the base for the most sophisticated functions and tasks of the body including memory and creativity, the tissue of the nervous system attracts highest attention since the earliest times of microscopy up to today. The complex functional processes in central and peripheral parts of the nervous system are not only continuously actual topics in life sciences and medical research. The recent identification and characterisation of neural stem cells within certain areas of the adult brain has revolutionised our knowledge and understanding of adult neurogenesis, and has increased hopes for possible treatment and repair of damages due to injuries or degenerative diseases.

At first glance, electron micrographs of nerve tissue originating from the central nervous system as shown in panels A and B, convey a confusing impression. It is difficult and may be impossible to identify every detailed structure visible, belonging to either of the two principal types of cells, nerve cells (neurons) or glial cells. At the left hand side of panel A, the large body of a nerve cell can be seen, as well as one of its processes, a dendrite, projecting toward the upper right corner of the picture. The neighbouring brain parenchyma consists of numerous, countless processes of both nerve and glial cells. The pictures in both panels make evident that the intercellular spaces are extremely narrow and intercellular materials are virtually lacking.

Neurons are the structural and functional units of the nerve tissue and show different functional domains. These include the cell body (soma), one process extending from the cell body and transmitting impulses away from the cell body, called neurite or axon, one or more processes transmitting impulses from the periphery toward the cell body, called dendrites, and the synaptic junctions, where another neuron or an effector cell is contacted and the impuls transmitted (cf. Figs. 158 and 159). The nerve cell body (soma, panel A) contains the nucleus and the surrounding perinuclear cytoplasm, the perikaryon, where all components of the biosynthetic apparatus, ribosomes, endoplasmic reticulum, Golgi apparatus and all organelles that maintain the cell, are present. A similar range of organelles is found in the dendrites. Although the axon hillock, the site where the axon (neurite) extends from the perikaryon, due to the plane of the section is not visible in panel A, multiple axons can be seen in both panels. The micrographs show characteristic ultrastructures of brain parenchyma. Numerous profiles of axons are visible. They are grouped and sheathed by fine processes of glial cells, and contain mitochondria and abundant neurofilaments and microtubules that function as tracks for axoplasmic transport. In the right hand segment of panel B, several axons are sheathed by myelin (asterisks; cf. Fig. 162), which in the central nervous system is produced by specialised neuroglial cells, the oligodendrocytes. Protoplasmic and fibrous astrocytes represent another class of glial cells. With their processes, they are intimately apposed to neurons, cover axons, build up superficial limiting layers at the boundaries to the pia mater, and form perivascular feet (cf. Fig. 158).

References

Bannai H, Inoue T, Nakayama T, Hattori M, and Mikoshiba K (2003) Kinesin dependent, rapid, bi-directional transport in dentrites of hippocampal neurons. J Cell Sci 117: 163

Hickey WF, Vass K, and Lassmann H (1992) Bone marrow-derived elements in the central nervous system: an immunohistochemical and ultrastructural survey of rat chimeras. J Neuropathol Exp Neurol 51: 246

Horne J (2004) New neurons? Nat Cell Biol 6: 287

Matteoli M, Coco S, Schenk U, and Verderio C (2004) Vesicle turnover in developing neurons: how to build a presynaptic terminal. Trends Cell Biol 14: 133

Priller L (2003) Grenzgänger: adult bone marrow cells populate the brain. Histochem Cell Biol 120: 85

Sanai N, Tramontin AD, Quninones-Hinojosa A, Barbaro NM, Gupta N, Kunwar S, Lawton MT, McDermott MW, Parsa AT, Verdugo JM-G, Berger MS, and Avlarez-Buylla A (2004) Unique astrocyte ribbon in adult human brain contains neural stem cells but lacks chain migration. Nature 427: 740

Song H, Stevens CF, and Gage FH (2002) Astroglia induce neuro genesis from adult stem cells. Nature 417: 39

Wisco D, Anderson ED, Chang MC, Norden C, Boiko T, Fölsch H, and Winckler B (2003) Uncovering multiple axonal targeting pathways in hippocampal neurons. J Cell Biol 162: 1317

Zimmermann DR and Dours-Zimmermann MT (2008) Extracellular matrix of the central nervous system: from neglect to challenge. Histochem Cell Biol 130: 635

Magnification: x 5,500 (A), x 26,500 (B)

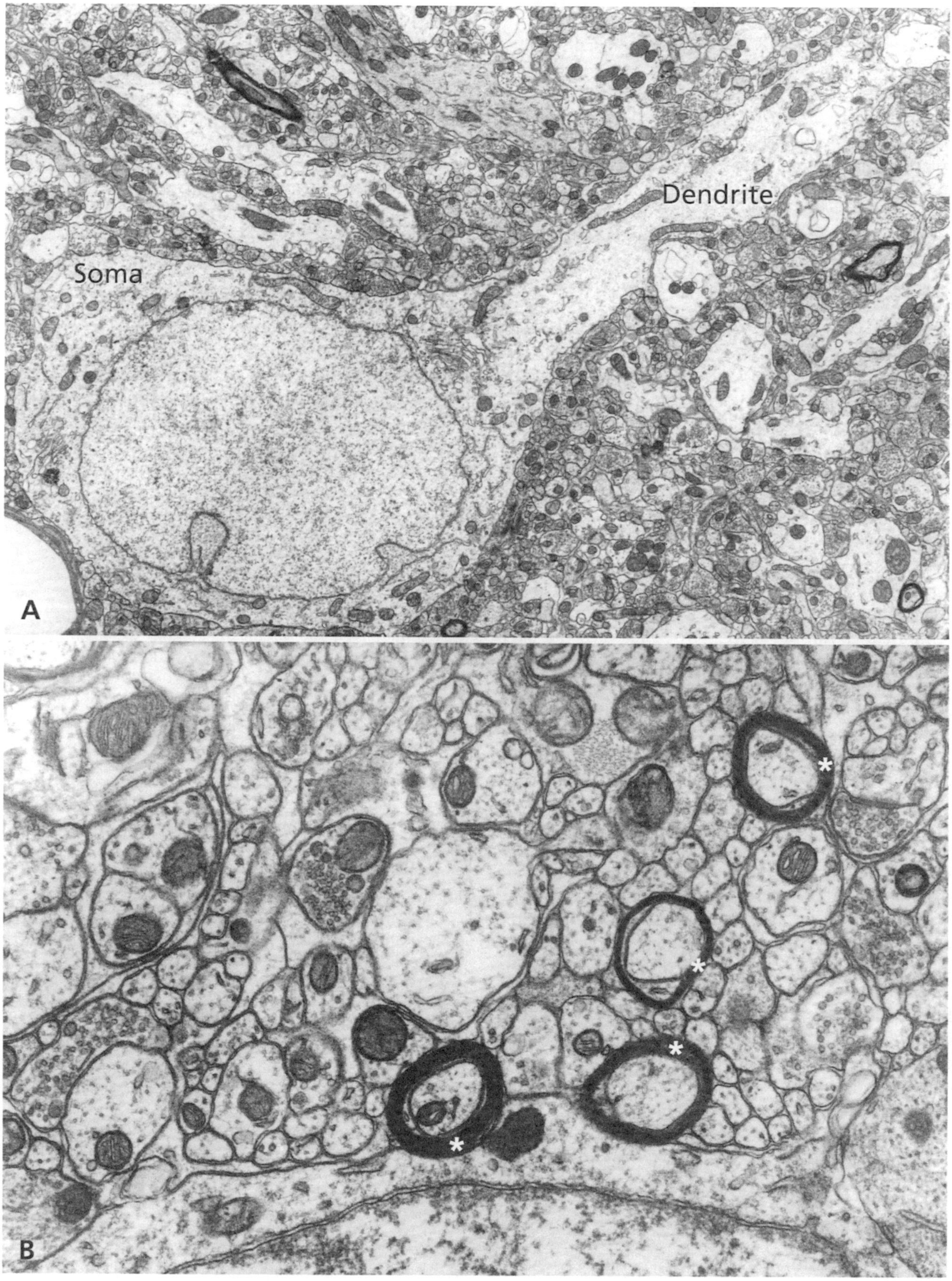
Dendrite
Soma
A
*
*
*
*
B

BLOOD-BRAIN BARRIER, SYNAPSES

Since more than 100 years, it is known that a specialised barrier protects the brain from harmful substances circulating in the blood. Different mechanisms contribute to the blood-brain barrier, which influence both paracellular and transcellular traffic. The endothelium is continuous, and extended tight junctions between the adjacent endothelial cells seal the intercellular spaces. Specific carrier mechanisms are involved even in the transport of small molecules, and the number of receptors is small. Furthermore, substances leaked into the brain parenchyma are transported back into the blood by multidrug resistance transporters present in the endothelial cells. The two neighbouring vessels (asterisks) shown in panel A are characteristic for the brain microvasculature. They are lined with a continuous layer of endothelial cells. Transport vesicles are sparse, reflecting the restricted transepithelial transport. Perivascular cells, known to be derived from the bone marrow, accompany the vessels. Within the basal lamina (arrowheads), a pericyte can be seen localised close to the endothelium. Several cell bodies of astrocytes are apparent near the vessels, and numerous profiles of glial cell processes can be seen closely apposed to the basal lamina, forming a perivascular layer.

Panels B and C show chemical synapses, specialised adhesive sites for communication between two neurons, where impulses are conducted by neurotransmitters (cf. Fig. 159). A synapse consists of a presynaptic component that often is a terminal knob of an axon where synaptic vesicles (arrows) containing the transmitters are accumulated, and a postsynaptic component that often is a spine-shaped process of a dendrite. The specialised pre- and postsynaptic membranes are separated only by a 20–30 nm space, the synaptic cleft. Triggered by Ca^{2+}-influx via voltage-gated channels, the transmitter is released into the synaptic cleft by fusion of the synaptic vesicle membrane with the presynaptic membrane (cf. Fig. 159). The transmitter diffuses across the synaptic cleft and is bound by the specific receptors concentrated in the postsynaptic membrane. As a consequence, ions are allowed to enter via ligand-gated channels, leading to changes of the polarisation and causing either excitation or inhibition at the postsynaptic membrane. At the presynaptic neuron, transmitter release is rapidly followed by, either "kiss and run" or clathrin-mediated, endocytosis, and transport to endosomal compartments. Under the electron microscope, layers of dense materials are visible, which are associated with the synaptic membranes, corresponding to their particular composition and organisation. According to the favourite localisations, asymmetric and symmetric types of synapses have been discriminated. Asymmetric synapses (panel B) that correspond mainly to excitatory synapses with glutamate used as transmitter, show particularly prominent postsynaptic densities (arrowhead in B). Asymmetric synapses are further characterised by predominance of round synaptic vesicles (arrows in B). By contrast, symmetric synapses (panel C), which often correspond to inhibitory synapses, show prominent densities associated with both the pre- and postsynaptic membranes. In symmetric synapses, flattened synaptic vesicles predominate (arrows in C).

References

Dresbach T, Nawrotzki R, Kremer T, Schumacher S, Quinones D, Kluska M, Kuhse J, and Kirsch J (2008) Molecular architecture of glycinergic synapses. Histochem Cell Biol 130: 617

Hickey WF, Vass K, and Lassmann H (1992) Bone marrow-derived elements in the central nervous system: An immunohistochemical and ultrastructural survey of rat chimeras. J Neuropath Exp Neurol 51: 246

Holburg A and Lippoldt A (2002) Tight junctions of the blood-brain barrier: development, composition, and regulation. Vascul Pharmacol 38: 323

Jarousse N and Kelly RB (2001) Endocytic mechanisms in synapses. Curr Opin Cell Biol 13: 461

Matteoli M, Coco S, Schenk U, and Verderio C (2004) Vesicle turnover in developing neurons: how to build a presynaptic terminal. Trends Cell Biol 14: 133

Micheva KD, Buchanan J, Holz RW, and Smith SJ (2003) Retrograde regulation of synaptic vesicle endocytosis and recycling. Nat Neurosci 6: 925

Nitta T, Hata M, Gotoh S, Seo Y, Sasaki H, Hashimoto N, Furuse M, and Tsukita S (2003) Size-selective loosening of the blood-brain barrier in claudin-5-deficient mice. J Cell Biol 161: 653

Sytnyk V, Leshchynška I, Dityatev A, and Schachner M (2004) *Trans* Golgi network delivery of synaptic proteins in synaptogenesis. J Cell Sci 117: 381

Yamagata M, Sanes JR, and Weiner JA (2003) Synaptic adhesion molecules. Curr Opin Cell Biol 15: 621

Magnification: x 6,600 (A); x 53,000 (B); x 82,000 (C)

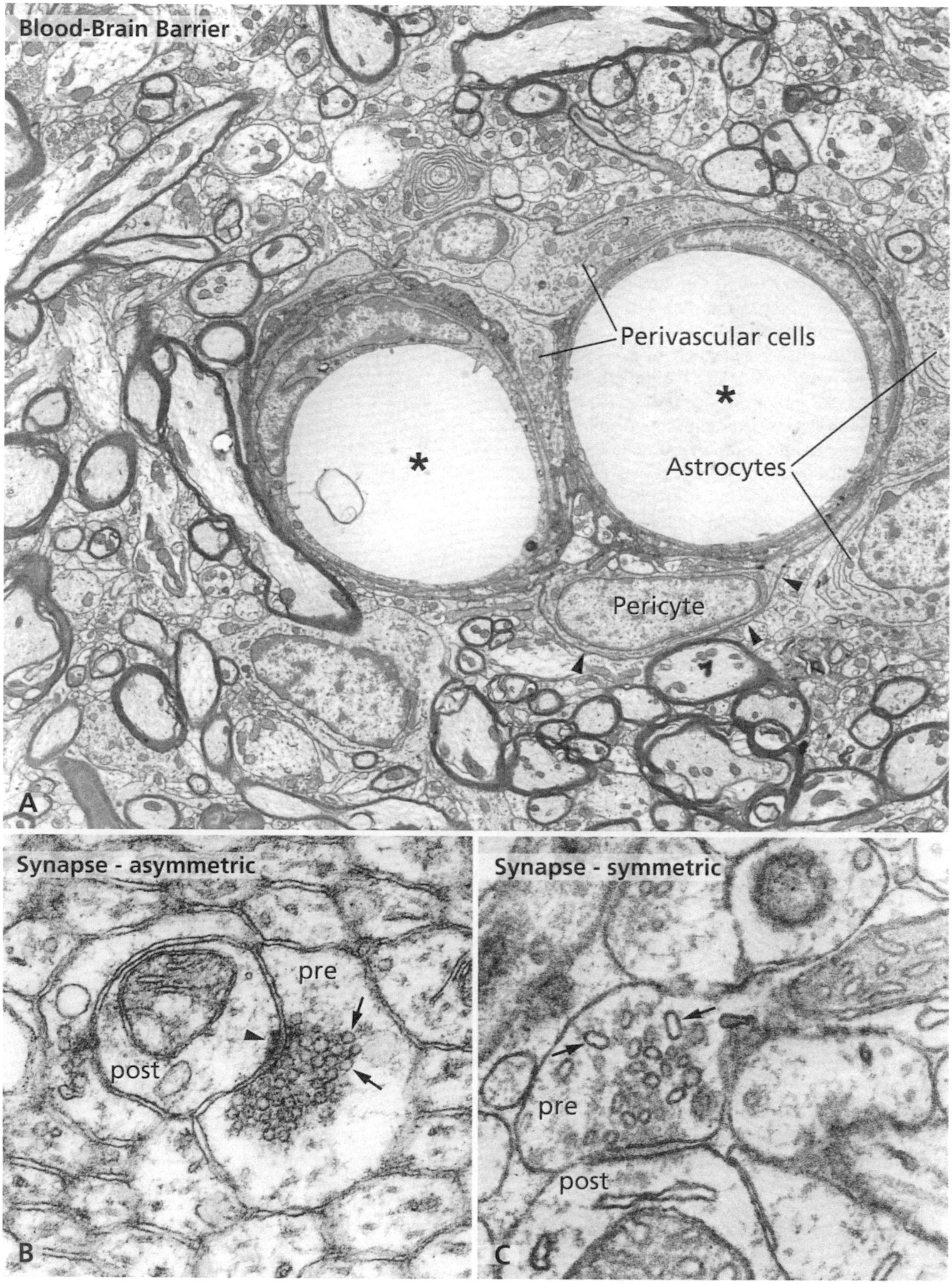
Blood-Brain Barrier
Perivascular cells
*
*
Astrocytes
Pericyte
A
Synapse - asymmetric
pre
post
B
Synapse - symmetric
pre
post
C

STRUCTURE OF THE SYNAPTIC TERMINAL

The transfer of information between neurons, or between neurons and other excitable cells, takes place at specialised junctions called *synapses.* In the mammalian central nervous system, most synapses function by converting electrical signals into chemical substances *(neurotransmitters).* Once released from the presynaptic terminal neurotransmitters bind postsynaptic receptors. Communication is unidirectional, resulting in the formation of very different structures at each side of the synapse (termed *pre-* and *postsynaptic* terminals, respectively). Synaptic terminals are separated by a ~ 25 nm gap known as the *synaptic cleft* (SC). Figure A shows a tomographic slice (2.7 nm in thickness) of a vitrified central nervous system synapse from the synaptosomal fraction.

Presynaptic terminals are specialised compartments that contain numerous neurotransmitter-filled synaptic vesicles (SV), as well as other organelles such as endoplasmatic reticulum and mitochondria (mit). Electrical signals *(action potentials)* are generated at the neuronal cell body and travel along the axon until they reach the presynaptic terminal. Action potential arrival causes a Ca^{2+} influx through voltage-gated Ca^{2+} channels that in turn stimulates synaptic vesicle fusion with the presynaptic membrane at the so-called *active zone,* a region of the membrane that directly faces the postsynaptic side and hosts the fusion machinery. After releasing the neurotransmitters, synaptic vesicles are endocytosed and prepared for the next round of release. Figure C3 shows a fusing vesicle, captured in the moment when its membrane is continuous with the plasma membrane.

As can be seen in Figure B, synaptic vesicles (yellow) are extensively interconnected via filamentous bridges (red), while those located in the vicinity of the active zone (grey) are most often tethered to it by similar strands (blue). Details of these structures and their corresponding 3D renderings can be seen in Figures C1 (connector) and C2 (tether).

The postsynaptic terminal is normally found on dendrites and it harbors a dense array of postsynaptic receptors and signalling and scaffolding proteins, the so-called *postsynaptic density* (PSD). The effects of the neurotransmitter binding to postsynaptic ionic channels and receptors can be broadly separated in two categories. First, the ionic flux at the postsynapse is altered, resulting in excitatory or inhibitory potentials (depending on the synapse type) that change the excitability of the postsynaptic cell. Second, various signal transduction pathways can be initiated by the activation of second messengers leading to diverse effects, from modulation of synaptic strength to gene transcription.

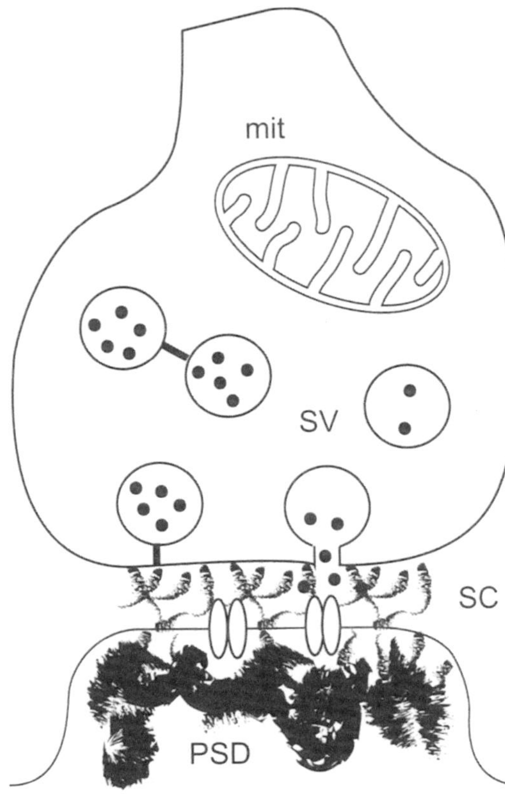

References

Dalva MB, McClelland AC, and Kayser MS (2007) Cell adhesion molecules: signalling functions at the synapse. Nat Rev Neurosci 8, 206

Kennedy MJ and Ehlers MD (2006) Organelles and trafficking machinery for postsynaptic plasticity. Ann Rev Neurosci 29: 325

Levitan EB and Kaczmarek LK (2002) The neuron. Oxford University Press.

Murthy VN and De Camilli P (2003) Cell biology of the presynaptic terminal. Ann Rev Neurosci 26: 701

Neher E and Sakaba T (2008) Multiple roles of calcium ions in the regulation of neurotransmitter release. Neuron 59: 861

Schweizer FE and Ryan TA (2006) The synaptic vesicle: cycle of exocytosis and endocytosis. Curr Opin Neurobiol 16: 298

Sheng M and Hoogenraad CC (2007) The postsynaptic architecture of excitatory synapses: a more quantitative view. Ann Rev Biochem 76: 823

Südhof TC (2004) The synaptic vesicle cycle. Ann Rev Neurosci 27: 509

Triller A and Choquet D (2005) Surface trafficking of receptors between synaptic and extrasynaptic membranes: and yet they do move! Trends Neurosci 28: 133

Yamagata M, Sanes JR, and Weiner JA (2003) Synaptic adhesion molecules. Curr Opin Cell Biol 15: 621

Magnification: x 142,000 (A), x 260,000 (C1, C2, C3)

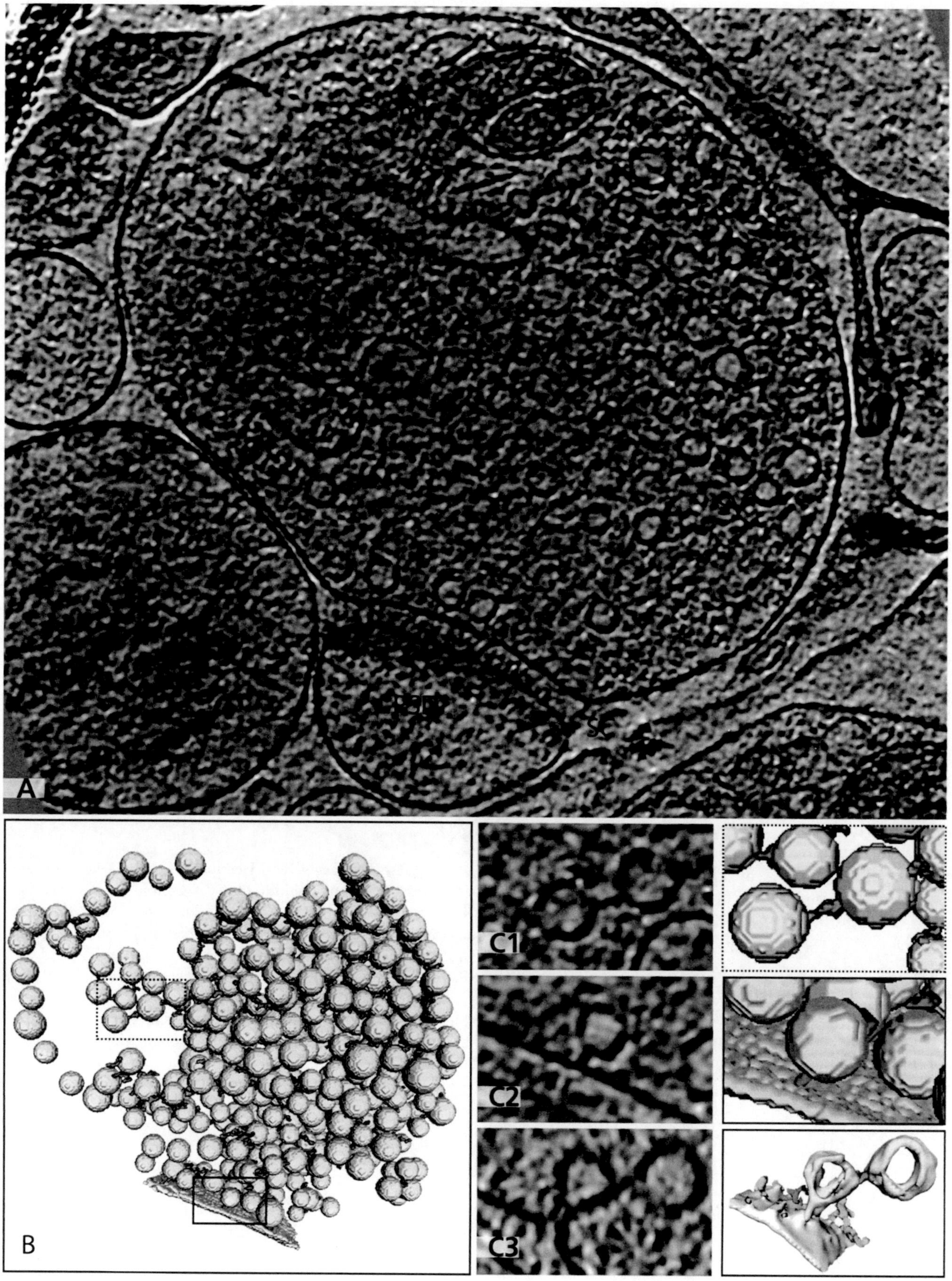
PSD
SC
A
B
C1
C2
C3

UNMYELINATED NERVE FIBRE

In the peripheral nervous system, nerve cell processes – both axons (neurites) and dendritic axons of sensory neurons – are surrounded by specialised glial cells, the cells of Schwann. The axon, its sheath of Schwann cells and the surrounding basal lamina form the impulse conducting structures, the nerve fibres, classified in myelinated and unmyelinated types with respect to the presence or absence of a myelin sheath (cf. Fig. 162). Along myelinated nerve fibres, conduction of impulses is fast, since the myelin sheath forms an insulating layer, and the impulse "jumps" from one intercellular region free of myelin to the next myelin-free region (cf. Fig. 163). In nerve fibres lacking a myelin sheath, the nerve impulses move continuously and less rapidly. Unmyelinated axons include most of the postganglionic axons from autonomic ganglia.

The unmyelinated nerve fibre shown in the electron micrograph belongs to the autonomic nervous system in the wall of the small intestine. Seven axons can be seen, either more superficially or more deeply invaginated in the cytoplasm of a cell of Schwann. The Schwann cell nucleus is visible in the right segment of the nerve fibre. In unmyelinated nerve fibres, one Schwann cell as a rule houses more than one axon. The Schwann cell cytoplasm forms thin envelopes around the axons, for each of which an own groove is provided. Schwann cells also produce the components of the basal lamina that covers the entire nerve fibre. Localised close to the axolemma (plasma membrane of the axon), the plasma membrane of the Schwann cell surrounds the axons, leaving a small but regular space. Most axons are completely engulfed by the Schwann cell. In these cases, the "lips of the grooves" are closed and a mesaxon (open arrows) is formed. By contrast, the axon at the left hand side of the nerve fibre containing microtubules (arrows), a mitochondrion (asterisk), and accumulations of small synaptic vesicles, is only partially enveloped by the cell of Schwann. Partly, its plasma membrane is devoid of a Schwann cell sheath and covered only by the basal lamina (arrowheads). Along unmyelinated nerve fibres, such areas with open lips and the axolemma exposed to the neighbourhood are particularly prominent close to regions of special synapses where transmitters are delivered to diffuse to the surfaces of target cells, such as smooth muscle or secretory cells. Such a "synapse á distance en passant" is also shown in panel B of Fig. 155, where the axon is completely free of a Schwann cell sheath and the axolemma covered only by the basal lamina.

In the axoplasm (the cytoplasm of the axons), in addition to microtubules called neurotubules (arrows), abundant filaments including 10 nm thick neurofilaments, are visible. Both microtubules and neurofilaments provide tracks for axonal transport. Fast axonal traffic is bidirectional, requires an intact microtubule cytoskeleton, and involves motor proteins, such as kinesin and dynein. Kinesin is implicated in the anterograde transport, by which newly synthesised molecules and membrane-bound organelles, such as mitochondria and multiple vesicles, including synaptic vesicles, are carried from the perikaryon to the axon periphery. Dynein is involved in retrograde traffic from terminal regions of the axon to the perikaryon, which in particular, is the route of substances and molecules internalised at the axon endings, and is misused by viruses and toxins.

The nerve fibre is surrounded by the loose connective tissue of the endoneurium (cf. Fig. 161), which contains multiple collagen fibrils.

References

Janmey PA, Leterrier J-F, and Herrmann H (2003) Assembly and structure of neurofilaments. Curr Opin Colloid Interface Sci 8: 40

Lariviere RC, and Julien JP (2004) Functions of intermediate filaments in neuronal development and disease. J Neurobiol 58: 131

Mirsky R, Jessen KR, Brennan A, Parkinson D, Dong ZP, Meier C, Panmantier E, and Lawson D (2002) Schwann cells as regulators of nerve development. J Physiol 96: 17

Smith RS and Snyder RE (1992) Relationship between the rapid axonal transport of newly synthesized proteins and membranous organelles. Mol Neurobiol 6: 285

Magnification: x 33,300

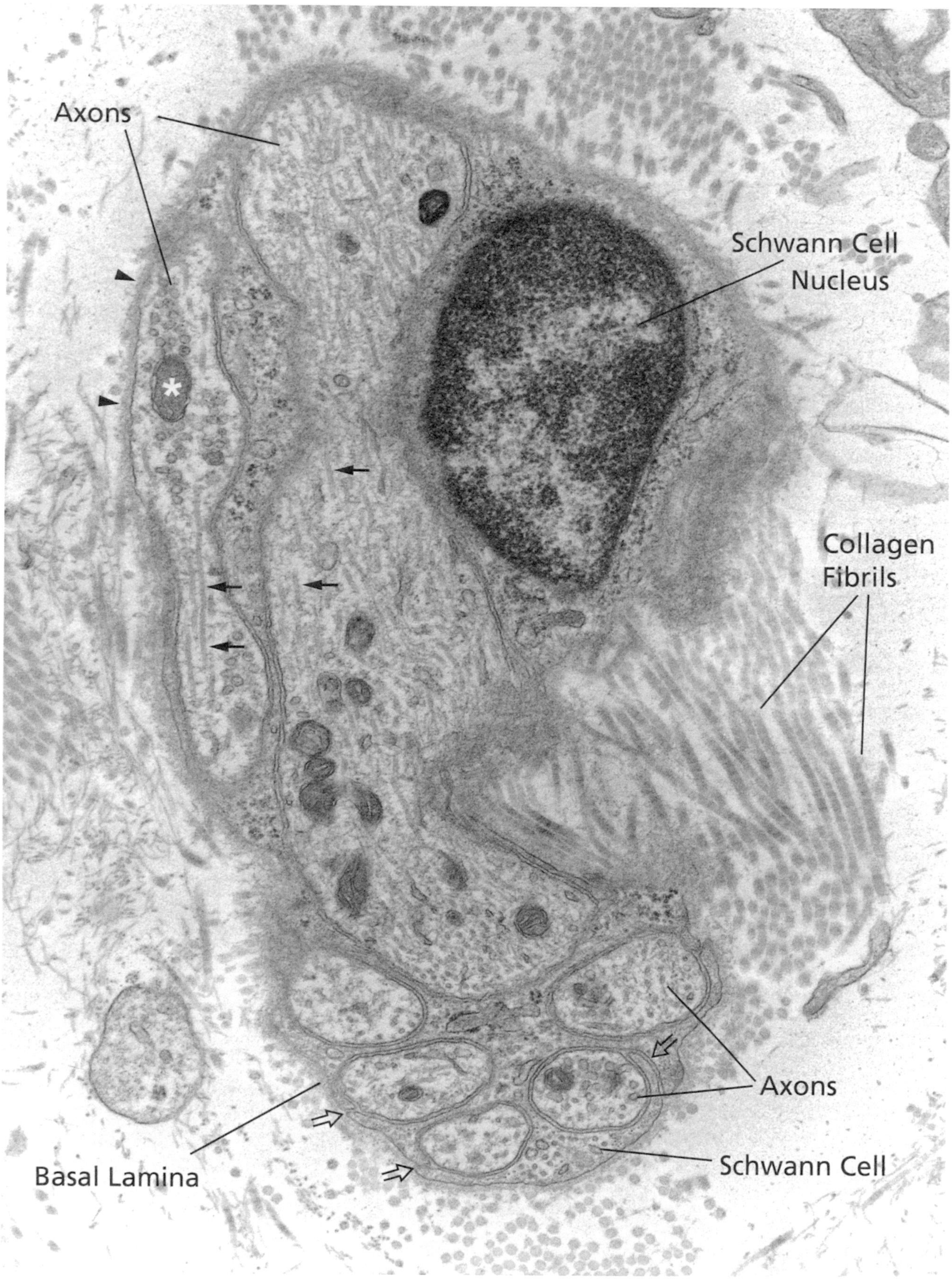
Axons
Schwann Cell Nucleus
Collagen Fibrils
Axons
Basal Lamina
Schwann Cell

PERIPHERAL NERVE: CONNECTIVE TISSUE COMPONENTS

In a peripheral nerve, the individual nerve fibres are organised by connective tissue that consists of three distinct components, called endoneurium, perineurium, and epineurium. Each of the three components has specific functional tasks and morphological characteristics. All three components can be seen in the survey electron micrograph in panel A. An unmyelinated nerve fibre is visible in the centre. Three axons containing a dense cytoskeletal network of neurotubules and neurofilaments are enveloped by thin cytoplasm sheaths of a cell of Schwann (S). The nerve fibre is surrounded by the endoneurium, consisting mainly of a fine network of collagen fibrils (C). Collagen fibrils also hold together nerve fibres and blood capillaries in larger nerve fibre bundles and are closely related to the basal laminae of both the Schwann cell of the nerve fibre and the perineurial cells. Within the endoneurium, fibroblasts are sparse. It is assumed that most of the collagen of the endoneurium is produced by the cells of Schwann.

The perineurium is a particularly specialised connective tissue that surrounds a nerve fascicle. The micrograph in panel A shows a small nerve fascicle, in which only one unmyelinated nerve fibre is visible. Large nerve fascicles contain multiple myelinated and unmyelinated nerve fibres held together by the endoneurium and surrounded by the perineurium. The perineurial sheath of a larger fascicle containing also myelinated nerve fibres is shown at higher magnification in panel B. The perineurium consists of flat, squamous cells that are composed in one or more layers, and are connected by junctional complexes (circle in panel A) that include a system of tight junctions preventing paracellular transport. In this aspect, the perineurium has the character of an epithelium, called "perineurial epithelium", and functionally contributes to the formation of a blood-nerve barrier (cf. Fig. 158, blood-brain barrier). Five to six cellular layers may be built. Cell branches contribute to the formation of multilayers, as is shown in the right hand part of the perineurial sheath in the survey micrograph in panel A, and at higher magnification in panel B. The two thin perineurial layers shown in panel B are connected by a fine cytoplasmic bridge, visible in the right segment of the picture. Each cellular layer is covered by a basal lamina (arrowheads) that is closely related to collagen fibrils of the endoneurium inside and the epineurium outside of the nerve fascicle. Collagen fibrils (C in panel B) are present between the individual perineurial cell layers but fibroblasts are absent. Perineurial collagen fibrils are assumed to be a product of the perineurial cells. In the cytoplasm, mitochondria and other cell organelles are visible, and numerous microtubules and filaments reflect active transport functions and high cell contractility. Most conspicuous are the numerous vesicles occupying extended areas of the cytoplasm. At places, the cytoplasm of the perineurial layers is extremely thin, measuring no more than 60–100 nm (panel B). In these areas, numerous vesicles are lined up, in part touching either of the two surfaces. Similarly to the endothelial cells forming the blood-brain barrier (cf. Fig. 158), the perineurial cells are equipped for serving as a diffusion barrier and possess molecular machineries for active transport of substances across the cellular layers.

The outermost component of the nerve-associated connective tissues is built by the epineurium, a more dense tissue with irregularly organised bundles of collagen fibrils (C). In a large nerve, the epineurium surrounds and holds together several or multiple nerve fascicles. The epineurium also leads the larger blood vessels, the branches of which, within a perineurium, penetrate into the individual fascicles.

References

Hirakawa H, Okajima S, Nagaoka T, Takamatsu T, and Oyamada M (2003) Loss and recovery of the blood-nerve barrier in the rat sciatic nerve after crush injury are associated with expression of intercellular junctional proteins. Exp Cell Res 284: 196

Smith CE, Atchabahian A, Mackinnon SE, and Hunter DA (2001) Development of the blood-nerve barrier in neonatal rats. Microsurg 21: 290

Terbo PM, Vuorinen VS, and Roytta M (2002) The endoneurium response to microsurgically removed epi- and perineurium. J Periph Nerve System 7: 155

Walbeehm ET, Afoke A, de Wit T, Holman F, Hovius SER, and Brown RA (2004) Mechanical functioning of peripheral nerves: linkage with the "mushrooming" effect. Cell Tiss Res 316: 115

Magnification: x 24,000 (A), x 28,800 (B)

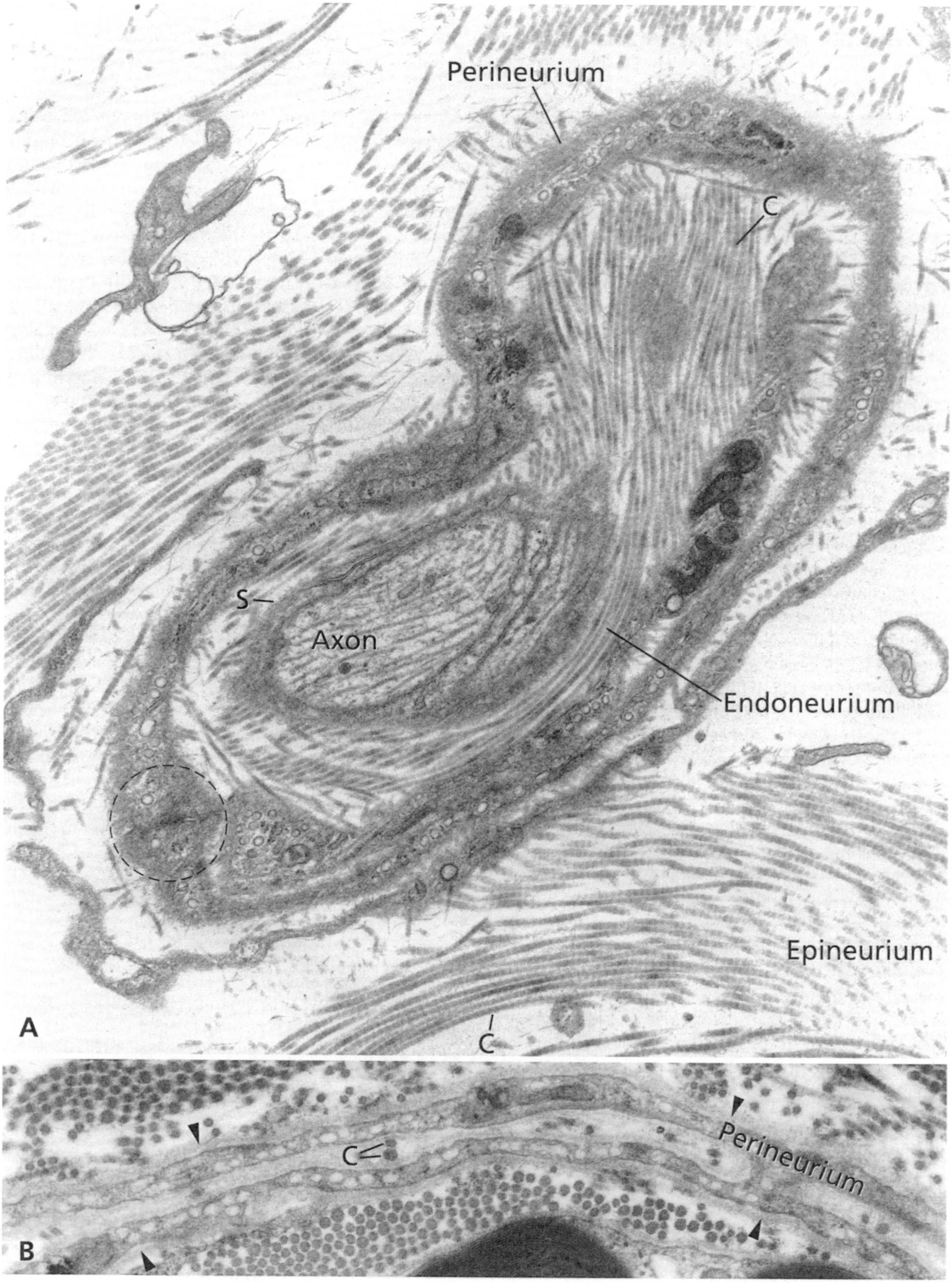
Perineurium
C
S
Axon
Endoneurium
Epineurium
A
C
C
Perineurium
B

MYELINATED NERVE FIBRE, MYELIN

Nerve fibres designed for particularly rapid and efficient conduction of action potentials are equipped with a myelin sheath, a lipid-enriched layer, produced by specialised glial cells, the oligodendrocytes in the central nervous system, and the Schwann cells in the peripheral nervous system. The myelin sheath isolates the axon from the surrounding compartments. It reduces the current flow across the axonal membrane allowing a saltatory conduction along the nerve fibre from one to the next area lacking a myelin sheath at the sites of the cell borders, called the nodes of Ranvier (cf. Fig. 163). A myelin sheath is absent from the axon hillock and the terminal ramifications of the axon where synapses are established. In the central nervous system, oligodendrocytes produce myelin sheaths for more than one axon, and the oligocyte cell body remains outside of the myelin sheaths. By contrast, in the peripheral nervous system the Schwann cells provide myelin only for one axon, and the entire cell including the nucleus becomes part of the myelinated nerve fibre.

The figures show a myelinated nerve fibre of the peripheral nervous system embedded in endoneurium and a high magnification micrograph of myelin of the central nervous system in the inset. The axon can be seen in the centre surrounded by the Schwann cell with its different parts including the sheath of myelin. Myelin is produced by a complex process including new production of membrane, redistribution of cytoplasm, cellular movement, and subsequent interactions between the myelinating glial cell and the axon. In the peripheral nervous system first, a segment of an axon is invaginated in a cell of Schwann and a mesaxon is formed. Subsequently, parts of the mesaxonal plasma membrane areas wrap in a spiral around the axon leading to a subdivision of the mesaxon in inner and outer parts, the inner and outer mesaxons, separated by the middle part, which produces the myelin. During the process of wrapping, the plasma membranes are closely packed. In the peripheral nervous system, protein zero is an essential mediator of membrane compaction. At the same time, the Schwann cell cytoplasm is redistributed. The main part of the Schwann cell body including most of the cytoplasm and the cell nucleus takes position outside the myelin sheath forming the outer collar of Schwann cell cytoplasm (OCS). Inside the myelin sheath, only a thin layer of cytoplasm remains surrounding the axon, the inner Schwann cell collar (asterisk). From the regions between the wrapping mesaxonial plasma membranes, most of the cytoplasm is extruded, and the inner leaflets of the plasma membrane fuse, accounting for the major dense lines visible in the myelin under the electron microscope. The closely apposed outer leaflets are visible as distinct intraperiod lines. The characteristic ultrastructural periodicity of compact myelin can be seen in the survey micrograph and is more clearly visible at a higher magnification in the inset. The major dense lines formed by the fused inner leaflets of the Schwann cell plasma membrane (arrows) alternate with the two intraperiod lines, corresponding to the outer plasma membrane leaflets, which reside closely apposed and are separated by the narrow extracellular space (arrowhead). In the myelin, only remnants of cytoplasm remain. Narrow cytoplasmic "streets" are formed throughout the myelin sheath, known as Schmidt-Lanterman clefts or incisures. Via these streets of cytoplasm, a short diffusion pathway is provided between the outer and the inner Schwann cell collars. Schmidt-Lanterman clefts are part of non-compact myelin, which also includes the cytoplasm at the nodes of Ranvier and the paranodal loops (cf. node of Ranvier, Fig. 163). During formation of the myelin sheath, autotypic junctions are built connecting membranes between themselves at the inner and outer mesaxons, in the Schmidt-Lanterman clefts, and in the paranodal regions. At the tips of the paranodal loops, specialised heterotypic junctions connect the cell of Schwann with the axon (cf. Fig. 163).

References

Agrawal D, Hawk R, Avila RL, Inouye H, and Kirschner DA (2009) Internodal myelination during development quantitated using X-ray diffraction. J Struct Biol (in press)

Arroyo EJ and Scherer SS (2000) On the molecular architecture of myelinated fibers. Histochem Cell Biol 113: 1

Balice-Gordon RJ, Bone L, and Scherer SS (1998) Junctional gap junctions in the Schwann cell myelin sheath. J Cell Biol 142: 1095

Lemke G (1996) Unwrapping myelination. Nature 383: 395

Peles E and Salzer JL (2000) Functional domains in myelinated axons. Curr Opin Neurobiol 10: 558

Spiegel I and Peles E (2002) Cellular junctions of myelinated nerves. Mol Membr Biol 19: 95

Magnification: x 80,000; x 200,000 (inset)

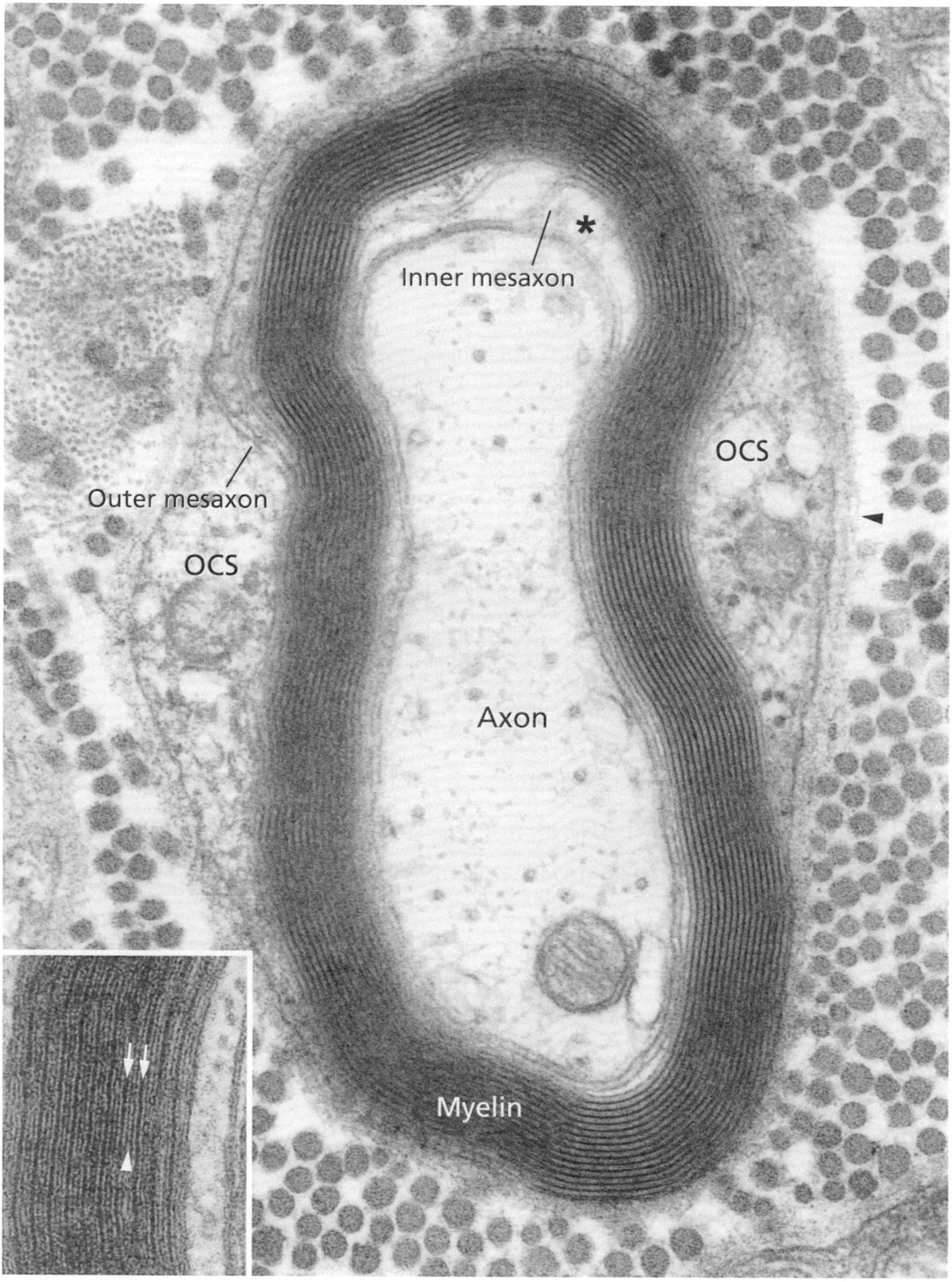
Inner mesaxon
*
Outer mesaxon
OCS
OCS
Axon
Myelin

NODE OF RANVIER

Along a myelinated nerve fibre, numerous Schwann cells are arrayed sequentially. At the cell borders where successive Schwann cells meet, the myelin sheath is interrupted, and the axon enveloped only by thin specialised Schwann cell extensions. These regions are called nodes of Ranvier. They are the sites where myelinated nerve fibres are excitable, whereas in the segments between the nodes, the internodia, due to the myelin properties, the axon is insulated. Together with adjacent paranodal and juxtaparanodal regions, the nodes form specialised areas along the nerve fibre where Na^+ and K^+ channels and various other specific proteins are concentrated in distinct domains. Since the compact myelin in the internodia reduces the current flow across the axonal membrane, the nerve impulse moves saltatory from one to the next node of Ranvier, skipping the internodial regions. The length of an internodium depends on the Schwann cell size and can reach 80–100 µm.

The electron micrographs in panels A, B, and C show a node of Ranvier along a myelinated nerve fibre of the peripheral nervous system and the adjacent paranodal (PN) and juxtaparanodal regions (JXP). A survey is presented in panel A in which the node region is marked by arrows. In the zone of the node shown at higher magnification in panel B, the axon is encapsuled only by flat extensions (microvilli) of the adjacent cells of Schwann and the basal lamina (arrowheads). The node region continues laterally into the paranodal regions where the compact myelin opens up, and the Schwann cells form characteristic loops. The paranodal loops are visible in all three panels. At the higher magnification in panel C, it can be seen that the peaks of the loops are closely attached to the axolemma. The intercellular gap is 2–3 nm, and specialised axo-glial junctions are formed appearing as dense areas (double arrows) under the electron microscope. They correspond to complexes of cell adhesion molecules including contactin and Caspr (contactin associated protein) in the axonal membrane connected to neurofascin 155 in the loop membrane. The paranodal axo-glial junctions at the peaks of the glial loops separate the membrane domains in the zone of the node, where Na^+ channels are concentrated, from the juxtaparanodal domains, which adjoin laterally and are enriched in K^+ channels and Caspr2. The juxtaparanodal regions are localised at the ends of the internodium below the compact myelin. Other functions attributed to the axo-glial junctions include attachment of the myelin sheath to the axon, signalling between glial cell and axon, and limitation of lateral diffusion of membrane constituents. In addition to the heterotypic junctions with the axon, the paranodal loops also contain autotypic tight, adherens, and gap junctions through which they are connected among themselves.

In the central nervous system (CNS), the central region of the nodes is not covered by oligodendrocytes but is occupied by processes of perinodal astrocytes. In overall lipid and protein contents, the compact myelin sheaths are similar in the CNS and peripheral nervous system (PNS). In the PNS, myelin is particularly enriched in sphingomyelin and glycoproteins. Myelin basic protein residing in the cytoplasmic leaflet of the plasma membrane is present in the myelin of both the CNS and the PNS. Via homophilic interactions, myelin-associated protein zero in the PNS stabilises adjacent membrane lamellae. Proteolipid protein is the main protein in the CNS compact myelin.

References

Arroyo EJ and Scherer SS (2000) On the molecular architecture of myelinated fibers. Histochem Cell Biol 113: 1

Fannon AM, Sherman DL, Ilyina-Gragerova G, Brophy PJ, Friedrich VL Jr, and Colman DR (1995) Novel E-cadherin-mediated adhesion in peripheral nerve: Schwann cell architecture is stabilized by autotypic adherens junctions. J Cell Biol 129: 189

Martin S, Levine AK, Chen ZJ, Ughrin Y, and Levine JM (2001) Deposition of NG2 proteoglycan at nodes of Ranvier in the peripheral nervous system. J Neurosci 21: 8119

Peles E and Salzer JL (2000). Functional domains in myelinated axons. Curr Opin Neurobiol 10: 558.

Spiegel I and Peles E (2002) Cellular junctions of myelinated nerves. Mol Membr Biol 19: 95

Wiley CA and Ellisman MH (1980) Rows of dimeric-particles within the axolemma and juxtaposed particles within glia, incorporated into a new model for the paranodal glial-axonal junction at the node of Ranvier. J Cell Biol 84: 261

Magnification: x 10,700 (A), x 28,000 (B), x 46,600 (C)

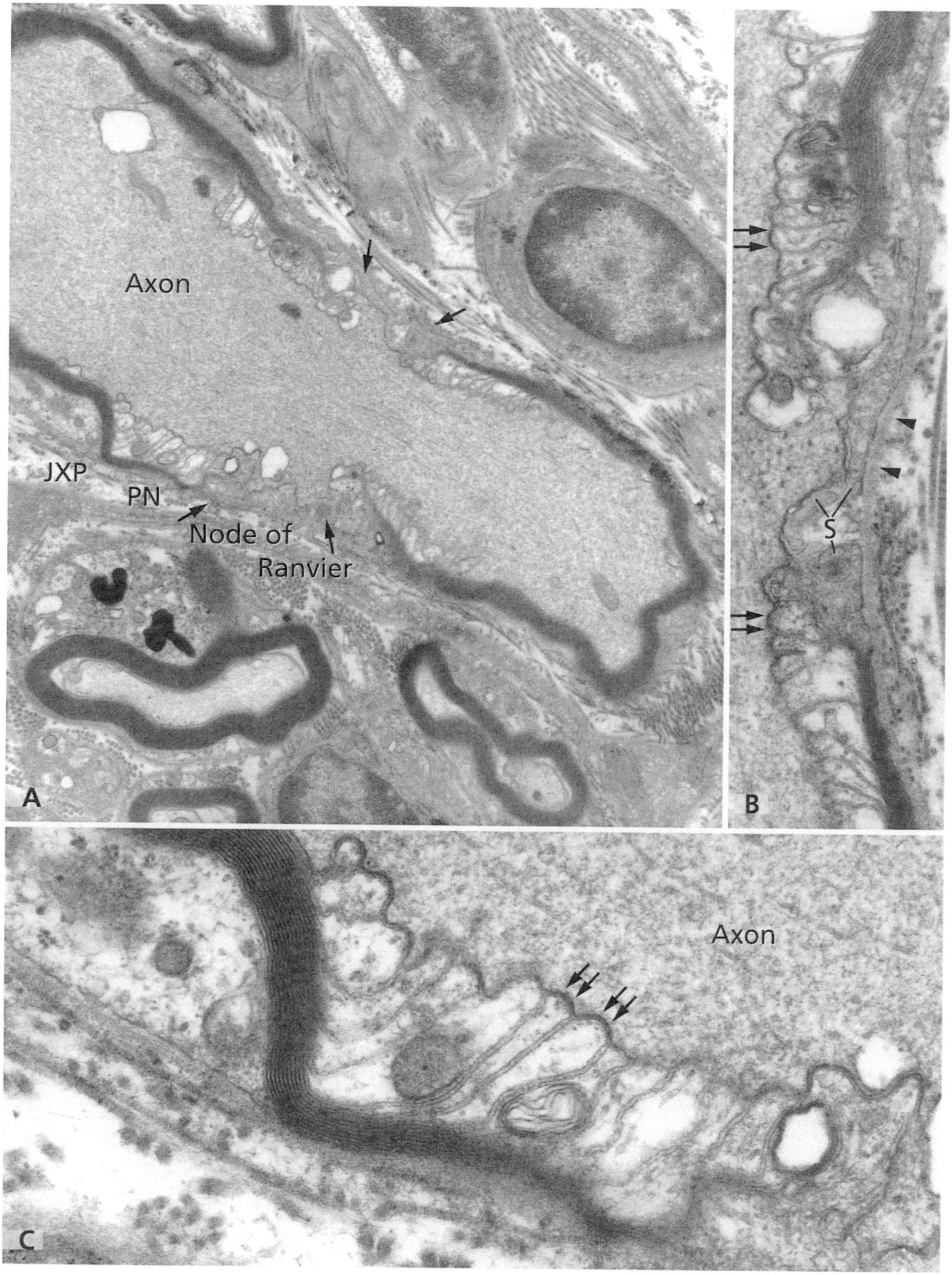
Axon
JXP
PN
Node of
Ranvier
A
S
B
Axon
C

AXONAL DEGENERATION

Various types of injury including mechanical transection, ischemia and compression result in axonal degeneration. The classical type is the Wallerian degeneration after experimental nerve transection and the structural alterations observed in the nerve distal to the injury occur also in a variety of pathological states. The degeneration starts at a single point and finally involves the entire axon distal of the injured part. This process is referred to as anterograde (Wallerian) degeneration. The one involving the part of the axon proximal to the injury is called retrograde degeneration. The structural changes follow a certain time course and develop gradually. Initially, axon swelling and disintegration of neurofilaments and microtubules occurs. The axolemma becomes discontinuous, and retraction and disintegration of myelin ensues. These changes result in the progressive disintegration of the axon and myelin. Fragments of the axon often contain large amounts of myelin debris and are called myelin ovoid. Schwann cells proliferate and ingest the degenerating axon. Finally, phagocytes eliminate the remnants of the axons. Such degenerative changes also occur in the axon proximal to the site of injury and chromatolysis takes place in the neuronal cell body.

Panel A shows the advanced disintegration of an axon and its myelin sheath. The axon contains many vesicular elements, vacuoles and highly electron dense myelin debris. The degenerating axon is surrounded by Schwann cells. In Panel B, a Schwann cell containing electron dense myelin ovoids is seen. The central axoplasm is almost completely disintegrated but the remaining myelin still shows some laminated structure. Although finally the axon and the myelin sheath break down completely, regeneration of the axon can occur in the peripheral nervous system from the intact soma. Axonal regeneration takes place by sprouting of the proximal nerve stump. Dividing Schwann cells are of great importance in this process since they arrange themselves in tubes (bands of Bünger) guiding the regenerating axon.

References

Chaudry V, Glass J, and Griffin J (1992) Wallerian degeneration in peripheral nerve disease. Neurol Clin 10: 613

Griffin J, George R, and Lobato C (1992) Macrophage responses and myelin clearance during Wallerian degeneration: relevance to immune-mediated demyelination. J Neuroimmunol 40: 153

Schröder J (2001) Pathology of peripheral nerves. An atlas of structural and molecular pathological changes. Berlin: Springer

Magnification: x 12,000 (A); x 9,500 (B)

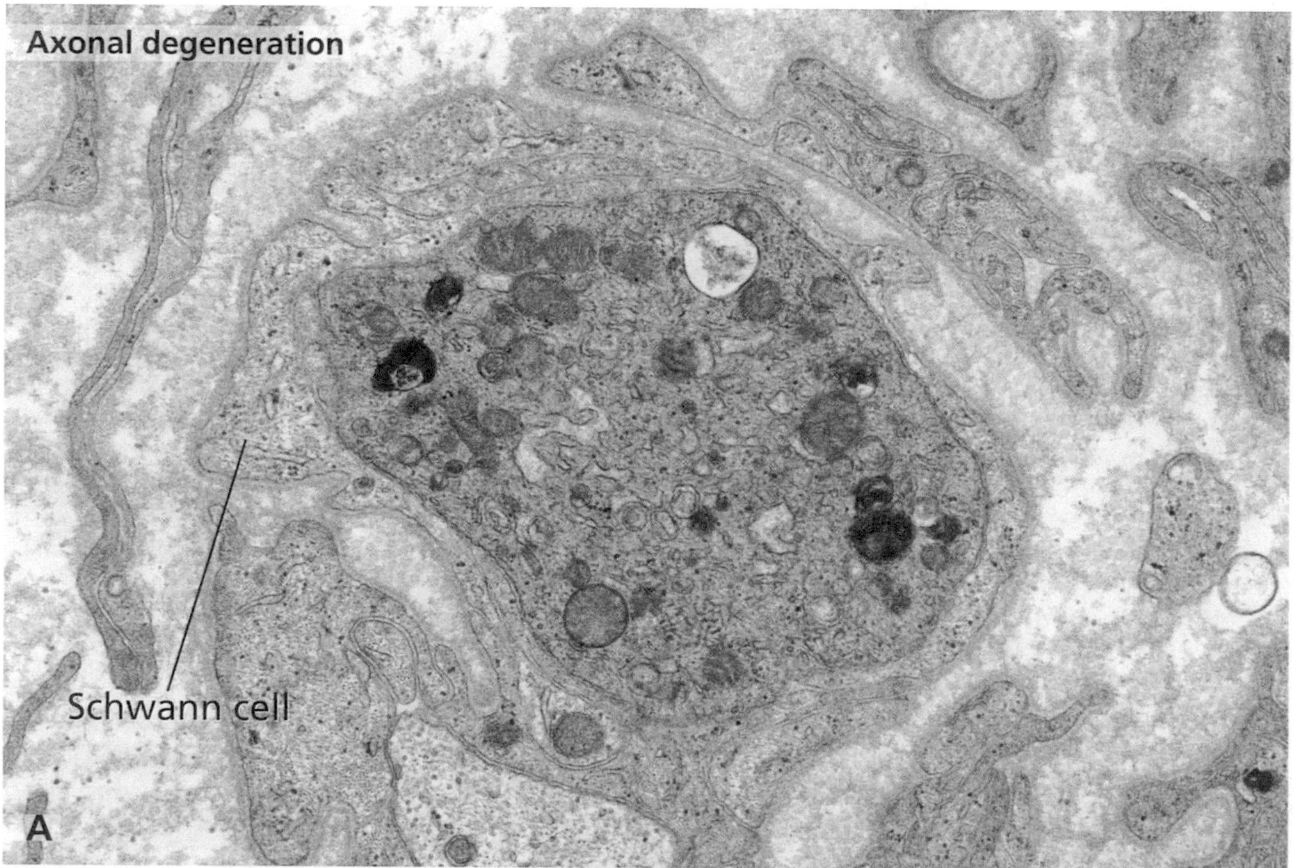
Axonal degeneration
Schwann cell
A

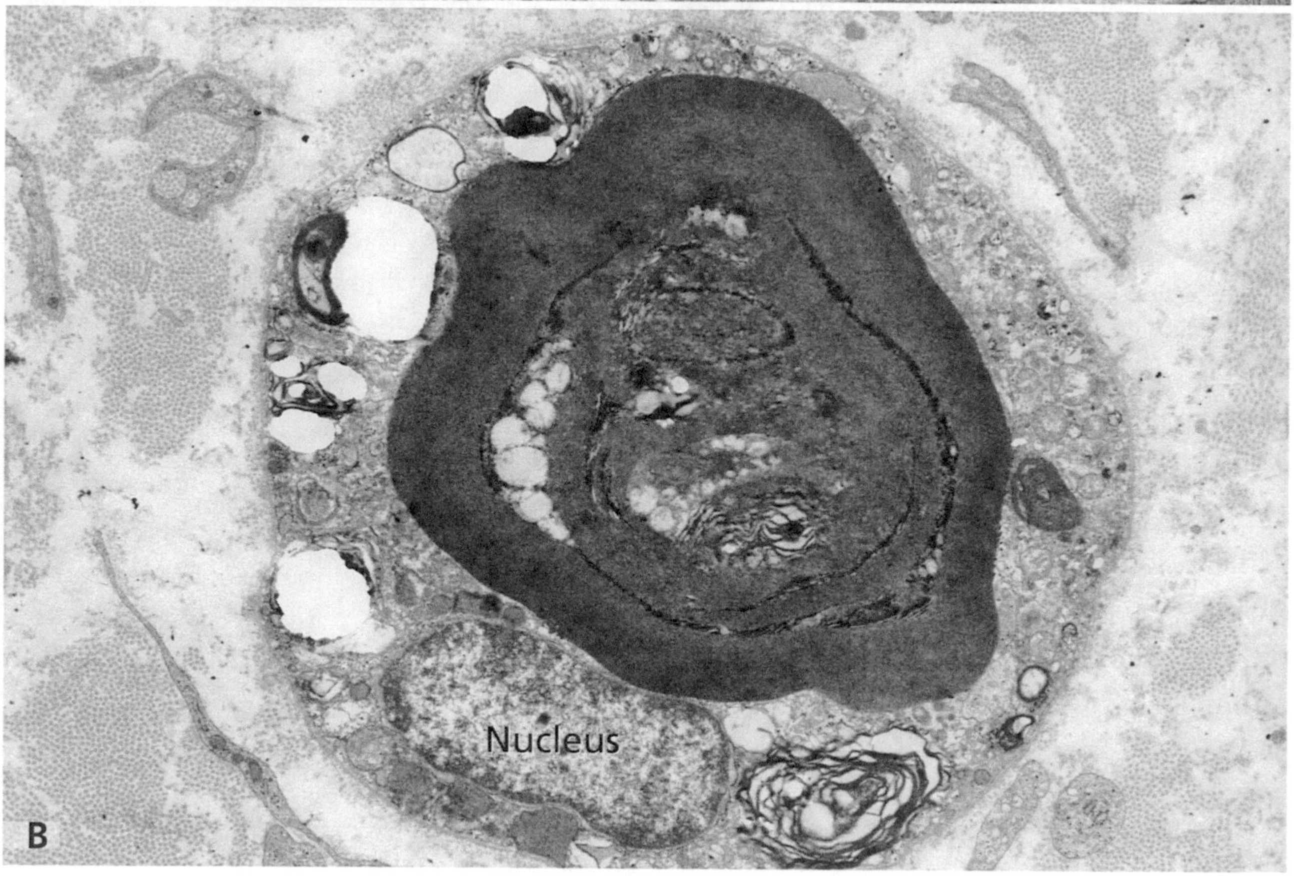
Nucleus
B

NEUROAXONAL DYSTROPHY

Neuroaxonal dystrophy (Schindler's disease) is a rare autosomal recessive disease caused by deficiency in the activity of lysosomal α-*N*-acetylgalactosaminidase (formerly termed α-galactosidase B). The gene for the enzyme has been mapped to chromosome 22q13.1-13.2 and missense mutations and a nonsense mutation have been shown to cause the disease. Due to deficient or missing α-*N*-acetylgalactosaminidase, different types of *N*- and *O*-linked glycoproteins as well as glycosphingolipids and proteoglycans are abnormally terminated with α-*N*-acetylgalactosamine residues and accumulate in various tissues and in body fluids.

Panel A shows an axon that exhibits fine structural changes typical of type-I infantile onset neuroaxonal dystrophy. A greatly distended axon contains prominent intra-axonal tubulovesicular arrays. At higher magnification, as shown in panel B, the enormous number of ramifying irregular tubules is evident. Secondary to these changes, axon demyelination can be observed (panel A). In other cases, the spheroids may be composed of lamelliform membrane arrays and exhibit acicular clefts. In type-I neuroaxonal dystrophy, no morphological evidence for lysosomal accumulation was found. However, this is the characteristic finding in type-II adult onset Schindler's disease. In the skin, various types of cells, among them most prominently endothelial cells, eccrine sweat gland cells, fibroblasts, and smooth muscle cells, were found to contain numerous cytoplasmic vacuoles. Such membrane limited vacuoles either appeared "empty" or contained a filamentous material.

Clinically, three major phenotypes of the disease can be distinguished, of which types I and II could be characterised in more detail. Type I is an infantile-onset classical neuroaxonal dystrophy. The accumulations occur in the nervous system and not in visceral organs and peripheral blood. The electron microscopic examination of neocortex, myenteric plexus from rectum biopsies and skin nerves showed highly characteristic structural features in preterminal and terminal axons. Axonal membranous spheroids of complex and heterogenous composition were observed in GABAergic neocortical neurons and autonomous myenteric axons.

References

De Leon G and Mitchel M (1985) Histological and ultrastructural features of dystrophic isocortical axons in infantile neuroaxonal dystrophy (Seitelberger's disease). Acta Neuropathol (Berlin) 66: 89

Desnick R and Schindler D (2001) α-*N*-acetylgalactosaminidase deficiency: Schindler's disease. In: The metabolic and molecular bases of inherited disease (Scriver C, Beaudet A, Valle D, Sly WS, Childs B, Kinzler K, and Vogelstein B, eds). New York: McGraw-Hill, pp 3483

Kanzaki T, Wang A, and Desnick R (1991) Lysosomal α-*N*-acetylgalactosaminidase deficiency, the enzymatic defect in angiokeratoma corporis diffusum with glycopeptiduria. J Clin Invest 88: 707

Schindler D, Bishop DF, Wolfe D, Wang A, Egge H, Lemieux R, and Desnick R (1989) Neuroaxonal dystrophy due to lysosomal α-*N*-acetylgalactosaminidase deficiency. N Engl J Med 320: 1735

Schröder J (2001) Pathology of peripheral nerves. An atlas of structural and molecular pathological changes. Berlin: Springer

Walkey S, Baker H, Rattazzi M, Haskins M, and Wu J-Y (1991) Neuroaxonal dystrophy in neuronal storage disorders: Evidence for major GABAergic neuron involvement. J Neurol Sci 104: 1

Wolfe D, Schindler D, and Desnick R (1995) Neuroaxonal dystrophy in infantile α-*N*-acetylgalactosaminidase deficiency. J Neurol Sci 132: 44

Magnification: x 9,500 (A); x 105,000 (B)

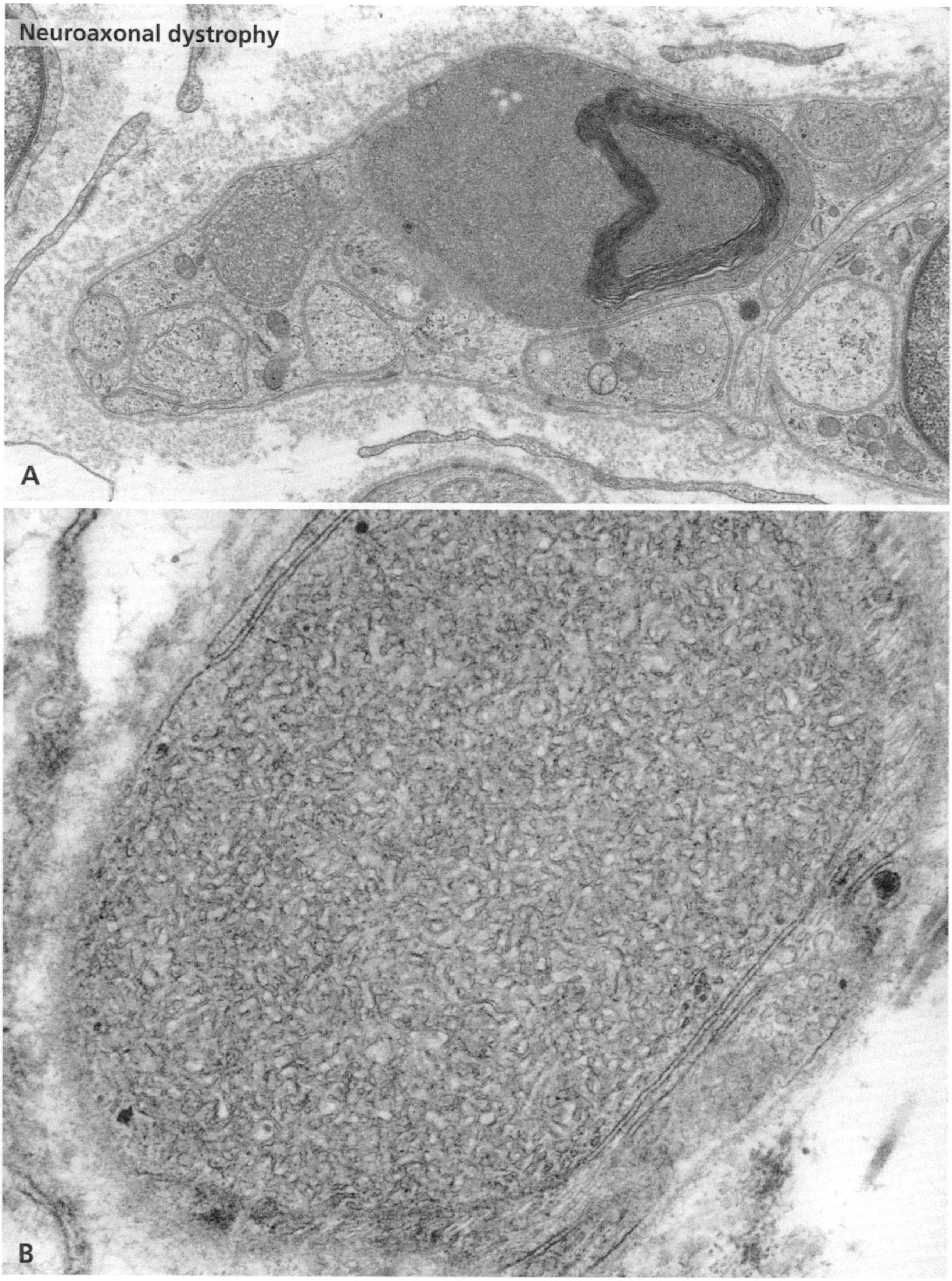
Neuroaxonal dystrophy
A
B

NEUROPATHIES ASSOCIATED WITH DYSPROTEINAEMIAS

Peripheral neuropathies can be divided into two groups: hereditary and acquired forms. Among the various acquired forms, immune-mediated neuropathies can be found that are associated with multiple myeloma (Waldenström's diseases), monoclonal gammopathies of unknown significance, and B-cell lymphoma.

Dysproteinaemic neuropathies result predominantly in structural disturbances of the myelin sheath. By electron microscopy, different degrees of separation of myelin lamellae with opening of the interperiod lines can be observed in cases of IgM monoclonal gammopathy. The changes occur mainly in the outer layers. This results in a decompaction of the myelin sheet as can be seen in the electron micrograph from a patient with dysproteinaemic neuropathy. Further progression of the disease results in demyelination.

The pathogenesis of dysproteinaemic neuropathies associated with IgM monoclonal gammopathies could be studied in some detail. The IgM was found deposited in the myelin sheaths and had a specificity for the myelin-associated glycoprotein. Probably, the presence of the antigen-antibody complexes initiated complement-mediated demyelination. Immunoglobulin depositions in the myelin were not observed in IgG and IgA gammopathy-associated neuropathies.

References

Bollensen E, Steck A, and Schachner M (1988) Reactivity with the peripheral myelin glycoprotein Po in serum from patients with monoclonal IgM gammopathy and polyneuropathy. Neurology 38: 1266

Braun P, Frail D, and Latov N (1982) Myelin-associate glycoprotein is the antigen for a monoclonal IgM in polyneuropathy. J Neurochem 39: 1261

Jacobs J and Scadding J (1990) Morphological changes in IgM paraproteinemic neuropathy. Acta Neuropathol (Berlin) 80: 77

Lach B, Rippstein P, Atack D, Afar D, and Gregor A (1993) Immunoelectron microscopic localisation of monoclonal IgM antibodies in gammopathy associated with peripheral demyelinating neuropathy. Acta Neuropathol (Berlin) 85: 298

Latov N, Hays A, and Sherman W (1988) Peripheral neuropathy and anti-MAG antibodies. Crit Rev Neurobiol 3: 301

Lupski J and Garcia C (2001) Charcot-Marie-Tooth peripheral neuropathies and related disorders. In: The metabolic and molecular bases of inherited disease (Scriver C, Beaudet A, Valle D, Sly WS, Childs B, Kinzler K, and Vogelstein B, eds). New York: McGraw-Hill, pp 5759

Meier C, Vandevelde M, Steck A, and Zurbriggen A (1984) Demyelinating polyneuropathy associated with monoclonal IgM-paraproteinemia. Histological, ultrastructural and immunocytochemical studies. J Neurol Sci 63: 353

Monaco S, Bonetti B, Ferrari S, Moretto G, Nardelli E, Tedesco F, Mollnes T, Nobile-Orazio E, Manfredini E, and Bonazzi L (1990) Complement-mediated demyelination in patients with IgM monoclonal gammopathy and polyneuropathy. N Engl J Med 322: 649

Schröder J (2001) Pathology of peripheral nerves. An atlas of structural and molecular pathological changes. Berlin: Springer

Smith I, Kahn S, and Lacey B (1983) Chronic demyelinating neuropathy associated with benign IgM paraproteinemia. Brain 106: 169

Vital C and Vallat J (1987) Ultrastructural study of the human diseased peripheral nerve, 2nd edn. New York: Elsevier

Magnification: x 27,000

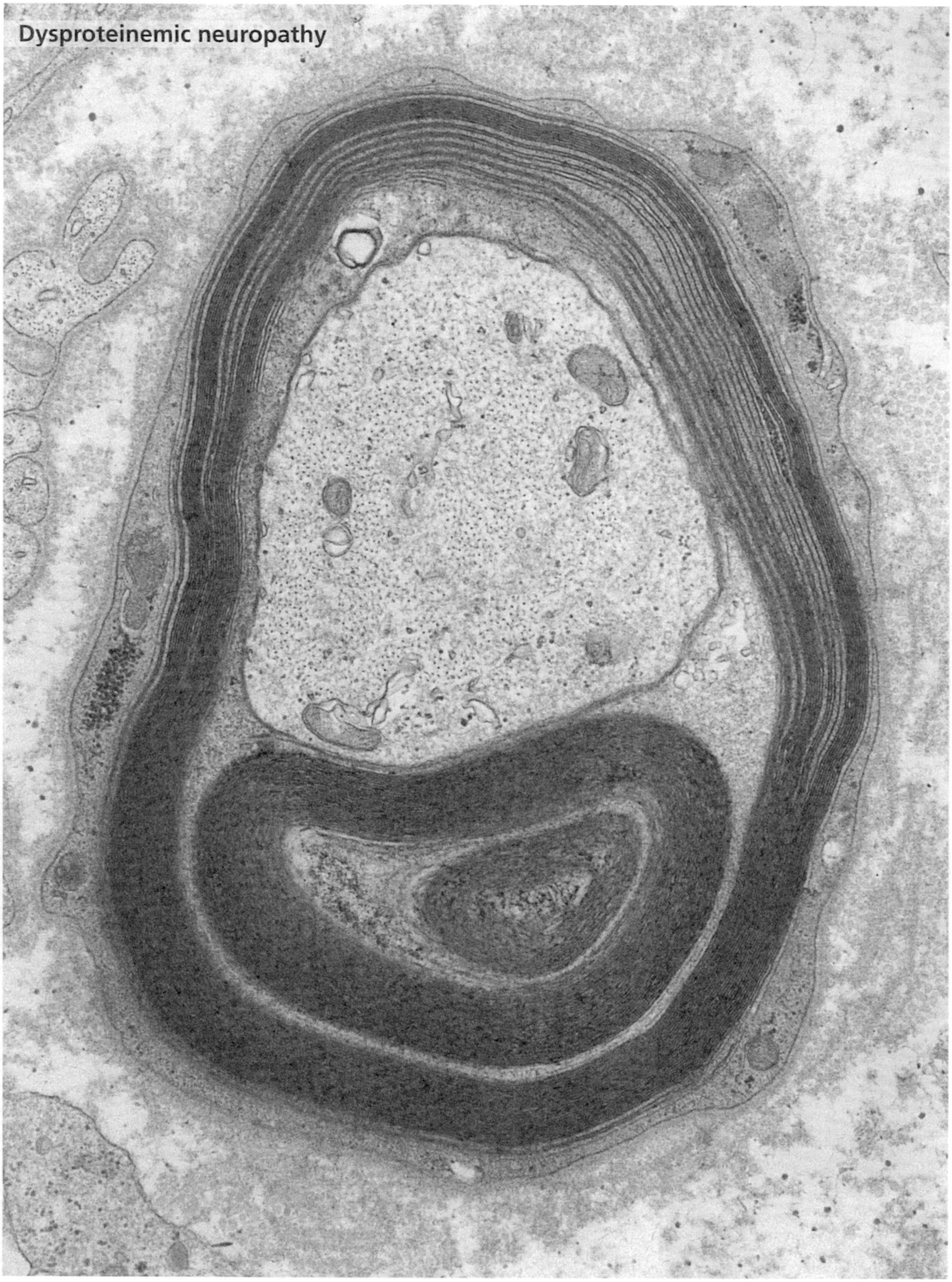
Dysproteinemic neuropathy

METACHROMATIC LEUKODYSTROPHY

Metachromatic leukodystrophy is a lysosomal storage disease that is inherited in an autosomal recessive manner. It is caused by a deficiency in catalytic activity or absence of arylsulfatase A and in some rare cases by absence of saposin B, a sphingolipid activator protein.

The principal structural alteration in metachromatic leukodystrophy consists of demyelination and the presence of metachromatic granules. Panel A illustrates an advanced demyelination of a nerve fibre of the skin with prominent irregularities of the myelin sheath (inset). This is accompanied with a proliferation of Schwann cells. In skin biopsies, the sweat glands show characteristic inclusions referred to as zebra bodies and tuff stone inclusions as well as prismatic inclusions (panel B). In adult metachromatic leukodystrophy, in addition to prismatic and tuff stone inclusions, composite inclusion bodies (asterisks in B) exist that are shown at higher magnification in the inset.

The gene locus for arylsulfatase A has been mapped to chromosome 22q13, and allelic mutations comprising splice donor site mutations and different amino acid substitutions are most commonly causative. A general, however somewhat imprecise, genotype/phenotype correlation has been established in that the severity of the disease is inversely correlated with residual activity of arylsulfatase A. Clinically, the disease can be subdivided in different forms according to the age of onset: infantile, juvenile, and adult.

The deficiency in arylsulfatase A activity results in an impaired desulfation of 3-*0*-sulfogalactosyl-containing glycolipids and additionally of sulfatides, sulfogalactoglycerolipids, and various other sulfated glycolipids. Sulfated glycolipids are important constituents of myelin sheaths. Together with galactosylceramides, they are involved in the maintenance of the insulator function of myelin membranes and may be important during brain maturation. Since sulfated glycolipids exist predominantly in the myelin sheaths of the central and peripheral nervous system and only in low amounts in visceral organs, the disease manifests preferentially in the nervous system. Here it results in demyelination of nerve fibres, with subsequent attenuation of the white matter. The function of visceral organs, with the exception of gall bladder, is not impaired by the deposition of metachromatic lipid.

References

Berger J, Loschl B, Bernheimer H, Lugowska A, Tylki-Szymanska A, Gieselmann V, and Molzer B (1997) Occurrence, distribution, and phenotype of arylsulfatase A mutations in patients with metachromatic leukodystrophy. Am J Med Genet 69: 335

Bosio A, Binczek E, and Stoffel W (1996) Functional breakdown of the lipid bilayer of the myelin membrane in central and peripheral nervous system by disrupted galactocerebroside synthesis. Proc Natl Acad Sci USA 93: 13280

Kappler J, Leinekugel P, Conzelmann E, Kleijer WJ, Kohlschutter A, Tonnesen T, Rochel M, Freycon F, and Propping P (1991) Genotype-phenotype relationship in various degrees of arylsulfatase A deficiency. Hum Genet 86: 463

Leinekugel P, Michel S, Conzelmann E, and Sandhoff K (1992) Quantitative correlation between the residual activity of beta-hexosaminidase A and arylsulfatase A and the severity of the resulting lysosomal storage disease. Hum Genet 88: 513

Phelan MC, Thomas GR, Saul RA, Rogers RC, Taylor HA, Wenger DA, and McDermid HE (1992) Cytogenetic, biochemical, and molecular analyses of a 22q13 deletion. Am J Med Genet 43: 872

Polten A, Fluharty AL, Fluharty CB, Kappler J, von Figura K, and Gieselmann V (1991) Molecular basis of different forms of metachromatic leukodystrophy. N Engl J Med 324: 18

von Figura K, Gieselmann V, and Jaeken J (2001) Metachromatic leukodystrophy. In: The metabolic and molecular bases of inherited disease (Scriver D, Beaudet A, Valle D, and Sly WS, eds). New York: McGraw-Hill, pp 3695

Magnification: x 13,000 (A); x 80,000 (inset); x 32,500 (B); x 84,500 (inset)

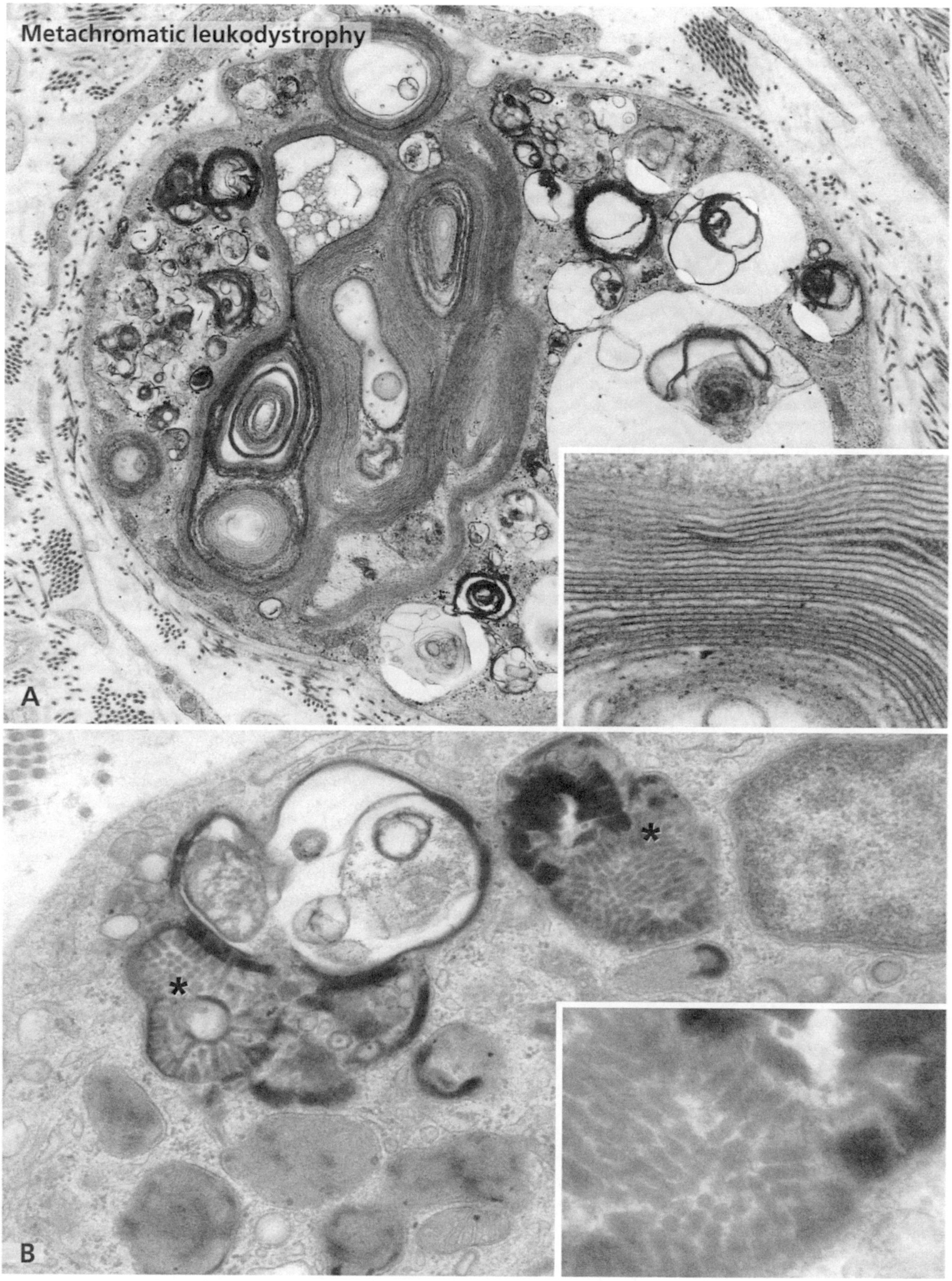
Metachromatic leukodystrophy
A
*
*
B

NEURONAL CEROID LIPOFUSCINOSIS

Neuronal ceroid lipofuscinoses (NCL) are lysosomal storage diseases comprising a group of progressive neurodegenerative disorders inherited in an autosomal recessive manner. In children, they represent the most common neurodegenerative disorder. Three major types of NCL are distinguished clinically: infantile, late onset infantile, and juvenile. They can also be distinguished by electron microscopy since the fine structure of the depositions is characteristic for each of them.

The various types of NCL are not limited to the nervous system. Therefore, the diagnosis is made primarily by electron microscopic investigation of skin biopsies or peripheral blood leukocytes. Panels A and B show numerous storage bodies (some marked by asterisks in A) in the cytoplasm of a capillary endothelial cell of a skin biopsy. At higher magnification (B), the stored material consists of curvilinear bodies (arrowhead) and fingerprint profiles (arrows) as well as granular osmiophilic material. This mixed type of inclusion material was observed in a case of juvenile NCL. However, only fingerprint profiles may be observed in juvenile NCL.

In infantile NCL, the stored material is mainly saposin A and D and fragments of glial fibrillar acidic protein. The defect lies in palmitoyl-protein thioesterase (PPT, CLN1) whose gene has been mapped to chromosome 1p32. The most commonly observed mutations are a premature stop codon at arginine 151 and a missense mutation (R122W). By electron microscopy, the lysosomal deposits consist of pathognomonic granular osmiophilic material.

In late onset infantile NCL, mainly the subunit c of mitochondrial ATP synthase is stored in lysosomes. The cause is a deficiency in pepinase (CLN2), a lysosomal protease, whose gene has been mapped to chromosome 11p15. By electron microscopy, typical curvilinear inclusions are found in lysosomes.

In juvenile NCL, a gene located at chromosome 16p12, which encodes a transmembrane protein that is referred to as battenin, is causative. Battenin most probably is a lysosomal membrane glycoprotein. The stored material is a lipopigment that shares similarities with the age-related pigment lipofuscin. A major component is the subunit c of mitochondrial ATP synthase complex and small amounts of saposins. By electron microscopy so-called fingerprint profiles are observed in lysosomes.

References

Eiberg H, Bisgaard ML, and Mohr J (1989) Linkage between alpha 1B-glycoprotein (A1BG) and Lutheran (LU) red blood group system: assignment to chromosome 19: new genetic variants of A1BG. Clin Genet 36: 415

Gonatas NK, Gambetti P, and Baird H (1968) A second type of late infantile amaurotic idiocy with multilamellar cytosomes. Neuropath Exp Neurol 27: 371

Haltia M, Rapola J, and Santavuori P (1973) Infantile type of so-called neuronal ceroid-lipofuscinosis. Histological and electron microscopic studies. Acta Neuropathol (Berlin) 26: 157

Hofmann S and Peltonen L (2001) The neuronal ceroid lipofuscinoses. In: The metabolic and molecular bases of inherited disease (Scriver C, Beaudet A, Valle D, and Sly WS, eds). New York: McGraw-Hill, pp 3877

Mitchison HM, Hofmann SL, Becerra CH, Munroe PB, Lake BD, Crow YJ, Stephenson JB, Williams RE, Hofman IL, Taschner PE, et al. (1998) Mutations in the palmitoyl-protein thioesterase gene (PPT; CLN1) causing juvenile neuronal ceroid lipofuscinosis with granular osmiophilic deposits. Hum Mol Genet 7: 291

Sharp JD, Wheeler RB, Lake BD, Savukoski M, Jarvela IE, Peltonen L, Gardiner RM, and Williams RE (1997) Loci for classical and a variant late infantile neuronal ceroid lipofuscinosis map to chromosomes 11p15 and 15q21-23. Hum Mol Genet 6: 591

Sleat DE, Donnelly RJ, Lackland H, Liu CG, Soha I, Pullarkat RK, and Lobel P (1997) Association of mutations in a lysosomal protein with classical late-infantile neuronal ceroid lipofuscinosis. Science 277: 1802

Vesa J, Hellsten E, Verkruyse LA, Camp LA, Rapola J, Santavuori, P, Hofmann SL, and Peltonen L (1995) Mutations in the palmitoyl protein thioesterase gene causing infantile neuronal ceroid lipofuscinosis. Nature 376: 584

Magnification: x 32,000 (A); x 63,000 (B)

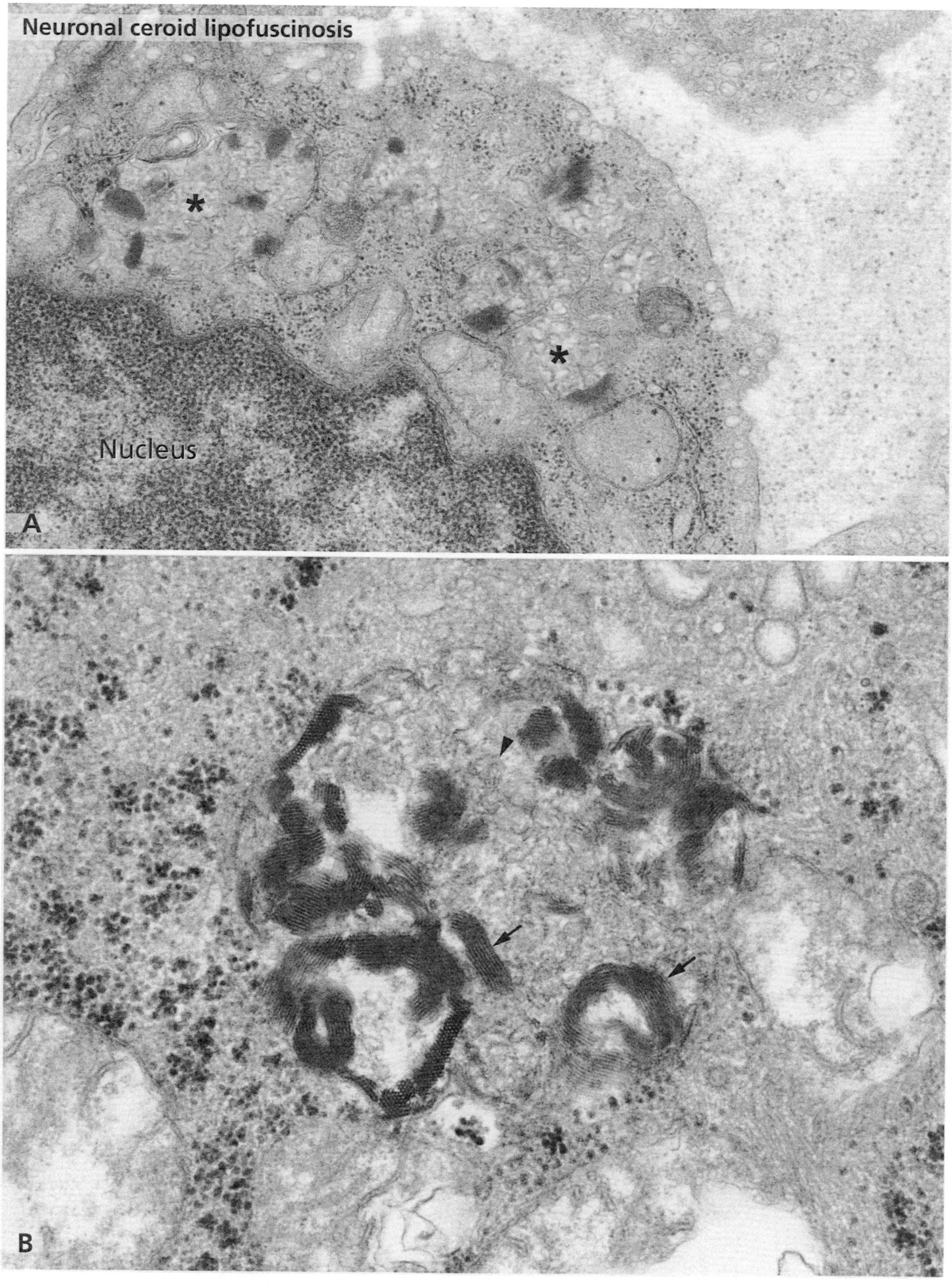
Neuronal ceroid lipofuscinosis
*
*
Nucleus
A
B

RED BLOOD CELLS AND CELLS OF THE ERYTHROID LINEAGE

The cells of the erythroid lineage develop from a multipotential myeloid stem cell under the influence of the major regulator erythropoietin, a glycoprotein hormone synthesised in the kidney. Erythropoiesis is stimulated in states of a disproportion of oxygen need and supply, e.g. hypoxia due to a decrease in oxygen level in the inspired air or a decreased number of erythrocytes in the circulating blood, due to bleeding. The erythropoietin-sensitive erythrocyte progenitor cell CFU-E (erythroid colony forming unit) differentiates to form a proerythroblast. Further differentiation stages include the basophilic, the polychromatophilic, and the orthochromatophilic erythroblast, as well as the reticulocyte, which already is able to leave the bone marrow to enter the circulating blood. Erythropoiesis takes place in a period of 4–6 days and is characterised by the synthesis and increased accumulation in the cytoplasm of haemoglobin, the protein specialised for transporting oxygen and carbon dioxide. Concomitantly, programmed cell changes take place, leading to the exclusion of the nucleus and disappearance of all cell organelles.

Panels A and B show electron micrographs of cells of the erythroid lineage taken by puncture of human bone marrow. Erythroblasts of earlier and later states of differentiation are shown in panel A. The electron density of the cytoplasm increases concomitantly with the increased accumulation of haemoglobin. The partially sectioned erythroblast on the right hand side shows a less dense cytoplasm compared with the two erythroblasts visible in the centre. The former presumably corresponds to an erythroblast of the polychromatophilic type (pE), the latter to orthochromatophilic erythroblasts (oE). Different parts of the cytoplasm are involved in haemoglobin production, including mitochondria (M) where protoporphyrin is synthesised and combined with iron to form the haem part of the haemoglobin molecule. Synthesis of the globin chains takes place on free ribosomes in the cytoplasm (arrows in the inset). By receptor-mediated endocytosis via coated vesicles (arrowhead), iron-transferrin complexes are taken up; after iron is released, transferrin recycles to the cell surface along with the receptor. In the nuclei, condensed chromatin (C) occupies extensive areas leading to a "checkerboard" pattern. Nuclei are delineated by distended perinuclear cisterns. At this stage of development, the cells are no longer able to divide. One of the orthochromatophilic erythroblasts shows a bizarre surface with irregular extensions. They reflect the surface dynamics occurring when the cells start to push out the nuclei. Orthochromatophilic erythroblasts extrude their nuclei giving rise to reticulocytes (panel B and inset in A). In panel B, a reticulocyte can be seen that shows the characteristic fimbrated processes that appear just after extrusion of the nucleus. Mitochondria and polyribosomes (arrows in the inset) are retained, endocytosis takes place (arrowhead in the inset), and haemoglobin is still synthesised. Within a few days, the reticulocytes mature to form erythrocytes. During this final stage, the reticulocytes remodel their surfaces and get rid of all intracellular compartments. By binding of ubiquitin to proteins of cell organelles, their destruction is initiated. Membrane proteins are endocytosed and sorted into special domains of endosomal membranes, which bud into the lumen. This leads to the formation of multivesicular or multilamellar organelles (arrowheads in panel B). By fusion with the plasma membrane, their contents are released, then called exosomes.

The mature erythrocyte (right segment of panel B) is a discoid biconcave "haemoglobin bag". It is devoid of cell organelles, and the cytoplasm shows a homogenous electron density caused by the high haemoglobin concentration. A unique membrane skeleton consists of spectrin, a heterodimeric protein interacting with actin and plasma membrane proteins, band 3 protein and glycophorin via ankyrin and protein 4.1. The membrane skeleton is responsible for the biconcave erythrocyte shape that provides a large surface-volume ratio facilitating gas exchange. The elastic submembraneous matrix formed by the spectrin lattice returns shear-deformed cells to the biconcave shape.

References

Géminard C, de Gassart A, Blanc L, and Vidal M (2004) Degradation of AP2 during reticulocyte maturation enhances binding of Hsc70 and Alix to a common site on TfR for sorting into exosomes. Traffic 5: 181

Magnification: x 10,500 (A); x 39,200 (inset upper right corner); x 17,200 (inset lower left corner); x 15,000 (B)

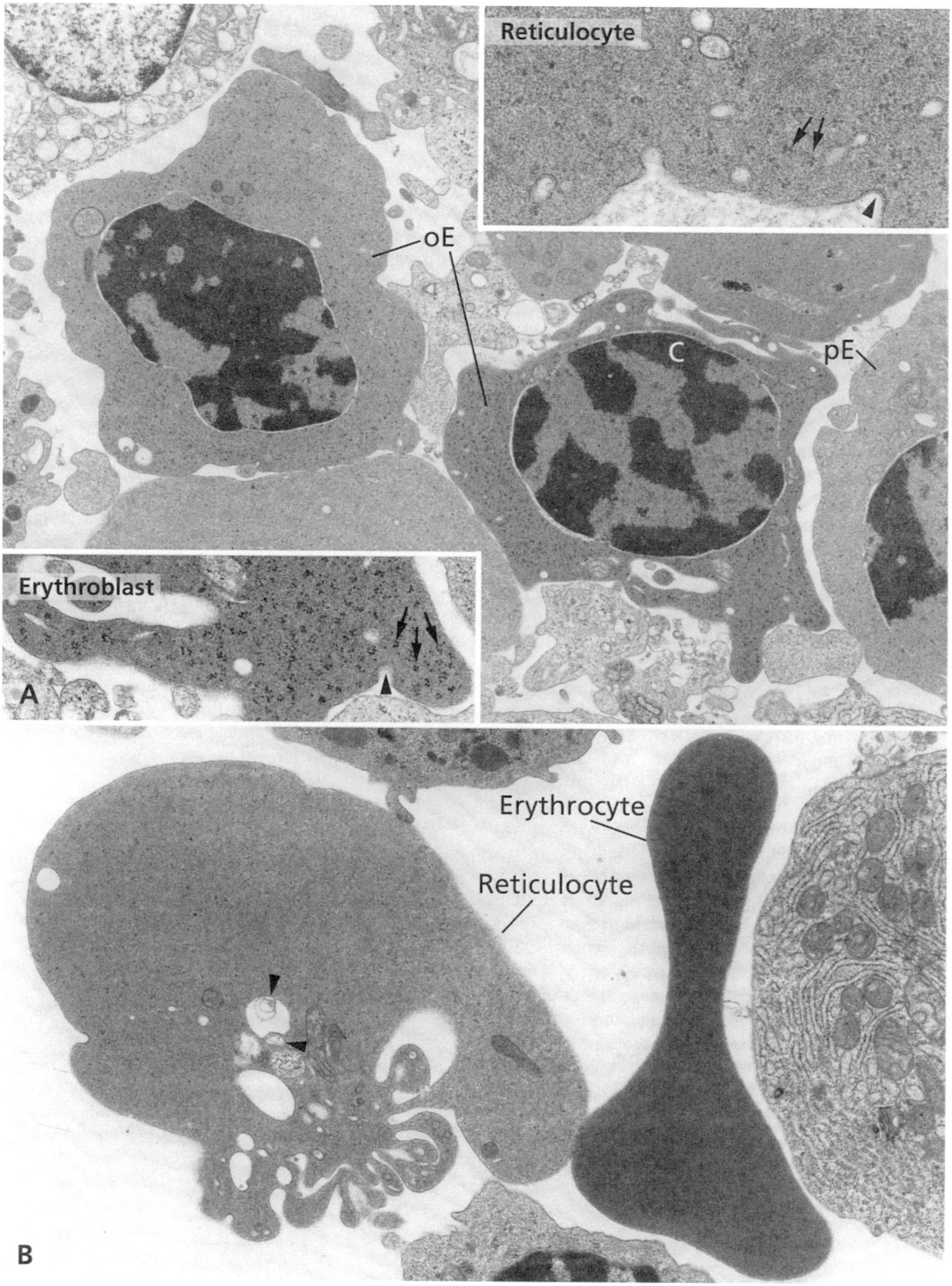
Reticulocyte
oE
C
pE
Erythroblast
A
Erythrocyte
Reticulocyte
B

NEUTROPHILIC GRANULOCYTE

Among the white blood cells (leukocytes), the neutrophilic granulocytes are the most numerous, and they also are the most common granulocytes. Neutrophilic granulocytes make up some 65% of all blood leukocytes. Because of their multilobed nuclei, they are also called polymorphonuclear neutrophils or polymorphs. Neutrophilic granulocytes are motile cells. By complex interactions with endothelial cells in postcapillary venules (homing), and binding of extracellular matrix and chemoattractant molecules (chemotaxis), neutrophils leave the circulation and migrate to the sites where their action is needed. Neutrophils are the initial cells at sites of infection. They are active phagocytes (microphages) capable of engulfing foreign material and organisms, mainly via Fc receptors present in the plasma membrane, and interacting with the Fc region of antibodies bound to antigen, e.g. antibodies decorating the surfaces of bacteria. Neutrophilic granulocytes have a central role in microbial defence and, together with other leukocytes – such as eosinophilic and basophilic granulocytes, monocytes/macrophages, lymphocytes, and fibroblasts – hold key functions in inflammation and wound healing.

The electron micrograph shows a neutrophilic granulocyte of the human blood. Two of the nuclear lobes can be seen. The condensed chromatin is apparent mainly in the nuclear periphery, and only small euchromatic regions are in contact with the nuclear envelope. The Golgi apparatus (Golgi) is in a central position between and close to the nuclear segments. Most of the cytoplasm is occupied by the densely packed granules the contents of which are responsible for the homing and antimicrobial functions of the neutrophils. Three main kinds of granules are accumulated in the neutrophilic granulocyte cytoplasm, the specific neutrophilic granules (arrowheads), azurophilic granules (arrows), and tertiary granules (open arrows), among which also different subtypes exist. It is difficult and may be impossible by ultrastructural examination to distinguish all classes of granules. The specific granules show globular or ellipsoidal shapes. Among the diverse granules, they are the smallest, and the most numerous, and contain a range of enzymes, such as phospholipases and type IV collagenase, bacteriostatic and bacteriocidal substances, such as lysozyme, as well as complement activators. The azurophilic granules are larger and less numerous than the specific granules. They appear most early during granulopoiesis (primary granules), but they are not confined to neutrophilic granulocytes. They arise in all types of granulocytes, and also are present in monocytes and lymphocytes. Azurophilic granules are lysosomes. They contain a range of typical acid hydrolases, antibacterial agents, including azurocidin, as well as myeloperoxidase, which is responsible for the generation of highly bactericidal hypochlorite and chloramines. Specific and azurophilic granules are designated as special types of secretory lysosomes (cf. Figs. 53 and 54). They fuse with phagosomes and release their contents, forming a phagolysosome. This process is called degranulation. After digestion, the degraded materials are either stored in residual bodies, or exocytosed. Subtypes of tertiary granules contain phosphatases and metalloproteinases, such as collagenases and gelatinases, which are assumed to facilitate the migration of the cells through the connective tissue.

Aside from the various granules, other cell compartments and organelles – including endoplasmic reticulum and mitochondria – are sparse, but the cytoplasm contains a considerable amount of glycogen, which can be seen in the form of numerous electron dense particles. Glycogen is broken down for yielding energy enabling neutrophilic granulocates to survive in an anaerobic environment.

References

Blott EJ and Griffiths GM (2002) Secretory lysosomes. Nat Rev Mol Cell Biol 3: 122

Jiang X, Kobayashi T, Nahirney PC, del Saz EG, and Seguchi H (2000) Ultracytochemical study on the localization of superoxide producing sites in stimulated rat neutrophils. Anat Rec 258: 156

Lowe JB (2003) Glycan-dependent leukocyte adhesion and recruitment in inflammation. Curr Opin Cell Biol 15: 531

Robinson JM (2008) Reactive oxygen species in phagocytic leukocytes. Histochem Cell Biol 130: 281

Stephens L, Ellson CC, and Hawkins P (2002) Roles of PI3Ks in leukocyte chemotaxis and phagocytosis. Curr Opin Cell Biol 14: 203

Weiner OD (2002) Regulation of cell polarity during eukaryotic chemotaxis: the chemotactic compass. Curr Opin Cell Biol 14: 196

Magnification: x 28,000

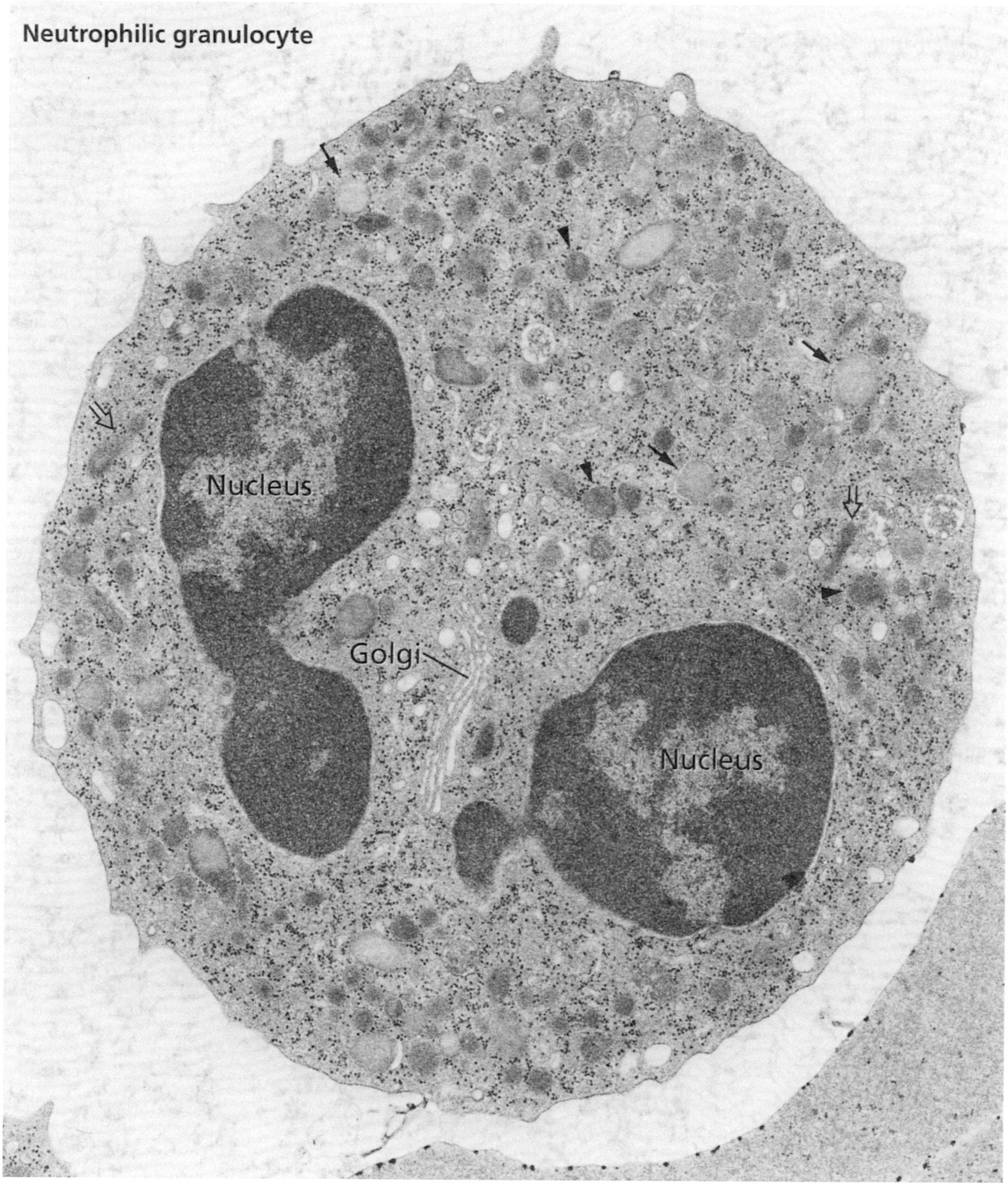
Neutrophilic granulocyte
Nucleus
Golgi
Nucleus

EOSINOPHILIC GRANULOCYTE

Eosinophilic granulocytes make up approximately 2-4% of the total leukocytes in the blood. The count of eosinophils in blood samples is usually increased in patients suffering from parasitic infections and allergies. Eosinophils are involved in immunological responses and have an important role in the host defence against helminthic parasites. Like neutrophils, eosinophilic granulocytes leave the circulation and populate connective tissues at sites of potential foreign invasion. They are particularly abundant in the lymphoreticular tissue of the intestinal mucosa. Eosinophilic granulocytes are named for their large eosinophilic granules occupying the cytoplasm. The nucleus is often bilobed and, as in neutrophils, condensed chromatin is concentrated mainly the periphery.

One of the nuclear segments is visible in the eosinophilic granulocyte of the human blood shown in the micrograph. The ultrastructure of the cytoplasm is dominated by the large specific eosinophilic granules, which contain a characteristic electron dense central crystalloid body that makes them easy to discern. The crystalloid body contains the major basic protein, which also accounts for the eosinophilia of the granules. The major basic protein, and other proteins residing in the matrix of the specific granules, such as the eosinophil cationic protein, are particularly toxic for helminthic parasites and protozoans. Eosinophil peroxidase, and eosinophil-derived neurotoxin are other contents of eosinophilic granules, both attacking parasitic organisms. After binding to the surface of parasites, the granule contents are released directly onto the parasites' membranes. Other enzymes contained in the specific granules neutralise and moderate potentially deleterious effects of vasoactive agents released during inflammation. These include cathepsins, arylsulfatase, and histaminase that are particularly important at sites of allergic reactions. Eosinophilic granulocytes also contain azurophilic granules. Some azurophilic granules are visible aside the nucleus and between the large specific granules. They are lysosomes containing a range of acid hydrolases, which are active in the destruction of parasites and capable of decomposing antigen-antibody complexes. Both specific and azurophilic granules of eosinophils belong to special classes of secretory lysosomes.

Close to the wavy surface of the eosinophilic granulocyte shown in the micrograph are vacuoles similar to those involved in macropinocytosis. Eosinophilic granulocytes are known to be particularly specialised for phagocytosis of antigen-antibody complexes.

References

Blott EJ and Griffiths GM (2002) Secretory lysosomes. Nat Rev Mol Cell Biol 3: 122

Magnification: x 29,000

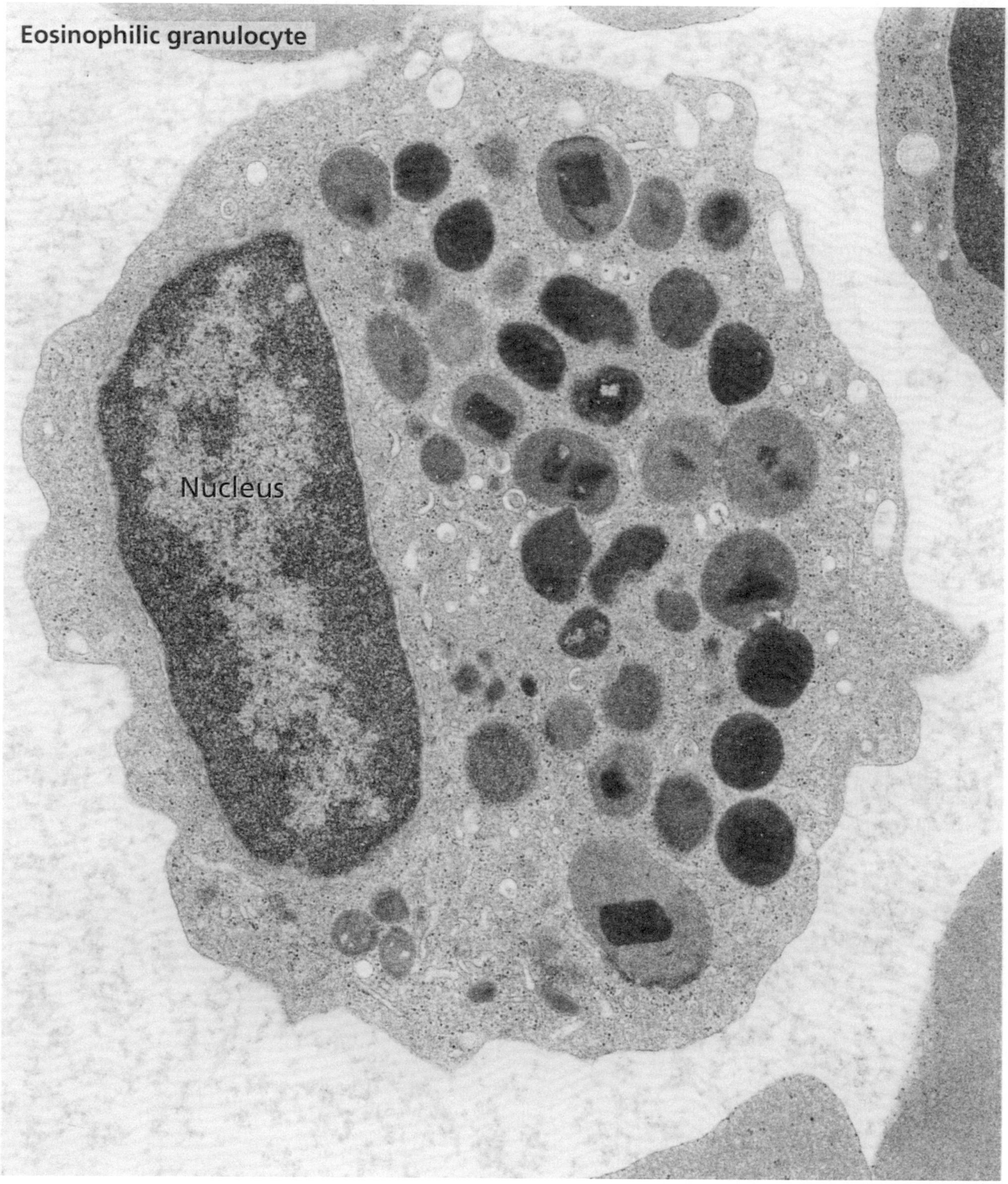
Eosinophilic granulocyte
Nucleus

BASOPHILIC GRANULOCYTE

Basophilic granulocytes are the least common granulocytes, making up only 0.5–1% of the total leukocytes. For this reason, they are difficult to find. The figure shows a basophilic granulocyte in the lumen of a blood vessel; the endothelium is visible at the right side of the picture. Four sections of the lobed nucleus can be seen. In the cytoplasm, mitochondria and lysosomes are visible; membranes of the rough and smooth endoplasmic reticulum are apparent but scarce. Most prominent are the numerous large specific granules (asterisks) exhibiting homogenously dense contents, which clearly can be discriminated from the granules of eosinophilic granulocytes with their crystalloid inclusions shown in Fig. 171.

Basophils derive from progenitor cells in the bone marrow; they are constituents of the circulating blood, but they may leave the circulation system and migrate into the connective tissue. During the past years, basophilic granulocytes have attracted continuously increasing attention, and they are being recognised as main cells of the immune system. Like mast cells in the connective tissue, they express IgE-receptors on their surfaces; they release histamine and cytokines, effect propagation of immune responses, and are involved in immediate and delayed hypersensitivity. Their number is increased in inflammatory diseases, and in acute and chronic hypersensitivity reactions. Recently, basophilic granulocytes have been shown to function as antigen-presenting cells, and novel roles for basophils in the induction of T-helper type 2 cell responses have been presented. New functional concepts about the roles of basophils pay increasing attention and interest in these rare cells.

References

Dvorak AM (1998) Cell biology of the basophil. Int Rev Cytol 180: 87

Falcone FH, Haas H, and Gibbs BF (2000) The human basophil: a new appreciation of its role in immune responses. Blood 96: 4028

Iikura M, Ebisawa M, Yamaguchi M, Tachimoto H, Ohta K, Yamamoto K, and Hirai K (2004) Transendothelial migration of human basophils. J Immunol 173: 5189

Perrigoue JG, Saenz SA, Siracusa MC, Allenspach EJ, Taylor BC, Giacomin PR, Nair MG, Du YR, Zaph C, van Rooijen N, Comeau MR, Pearse EJ, Laufer TM, and Artis D (2009) MHC class II-dependent basophil-CD4(+) T cell interactions promote T (H)2 cytokine-dependent immunity. Nat Immunol 10: 697

Schroeder JT (2009) Basophils: Beyond effector cells of allergic inflammation. Adv Immun 101: 123

Sokol CL, Chu NQ, Yu S, Nish SA, Laufer TM, and Medzhitov R (2009) Basophils function as antigen-presenting cells for an allergen-induced T helper type 2 response. Nat Immun 10: 713

Valent P, and Bettelheim P (1990) The human basophil. Crit Rev Oncol/Hematol 10: 327

Yoshimoto T, Yasuda K, Tanaka H, Nakahira M, Imai Y, Fujimori Y, and Nakanishi K (2009) Basophils contribute to T (H)2-IgE responses in vivo via Il-4 production and presentation of peptide-MHC class II complexes to CD4(+) T-cells. Nat Immun 10: 706

Magnification: x 30,000

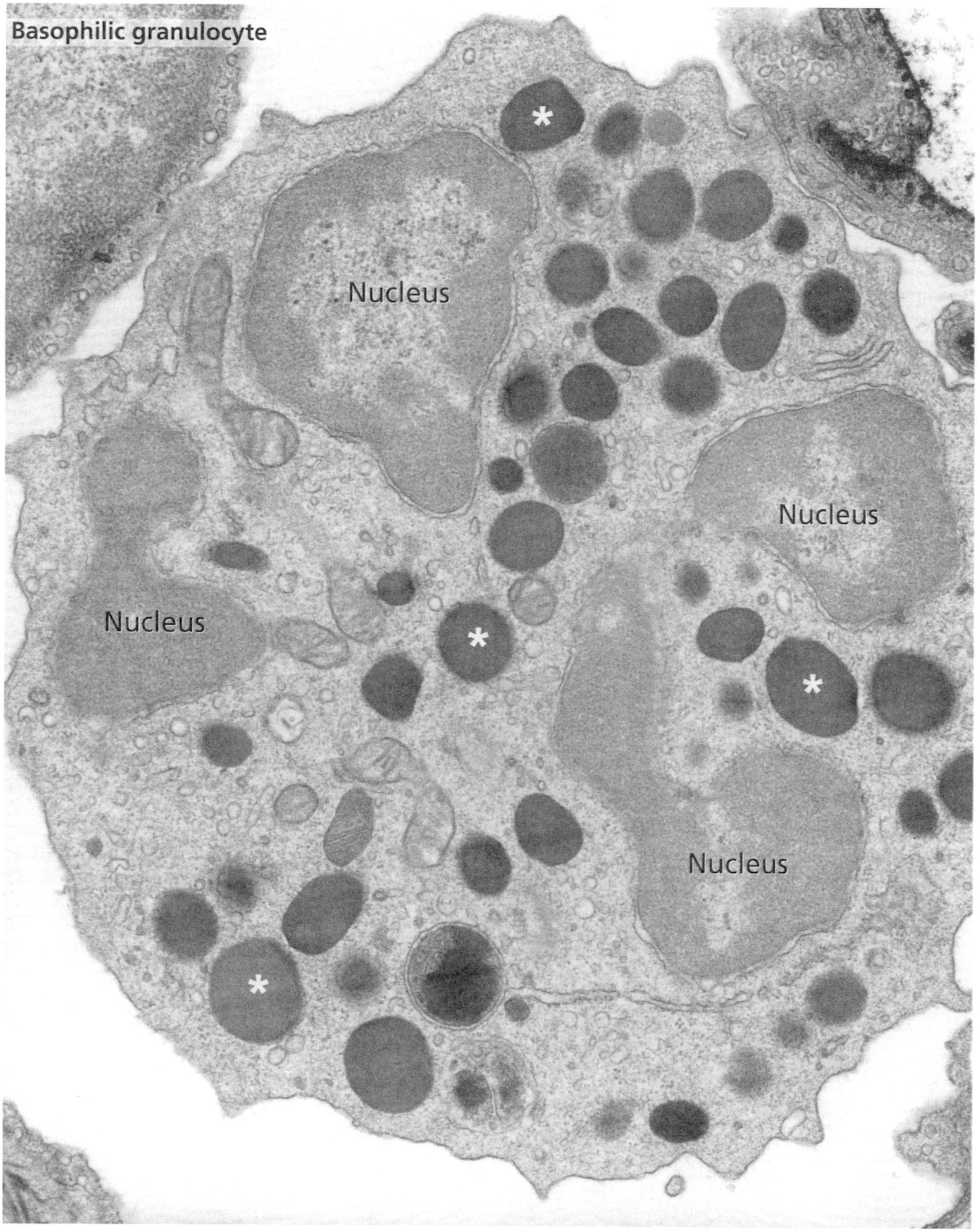
Basophilic granulocyte
Nucleus
Nucleus
Nucleus
Nucleus
*
*
*
*

MONOCYTE

Monocytes are the largest in the class of leukocytes, and make up 2–8% of total leukocytes in the blood. They are bone marrow-derived and are the precursor cells of macrophages, and phagocytes of the mononuclear phagocytic system that are present in many tissues and organs, such as the alveolar phagocytes in the lung, the Kupffer cells in the liver (cf. Fig. 106), the histiocytes in the connective tissues, the microglial cells in the brain, and the phagocytes in the red pulp of the spleen that remove the senescent erythrocytes. Among the blood monocytes, also precursors of osteoclasts (cf. Fig. 148) are counted. Monocytes travel from the bone marrow to the diverse organs. They are transported in the blood, where they still remain a few days. Further differentiation of monocytes and transformation into phagocytic cells occurs outside the blood vessels. They infiltrate almost every tissue of the body. Monocytes-macrophages have pivotal roles in the physiological turnover of tissue constituents, including the degradation of extracellular matrix materials and senescent connective tissue fibres, and are major cells of the specific and non-specific immune systems. Attracted by components of bacteria, and products developing during injury of the tissue, phagocytic cells also migrate to sites of inflammation, and regions of pathologic tissue remodelling. They are involved in phagocytosis of cells, and microorganisms, and clean up the tissue from dead cell and tissue debris. Monocytes also develop into antigen-presenting cells, which partially degrade antigens and present the fragments associated with major histocompatibility II molecules located in the plasma membrane to helper $CD4^{+}T$ lymphocytes (cf. Fig. 174).

The electron micrograph shows a monocyte of the human blood with a typically indented nucleus. In the cytocentre, parts of centrioles can be seen (arrows) and small stacks of Golgi cisternae are localised nearby. Although monocytes are counted to the class of "agranular" blood cells, azurophilic granules are present. Azurophilic granules correspond to lysosomes (Ly) and contain a range of typical lysosomal enzymes. They fuse with phagosomes to form phagolysosomes, where degradation of the engulfed materials takes place. Mitochondria are abundant. Rough endoplasmic reticulum, multiple transport vesicles near the Golgi apparatus, and endocytic vesicles budding from the plasma membrane indicate intense activities of the secretory and endocytosis systems.

References

Zhu Ch, Ying DJ, Mi JH, Zhu XH, Sun JS, and Cui XP (2004) Low shear stress regulates monocyte adhesion to oxidized lipid-induced endothelial cells via an IkappaBalpha dependent pathway. Biorheol 41: 127

Ishibashi M, Hiasa K, Zhao QW, Inoue S, Ohtani K, Kitamoto S, Tsuchihashi M, Sugaya T, Charo IF, Kura S, Tsuzuki T, Ishibashi T, Takeshita A, and Egashira K (2004) Circ Res 94: 1203

Löms Ziegler-Heitbrock H-W (1989) The biology of the monocyte system. Eur J Cell Biol 49: 1

Takahashi A, Kono K, Ichihara F, Sugai H, Fujii H, and Matsumoto Y (2004) Vascular endothelial growth factor inhibits maturation of dendritic cells induced by lipopolysaccharide, but not by proinflammatory cytokines. Cancer Immunol Immunother 53: 543

Vogl-Willis CA and Edwards IJ (2004) High glucose-induced alterations in subendothelial matrix perlecan leads to increased monocyte binding. Aterioscl Thromb Vasc Biol 24: 858

Magnification: x 34,000

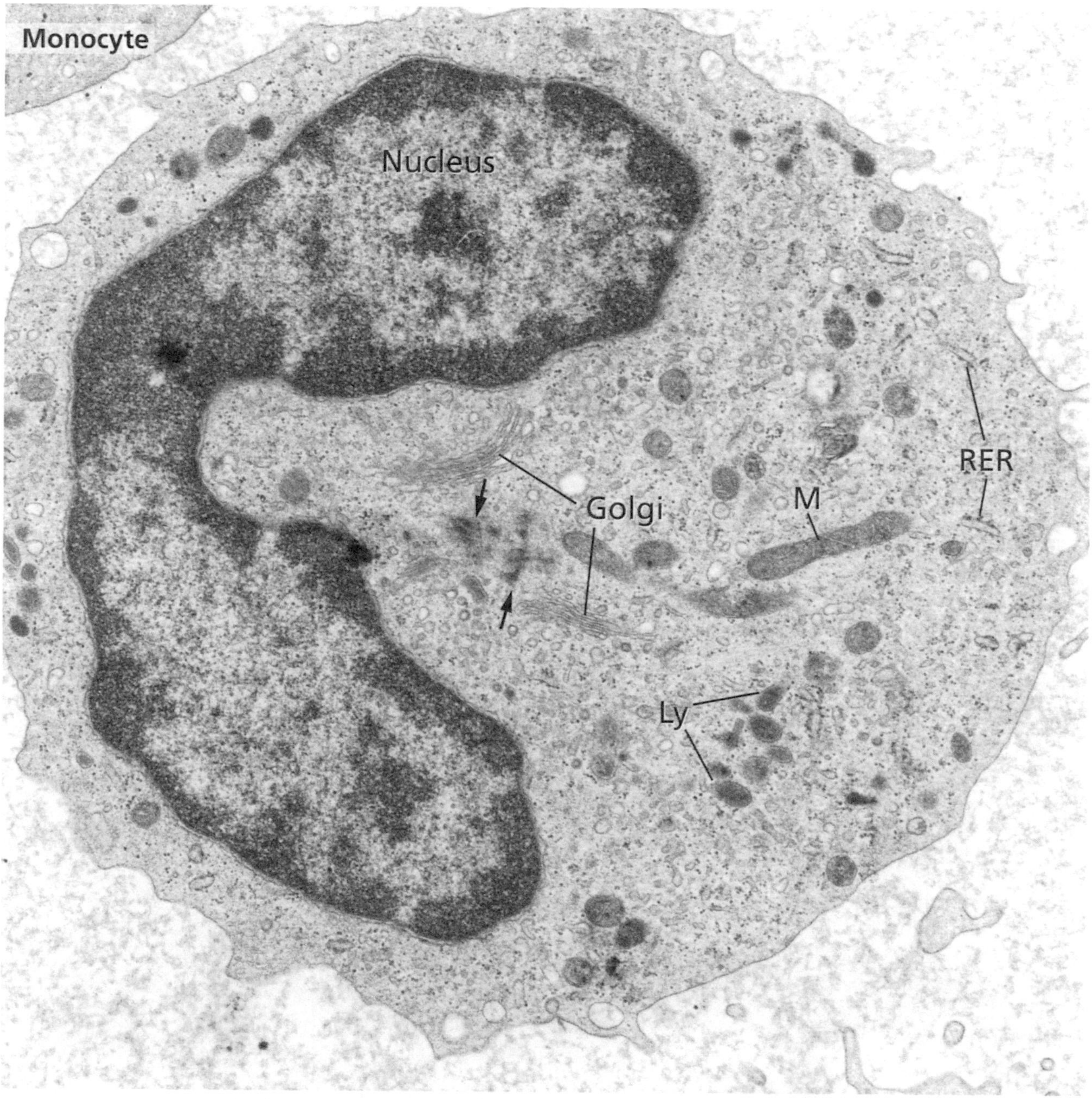
Monocyte
Nucleus
Golgi
M
RER
Ly

LYMPHOCYTE

Lymphocytes are the immunocompetent cells of the lymphatic or immune system, which have developed the ability to recognise and respond to antigens. They account for some 30% of the total blood leukocytes. In blood and lymph, they are in transit, re-circulating between different lymphatic tissues. Although they are morphologically similar, they represent a heterogenous population of cells differing from each other in various terms including origin and sites of maturation, cell surface markers, specific functions, localisations within the lymphoid tissues, and life span. Three functionally distinct types exist in the body: T lymphocytes (T-cells), B lymphocytes (B-cells), and natural killer cells (NK-cells).

The name of the T-cell is derived from the organ where they undergo differentiation, the thymus. T-lymphocytes are responsible for cell-mediated immunity and are further classified according to the presence of CD4 or CD8 proteins, recognising antigens bound to the major histocompatibility complex (MHC) II or MHC I molecules, respectively. $CD4^{+}$ helper lymphocytes have a central role in inducing of an immune response to a foreign antigen. They are activated by binding of their receptors to antigen-MHC II complexes present on the surfaces of antigen-presenting cells, which induce proliferation and differentiation of more T- and NK-cells and stimulate differentiation of B cells into plasma cells capable of synthesising and secreting antibodies. $CD8^{+}$-cells are primary effector cells. Stimulated by binding of their receptors to antigens associated with MHC I-molecules on the surface of a virus infected or neoplastic cell, they secrete perforins that assemble to form ion channels in the plasma membranes of transformed cell, leading to its lysis.

B-cells are named after the Bursa of Fabricius in birds where they were first recognised, or the bursa equivalents, such as the bone marrow, in mammals. B lymphocytes represent the major cells of the antibody-mediated humoral immunity. Mature B-cells express MHC II molecules and antibodies on their surface, and after activation, differentiate into antibody-secreting plasma cells.

NK cells are capable of killing cells, such as certain virus-infected cells and types of tumour cells but their activity is independent of antigen activation.

The electron micrograph shows a lymphocyte of the human blood. Most of the blood lymphocytes are small or medium-sized, measuring 6–15 µm in diameter. In the cytoplasm, abundant free ribosomes are visible. Several mitochondria (M) can be seen, as well as few cisternae of the rough endoplasmic reticulum (RER). Short stacks of the Golgi apparatus (Golgi) accompanied by multiple coated and non-coated vesicles are localised in the cytocentre close to the nucleus. A multivesicular body (MVB) residing in the Golgi apparatus neighbourhood may be part of the endosomal system where loading of MHC II molecules with antigenic peptide takes place.

Lymphocytes circulating in the blood represent mainly mature T-cells, which make up 60–80% of the total number of blood lymphocytes. 29%–30% are mature B-cells. Approximately 5–10% of the cells diagnosed as lymphocytes do not exhibit markers of either T- or B-cells. These include NK-cells as well as rare circulating hemopoietic stem cells.

References

Boes M, Cuvillier A, and Ploegh H (2004) Membrane specializations and endosome maturation in dendritic cells and B cells. Trends Cell Biol 14: 176

Chin YH, Cai J-P, and Xu X-M (1991) Tissue-specific homing receptor mediates lymphocyte adhesion to cytokine-stimulated lymph node high endothelial venule cells. Immunol 74: 478

Ikuta K, Uchida N, Frieman J, and Weissman IL (1992). Lymphocyte development from stem cells. Annu Rev Immunol 10: 759

Yagita H, Nakata M, Kawasaki A, Shinkai Y, and Okumura K (1992) Role of perforin in lymphocyte-mediated cytolysis. Adv Immunol 51: 215

Magnification: x 37,000

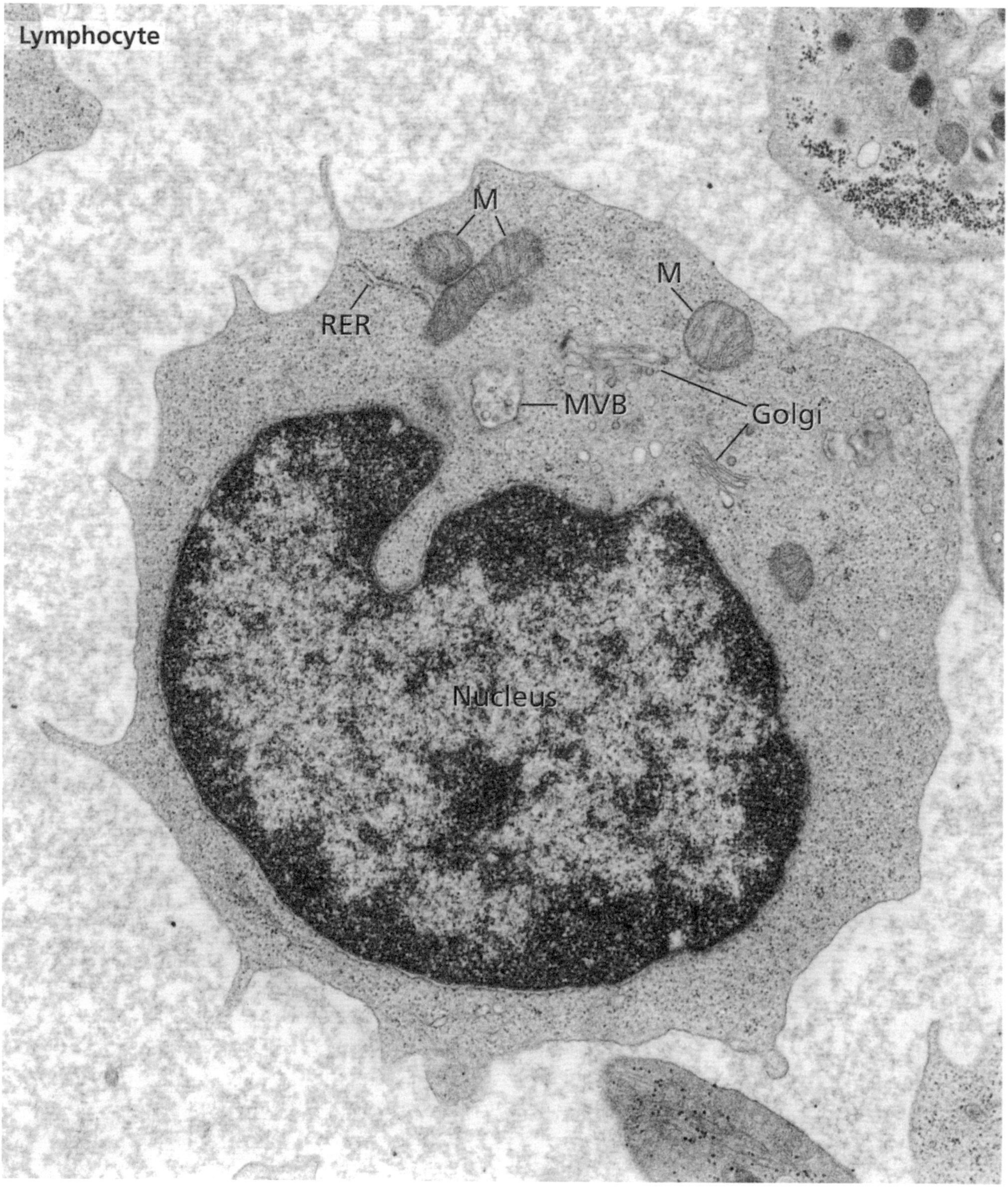
Lymphocyte
M
M
RER
MVB
Golgi
Nucleus

MEGAKARYOCYTE AND THROMBOCYTE

Megakaryocytes are large polyploid cells present of the bone marrow, which produce the blood platelets or thrombocytes, small disk-shaped cytoplasmic fragments that circulate in the blood and are essential components of the haemostasis system. Formation and release into the blood is under the primary regulation of thrombopoietin, a glycoprotein produced in the kidney and liver. During differentiation of megakaryocytes, continued DNA replication in the presence of abortive mitosis leads to the formation of polyploid nuclei, a process called endomitosis. It has been shown that polyploidisation is connected with the suppression of stathmin, a microtubule-regulatory protein that has a major role in the formation of the mitotic spindle. Platelets are pre-formed in the cytoplasm of megakaryocytes at sites of protrusions, called proplatelets. A megakaryocyte may form 10–20 proplatelets from the tips of which the platelets are released. Maturing platelets become filled with organelles and granules, which move from the megakaryocyte cell body to the tips of the proplatelets. Extensive platelet demarcation channels are formed by invaginations of the plasma membrane delineating the future platelets. The platelet demarcation system is continuous with the extracellular space. Once the thrombocyte has been filled with organelles, granules, and cytoskeletal proteins, a marginal band is formed consisting of a single peripheral microtubule wound in 8–12 coils. Proplatelets protrude through apertures of the endothelial lining of the bone marrow sinusoids allowing the platelets to be released directly into the circulating blood.

Panels A and C show segments of the superficial cytoplasm of a megakaryocyte of the human bone marrow. Future platelets are roughly delineated by membranes of the demarcation system. Delineated cytoplasmic areas, in part, are still devoid of organelles, in part they are already "filled" containing a range of compartments, organelles, and granules, such as large alpha granules and dense core granules, corresponding to the equipment of the future thrombocytes. After release, parts of the demarcation membrane system that did not participate in subdividing the megakaryocyte cytoplasm, mostly remain forming the open canalicular system of the mature platelets (panel B). The open canalicular system is in continuity with the extracellular space. Parts of the second type of the platelet membrane channel systems, the dense tubular system, can be seen vaguely in the left segment of the thrombocyte shown in panel B (asterisk, cf. also Fig. 176). The dense tubular system derives from the endoplasmic reticulum of the megakaryocyte, and serves as a calcium storage compartment.

Two phases of platelet formation can be differentiated. The first phase is characterised by polyploidisation, enlargement of the cytoplasm, and synthesis of the platelet-specific membranes, compartments and granules. This is induced by megakaryocyte-specific growth factors and takes place over a period of days. In the second phase, proplatelets are formed, filled with the bulk of organelles and granules, and the platelets are released into the blood. The events of the second phase are completed within hours.

References

Ebbeling L, Robertson C, McNicol A, Gerrard JM (1992) Rapid ultrastructural changes in the dense tubular system following platelet activation. Blood 80: 718

Fuhrman B, Brook GJ, and Aviram M (1992) Proteins derived from platelet alpha granules modulate the uptake of oxidized low density lipoprotein by macrophages. Biochim Biophys Acta 1127: 15

Hartwig J and Italiano J (2003) The birth of the platelet. J Thromb Haemost 1: 1580

Italiano JE and Shivdasani RA (2003) Megakaryocytes and byond: the birth of platelets. J Thromb Haemost 1: 1174

Kaushansky K (2003) Thrombopoietin: a tool for understanding thrombopoiesis. J Thromb Haemost 1: 1587

Mahaut-Smith MP, Thomas D, Higham AB, Usher-Smith JA, Hussain JF, Martinez-Pinna J, Skepper JN, and Mason MJ (2003) Properties of the demarcation membrane system in living rat megakaryocytes. Biophys J 84: 2646

Rubin CI, French DL, and Atweh GF (2003) Stathmin expression and megakaryocyte differentiation: A potential role in polyploidy. Exp Hematol 31: 389

Magnification: x 15,000 (A), x 12,000 (B), x 34,500 (C)

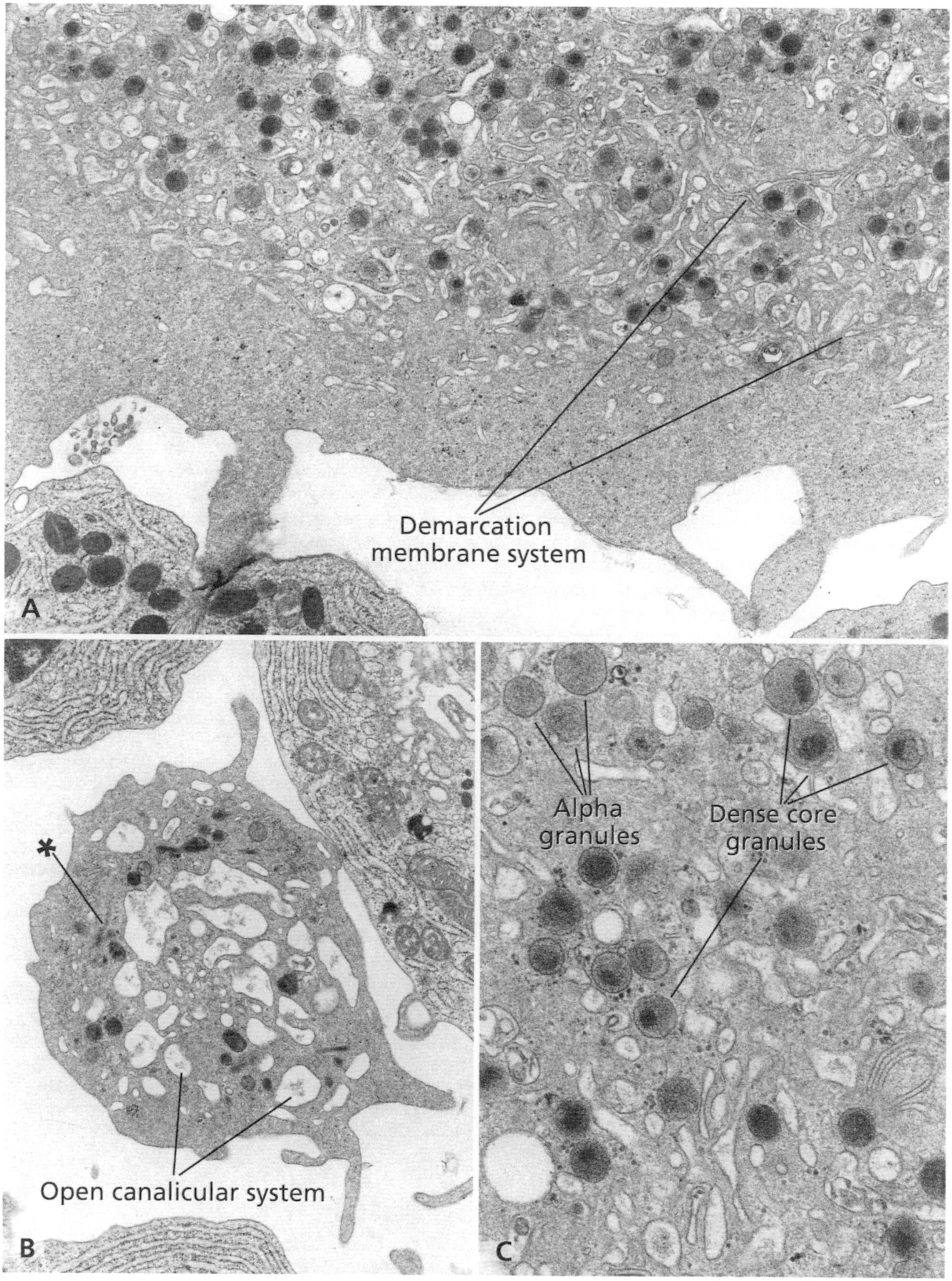
Demarcation membrane system
A
*
Open canalicular system
B
Alpha granules
Dense core granules
C

THROMBOCYTES

Thrombocytes (blood platelets) are cytoplasmic fragments of the bone marrow megakaryocytes (cf. Fig. 175). After release into the blood, the thrombocytes circulate as small cellular disks measuring 3–4 μm in diameter, with a life span of about 10 days. Platelets have a pivotal role in hemostasis, where they are involved in several aspects including blood clot formation, clot retraction, and repair of the injured tissue. Platelets also continuously survey the endothelial lining of the blood vessels for possible breaks or damages, and are among the first cells to be recruited to sites of injuries.

Panels A and B show platelets in the lumen of a blood vessel in the wall of the small intestine. In the cytoplasm, organelles, granules, and parts of the membrane systems (cf. also Fig. 175) are visible. Both panels show the prominent microtubule band (MT) in the platelet marginal structural zone. The band consists of one single microtubule polymer that is wound in 8-12 coils, which are particularly well visible in the cross section of the band shown in the upper part of panel B. About 90% of β-tubulin in the marginal microtubule band is of the β1-isoform that is restricted to cells of the megakaryocyte lineage. The marginal microtubule band maintains the discoid cell shape of the platelets, and has also an essential role in the microtubule-driven processes of formation of the proplatelets in the megakaryocyte cytoplasm. Aside the microtubule coil, the platelet cytoskeleton includes crosslinked actin filaments connected to spectrin and associated proteins of the membrane skeleton. While circulating in the blood stream, platelets have to withstand high shear stresses. Rapid cytoskeletal remodelling is required when platelets, in response to vascular damages, change their shapes, round up, and extend filopodia and lamellae. The marginal structural zone of the platelets is also referred to as hyalomer.

The centre of the platelets, called organelle zone or granulomer, contains mitochondria, peroxisomes, glycogen, and different types of granules. The large alpha granules (asterisks in panel B, cf. also Fig. 175) contain mainly coagulation factors, fibrinogen, the platelet-derived growth factor, plasminogen, and plasminogen activator inhibitor. The dense core granules (delta granules, cf. Fig. 175) store mainly serotonin, adenosine diphosphate, adenosone triphosphate, and histamine, and the lamda granules, corresponding to lysosomes, contain several hydrolases. The contents of the alpha and delta granules are particularly important in the initial phases of haemostasis, platelet adhesion and aggregation, blood coagulation, and vasoconstriction in the area of the injured vessel. They have a major role also in the initial stages of vessel repair. The lamda granule contents are involved mainly in the resorption of the clot during the later stages of vessel repair.

Two classes of membrane systems are present in the thrombocyte cytoplasm: Membrane channels and tubules of the open canalicular system, which are remnants of the demarcation channels (cf. Fig. 175), and the dense tubular system derived from the megakaryocyte endoplasmic reticulum (white arrows in panel A), which serves as a storage compartment for calcium ions. It has a major role in the regulation of the platelet calcium concentration.

References

Barkalow KL, Italiano JE Jr, Chou DE, Matsuoka Y, Bennett V, and Hartwig JH (2003) Alpha – adducin dissociates from F-actin and spectrin during platelet activation. J Cell Biol 161: 557

Colman RW (1990) Aggregin: a platelat ADP receptor that mediates activation. FASEB J 4: 1425

Elrod JW, Park JH, Oshima T, Sharp CD, Minagar A, and Alexander JS (2003) Expression of junctional proteins in human platelets. Platelets 14: 247

Italiano JE Jr, Bergmeier W, Tiwari S, Falet H, Hartwig JH, Hoffmeister KM, André P, Wagner DD, and Shivdasani RA (2003) Mechanisms and implications of platelet discoid shape. Blood 101: 4789

Mangin P, Ohlmann P, Eckly A, Cazenave JP, Lanza F, and Gachet C (2004) The P2Y(1) receptor plays an essental role in the platelet shape change induced by collagen when TxA2 formation is prevented. J Thromb Haemost 2: 969

Quinton TM, Murugappan S, Kim S, Jin J, and Kunapuli SP (2004) Different G-protein-coupled signaling pathways are involved in alpha granule release from human platelets. J Thromb Haemost 2: 978

Veljkovic.DK, Cramer EM, Alimardani G, Fichelson S, Masse JM, and Hayward CPM (2003) Studies of alpha-granule proteins in cultured human megakaryocytes. Thromb Hemost 90: 844

Magnification: x 63,500 (A), x 43,500 (B)

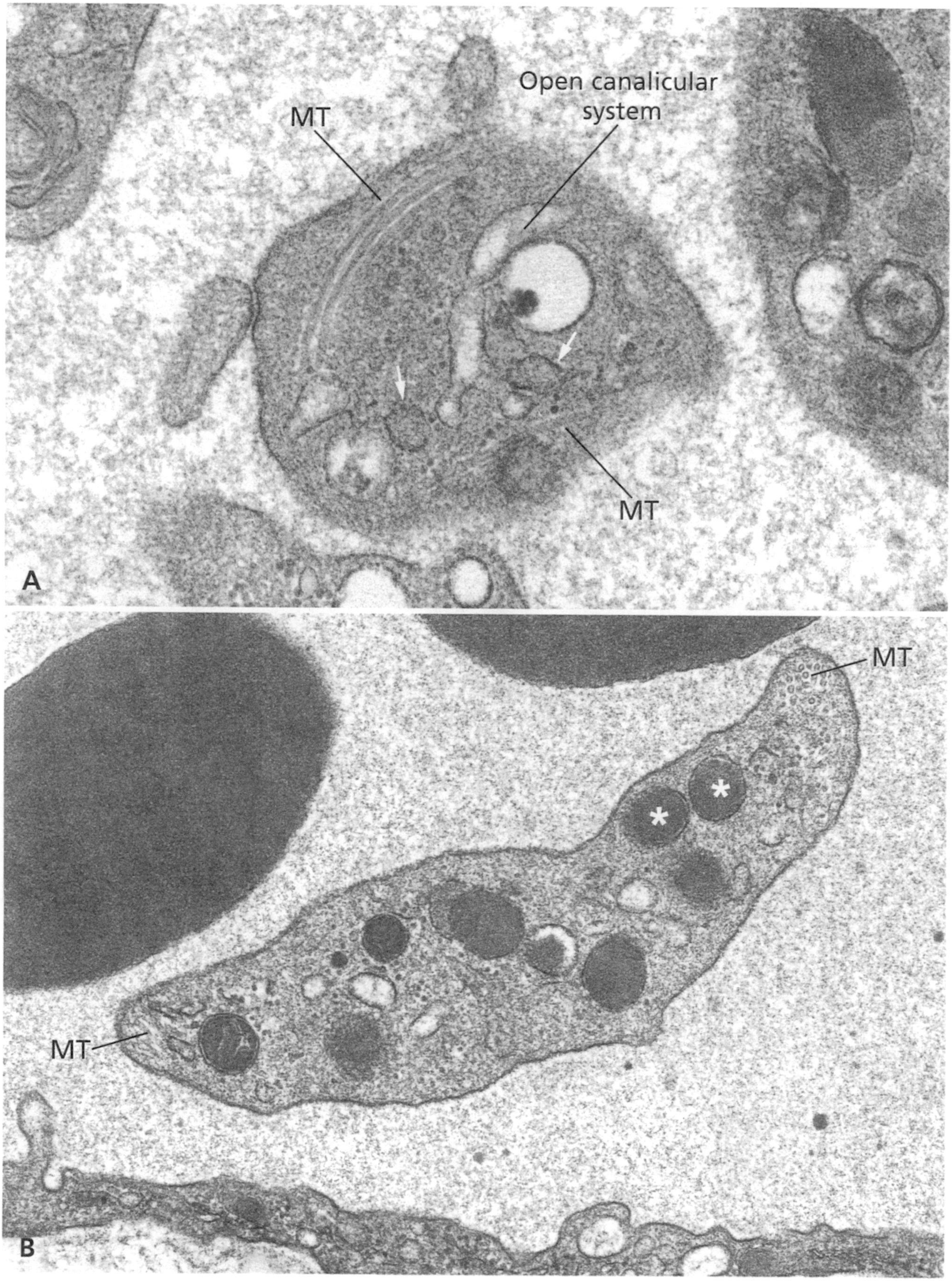
MT
Open canalicular system
MT
A
MT
MT
B

SUBJECT INDEX

A

B

C

D

E

F

G

H

I

J

K

L

M

N

R

S

Kolibri 360 GmbH
Ferdinand-Jühlke-Straße 7
99095 Erfurt, Deutschland
produktsicherheit@kolibri360.de